```
QR              Developmental
184.5              immunology.
.D49
1993


$95.00
```

DATE			

DEVELOPMENTAL IMMUNOLOGY

Developmental Immunology

Edited by

EDWIN L. COOPER, PH.D.

Department of Anatomy and Cell Biology
School of Medicine
University of California, Los Angeles

ERIC NISBET-BROWN, M.D., PH.D.

Departments of Immunology and Medicine
University of Toronto

New York   Oxford
OXFORD UNIVERSITY PRESS
1993

Oxford University Press

Oxford   New York   Toronto
Bombay   Calcutta   Madras   Karachi
Kuala Lumpur   Singapore   Hong Kong   Tokyo
Nairobi   Dar es Salaam   Cape Town
Melbourne   Auckland   Madrid

and associated companies in
Berlin   Ibadan

Published by Oxford University Press, Inc.,
200 Madison Avenue, New York, New York 10016

Oxford is a registered trademark of Oxford University Press

Library of Congress Cataloging-in-Publication Data
Developmental immunology / edited by Edwin L. Cooper,
Eric Nisbet-Brown.
p. cm.   Includes bibliographical references and index.
ISBN 0-19-504353-7
1. Developmental immunology.   I. Cooper,
Edwin L. (Edwin Lowell), 1936-   .   II. Nisbet-Brown, Eric.
[DNLM: 1. Developmental Biology.   2. Immune System.
3. Immunity.   QW 504 D48961]
QR184.5.D49   1993
616.07′9—dc20   DNLM/DLC
for Library of Congress   91-39118

1 3 5 7 9 8 6 4 2

Printed in the United States of America
on acid-free paper

PREFACE

Developmental Immunology is a comprehensive review of this rapidly developing field, which has not been summarized in the past few years. Recent advances in molecular and cellular biology and their application to immunology have led to a marked increase in our understanding of developmental and regulatory processes, and they make this synthesis both appropriate and timely. The book is intended to progress sequentially from "basic" molecular and cellular immunology, to tissue and morphogenetic structure, and ultimately to the regulation of the immune system in health and disease. Information has been drawn from both recent and historical works in evolution, ontogeny, and phylogeny, and from studies of nonmammalian and mammalian organisms, including humans.

The book is intended for advanced students and active workers in experimental biology, including immunologists, developmental biologists, and cell and molecular biologists. Given the great diversity of background among our potential audience, we have attempted to summarize pertinent basic information in each chapter, before describing recent advances or areas of controversy. Throughout, we have tried to integrate information derived from different systems and using a variety of experimental approaches to achieve a balanced view of the field—unlike the majority of works in this field, which have generally been symposium or congress proceedings.

Developmental Immunology is divided into five parts, of which the first serves as an Introduction to the book. Part II, Phylogeny of the Immune System, comprises six chapters that focus on the varied roles of members of the immunoglobulin gene superfamily (including MHC molecules, immunoglobulins, and T cell receptors) in the development and regulation of the immune system. Of necessity, the majority of information in this section is drawn from studies of vertebrates, with the exception of Chapter 2, which deals with invertebrate recognition systems and is largely theoretical.

Part III, Ontogeny of the Immune System, includes four chapters on the development of stem cells, monocytes, macrophages, accessory cells, and T and B lymphocytes. Data from both vertebrates and invertebrates are presented, particularly in regard to the development of stem cells and monocytes. One need only reflect on the enormous impact of Metchnikoff's description of phagocytic cells on the subsequent development of immunology to appreciate its relevance to this entire discipline.

Molecular aspects of immune recognition and regulation are covered in the five chapters in Part IV. These chapters cover a diverse range of topics, including antigen processing and presentation, the complement system, the role of cytokines in the regulation and development of the immune system, molecular mechanisms of T cell activation, and regulatory processes for B cell responses. All of these areas reflect the critical importance of recent advances in molecular biology to our understanding of regulatory and developmental processes in immunology. The reviews of complement proteins and cytokines have particular relevance to phylogenetic analyses, since homologous substances have been described in ectothermic vertebrates.

Finally, Part V reviews immunological function in health and disease, with particular emphasis on the human immune system. Individual chapters deal with the ontogeny and phylogeny of the thymosins; immune surveillance and its relationship to natural immunity; immunobiology of aging; and the development of normal immune responses and primary immunodeficient states in humans.

Developmental immunology gained recognition as a crucial area for study when it was included at the First International Congress of Immunology, in 1971, and the field comes of age last year with the Eighth International Congress, in Budapest. Over the years, there has been much confusion over what precisely is encompassed within the discipline of "developmental immunology"; despite this, the area has become well-established and seems assured of a long and vigorous maturity. Among the important steps in the development of the field were the establishment, in 1977, of the international journal *Developmental and Comparative Immunology,* and publication of a number of symposium proceeding volumes. This ongoing interest led to the founding of the International Society of Developmental and Comparative Immunology, which is affiliated with the International Union of Immunological Societies and the International Union of Biological Societies. National organizations have also been founded to promote developmental immunology as an independent discipline, notably the Division of Comparative Immunology of the American Society of Zoologists, and similar groups in Japan, Great Britain, and France. Recent years have seen the proliferation of laboratories working in the area, the setting-up of international courses, and the founding of new journals, notably *Developmental Immunology,* all of which confirm the vigor and level of activity of the field.

Given this background we present *Developmental Immunology* to you in the hope that it will fulfill its goal of providing a cohesively written, factual, and current overview of an exciting field. We would be particularly pleased if our book inspires at least some young investigators to take up the study of developmental aspects of the immune system.

Los Angeles   E.L.C.
Toronto   E.N-B.

ACKNOWLEDGMENTS

Developmental Immunology had its origins a number of years ago, with a meeting between one of us (ELC) and Jeffrey House of Oxford University Press, at a FASEB meeting. Out of this, and many subsequent meetings, has come the present book. Jeffrey House deserves a special note of gratitude for his patience and persistence during the preparation and production of *Developmental Immunology.* We also express our appreciation to others at Oxford University Press for their assistance during production of this volume—most notably Stanley George and his colleagues.

After its first review, the manuscript was felt to need significant expansion, a reflection of the rapid developments in the field, and the growing importance of molecular biology to our understanding of the discipline. The time required to recruit new contributors meant that parts of the initial volume became dated, and we have had to depend on the goodwill of our authors to keep the book current. We are most grateful to all our contributors for their patience, forbearance and sustained interest as the book took its final form, despite the competing pressures of research, teaching and administration.

CONTENTS

IV   MOLECULAR ASPECTS OF RECOGNITION
            AND REGULATION

V   DEVELOPMENT OF THE IMMUNE SYSTEM IN HEALTH
            AND DISEASE

CONTRIBUTORS

Matteo Adinolfi
Paediatric Research Unit
Guy's Hospital Medical School
London, England SE1 9RT

Chris T. Amemiya
Lawrence Livermore National
*   Laboratory*
Division of Biomedical Science L-452
P.O. Box 5507
Livermore, California 94551

Roger J. Booth
Department of Molecular Medicine
School of Medicine
University of Auckland
Auckland, New Zealand

Robert Brines
Department of Microbiology and
*   Immunology*
School of Medicine
University of California, Los Angeles
Los Angeles, California 90024

Edwin L. Cooper
Department of Anatomy and Cell
*   Biology*
School of Medicine
University of California, Los Angeles
Los Angeles, California 90024

Kathryn E. Crosier
Department of Molecular Medicine
School of Medicine
University of Auckland
Auckland, New Zealand

Terry L. Delovitch
Banting and Best Department of
*   Medical Research and Department*
*   of Immunology*
University of Toronto
Toronto, Canada M5G 1L6

Stephen V. Desiderio
Department of Molecular Biology and
*   Genetics*
and Howard Hughes Medical Institute
*   Laboratory of Genetics*
Johns Hopkins University School of
*   Medicine*
725 North Wolfe Street
Baltimore, Maryland 21205

Erwin Diener
Department of Immunology
University of Alberta
Medical Sciences Building
Edmonton, Canada T6G 2H7

Paul J. Doherty
Immunology Laboratory
The Toronto Hospital
399 Bathurst Street
Toronto, Canada M5T 2S8

Hans-Michael Dosch
Department of Immunology and
*   Cancer Research*
The Hospital for Sick Children and
*   University of Toronto*
555 University Avenue
Toronto, Canada M5G 1X8

Rati Fotedar
Divison of Basic Sciences
Fred Hutchinson Cancer Research
Center
1024 Columbia Street
Seattle, Washington 98104

Allan L. Goldstein
Department of Biochemistry and
Molecular Biology
George Washington University School
of Medicine and Health Sciences
2300 I Street NW
Washington, DC 20037

Julia Green-Johnson
Department of Microbiology and
Immunology
Queen's University
Kingston, Canada K7L 3N6

Robert N. Haire
University of South Florida
All Children's Hospital
801 Sixth Street South
St. Petersburg, Florida 33701

Fiona A. Harding
Department of Immunology
University of California, Berkeley
415 LSA
Berkeley, California 94720

Kristin R. Hinds-Frey
University of South Florida
All Children's Hospital
801 Sixth Street South
St. Petersburg, Florida 33701

May F. Hui
Department of Pediatrics and
Immunology
The Hospital for Sick Children and
University of Toronto
555 University Avenue
Toronto, Canada M5G 1X8

Jim Kaufman
Basel Institute for Immunology
Grenzacherstraße 487
CH-4058 Basel, Switzerland

Fumio Kokubu
Department of Medicine
Showa University
Tokyo, Japan

Mitchell Kronenberg
Department of Microbiology and
Immunology
School of Medicine
University of California, Los Angeles
Los Angeles, California 90024

Kenneth S. Landreth
Department of Microbiology and
Immunology
and Mary Babb Randolph Cancer
Center
West Virginia University Health
Science Center
Morgantown, West Virginia 26506

Gary W. Litman
University of South Florida
All Children's Hospital
801 Sixth Street South
St. Petersburg, Florida 33701

Ronda T. Litman
University of South Florida
All Children's Hospital
801 Sixth Street South
St. Petersburg, Florida 33701

Eva Lotzová
Department of General Surgery
M. D. Anderson Hospital and Tumor
Institute
6723 Bertner Avenue
Houston, Texas 77030

Mohamed H. Mansour
Department of Zoology
Faculty of Science
Cairo University
Giza, Cairo, 12613 Egypt

Gordon B. Mills
Oncology Research
Toronto General Hospital
200 Elizabeth Street, Eaton Wing
Toronto, Canada M5G 2C4

Christa E. Müller-Sieberg
Medical Biology Institute
11077 North Torrey Pines
La Jolla, California 92037

Shigeru Muramatsu
Department of Zoology
Faculty of Science
Kyoto University
Kyoto 606, Japan

Eric Nisbet-Brown
Immunology Laboratory
The Toronto Hospital
399 Bathurst Street
Toronto, Canada M5T 2S8

Karen K. Oates
Department of Biology
George Mason University
4400 University Drive
Faifax, Virginia 20030

David A. Raftos
UTS Immunobiology Unit
University of Technology, Sydney
P.O. Box 123 Broadway
N.S.W., 2007 Australia

John W. Semple
Division of Hematology
St. Michael's Hospital
30 Bond Street
Toronto, Canada M5B 1W8

Michael J. Shamblott
C.O.M.B.
University of Maryland
600 East Lombard Street
Baltimore, Maryland 21202

Animesh Sinha
Department of Medical Microbiology
Stanford University School of
Medicine
Stanford, California 94305

Gerald J. Sprangrude
NIH-NIAID Rocky Mountain
Laboratories
903 South Fourth Street
Hamilton, Montana 59840

Daijiro Suzuki
Department of Surgery
Showa University
Tokyo, Japan

Myron R. Szewczuk
Department of Microbiology and
Immunology
Queen's University
Kingston, Canada K7L 3N6

Hung-Sia Teh
Department of Microbiology
University of British Columbia
Vancouver, Canada V6T 1W5

Andrew W. Wade
Department of Microbiology and
Immunology
Queen's University
Kingston, Canada K7L 3N6

James D. Watson
Department of Molecular Medicine
School of Medicine
University of Auckland
Auckland, New Zealand, 379716

I

INTRODUCTION

1

Basic Concepts and the Functional Organization of the Immune System

EDWIN L. COOPER

BASIC CONCEPTS: PHYLOGENY AND ONTOGENY

Introduction to Developmental Immunology

Developmental immunology is a subdiscipline concerned with mechanisms that emphasize ontogenetic events during immunogenesis (Cooper, 1975; Smith et al., 1967). To critically dissect its ontogeny, immunologists must resolve the immune system into its contituents, analyze the individual genes, cells, organs, and products, and then reintegrate it (Solomon and Horton, 1977). This approach will reveal, first, the developmental complexity of immunologic mechanisms and, second, the pervasiveness of immunity as an adaptive characteristic, one that is as necessary for survival as are digestion, respiration, reproduction, excretion, and other processes. Developmental immunology is a derivative of the parent field, immunology; however, its roots can be traced in relation to biology and, in turn, back to embryology and comparative anatomy.

The developmental immunologist can now retrace by various techniques (for example, those of the molecular biologist), the origins of the immune system to early events in embryological development. After fertilization, the contributions of each germ layer, especially mesoderm and endoderm, to the immune system's development can be analyzed with respect to the later movements and interactions of various cellular components. Once the immune system's basic plan has been laid out and normal processes of development have given rise to a fully functioning, integrated organism, the pursuit of the developmental immunologist can also extend to later periods as sources of new cells are sought during an individual's lifetime. Developmental immunology therefore deals with the maturation, aging, and ultimate decline of the immune system and, in so doing, considers the totality of events during the immune system's genesis and its senescence (Cooper and Brazier, 1982). Certain diseases that affect the normal functioning of the immune system may develop; at the moment these touch our daily lives. Causes of failure of the immune system include acquired immunodeficiency syndrome (AIDS) and severe combined immunodeficiency (SCID), which result from acquired and congenital modifications, respectively. Autoimmune diseases, by contrast, represent the paradoxical turning of the immune system on itself with the loss of clear acceptance of self, resulting in an autoaggressive attack on various organs or tissues (e.g., Hashimoto's thyroiditis, myasthenia gravis).

Developmental immunology, to borrow older terms, is concerned with the ontogeny of the immune system, including its origins and development during embryonic and immediate perinatal/hatching periods. The meaning of developmental immunology, used in this sense, is a gradual bringing to completion of the immune system, both in form and function, including cumulative progressive change (Solomon, 1981). This view of development, which emphasizes the components and responses of the immune system, allows comparison with earlier and later stages, and is particularly informative and significant in that it allows an integrated approach. An embryo is a functioning organism, adequately adapted to its needs and environment at all stages of development, whether in aquatic conditions (fishes, amphibians), within the shell of an egg (reptiles, birds, monotreme mammals), within a marsupial pouch (kangaroos, opossums), or within the uterus (mice, humans). The state of differentiation at any given period anticipates immune responses that appear later. Our understanding of these events has been very much dependent on the availability of pertinent and useful model systems (Cooper, 1976 a, b, c; Cooper et al., 1985; Gershwin and Cooper, 1978; Rowley et al., 1983; Van Muiswinkel, 1982). Thus, early developmental events involving the immune system are generally in sharp contrast to the recurring, nonprogressive, physiological changes involved mainly in the maintenance of life and the immune system during adult periods. Some of these post-ontogenetic immunologic changes may involve what has already been programmed in immune response genes, whereas other changes may represent the natural consequences of interactions between the immune system and other body systems, for example the endocrine and nervous systems. Investigators might argue that the life span of any organism, from conception through adulthood, to aging and the inevitable death of the individual in fact represents continuous, overlapping, or an extended period of development.

When Does the Immune System Develop?

Most research in immunology has focused on events taking place after birth, when the basic plan has already developed and the new individual's immune system is more or less prepared to deal with a hostile environment, including foreign material and pathogenic microorganisms (Snieszko et al., 1965). However, recent advances in cellular and molecular immunology, with broad conceptual and technical bases, have led to sustained interest in events occurring before birth, which establish the plan and framework for subsequent development of the immune system. Immunologists have only recently recognized, however, that crucial changes, other than mere growth, take place well into adult stages, and even into senescence. For example, events that include the continuous generation of stem cells in bone marrow or equivalent sites can be considered ontogenetic, even though they take place in adult mammals.

Comparative Immunology

Comparative immunology, by contrast, takes a phylogenetic view and is concerned with the diversity of immune reactions and underlying mechanisms found throughout the animal kingdom (Azzolina et al., 1981; Cooper, 1976 a, b, c; Langman, 1989; Manning and Turner, 1976; Marchalonis, 1976, 1977; Sigel, 1968). In what may seem to be an endless search, comparative immunologists analyze how immune systems have evolved over millenia (Liacopoulos and Panijel, 1972; Manning, 1980; Smith et al., 1966). Given the fundamental characteristics of immunity, including specificity, memory, and synthesis of components that sequester antigen, comparative immunologists attempt to define the times when these characteristics may have first appeared. To analyze such time spans requires comparative studies of immunologic events in representative living animals; some are related to extinct animals that developed and lived during earlier geologic eras. Comparative immunology, in contrast to developmental immunology, is therefore broader and more complex. Understanding the totality of responses throughout all phyla is the goal of the phylogenetic approach. Because of the complexities of individual phyla due to their enormous diversity, comparative immunology can have as its goal not only comparison of immune mechanisms between various groups (e.g., fish and mammals) but also comparisons within major groups (e.g., marsupial and placental mammals), thus allowing a certain perspective to be attained.

Comparative analyses may thus contribute greatly to our understanding of human immunity through what are viewed (often mistakenly) as simpler animal models, as well as themselves being of intrinsic interest. Yet, from a practical viewpoint, there is also the contrary, teleologically based view that the vertebrate immune system represents the highest state of development attained to date. This approach emphasizes a more restricted study of deuterostome rather than protostome invertebrates. Vertebrates are generally believed to have evolved from primitive chordates, which belong to the deuterostome line and share certain characteristics with vertebrates, including (1) a notochord or its equivalent, (2) a dorsal nerve cord, and (3) gill slits. A link to the vertebrate immune system is seen through the presence of lymphocyte-like cells in close proximity to the gill region of tunicates, an anatomical site equivalent to the vertebrate thymus. Thus, vertebrate lymphocytes probably evolved from cells present in early chordates, suggesting homology. Conversely, analysis of the origins of lymphocyte-like cells in protostomes suggests that they are analogous to vertebrate lymphocytes.

Some Biological Principles and the Role of Invertebrate Immunity

When we describe the immune responses of different animal groups, we are often forced to use terms such as "primitive" and "advanced," "lower" and "higher," or "simple" and "specialized." Such classifications should not imply, however, that the immune systems of some animals are "better" than others. General concepts have been developed, which seek to explain the evolution of animals and, in turn, of their immune systems. Based on structural and functional studies, it appears that lymphocyte populations may have evolved from a common lymphocyte precursor present in amphibians. These lines may have undergone divergent evolution along various pathways, becoming distributed among various organs and ultimately giving rise to T and B lymphocytes. The development of similar characteristics by lymphocytes from quite different evolutionary ancestries represents an example of convergent evolution. The structural and functional development of lymphocytes in mammals and lymphocyte-like cells in advanced invertebrates may therefore be an example of convergent evolution.

Descriptions of the evolution of immune systems must make distinctions between resemblances—and whether they represent homology or analogy. Homology may apply to leukocytes or their products in different animals, which share structural and functional similarities due to inheritance of features from a common ancestor. However, cells or their products (such as agglutinins), which are functionally similar in different animals but which have little gross structural resemblance and lack a common ancestor, are analogous, non-homologous entities. To take an extreme example of analogy, the coelomic cavity of the earthworm and the bone marrow cavity of tetrapod vertebrates are both hemopoietic sites and lack a common ancestor, but they are functionally similar in that they produce and house leukocytes.

Adaptive Immunity

Adaptive immunity is traditionally defined on the basis of vertebrate cellular and humoral responses, supporting the erroneous assumption made by classical immunologists that invertebrates possess no form of adaptive immunity. Throughout ontogeny and phylogeny, cell-mediated reactions appear to be the earliest specific immune responses, and are therefore common to all multicellular animals (Cohen and Sigel, 1982; Fudenberg et al., 1978; Good and Papermaster, 1964). Graft rejection in multicellular invertebrates is characterized by infiltration of the grafts by leukocytes strongly resembling those of vertebrates.

Invertebrate immunity is a vast, relatively unexplored, and exciting area, which offers many opportunities for sustained analysis (Anderson and Sigel, 1983; Bang, 1967; Brehelin, 1986; Burnet, 1974; Coombe et al., 1984; Cooper, 1974; Ermak, 1976; Greenberg, 1987; Ratcliffe et al., 1985). These studies may be expected to confirm that immune reactions are biological phenomena characteristic of all living species. As well, they will contribute to our understanding of mechanisms—the quest for cause-and-effect relationships, including the evolutionary pressures that cause immunogenesis. We can also more easily identify the origins of vertebrate immune responses through experimentation with ancestral animals. A final and perhaps more practical reason for these studies is to define the utility of any immune system—which we can do by comparing immune capabilities, whether primitive or advanced, invertebrate or vertebrate. In other words, the question of how the immune system evolved to its present state in humans can be answered only if we understand less complex immune systems. It is useful to speculate about the nature of the evolutionary pressures that were responsible for development of the immune system. A major pressure would be the need to maintain homeostasis in the face of threats of extinction by external pathogens. A second selective force might have been the need to maintain the internal milieu as a safeguard against altered self components, including the need for an internal surveillance and regulatory system to guard against the development of cancer or degenerative diseases. It is also of heuristic value to speculate that development of the vertebrate immune system was paralleled by an increasing diversity of leukocyte types.

Relationship Between the Evolution of Immunity and Cancer

Cancer in invertebrates, fish, amphibians, and reptiles has been the subject of two early reviews. Much controversy and speculation once centered around the question of whether tumors could in fact develop in animals other than birds and mammals; these speculations now seem similar to those that concerned the existence of immunity in more primitive creatures. Whereas comparative immunology is beginning to keep pace with its parent discipline, however, comparative oncology still lags behind (Dawe and Harshbarger, 1968; Mizell, 1969). The problem of immunity and neoplasia in invertebrates, as in vertebrates, has a central characteristic—namely, the capacity to distinguish between *self* and *non-self* material and to display specific recognition (Marchalonis and Cohen, 1980; Mizuno et al., 1982; van de Vyver and Herlant-Meewis, 1974). In an historical context this potential was recognized nearly 50 years ago as "individuality differentials" (Loeb, 1945). Exogenous infectious material introduced into invertebrates is sequestered and induces a state of rapid immunity, but neoplastic cells that arise from within (assuming they are present in invertebrates) may not elicit an immune response. Assuming a cause-and-effect relationship, animals would then possess a cell-mediated immune system as defense against neoplasia. Thus, animals with immune systems could be expected to have neoplasia, and immunological deficiency caused by breakdown of the immune system would allow neoplasms to develop (Burnet, 1970). Studies of neoplasia and immunity in primitive animals such as mollusks (snails, clams), in which there are more data on both phenomena, are important for understanding how these two phenomena evolved. Some form of immunity with both specific and nonspecific components exists among invertebrates. The mollusks and arthropods are the only groups, however, in which there are unequivocal cases of neoplastic growths with characteristics similar to those in mammals. The major problem that must be resolved is whether there is a causal association between the amount and quality of invertebrate immune reactions and the capacity to resist the development of abnormal growths analogous or homologous to vertebrate neoplasia. Perhaps it is safer to conclude simply that neither neoplasia nor immunity is a purely vertebrate attribute.

The Receptor Problem

Whether dealing with simpler animals or advanced protostomes, deuterostomes, and vertebrates, the ability to distinguish between *self* and *non-self* is a universal attribute that requires expression of a receptor or receptors. Indeed, recognition mechanisms can be identified in even the simplest protozoans, in which transplantation of cell organelles causes incompatibility reactions as basic to immunity as primitive irritability is to the nervous system. *Quasi-immunorecognition* is the capacity for recognition of non-self tissue followed by incompatibility reactions. Regardless of the experimental system, recognition is probably the fundamental event that initiates the immune response, and comparative phylogenetic analysis is essential to understanding the basic mechanisms involved. Two examples that support the presumed existence of receptors in invertebrates are worth reviewing for two reasons. First, there is a comparison between the protostome and deuterostome lines; second, both putative cell receptor structures are related to humoral substances. This raises the possibility that invertebrate receptors may turn out to be humoral products similar to leukocyte receptors in vertebrates. Finally, these receptors appear to share many similarities even though the animals and assays studied have been quite different.

Antibodies or immunoglobulins can serve as cell surface receptors for antigens, but invertebrate humoral products, including agglutinins, have not yet been clearly and universally shown to be receptors, although excellent preliminary work has suggested they have such properties. Unlike vertebrate antibodies, the humoral substances of invertebrates still remain to be fully biochemically characterized. Progress has been slow, owing to technical and conceptual limitations in our understanding of the synthesis of these humoral mediators. Despite the debate over precursor T cells (cell-mediated immunity) and the question of humoral substances as receptors (humoral immunity), it should be possible to arrive at a definition of adaptive immunity that encompasses both vertebrate and invertebrate characteristics. Thus, at a minimum, it should be assumed that certain characteristics of the leukocytes or humoral substances of an animal can be altered after antigenic stimulation has triggered a receptor.

MOLECULAR EVENTS IN THE EVOLUTION OF THE IMMUNE SYSTEM

Ancestors of the Major Histocompatibility Complex (MHC) and MHC-Like Molecules in Invertebrates

The major histocompatibility complex (MHC) is an extensive genetic locus that encodes numerous different proteins and determines various biological functions, most importantly those related to immune responses (Kelsoe and Schulze, 1987; Klein, 1986). The MHC is characterized by the expression of polymorphic cell surface heterodimeric glycoproteins, corresponding to classical MHC transplantation antigens. These molecules are divided into two multigene families termed class I and class II, which are related in structure, function, and presumably evolution. MHC antigens can be recognized by certain thymus-derived (T) lymphocytes via clonally expressed T cell receptors (TCR). These cells mediate tissue graft rejection and various other cell-mediated immune phenomena. T lymphocytes that recognize classical transplantation MHC molecules are an important component of the immune system, both in resistance to disease and involvement in the pathogenesis of autoimmune disease.

In searching for the origins of MHC antigens in invertebrates, comparative immunologists have taken similar approaches to those used to identify MHC molecules in mammals. Allorecognition responses in many invertebrates resemble vertebrate graft rejection reactions. These allorecognition phemonena have been extensively documented in a wide range of colonial marine invertebrates, from metazoans as primitive as sponges (Phylum Porifera) to tunicates (Subphylum Urochordata), which are thought to be the closest ancestors of vertebrates. Presumably, the primary function of polymorphic allorecognition systems in invertebrates, like those of vertebrates, is to recognize pathogens.

Allorecognition responses in vertebrates and invertebrates may have evolved independently, leaving invertebrate systems based on molecular families unrelated to the MHC (Cooper, 1977; Hildeman and Cooper, 1970). In plants, there is a superficially similar system. During fertilization, the incompatibility system of some plants is determined by the highly polymorphic S locus, which evidently encodes proteins with no similarity to MHC molecules. At the molecular level, allorecognition in vertebrates is based on interaction between peptides bound in the groove of the MHC molecule and the T cell receptor, rather than on direct binding of polymorphic residues on MHC molecules. Invertebrate allorecognition reactions may function similarly, in which case a polymorphic histocompatibility molecule is not required; the polymorphisms of bound peptides (like minor antigens) could be sufficient to mediate allorecognition. One final point deserves mention. There are at least three alternative mechanisms that may account for the origins of vertebrate MHC molecules but not for the development of polymorphism. These include facilitation of cell–cell interactions, peptide binding, or intracellular transport. These three possibilities may be more important than polymorphism for the functions of ancestral MHC molecules.

The Ig Superfamily

The Importance of Flexibility in Generating Diversity

To initiate an immune response, an antigen must interact with an immunocompetent cell by perturbation of its cell surface receptor. There are specific cell surface molecules that are involved in antigen recognition, as well as other functions mediated by T (thymus-derived) and B (bursa or bone marrow–derived) lymphocytes. All these molecules, despite their apparent heterogeneity, are made up of variable numbers of homologous units, which correspond in three-dimensional structure to a single immunoglobulin (Ig) domain. This structure, according to present interpretations, may have evolved to serve as a basic structural and functional unit for recognition. During evolution, it has been used to generate numerous different molecules, which participate not only in immune responses but also in aspects of recognition within the nervous system. Taken together, this collection of related molecules constitutes the Ig superfamily.

The immune system is phenotypically complex, and its evolution may have proceeded within limits imposed by this superfamily of molecules. The common evolutionary and structural denominator of these molecules is probably a highly conserved, yet extremely versatile, three-dimensional structure that supports two important features. First, the presence of a highly conserved Ig fold may accommodate many different sequences, and, hence, may have evolved to efficiently generate diverse structures, for example, V (variable) domains in Ig and T cell receptors. Second, this domain structure contains a highly conserved feature, namely the ability to interact with other structurally related sequences; thus, duplication of these domains may have led to the development of new sets of molecules with an inherent ability to interact (e.g., T cell receptor–MHC polypeptide; poly Ig receptor–IgA and IgM; CD8–MHC class I polypeptide, etc.).

Considering the marked diversity of sequences among Ig-related domains within the superfamily and consistent with the fact that complex molecules like Ig and MHC are only found in vertebrates, the evolutionary starting point of the superfamily (the primordial domain) may be found before the emergence of vertebrates. The gene that encoded the primordial domain may have also encoded a cell surface protein with an Ig-like fold having both V- and C (constant)-domain characterstics. The general tendency for these domains to interact suggests that this primitive molecule may have been a homodimer or may have interacted with similar units expressed on different cell types.

Partial duplication and independent divergence taking place within a number of these primitive molecules may have led to the development of more specialized V-like and C-like domains, and this probably was a critical early event in the evolution of the superfamily. This would have made it possible to generate proteins with separate regions for recognition and for membrane-anchoring or effector functions. At some point early in the evolution of the Ig superfamily, genes encoding Thy-1, β_2-m, MRC-OX2, and the poly Ig receptor may have begun to diverge, and these may have been followed by the development of precursors to MHC-encoded class I and class II polypeptides. The Thy-

1 gene may have diverged first as it encodes a single domain that is intermediate in structure between the V and C domains (Cooper and Mansour, 1989).

Acquisition of mechanisms for DNA rearrangement was a second critical event in the evolution of the Ig superfamily. Development of the rearrangement mechanism would provide new opportunities for molecules—including immunoglobulins, T cell receptors, and their accessory molecules—to diversify through (a) an extensive duplication of gene segments, (b) combinatorial joining of these gene segments, and (c) somatic mutation arising from joining imprecisions.

Evolution of an Immune System

Immunoglobulin superfamily evolution probably stems from the development of heterophilic recognition between related molecules (Williams, 1987). A heterophilic recognition system that had undergone extensive diversification would allow sophisticated signaling in cell interactions. It is necessary to consider how recognition-mediating interactions within an organism could be externalized to produce an immune system. According to one view, external recognition could have been derived from a cytotoxicity system that involved programmed cell death. These early cytotoxic cells could have provided a precursor for the vertebrate immune system. Subsequent integration of functional and structural aspects of recognition and killing could occur if the specificity of a killer cell for programmed cell death was conferred by Ig-related molecules. Finally, if this recognition specificity underwent modification such that it incorporated a determinant expressed by a virus or parasite, it would then possess features similar to those of T lymphocyte antigen receptors. Increasing evidence suggests that natural cell cytotoxicity is a possible link between invertebrate and vertebrate cell-mediated responses, and this speculation thus appears to merit further theoretical and experimental analysis.

Early Development of Immunoglobulins

Uniqueness of Immunoglobulins

The immunoglobulins are unique among members of multigene families. In addition to representing the most complex gene system described to date, immunoglobulins and T cell receptor (TCR) genes undergo somatic reorganization in individual cells committed to immunological recognition. In mammals, these rearrangements take place over extended chromosomal distances by a mechanism that involves sequence-specific recombination. The Ig heavy chain variable region (V_H) genes of mice and humans are the most complex that have been defined among variable gene systems with respect to the number and relative diversity of individual segmental components. In addition, they are distributed along much larger chromosomal regions than those for the light chain variable region (V_L) and the TCR variable region (V_T) genes (Yancopoulous and Alt, 1986). Considering the unusual nature of the Ig gene system and the complex control mechanisms that govern its expression, specific strategies have been developed to elucidate the bases for its evolutionary diversification. This includes the study of antibody gene structure and function in unique animal models that occupy critical and distant phylogenetic positions. The primary emphasis thus far has focused on the V_H system by direct comparative analysis of gene structure and organization in living species that represent critical points in the evolutionary radiation of vertebrates.

Evolution of the V_H System

An Ig gene system that resembles the system observed in higher vertebrates arose early in evolution (Litman, 1984). When and how it diverged from the TCR system is unclear, although there is little question that these gene families are closely related in structure and function. Even at distant levels of evolutionary development, the V_H gene family is complex and appears to undergo somatic rearrangement by mechanisms that resemble those of higher vertebrates. It is difficult to reconcile the genetic complexity of nonmammalian vertebrate immunoglobulins with the apparent species restrictions in antibody diversity. Diverse somatic effects, unusual cell population kinetics and different cell–cell interactions could nonetheless explain many of these biological observations.

The Ig gene system of one nonmammalian vertebrate *(Heterodontus)* is not merely equivalent to those in higher mammals. The V_H gene family in *Heterodontus* shows extreme differences in organization as compared to mammals, making this genetic system unique. The existence of multiple, closely linked gene clusters has necessitated the maintenance and regulation of a series of individual segmental elements throughout evolution. A major biological point that has emerged from these studies has suggested the presence of multiple C_H genes and the possibility that they may subserve different biological functions. Although the recent finding of VD- and VDJ-joined germline genes is without precedent in any other immune system, the functional consequences of this arrangement are not clear. The mechanisms by which this extended gene family is regulated in a coordinated manner are of considerable interest and enormous potential importance as they may help link vertebrate evolutionary development with the genetic basis for specific immunologic recognition.

How Ig Genes Are Organized and Assembled

Antibody Response in Relation to B Cells

Soluble antigens bind to immunoglobulins on the surfaces of only those B lymphocytes that express appropriate receptor specificities. The B cells that have bound antigen are induced to proliferate and to differentiate into cells that secrete Ig of the same or similar specificities as the parental cells; this is a manifestation of the *clonal selection theory.* In the years since this theory was first proposed, a large body of experimental evidence has been amassed in its support. The clonal selection theory required that (1) diversification of the antibody repertoire precedes initial exposure to antigen; (2) a given B lymphocyte expresses Ig of only one antigenic specificity; and (3) the antibody present at the cell surface must have the same specificity as the antibody that is secreted later in response to antigen. The theory therefore places unusual constraints on Ig structure and expression. How these constraints are met can best be understood by examining the molecular genetic basis of antibody diversity.

Significance to Developmental Immunology:
Structure of Ig Genes and Gene Rearrangement

The structure of immunoglobulins poses several unusual genetic problems, the two most obvious being (1) how the genome encodes such a large number (10^7 10^8) of variable regions and (2) how different variable regions come to be associated with the same constant region. Immunoglobulin gene rearrangement is an essential feature of B cell differentiation. Two recombinational mechanisms participate in the assembly of Ig genes (Honjo, 1983). The first generates DNA sequences that encode the V regions of heavy and light chains, and it occurs via a developmentally ordered series of rearrangements. Regulation of this process underlies the phenomena of allelic and isotypic exclusion (the expression of only one allelic form of an Ig heavy chain and one light chain isotype, respectively, by a single B lymphocyte). The second type of rearrangement is responsible for Ig class switching. An increasing body of evidence suggests that class switching, like other aspects of B cell development, can be modulated by specific lymphokines. Thus, both types of immunoglobulin gene rearrangements are subject to developmental control.

At about day 12 of murine development, B cell precursors that express cytoplasmic μ chain, but not light chain ("pre-B cells"), appear in the fetal liver. This event precedes the appearance of surface IgM^+ B cells in fetal liver by several days. Analyses of rearrangements in normal and immortalized B-lineage cells reveals a defined developmental order in the assembly of heavy and light chain genes, although the anatomical site where regulation of rearrangement is effected is not yet known. The assembly of heavy and light chain genes is mediated by a common recombinational apparatus that is active in immature B cells but not in mature B cells or plasma cells (Rogers and Wall, 1984). Whereas the suppression of recombination that follows productive light chain rearrangements may involve the shutting off of a recombinational activity, regulation of heavy chain recombination requires more subtle and selective processes.

The T Cell Antigen Receptor Gene: Cell-Mediated Immunity in Relation to T Cells

T lymphocytes effect cell-mediated immunity and may be divided into at least two main functionally defined subcategories: those that provide help for other responses, and those that kill. Helper T cells act on B cells to stimulate Ig production or may be involved in induction of cytotoxic T cells. Cytotoxic (killer) cells directly eliminate virus-infected or other cells that may be antigenic. Both helper and cytotoxic cells recognize only antigens that have been processed and correspond to degraded peptide fragments of larger protein antigens. These peptides are bound to the antigen-binding groove of MHC-encoded cell surface molecules and can only then be recognized by T cells through specific cell surface receptors. The T cell antigen receptor (TCR) that mediates recognition of antigen in the context of MHC determinants has been defined in molecular and biochemical terms (Kronenberg et al., 1986; Lanier and Allison, 1987). It is a heterodimeric structure made up of disulfide-bonded α (acidic isoelectric point) and β (basic isoelectric point) chains. Another species of T cell receptor has been identified, which consists of a γ-chain disulfide-bonded to a TCR δ polypeptide (Janeway et al., 1988). The expression of TCR is subject to isotypic exclusion. This means that nearly all T cells express either α/β TCR or γ/δ TCR, but not both. It is not yet certain whether γ/δ TCRs recognize antigens bound to MHC molecules in a fashion similar to α/β TCR, and there remain many other unresolved questions about these receptors. However, their origin early in ontogeny, as well as their presence in sites such as skin and the intestinal epithelium where contact with pathogens first occurs, has sparked new hypotheses about the developmental physiology of the immune system.

CELLS OF THE IMMUNE SYSTEM:
THEIR DEVELOPMENT

Stem Cells

Evolutionary Aspects of Hemopoiesis: the Condition in Ectothermic Vertebrates

The body can regenerate blood cells, including cells of the immune system. The hemopoietic system is unique in that it is developmentally active throughout an animal's life span and constantly generates a large variety of functionally diverse cells (LeDouarin et al., 1976). By possessing the ability to regenerate its own kind, the hemopoietic system is particularly suitable for analyzing problems related to developmental biology, including lineage relationships and regulation of differentiation (Hadjii-Azimi et al., 1987).

Almost all terrestrial vertebrates possess hemopoietic bone marrow (Campbell, 1970; Cooper, 1976 a, b, c; Cooper et al., 1980). Beginning with amphibians, this tissue becomes progressively more specialized during evolution, through reptiles, birds, and mammals (Wright and Cooper, 1976). One may question why this tissue is present only in terrestrial vertebrates and why hemopoiesis is confined to the marrow. A simple explanation would include the presence of long bones in tetrapods, although this does not account for the presence of hemopoietic sites in animals without marrow or long bones.

Hemopoiesis in Invertebrates

All coelomate invertebrates possess hemopoietic/lymphoid tissues and can generate cells that participate in immune responses. Comparative immunologists are only now beginning to recognize that numerous blood cell types participate in invertebrate defense reactions (Ratcliffe and Rowley, 1981). Cells of invertebrate immune systems can be subdivided into two main groups: the freely circulating blood cells, or coelomocytes, and the various fixed cells. Fixed cells may be either scattered throughout the tissues or localized in hemopoietic/phagocytic organs. As in vertebrates, cell nomenclature is complicated; often terms overlap, and many descriptions have been based on supposed functions of particular cells or populations. Because of the morphological similarity between invertebrate progenitor cells and vertebrate lymphocytes, there has been considerable controversy over the evolutionary origins of lymphocytes. In tunicates, for example, stem cells and lymphocyte-like cells are apparently morphologically distinct.

Monocytes, Macrophages, and Accessory Cells

Relevant History to the Entire Animal Kingdom

Metchnikoff was the first investigator to identify phagocytic cells, based on experiments in which rose thorns were inserted into the blastcoel of transparent starfish gastrulae. The thorns rapidly became encapsulated by a large number of free amoeboid cells. Subsequently, it became apparent that functionally identical cells are present throughout the animal kingdom and that they act to engulf or isolate foreign materials, including inorganic and organic particulates, microbes, effete cells, and cell debris. The term macrophage has come to designate the free or fixed large, mononuclear phagocytes that are found in tissues, while smaller, polymorphonuclear phagocytes or neutrophils, known originally as microphages, circulate in the blood.

The evolutionarily oldest cells in the animal kingdom are the macrophages, or mononuclear phagocytes, whose origins can be traced to protozoans, the evolutionary progenitors of existing animals. Protozoans are unicellular organisms and are capable of ingesting particulate matter, some of which is nutritious, from the environment. Metazoans (multicellular animals) probably originated from colonies of ancient protozoans that were composed of food-getting outer zooids, and non-food-getting, but diversely specialized, inner zooids. In existing metazoans, the outer zooids are analogous to the gut-associated, food-ingesting epithelial cells of poriferans, coelenterates, and platyhelminths, whereas the inner zooids correspond to the differentiated cells that make up various specialized tissues. Among these specialized cells, one population has conserved its original protozoan characteristics. Such cells therefore probably serve as the progenitors of macrophages because they do not participate directly in food-gathering but are fixed in tissues where they eliminate or isolate exogenous, "foreign" materials, as well as "denatured" endogenous ones.

Macrophages in Vertebrates

The reticuloendothelial system (RES) theory suggests that cells that strongly ingest vital dyes, such as histiocytes and monocytes (endothelial leukocytes or blood histiocytes), may be derived from cells that take up vital dyes weakly, including reticulum cells of lymphoid tissues or organs, and from RES cells of lymph node sinuses and sinusoids in the liver, spleen, bone marrow, adrenal cortex, and pituitary (van Furth et al., 1982). Historically, macrophages were assumed to be derived from mesenchymal cells and to be independent of the hemopoietic system. In the 1960s, technical advances in cell biology, including cytochemistry, autoradiography, and utilization of chromosomal markers, contributed greatly to analyses of cell lineages; based on these studies monocytes were identified as the precursors of macrophages.

RES theory provides a unified interpretation of the cellular lineages of mononuclear phagocytes. All highly phagocytic mononuclear cells and their precursors can be placed in one system, termed the mononuclear phagocyte system (MPS). According to this view, all mononuclear phagocytes, or macrophages, in adult animals are derived solely from blood monocytes, which are the progeny of promonocytes and monoblasts and which, in turn, originate from hemopoietic stem cells in the bone marrow. Monoblasts and promonocytes are morphologically identifiable in the bone marrow, and they share various characteristics with blood monocytes. Thus, they express receptors for the Fc region of IgG (FcR) and for the activated third component of complement (C3bR). They also possess phagocytic and pinocytic capacities. Most importantly, macrophages play a crucial role in the presentation of antigens after initiation of an adaptive immune response (Unanue, 1984).

Natural Killer (NK) Cells

Characteristics of NK Cells and Their Surface Phenotype

Natural killer (NK) cells belong to the lymphocyte lineage; however, they differ from classical B cells and T cells by several functional, phenotypic, and morphological characteristics (Brion and Welsh, 1986). NK cells are defined by their ability to destroy a variety of tumor cells rapidly (within the first hour after contact with the target) without prior stimulation and/or arming with antibody (Herberman, 1986; Lotzová and Herberman, 1987). NK cells are derived from bone marrow (Hackett et al., 1986), and their activity is mediated independently of the thymus and is not restricted by MHC class I or class II antigens. In contrast to T and B lymphocytes, specific secondary (memory) responses have not been shown to be mediated by NK cells; thus, their responses are not "learned" or adaptive. In contrast to B cells, NK cells do not express surface immunoglobulin, and they do not bear classical T cell markers. Unlike macrophages, NK cells are not phagocytic and they are nonadherent. However, some NK cells do adhere to nylon wool or to glass or plastic surfaces, especially after activation.

Most human NK cells react with Leu-11, B73.1, and 3G8 antibodies, which recognize epitopes on the Fc gamma receptor (CD16 antigen). Although treatment with antibodies against this structure largely abrogates typical NK cell activity, the CD16 molecule is not NK cell-specific, since it is also present on granulocytes and some macrophages. Most NK cells react with the OKM1 and Leu-15 antibodies, which detect the CD11b antigen. This antigen is expressed at high levels on monocyte/macrophages, polymorphonuclear leukocytes, and certain subsets of T cells. However, CD11b is present only at low density on NK cells and does not appear to be expressed on NK cells maintained in culture.

Phylogeny of NK Cells

NK cell function seems to precede T and B cell-mediated responses in phylogeny, and NK cell-like activity has been detected in invertebrates. Although invertebrate cytotoxic cells have not been characterized with respect to surface phenotype, they resemble NK cells functionally and morphologically. Thus, cytotoxic effector cells are nylon wool–nonadherent in echinoderms and mollusks and are nonspecific esterase-negative. Their cytotoxic function is inhibited by specific monosaccharides. Ectothermic vertebrates, including fish and amphibians, also possess cells that express NK-like cytotoxic activity. For instance, freshwater fish possess cells that are cytotoxic against a range of established normal and malignant mammalian cell lines. The highest levels of cytotoxic activity have been found in the pronephros (a hemopoietic organ in fish); lower and variable activity is present in peripheral blood and spleen. Cytotoxicity is qualitatively similar to that displayed by mammalian NK cells in terms of its rapidity (lysis within 6 hours) and lack of species restriction. In amphibia, splenocytes are the main source of NK-like cells, whereas bone marrow and thymus exhibit negligible cytotoxic levels. Interestingly, there is clear evidence for preservation of NK-like cells during phylogeny. As an example, a conserved NK receptor-like molecule has recently been shown to be expressed on fish, mouse, and rat NK cells. This conservation of NK cell function suggests that NK cells play an important role throughout the animal kingdom.

T Lymphocytes: Activation and Characterization

The Environment

We are only now beginning to understand the mechanisms by which lymphocytes interact with their environment, particularly how contact with foreign antigen initiates cell activation. Regulation of the signaling networks, which determine how contact with a foreign antigen leads to an appropriate cellular response, is less clear. Regulatory events in immune responses are complicated, particularly when one considers the requirements for maturation and differentiation of both T and B lymphocytes. Although the transmembrane signals that are generated in mature and immature cells are apparently qualitatively similar, their physiologic role in the development of receptor repertoires, in self recognition, and in tolerance in immature cells is unknown. Cells interact with their environment either through direct cell–cell contact or through the action of intercellular mediators such as steroids, polypeptide hormones, or neurotransmitters. Steroid hormones penetrate the cell membrane and may interact directly with intracellular regulatory elements. Steroid hormones are known to influence immune function both directly and indirectly, but little is known about the role of these mediators in normal immune responses.

Activation of Lymphocytes

Resting T and B cells display restricted patterns of expression of cell surface receptors. After activation mediated by binding of ligand to the antigen receptor, receptors for several lymphokines are expressed on the cell surface. The interaction of these newly expressed receptors with their specific ligands leads to proliferation and differentiation of responding cells. Therefore, activation of particular antigen receptors may determine the specificity of the induced response. However, the cell's intracellular machinery, as determined by its lineage and state of differentiation, establishes the types of responses that are generated. The array of lymphokine receptors expressed, together with the pattern of lymphokines produced, regulates the magnitude of the response. The duration of responses is determined both by the kinetics of local lymphokine production and by the range of lymphokine receptors on cell surfaces.

Activation of lymphocytes normally occurs within the content of a recognition complex, which includes antigen presenting cells, helper and effector T lymphocytes, and B lymphocytes (MacDonald and Nabholz, 1986). The interaction between TCR and antigen in the context of MHC has a relatively low affinity. However, the complex is stabilized by the presence of several adhesion molecules such as CD2, CD4, and N-CAM, and of lymphocyte function-associated antigens (LFAs) expressed on the surface of the responding and stimulating cells. The adhesion molecules do not appear to function solely to stabilize the complex but may also be involved in signal transduction. This may result in amplification of signals transduced through the TCR or may communicate negative signals in the absence of effective TCR ligation.

Within this activation complex, local concentrations of lymphokines are sufficient to allow binding to high-affinity receptors. In some cases, lymphokines are secreted specifically from the pole of the cell involved in the complex, thus increasing still further their local concentrations. Since lymphokines bind to their receptors and are then endocytosed and degraded, most of the secreted lymphokines never leave the activation complex. Dilution in interstitial fluid ensures that lymphokine concentrations decrease rapidly as the distance from the complex increases. Furthermore, systemic-free lymphokine concentrations are limited by the presence of binding proteins, which in the case of interleukins 2 and 4 (IL-2 and IL-4) may represent secreted forms of the receptor itself.

Activation of cell surface receptors can lead to additive, synergistic, or antagonistic effects, depending on the cell type and on the nature of response. Although activation via the T cell receptor is required to induce optimal expression of most lymphokine receptors, concurrent engagement of lymphokine receptors and the TCR may in fact decrease the ultimate response, possibly through attenuation of transmembrane signaling. Similarly, IL-2 and IL-4 synergize in the proliferative response of some T lymphoid cell lines, whereas IL-4 can inhibit the effect of IL-2 on B cell activation and proliferation. Therefore, receptor-mediated activation should not be considered in isolation, and the ultimate response includes a component of crosstalk between signals transduced through an array of receptors on the cell surface. Moreover, a given biochemical event should not be considered to have solely stimulatory (positive) or negative effects. The final effect will reflect the character, magnitude, and duration of the various signals, together with the lineage and differentiation state of the responding cell.

Characteristics

T lymphocytes are morphologically similar to B lymphocytes and express clonally-distributed antigen receptors. During the development of B and T lymphocytes, DNA gene segments that are required for the synthesis of receptor molecules for both lymphocyte classes undergo somatic rearrangement. However, T lymphocytes differ from B lymphocytes in two aspects: (1) T lymphocytes do not directly bind soluble antigens; (2) T lymphocytes recognize foreign antigens only in the context of molecules that are encoded by MHC and expressed on the surface of viable cells (Swain, 1983). This form of antigen recognition, referred to as MHC restriction, enables T lymphocytes to perform two vital immune functions, namely the destruction of pathogen-infected host cells by effector T lymphocytes, and the delivery of signals to B lymphocytes by helper T lymphocytes (T_H). In this manner, T_H cells regulate both the quantity and class of specific antibodies made by B cells.

Despite the requirement for self MHC molecules in antigen recognition by T cells, T cells do not normally respond to self MHC antigens alone or to self MHC plus self antigens (Wagner et al., 1981), a manifestation of self tolerance. However, T cells show a marked propensity to respond to non-self MHC molecules, a phenomenon referred to as alloreactivity. The mechanisms by which T cells learn to recognize foreign MHC antigens while remaining tolerant of self MHC and other self antigens is an important biological question and is of much interest to immunologists concerned with the developmental biology of T cells (Nossal, 1983; von Boehmer, 1988).

Surface Markers That Define T Lymphocyte Subsets

All murine T lymphocytes express the Thy-1 antigen, a 24-kD glycoprotein that is recognized by a number of monoclonal antibodies (mAb). Recently, the CD3 complex has been shown to be a more definitive marker for T lymphocytes. The complex comprises at least six distinct proteins, which are noncovalently associated with the TCR and which appear to be involved in signal transduction in cell activation. Two other surface markers, designated CD4 (previously L3T4 in the mouse) and CD8 (formerly Lyt-2 in the mouse), serve to identify two major T lymphocyte subpopulations with the $CD4^+8^-$ or $CD4^-8^+$ cell surface phenotypes.

The murine Lyt-2 molecule is the homologue of the human CD8 molecule. In mice, the Lyt-2 molecule is normally coexpressed on T lymphocytes with the Lyt-3 molecule. The Lyt-2/3 molecular complex is a 70-kD heterodimer, made up of two covalently linked chains, an α chain (35–38 kD), which expresses epitopes detected by anti-Lyt-2 mAb, and a β chain (30–34 kD), which expresses Lyt-3 epitopes. The Lyt-2 and Lyt-3 molecules are transmembrane glycoproteins and are both encoded by closely linked genes on chromosome 6. There are two variants of Lyt-2 chains, designated α and α'. These products are of slightly different molecular weight and probably arise from alternative mRNA splicing. The biological function of the Lyt-2/3 heterodimer is largely determined by the Lyt-2 chain, a conclusion that has been confirmed by transfection of the Lyt-2 gene into CD8$^-$ T cells. The murine L3T4 (CD4) molecule exists as a monomeric cell surface glycoprotein with a molecular mass of about 55 kD.

What Does the Future Hold with Respect to T Lymphocytes?

Dramatic advances have recently occurred in our understanding of the mechanisms that lead to T cell tolerance of self antigens and T cell repertoire selection. These have come about through the use of TCR transgenic mice and specific antigenic systems in which large numbers of specifically reactive T cells are present and can be detected with TCR-specific mAbs. However, the cells within the thymus that mediate positive or negative selection and the intracellular signals for these processes have yet to be determined. TCR transgenic mice also have allowed immunologists to obtain large numbers of homogeneous antigen-specific T cells in vivo. These cells have been particularly useful for (1) analyzing the mechanisms by which MHC-restricted T cells are activated; (2) understanding the mechanisms by which mature CD4$^+$8$^-$ or CD4$^-$8$^+$ T cells become tolerant of their specific antigens; and (3) analyzing the various mechanisms that lead to autoimmunity.

B Lymphocytes and Regulation of Their Responses

The Situation in Mammals and Birds

Vertebrates are able to mount specific antibody responses to numerous antigens, including those to which they have never been exposed. Because of this unique trait, the humoral immune response is said to be adaptive. In an individual who has previously encountered a particular antigen, a second exposure generally elicits an antibody response that is more rapid, and of higher titer, than the original response. In this respect, the humoral response is said to be anamnestic—that is, it possesses a memory component. The adaptive and anamnestic features of humoral immune responses have elegantly been explained at a cellular level by the clonal selection theory of MacFarlane Burnet (1959). According to this theory, an organism generates large numbers of B lymphocytes, each expressing immunoglobulin of a single specificity. The clear demonstration that "bursa-derived cells" in birds and B cells in mammals could be identified by the presence of cell surface immunoglobulin (sIg) fulfilled an important prediction of Ehrlich, made some 70 years earlier.

B Lymphocyte Development in Poikilothermic Vertebrates

B-lineage precursors appear sequentially in developing urogenital tissues, gut-related tissues, and the bone marrow throughout phylogeny (Kincade, 1981). B lymphopoiesis in endothermic vertebrates occurs not only during embryonic development, but also continues postnatally, when it is associated with organized gut-derived sites or with the bone marrow. The striking differences in the anatomical location of B lymphopoiesis in birds and mammals presents a fascinating evolutionary puzzle, the understanding of which may teach us much about the selective pressures that have affected the evolution of lymphopoiesis in ectothermic vertebrates (fish, amphibians, and reptiles) (Cuchens et al., 1976).

Regulation of B Cell Responses

B lymphocytes have for many years been the major focus of immunologists for studies of the mechanisms that control cellular responses to external stimuli. A number of different techniques have been used to identify distinct B cell activation and differentiation stages. These include antigen binding to cell surface Ig receptors, measurements of cell proliferation as detected by isotope incorporation, and analysis of population kinetics for antibody-secreting cells. The external stimuli involved in B cell activation have been gradually defined, in parallel with progress in identifying the various steps in antibody synthesis. It has been shown that an immunocompetent B cell must not only bind antigen but must also interact with co-stimulators derived from non-B cells in order to achieve a full immune response, including the secretion of antibody.

Antigen-specific helper T cells might increase their potential for interaction with B cells through secretion of antigen-specific helper factors. However, molecular and biochemical analyses have so far failed to provide evidence for the existence of such helper factors, which were thought to correspond to secretory forms of T cell antigen receptors. Most available evidence, taken together, suggests that the initial triggering stimulus in T–B cell interactions involves a direct cooperative contact between the cooperating cells via an antigen bridge. Following this initial contact, antigen nonspecific lymphokines, which are secreted by T cells and act locally on B cells, may contribute additional stimuli. Analysis of B cell growth and differentiation has, for the most part, been carried out using experimental systems in which antigen-specific helper T cell function is replaced by any of a number of ligands, including T cell–independent (TI) antigens, polyclonal B cell activators, and antibodies directed at antigen receptors.

The lack of a consensus on mechanisms of B cell activation by cytokines and other factors may in part reflect the various experimental models that have been used by different investigators. Many of these models have failed to take sufficient account of B cell heterogeneity, possible contamination of B cells by other cell types, stimulatory artifacts due to the presence of foreign serum proteins in culture media, the use of mycoplasma-infected cell lines in reconstitution experiments, and the use of insufficiently defined cytokine preparations. Several distinct signals have nonetheless been identified and shown to mediate the functionally distinct steps that make up the B cell response. These include activation from the (resting) G_0 to the G_1 state, followed by proliferation and differentiation into the antibody-secreting effector cells (Leanderson et al., 1987). Lipopolysaccharide (LPS)-stimulated B cells have been shown to mature into antibody-forming cells at a much higher rate in the presence of IL-2 and IL-1 than in the presence of IL-2 alone. This marked synergism between IL-2 and IL-1 critically depends on the initial LPS-induced proliferative stimulus for the B cells (Fotedar and Diener, 1988).

Inactivation of B Lymphocytes

Studies on antigen-specific B cell inactivation, also referred to as B cell tolerance or unresponsiveness, have paralleled work on B cell activation. Engagement of B cell surface immunoglobulin (sIg) receptors with appropriate ligands, such as antigen or anti-Ig anti-bodies, may under certain conditions cause the cells to become refractory to a subsequent immunogenic challenge. Various theories have been advanced to account for the mechanisms of this unresponsiveness; in general, they argue that induction of unresponsiveness depends either on quantitative parameters of portions of the ligand or on the physiological state of the B cell. In either case, B cell unresponsiveness results from the interaction of B cell receptors with a ligand in the absence of additional cellular or humoral mediators. Thus, various versions of clonal deletion/abortion hypotheses seem irreconcilable with theories that postulate the involvement of suppressor T cells as the chief mechanism that controls tolerance to "self" (Diener and Waters, 1986).

MOLECULAR ASPECTS OF RECOGNITION AND REGULATION

Antigen Processing and Presentation

For an antigen to be processed by an antigen presenting cell (APC), it must first bind either specifically, via sIg on B cells, or nonspecifically to a component on the plasma membrane. Subsequently, multiple biochemical events occur within the APC that lead to the production of proteolytic peptide fragments which are immunogenic. Proteolysis may be mediated by enzymes located either at the plasma membrane or within the cell. The final and most crucial stage for effective antigen presentation involves association between the antigenic peptide and the antigen-binding groove of MHC class I or class II molecules. This process then gives rise to a specific peptide/MHC complex that can then be recognized by the T cell receptor (Berzofsky et al., 1988). The factors that regulate the magnitude of T cell stimulation and the net immune response remain uncertain, although they probably are related to properties of the peptides that interact with MHC antigens.

Developmental Aspects of Antigen Processing and Presentation in Invertebrates and Ectothermic Vertebrates

Antigen receptors on B and T cells share certain genetic and structural similarities, yet antigen recognition by T cells differs qualitatively from that by B cells. T cell receptors generally do not recognize native antigen, but instead interact with a degraded or processed form of native antigen associated with MHC-encoded molecules expressed on the surface of APCs. B cell sIg's, on the other hand, can recognize epitopes on native antigens and/or on degraded forms of the antigen.

There is little information concerning antigen processing by invertebrates and no firm evidence that molecules homologous to those encoded by the mammalian MHC exist within invertebrate species, although there is serological evidence for the existence of Ig gene superfamily members (β_2-microglobulin and Thy-1). It is therefore unclear whether antigen processing and presentation mechanisms are required in invertebrates. However, vertebrate antigen processing mechanisms may be assumed to have evolved from mechanisms in lower vertebrates and invertebrates, and some speculations therefore seem justified. Engulfment reactions and phagocytosis occur in almost all invertebrates. These reactions probably represent the precursors of the antigen processing mechanisms seen in vertebrates. Phagocytic cells that resemble macrophages, the classical mammalian APC, are found in most invertebrates and contain cytoplasmic vacuoles. In higher forms, enzymes are present within the vacuole which can degrade experimentally administered dyes with subsequent disappearance of the vacuole. Such degradation pathways are characteristic of mammalian macrophages but usually correlate with a terminal degradation event rather than with antigen processing (Unanue, 1984).

Let us now move to the vertebrates. Beginning in the fish, molecules with specific immune-associated functions are consistently present, including MHC-encoded antigens. For example, molecules homologous to MHC antigens have been discovered in advanced amphibians (anurans), fishes (teleosts), reptiles, and birds, and expression of these antigens correlates with the presence of mixed lymphocyte culture (MLC) reactivity and allograft reactions. The clawed toad *Xenopus* expresses both class I and class II MHC molecules that are polymorphic and have similar structural characteristics to mammalian MHC molecules (see Kaufman, Chapter 3 in this volume). The class II molecules of *Xenopus* are made up of α and β chains and have a molecular mass of 30–35 kD. In addition, toads contain helper and cytotoxic T cell subsets that are similar to mammals and show MHC restricted patterns of antigen recognition. These T cells can secrete growth factors similar to mammalian IL-2 when stimulated with the T cell mitogen phytohemagglutinin (PHA). The thymus of *Xenopus* contains both stromal and epithelial cells, which are probably involved in the education of T cell precursors and may also act as APC. Because of the anatomical and functional similarities that exist between *Xenopus* and mammals, it seems almost certain that the antigen processing and presentation pathways necessary for generating an immune response exist in ectothermic vertebrates.

Antigen Processing in Mammals

The generation of an immune response to a soluble protein antigen begins with binding of the antigen by an APC (Delovitch et al., 1988). The APC subsequently degrades (processes) the antigen into immunogenic peptides that associate with MHC antigens and can subsequently be recognized by antigen-specific T cells. Activated T cells in turn secrete lymphokines, which then stimulate various immune responses, including the production of immunoglobulins and/or the generation of cytotoxic lymphocytes. The basic requirements for cells to provide APC activity include (a) expression of class I or class II MHC molecules on their surface; (b) expression of biochemical pathways that degrade antigen; and (c) ability to synthesize and secrete cytokines such as IL-1. The classical APC is the macrophage; however, various cell types that possess the above characteristics can also function as APC and may therefore be important for generating immune responses in vivo. Such additional APC types may include dendritic cells, epithelial cells, monocytes, and B cells.

Complement

The Components of C and Their Mechanisms of Activation

The complement (C) system was discovered when it was observed that fresh serum would kill bacteria in the presence of antibodies. We now know much more about complement, including the close connections between activated complement proteins and biological functions related to inflammation. Thus, the complement system plays a major role in protection against infection by both specific and nonspecific mechanisms. In addition to mediating cell lysis, other biological activities of complement include immune adherence, viral neutralization, chemotaxis, and the release of lysosomal enzymes and histamine from basophils and mast cells (Müller-Eberhard, 1986). Other functions include involvement in bone metabolism, the production of antibodies, and cell proliferation. Finally, several components of complement behave as so-called "acute phase proteins" as their serum levels increase markedly during infection and inflammation.

The complement system consists of about 20 plasma and cell membrane proteins that undergo extensive mutual interactions on activation and thus acquire new biological activities. A unique characteristic of many components of complement is their intrinsic ability to be transformed from soluble molecules to membrane-associated species through the generation of proteolytic fragments that express specific binding activity. In fact, cleavage of many complement proteins results in the formation of fragments that are capable, at least for a short period of time, of binding to specific receptors on cells or bacteria or to other complement molecules.

Phylogeny of Complement

Studies on the evolution of complement have shown that complex systems of interacting proteins with biological and physicochemical properties similar to human complement are present in several mammalian vertebrates. In species such as guinea pigs, mice, rabbits, sheep, and cattle, numerous proteins have been identified that are involved in both classical and alternative pathways of complement activation. Both pathways generate C5 convertase and lead to the same membrane attack or lytic stage of complement activity. In many instances, components of complement from one species can interact with those of other species, although with variable efficiency.

Relatively little information is available on the complement system in ectothermic vertebrates (Koppenheffer, 1987). Results from early studies suggested that the vertebrate complement systems share a common evolutionary origin, but that differences exist between mammalian and ectothermic complement. Lamprey serum contains a factor that can lyse rabbit erythrocytes in the presence of Mg^{2+} but in the absence of specific antibodies. The lytic activity of lamprey sera is abolished by treatment with zymosan. These and similar observations in other invertebrate species have led to the suggestion that the primary evolutionary role of complement was to promote phagocytosis rather than to mediate cell lysis and that the alternative pathway, involving properdin and Factor B, represents the most primordial form of complement activation.

Cytokines

General Features

Cytokines are polypeptides that regulate the development and functions of cells that constitute the hemopoietic and immune systems. All cytokines that have been discovered up to now are pleiotropic in their activities and have effects on several different cell types. The regulatory mechanisms that influence the development and function of different cell lineages are clearly closely interwoven. The immune system is normally activated by the entry of pathogens (antigens) into the body, and their subsequent processing by antigen presenting cells. Cellular processes mediated by T lymphocytes after binding of these processed antigens initiate the synthesis of several cytokines. Cytokines form a communication network that stimulates lymphocyte development and function as it influences cell differentiation and renewal throughout the hemopoietic system. It now seems likely that these regulatory molecules can be used to selectively manipulate immune and hemopoietic functions.

Cytokines and Phylogeny

Studies of cytokines to date have concentrated on mammalian species and have revealed a considerable degree of sequence conservation and functional cross-reactivity among molecules derived from different members of this family. New information suggests that cytokines or cytokine-like molecules probably also exist in ectothermic vertebrates and invertebrates and we are only now beginning to appreciate their possible evolutionary significance. An IL-1–like protein has been isolated from the coelomic fluid and coelomocytes of an echinoderm, the starfish *Asterias forbesi,* and IL-1–like activity has been discovered in eight North American species of tunicates. Tunicate IL-1 has physical and biochemical properties that are strikingly similar to those of mammalian and echinoderm IL-1, and its activity in a thymocyte proliferation assay is inhibited by rabbit antibodies to human IL-1; this indicates considerable functional and structural homology (Beck and Habicht, 1986; Beck et al., 1989). Among ectotherms, IL-1–like activity has been identified in fish, and their lymphocytes respond to human IL-1. These findings support the view that IL-1 is an important molecule and is present from early in evolution. That it has been conserved in invertebrates and vertebrates underscores its important role during the evolution of host defenses.

　　IL-2–like activity has been identified in supernatants of mitogen-activated lymphocytes from amphibians *(Xenopus)* and fish *(Cyprinus).* Human, gibbon, and rat (but not mouse) IL-2 can stimulate carp lymphocytes, although cross-reactivity between *Xenopus* and human or murine IL-2 has not been observed (Hamby et al., 1986; Watkins and Cohen, 1985). It is therefore conceivable that IL-2 has been less conserved throughout evolution than has IL-1. Because gene probes for all known murine and human cytokines are now available, we can begin to systematically investigate the phylogeny of cytokines in order to better understand the significance of these molecules in evolution and development.

DEVELOPMENT OF THE IMMUNE SYSTEM
IN HEALTH AND DISEASE

Thymosins

The thymus is an endocrine organ that synthesizes several proteins having unique chemical structures and hormonal activities. The thymus and its hormones are responsible for the development and function of the T cell compartment of the immune system (Doria and Frasca, 1985). Implantation of a cell impermeable diffusion chamber that contains thymus into a neonatally thymectomized animal can prevent wasting, can reverse lymph node atrophy, and can establish immunological competence in these animals. Because the chamber prevents cell–cell contact between the implanted thymus and circulating host lymphocytes, its reconstitutive properties must be the result of humoral rather than cellular products. Beginning in the late 1960s and continuing in the present, efforts to purify thymic extracts and to identify which of the many "humoral" compounds produced within the thymus are responsible for restoring immunocompetency have been pursued with vigor.

Mechanisms of Action

The mechanisms of action of thymic hormones are currently being investigated by pharmacologic analyses, together with direct measurements of cyclic nucleotides and cytosolic calcium concentrations. Results to date indicate that thymic hormones act in lymphocyte activation and maturation via a second messenger mechanism. One of the earliest detectable effects of thymosin fraction 5 (TF5), a unique thymic peptide, on thymocytes was to increase their intracellular cyclic GMP (cGMP) levels. Stimulation of cGMP levels can be seen after as little as 1 minute of in vitro incubation of thymocytes with TF5, and maximum effects occur at between 5 and 10 minutes. TF5 also causes influx of calcium into thymocytes, consistent with a role of calcium in mediating the increases in cGMP. TF5 has not been shown to stimulate cyclic AMP (cAMP) levels. Based on studies of cellular subpopulations, cGMP may be primarily involved in maturation and activation of more mature or committed lymphocytes in the thymus and spleen.

Phylogenetic Distribution of Thymosin α

Thymosin α_1 ($T\alpha_1$), which is one of the principal thymic humoral factors, circulates in the blood of mammals (human, mouse, rat, cat, dog, monkey, baboon) and has been localized to the epithelial cells of the thymic subcapsular cortex and medulla. Using a specific radioimmunoassay, the presence of immunoreactive $T\alpha_1$ in body fluids or culture supernatants from fish, invertebrates, and bacteria has been reported. Certain hormones and related biochemical substances may have initially developed during evolution and ontogeny as local tissue factors. Later, as morphogenesis proceeds, a subset of these messenger molecules may develop into hormones and neurotransmitters. This may account for the observed biological effects of many mammalian hormones in unicellular organisms.

A $T\alpha_1$-like peptide has been found in body fluids and blood from a wide range of organisms, ranging from annelida (earthworms) to primates. The presence of $T\alpha_1$ has also been evaluated in unicellular prokaryotic organisms. It is not clear why a $T\alpha_1$-like peptide should be present in unicellular organisms (Oates et al., 1988). The presence of $T\alpha_1$ in extra thymic sites, including the spleen and brain, and in the blood suggests both endocrine and/or autocrine roles in humans. Thus, $T\alpha_1$ is highly conserved during evolution and should be added to the growing list of mammalian hormones that are secreted by lower forms of life, including bacteria and plants. The occurrence of a $T\alpha_1$-like peptide in unicellular organisms suggests that these cells express autocrine or primitive neuromodulating roles.

Therapeutic Applications of Thymosin

The successful use of thymic hormones in reconstituting immunodeficient animals supports the application of thymosins to the treatment of human diseases. Significant toxicity has not been reported in any clinical study to date with TF5 or $T\alpha_1$, although an occasional patient may develop an allergic reaction to TF5. Immunodeficient patients who have had the greatest responses to thymosin include those who suffer from primary thymic immunodeficiences, such as children with DiGeorge syndrome or thymic aplasia.

Aging and the Immune System

What Are Some General Characteristics?

Aging is a relatively ill-defined process involving complex multisystem changes that occur continuously from the fetus through to the final stages of life. Increasing age generally leads to diminished humoral and cell-mediated immune responses, as suggested by the higher incidence of malignancies and of degenerative and infectious diseases. The first evidence of age-related change directly related to the immune system is physiological— thymic involution begins early in life and has far-reaching consequences (Steinman, 1986). The most obvious and direct effect is the diminished function of T lymphocytes in senescent animals, which in turn influences other components of the immune system.

Decline and loss of immune function may not be equally severe throughout the whole organism. Evidence is accumulating which suggests that the cellular microenvironment plays a major role in regulating the effects of aging on cell function. Clearly, changes in cell function can ultimately be traced back to a basic age-related induction of intracellular defects, particularly the ability to transduce or respond to signals. The age-related loss of function of the immune system apparently occurs cumulatively, as increasing numbers of cells display these defects. Undoubtedly some individual cells remain functional and are as capable of responding and reacting as "young" cells. For example, recent evidence suggests that the mucosal immune system is less susceptible to the effects of age by virtue of its unique cellular composition (Wade and Szewczuk, 1984). Stem cells in old individuals also retain the potential to produce fully functional cells. The microenvironment in which cells mature thus may influence their susceptibility to age-related changes in immune function.

Aging and the Immune System: More Specific Characteristics

The definition of age-related changes in immune function remains controversial. Age-related effects have been noted in many parameters, including serum protein concentrations, lymphocyte numbers, serum antibody levels, and autoantibody production. Much of the controversy has probably arisen from difficulties in standardizing human populations for aging studies. For example, certain analyses have used different criteria for defining "old" and "healthy," which may account in part for variations in interpretation. Changes in minor cell subpopulations also cannot be ruled out. In addition, circannual and circadian rhythms in mice may contribute to the controversy about changes in cell numbers. Although attempts have been made to correlate the appearance of various autoantibodies in aged humans and mice with the development of clinical autoimmune disease, these correlations are not definitive. Instead, autoantibodies may simply indicate reduced efficiency of the regulatory systems that normally control their expression.

Young and old mice have comparable numbers of B cell clones capable of producing autoreactive antibodies, supporting the idea that autoantibody production results from dysregulation of such clones rather than an age-related increase in their frequency. Levels of antibody produced by elderly humans in response to immunization with flagellin, influenza, pneumococcal vaccine, or tetanus toxoid are lower than levels observed in young adults. Aged mice also have lower serum antibody responses than young mice after immunization and show decreased numbers of splenic antibody-producing cells. Responses to T-independent antigens decline only later in life than responses to T-dependent antigens, suggesting that the aging process affects T cells earlier than B cells.

Immunodeficiency

We have come a long way since beginning this chapter, defining developmental immunology—its ontogeny and phylogeny, or its embryology and evolution. We have sought to analyze regulatory mechanisms at a genetic level and to describe the various constituent cell populations and their interactions. We are then left with integrating these normal cells, controlled by normal genetic mechanisms into the homeostasis of the intact organism. However, the developing immune system does not always remain normal: Defects can arise at many stages during the transition from embryo to adult. Although AIDS is the most popular and feared of the diseases that may affect the immune system, many other primary immunological defects, including severe combined immunodeficiency disease, isolated T and B cell deficiencies, and presumed regulatory cell abnormalities, have been described in human and experimental systems. Analysis and understanding of the genetic bases and pathophysiology of these disease states can contribute greatly to our knowledge of the normal organization, function, and regulation of the immune system.

References

Anderson, R. S., and Sigel, M. M., eds. (1983). Comparative aspects of inflammation. *Am. Zool.* 23:129–227.

Azzolina, L. S., Tridente, G., and Cooper, E. L. (1981). Comparative immunology. *Dev. Comp. Immunol. Suppl. 1.* 180 pp.

Bang, F. B., ed. (1967). Defense reactons in invertebrates. *Fed. Proc. 26*:1664–1715.

Beck, G., and Habicht, G. (1986). Isolation and characterization of a primitive IL-1–like protein from invertebrate *Asterias forbesi. Proc. Natl. Acad. Sci. USA 83*:7429–7433.

Beck, G., Vasta, G. R., Marchalonis, J. J., and Habicht, G. S. (1989). Characterisations of interleukin 1 activity in tunicates. *Comp. Biochem. Physiol. B92*:93–98.

Berzofsky, J. A., Brett, S. J., Streicher, H. Z., and Takahasi, H. (1988). Antigen processing for presentation to lymphocytes: Function, mechanisms, and implications for the T cell repertoire. *Immunol. Rev. 106*:5–31.

Brehelin, M., ed. (1986). *Immunity in Invertebrates. Cells, Molecules, and Defense Reactions.* Berlin: Springer Verlag, 233 pp.

Brion, C. A., and Welsh, R. M. (1986). Antigenic distinctions and morphological similarities between proliferating natural killer and cytotoxic T cells. In *Natural Immunity, Cancer and Biological Response Modification* (Lotzová, E., and Herberman, R. B., eds.) Basel: S. Karger. pp. 289–297.

Burnet, F. M. (1959). *The Clonal Selection Theory of Acquired Immunity.* Nashville: Vanderbilt University Press, 208 pp.

Burnet, F. M. (1970). *Immunological Surveillance.* Oxford: Pergamon Press, 280 pp. (See Chapter V, The Evolution of the Immune Process, pp. 101–120.)

Burnet, F. M. (1974). Invertebrate precursors to immune response. In *Contemporary Topics in Immunobiology,* vol. 4 (Cooper, E. L., ed.) New York: Plenum Press, pp. 13–24.

Campbell, F. (1970). Ultrastructure of the bone marrow of the frog. *Am. J. Anat. 129*:329–356.

Cohen, N., and Sigel, M. M., eds. (1982). *The Reticuloendothelial System: A Comprehensive Treatise: Vol. 3. Phylogeny and Ontogeny.* New York: Plenum Press.

Coombe, D. R., Ey, P. L., and Jenkin, C. R. (1984). Self/non-self recognition in invertebrates. *Quart. Rev. Biol. 59*:231–256.

Cooper, E. L., ed. (1974). *Invertebrate Immunology. Contemporary Topics in Immunobiology: Vol. 4.* New York: Plenum Press, 299 pp.

Cooper, E. L., ed. (1975). Developmental immunology. *Am. Zool. 15*:1–213.

Cooper, E. L. (1976a). *Comparative Immunology.* Englewood Cliffs, N.J.: Prentice-Hall, 338 pp.

Cooper, E. L. (1976b). Evolution of blood cells. *Ann. Inst. Pasteur 127C*:817–825.

Cooper, E. L. (1976c). Immunity mechanisms. In *Physiology of the Amphibia* (Lofts, B., ed.). New York: Academic Press, pp. 163–272.

Cooper, E. L. (1977). Phylogenetic aspects of transplantation. In *Transplantation* (Masshoff, J. W., ed.). Berlin: Springer Verlag, pp. 139–167.

Cooper, E. L. (1981). Immunity in invertebrates. *CRC Crit. Rev. Immunol. 2*:1–32.

Cooper, E. L., and Brazier, M.A.B., eds. (1982). *Developmental Immunology: Clinical Problems and Aging.* New York: Academic Press, 321 pp.

Cooper, E. L., Klempau, A. E., Ramirez, J. A., and Zapata, A. G. (1980). Source of stem cells in evolution. In *Development and Differentiation of Vertebrate Lymphocytes* (Horton, J. D. ed.). Amsterdam: Elsevier/North-Holland, pp. 3–14.

Cooper, E. L., Klempau, A. E., and Zapata, A. G. (1985). Reptilian immunity. In *Biology of the Reptilia,* vol. 14. (Gans, C., Billett, F., and Maderson, P., eds.). New York: Wiley, pp. 600–678.

Cooper, E. L., and Mansour, M. H. (1989). Distribution of Thy-1 in invertebrates and ectothermic vertebrates. In *Cell Surface Antigen Thy-1: Immunology, Neurology, and Therapeutic Applications* (Reif, A. E., and Schlesinger, M., eds.). New York: Marcel Dekker, pp. 197–219.

Cooper, E. L., and Wright, R. K., eds. (1984). *Aspects of Developmental and Comparative Immunology,* vol. 2. New York: Pergamon Press, 280 pp.

Cuchens, M., McClean, E., and Clem, L. W. (1976). Lymphocyte heterogenity in fish and reptiles. In *The Phylogeny of Thymus and Bone-Marrow-Bursa Cells* (Wright, R. K., and Cooper, E. L., eds.). Amsterdam: Elsevier/North Holland, p. 205.

Dawe, C. J., and Harshbarger, J. C., eds. (1968). *Neoplasms and Related Disorders of Invertebrate and Lower Vertebrate Animals.* (National Cancer Institute Monograph 31). Washington, D.C.: U.S. Government Printing Office, 772 pp.

Delovitch, T. L., Semple, J. W., and Phillips, M. L. (1988). Influence of antigen processing on immune responsiveness. *Immunol. Today 9*:216–218.

Diener, E., and Waters, C. A. (1986). Immunological quiescence towards self: Rethinking the paradigm of clonal abortion. In *Paradoxes in Immunology* (Hoffman, G. W., Levy, J. G., and Nepom, G. T., eds.). Boca Raton, Fla.: CRC Press, pp. 27–40.

Doria, G., and Frasca, D. (1985). Effects of thymosin α_1 on immunoregulatory T-lymphocytes. In *Thymic Hormones and Lymphokines* (Goldstein, A. L., ed.). New York: Plenum Press. pp. 445–454.

Ermak, T. H. (1976). The hematogenic tissues of tunicates. In *The Phylogeny of Thymus and Bone-Marrow-Bursa Cells* (Wright, R. K., and Cooper, E. L., eds.). Amsterdam: Elsevier/North Holland, 325 pp.

Fotedar, R., and Diener, E. (1988). The role of recombinant IL-2 and IL-1 in murine B cell differentiation. *Lymphokine Res. 7*:393–402.

Fudenberg, H. H., Stites, D. P., Caldwell, J. L., and Wells, J. V. (1978). *Basic and Clinical Immunology.* Los Altos, Calif.: Lange Medical Publications, 758 pp. (See Section II, Immunobiology, Chapter 13, Phylogeny and Ontogeny of the Immune Response, pp. 141–154.)

Gershwin, M. E., and Cooper, E. L., eds. (1978). *Animal Models of Comparative and Developmental Aspects of Immunity and Disease.* New York: Pergamon Press, 396 pp.

Good, R. A., and Papermaster, B. W. (1964). Ontogeny and phylogeny of adaptive immunity. *Adv. Immunol. 4*:1–115.

Greenberg, A. H. (1987). *Invertebrate Models, Cell Receptors and Cell Communication.* Basel: S. Karger, 269 pp.

Hackett, J. Jr., Bennett, M., and Kumar, V. (1986). Natural killer cell precursors in the bone marrow are distinct from lymphoid and myeloid progenitors. In *Natural Immunity, Cancer and Biological Response Modification* (Lotzová, E., and Herberman, R. B., eds.). Basel: S. Karger, pp. 40–49.

Hadjii-Azimi, I., Coosemans, V., and Canicatti, C. (1987). Atlas of adult *Xenopus laevis* hematology. In *Developmental and Comparative Immunology,* vol. 11 (Cooper, E. L., ed.). New York: Pergamon Press, pp. 807–874.

Hamby, B., Huggins, E. E., Lachman, L., Dinarello, C., and Sigel, M. (1986). Fish lymphocytes respond to human IL-1. *Lymphokine Res. 5*:157–162.

Herberman, R. B. (1986). Natural killer cells. *Annu. Rev. Med. 37*:347–352.

Hildemann, W. H., and Benedict, A. A., eds. (1975). *Immunologic Phylogeny.* New York: Plenum Press, 485 pp.

Hildemann, W. H., and Cooper, E. L., eds. (1970). Phylogeny of transplantation reactions. *Transpl. Proc. 2*:179–340.

Honjo, T. (1983). Immunoglobulin genes. *Annu. Rev. Immunol. 1*:499–528.

Janeway, Jr. C. A., Jones, B., and Hayday A. (1988). Specificity and function of T cells bearing receptors. *Immunol. Today 9*:73–76.

Kelsoe, G., and Schulze, D. H., eds. (1987). *Evolution and Vertebrate Immunity. The Antigen-Receptor and MHC Gene Families.* Austin: University of Texas Press, 459 pp.

Kincade, P. W. (1981). Formation of B lymphocytes in fetal and adult life. *Adv. Immunol. 31*:177–245.

Klein, J. (1986). *Natural History of the Major Histocompatibility Complex.* New York: Wiley, 775 pp.

Koppenheffer, T. L. (1987). Serum complement systems of ectothermic vertebrates. *Dev. Comp. Immunol. 11*:279–286.

Kronenberg, M., Siu, G., Hood, L. E., and Shastri, N. (1986). The molecular genetic of the T cell antigen receptor and T cell antigen recognition. *Annu. Rev. Immunol. 4*:529–591.

Langman, R. E. (1989). *The Immune System. Evolutionary Principles Guide Our Understanding of This Complex Biological Defense System.* New York: Academic Press, 209 pp.

Lanier, L. L., and Allison, J. P. (1987). The T cell antigen receptor gamma gene: Rearrangement and cell lineages. *Immunol. Today 8*:293–296.

Leanderson, T., Andersson, J., and Rajasekar, R. (1987). Clonal selection in B cell growth and differentiation. *Immunol. Rev. 99*:53–69.

LeDouarin, N., Bach, J. F., and Dieterlin-Lièvre, F., eds. (1976). Hemopoiesis in vertebrate embryos. *Ann. Immunol. 127c*:811–815.

Liacopoulos, P., and Panijel, J. eds. (1972). *Phylogenic and Ontogenic Study of the Immune Response and its Contribution to the Immunological Theory.* Paris: Colloque INSERM.

Litman, G. W. (1984). Phylogeny of immunoglobulins. In *Molecular Immunology* (Atassi, M. Z., Van Oss, C. J., and Absolom, D. R., eds.). New York: Marcel Dekker, pp. 215–230.

Loeb, L. (1945). *The Biological Basis of Individuality.* Springfield, Ill.: C. C. Thomas, 711 pp.

Lotzová, E., and Herberman, R. B., eds. (1987). *Natural Immunity, Cancer and Biological Response Modification.* Basel: S. Karger.

MacDonald, H. R., and Nabholz, M. (1986). T-cell activation. *Annu. Rev. Cell Biol. 2*:231–253.

Manning, M. J., ed. (1980). *Phylogeny of Immunological Memory.* Amsterdam and New York: Elsevier/North Holland, 318 pp.

Manning, M. J., and Turner, J. (1976). *Comparative Immunobiology.* London: Blackwell Scientific, 184 pp.

Marchalonis, J. J., ed. (1976). *Comparative Immunology.* London: Blackwell Scientific, 470 pp.

Marchalonis, J. J. (1977). *Immunity in Evolution.* London: Edward Arnold, 316 pp.

Marchalonis, J. J., and Cohen, N., eds. (1980). *Contemporary Topics in Immunobiology: Vol. 9. Self/Non-Self Discrimination.* New York: Plenum Press, 293 pp.

Mizell, M. ed. (1969). *Biology of Amphibian Tumors.* New York: Springer Verlag, 484 pp. (See Section III, Immunity and Tolerance in Amphibia, pp. 130–183.)

Mizuno, D., Cohn, Z. A., Takeya, K., and Ishida, N., eds. (1982). *Self-Defense Mechanisms: Role of Macrophages.* Tokyo and Amsterdam: University of Tokyo and Elsevier Biomedical, 343 pp. (See Section I, Phylogenic and Ontogenic Aspects of Macrophages, pp. 3–82.)

Müller-Eberhard, H. J. (1986). The membrane attack complex of complement. *Annu. Rev. Immunol. 4*:503–508.

Nossal, G.J.V. (1983). Cellular mechanisms of immunological tolerance. *Annu. Rev. Immunol. 1*:33–62.

Oates, K. K., Grisburg, G. T., Naylor, P. H., Afronti, L. F., and Goldstein, A. L. (1988). Identification and distribution of thymosin α_1–like immunoreactivity. *Dev. Comp. Immunol. 12*:397–402.

Ratcliffe, N. A., and Rowley, A. F., eds. (1981). *Invertebrate Blood Cells,* vols. 1 and 2, New York: Academic Press, 641 pp.

Ratcliffe, N. A., Rowley, A. F., Fitzgerald, S. W., and Rhodes, C. P. (1985). Invertebrate immunity: Basic concepts and recent advances. *Int. Rev. Cytol. 97*:183–350.

Rogers, J., and Wall, R. (1984). Immunoglobulin RNA rearrangements in B lymphocyte differentiation. *Adv. Immunol. 35*:39–88.

Rowley, A. F., Ratcliffe, N. A., and Ambrosius, H., eds. (1983). Invertebrate and vertebrate immunity, conferences Swansea and Masserberg. *Dev. Comp. Immunol. 7*:599–804.

Sigel, M. M., ed. (1968). Differentiation and defense mechanisms. In *Lower Organisms.* Baltimore: Williams and Wilkins, 211 pp.

Smith, R. T., Good, R. A., and Miescher, P. A. (1967). *Ontogeny of Immunity.* Gainesville: University of Florida Press, 208 pp.

Smith, R. T., Miescher, P. A., and Good R. A., eds. (1966). *Phylogeny of Immunity.* Gainesville: University of Florida Press, 276 pp.

Snieszko, S. F., Nigrelli, R. F., and Wolf, K., eds. (1965). Viral diseases in poikilothermic vertebrates. *Ann. N.Y. Acad. Sci. 126*:1–680. (See Section VII, Immune Reaction in Cold Blooded Vertebrates, pp. 629–677.)

Solomon, J. B. ed. (1981). *Aspects of Developmental and Comparative Immunology,* vol. 1. New York: Pergamon Press, 572 pp.

Solomon, J. B., and Horton, J. D., eds. (1977). *Developmental Immunobiology.* Amsterdam: Elsevier/North Holland, 456 pp.

Steinmann, G. G. (1986). Changes in the human thymus during aging. *Curr. Top. Pathol. 75*:43–88.

Swain, S. L. (1983). T cell subsets and the recognition of MHC class. *Immunol. Rev. 74*:129–142.

Unanue, E. (1984). Antigen-presenting function of the macrophage. *Annu. Rev. Immunol. 2*:395–428.

van de Vyver, G., and Herlant-Meewis, H., eds. (1974). Les phénomènes de reconnaissance cellulaire. *Arch. Biol.* (Bruxelles) *85*:1–150.

van Furth, R., van der Meer, J.W.M., Blussé van Oud Albas, A., and Sluiter, W. (1982). Development of mononuclear phygocytes. In *Self-Defense Mechanisms. Role of Macrophages* (Mizuno, D., Cohn Z. A., Takeya, K., and Ishida, N., eds.). Amsterdam: Elsevier, pp. 25–41.

van Muiswinkel, W. B., ed. (1982). Immunology and immunization of fish. *Dev. Comp. Immunol. Suppl. 2,* 255 pp.

von Boehmer, H. (1988). The developmental biology of T lymphocytes. *Annu. Rev. Immunol. 6*:309–326.

Wade, A. W., and Szewczuk, M. R. (1984). Aging, idiotype repertoire shifts, and compartmentalization of the mucosal-associated lymphoid system. *Adv. Immunol. 36*:143–188.

Wagner, H., Hardt, C., Stockinger, H., Pfizemaier, K., Bartlett, R., and Rollinghoff, M. (1981).

Impact of thymus on the generation of immunocompetence and diversity of antigen-specific MHC-restricted cytotoxic T lymphocyte precursors. *Immunol. Rev. 58*:95–129.

Watkins, D., and Cohen, N. (1985). The phylogeny of interleukin-2. *Dev. Comp. Immunol. 9*:819–824.

Williams, A. F. (1987). A year in the life of the immunoglobulin superfamily. *Immunol. Today 8*:298–303.

Wright, R. K., and Cooper, E. L., eds. (1976). *The Phylogeny of Thymus and Bone Marrow-Bursa Cells.* Amsterdam: Elsevier/North Holland, 325 pp.

Yancopoulous, G. D., and Alt, F. W. (1986). Regulation of the assembly and expression of variable-region genes. *Annu. Rev. Immunol. 4*:339–368.

II

PHYLOGENY OF THE IMMUNE SYSTEM

2

Development of Primitive Recognition Systems in Invertebrates

DAVID A. RAFTOS

The concept of homology remains central to modern biology. It suggests that "nothing in evolution is created *de novo* . . . each new gene must have arisen from an already existing gene" (Ohno, 1970). Hence, the expression of a given characteristic is constrained by its ancestry. This corresponds with the original Darwinian view of descent with modification and suggests that characteristics may be elucidated by tracing their evolution as well as by analyzing their immediate functions. Indeed, primitive systems may express functional activities that are unlikely to be disguised by progressive adaptation. Such conserved homologies provide the focus for this chapter. The following discussion describes basic functions of immunological recognition as they are expressed in relatively simple organisms.

Despite considerable resistance to the concept, a continuous evolution of immunological reactivity may be demonstrable at a functional level. Here, I present an interpretation of the available evidence which reflects a cohesive diversification of immunological recognition throughout the animal kingdom. This approach allows us to explain some of the confusing or contradictory aspects of immunological reactivity in higher organisms in terms of the system's evolutionary past. It must be stressed, however, that this chapter presents only a few of the many possible interpretations of the available data. Moreover, experimental findings pertaining to invertebrate immunity are extremely diverse and are subject to substantial variations arising from the biological and phylogenetic differences between experimental subjects. For more detailed accounts of these various studies, the reader is referred to the excellent and comprehensive reviews by Coombe et al. (1984) and Ratcliffe et al. (1985).

PRIMITIVE ORIGINS OF IMMUNOLOGICAL RECOGNITION

Molecular Homologies

The fundamental requirements for recognitive activity in cell adhesion, differentiation, cellular regulation, and defense must be common to all metazoans. A need to develop such adapted systems from ancestral protozoan mechanisms probably represents the primitive and common basis for immunological reactivity and other forms of cellular recognition. However, the divergence of these essential recognition systems is no doubt so ancient that it precludes useful functional comparisons. Over such extreme phylogenetic distances, molecular analysis has proven to be a viable alternative to functional studies.

In mammals, sequence homologies are evident between immunoglobulins (Ig), class I and II major histocompatibility complex (MHC) glycoproteins, β_2-microglobulin, Thy-1 antigens, T cell antigen receptors, and a variety of other immune-related molecules (Hunkapiller and Hood, 1986). These similarities are generally believed to reflect homology and not evolutionary convergence arising from functional constraint. Hence, they are indicative of a homologous "Ig superfamily" derived from gradual evolutionary divergence. A somewhat cladistic view of that radiation is provided in Figure 2–1.

The degree of variation in structure and function evident among Ig-related molecules suggests that the divergence within this superfamily is an ancient one. A primitive origin is supported by the identification of Ig-like determinants in invertebrates. For instance, epitopes that cross-react with anti-β_2-m antisera or monoclonal antibodies have been identified in annelids (Roch and Cooper, 1983), crustaceans (Shalev et al., 1981), and dipterans (Shalev et al., 1983). Similar serological evidence for the existence of Thy-1 analogues has been reported in mollusks (Williams and Gagnon, 1982) and tunicates (Mansour et al., 1985). Moreover, biochemical analyses have been used to infer homology between serologically cross-reactive tunicate antigens and vertebrate Thy-1, although confirmatory sequence analyses are not available. The existence of such primitive epitopes, together with intraspecific molecular comparisons (Fig. 2–1), indicates that the Ig superfamily arose from a single primordial gene, which probably encoded a polypeptide domain of approximately 110 amino acids analogous to Thy-1 or β_2-m.

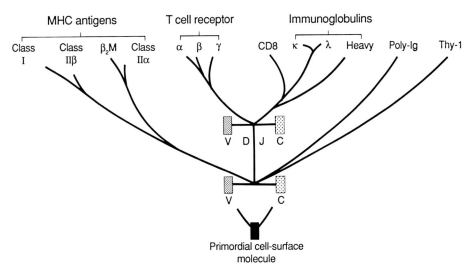

Figure 2–1.   Postulated evolution of Ig superfamily molecules. (Modified from Hood et al., 1985. *Cell 40*:225.)

Cellular Recognition

While the functions and evolutionary homologies of invertebrate Ig-like molecules have yet to be resolved, it is tempting to speculate that the original Ig genes were involved in a wide range of recognitive processes. Circumstantial evidence that supports a broad involvement has come from the recent identification of cell adhesion molecules as members of the extended Ig superfamily. The chicken neural cell adhesion molecule, N-CAM, exhibits substantial amino acid sequence homology with the Ig V_H and $C_\mu 4$ domains, as well as the CD4 and CD8 lymphocyte antigens (Matsunaga and Mori, 1987). Sequence homologies are also found in the cs-A (cell adhesion) molecules of the slime mold, *Dictyostelium discoideum*. These results suggest a primordial relationship between cell adhesion and defensive reactivity (Matsunaga and Mori, 1987). Perhaps ancestral members of the Ig superfamily arose among unicellular organisms and subsequently became committed to a number of recognition responses that are prerequisites for the development of multicellularity.

PRIMITIVE RECOGNITION MECHANISMS

Phagocytosis

Among recognition responses that are implicitly associated with the origins of multicellularity, phagocytosis represents the most readily identified functional progenitor of immunological reactivity. Phagocytosis is the most potent and extensively studied defense system in invertebrates. Its ubiquity suggests a continuous evolutionary history. Phagocytic activity was probably derived from the ingestive systems of protists and later assumed its defensive role. In metazoans, phagocytic cells prevent the proliferation of pathogens in homeostatic extracellular environments and act to sequester damaged tissues.

Discriminatory Capacity of Phagocytosis

Phagocytosis and its related effector responses, while clearly primitive, are not simply an automatic consequence of contact with foreign material (Coombe et al., 1984). To prevent autoreactivity, defensive reactions, including phagocytosis, must incorporate the capacity to discriminate between "self" and "non-self" (Burnet, 1974). It is therefore not surprising that discriminative phagocytic recognition has been demonstrated among a number of invertebrate phyla (Ratcliffe et al., 1985). For instance, molluscan phagocytes can distinguish between closely related globular proteins prior to pinocytic ingestion, while sea urchins exhibit differential phagocytic responses to gram positive and negative bacteria. Similarly, echinoid and molluscan phagocytes that are normally insensitive to vertebrate erythrocytes or to yeast will avidly ingest the same materials that have been pretreated with glutaraldehyde or formaldehyde.

Opsonic Recognition

The recognition mechanisms that confer this form of discrimination on invertebrate phagocytes have yet to be fully defined. However, many authors suggest a central role for opsonic factors. Prowse and Tait (1969) found that absorption of hemolymph from the snail, *Helix aspersa,* with yeast significantly inhibited the subsequent in vitro phagocytosis of yeast but not of erythrocytes. Similar observations have now been made in crustaceans, echinoderms, and mollusks, indicating a broad distribution of humoral opsonic functions (Coombe et al., 1984).

Opsonization in invertebrates has most often been attributed to the action of agglutinin molecules. Agglutinins are characterized by a broad range of target specificity and susceptibility to inhibition by various mono- and polysaccharides. This suggests that they act as lectins with the capacity to bind carbohydrate moieties on the surface of target cells. Specificity for carbohydrates has been correlated with the ability of agglutinins to induce phagocytic activation (Renwrantz et al., 1981). These authors showed that pretreatment of erythrocytes with N-acetylgalactosamine and N-acetylglucosamine inhibited both agglutination and phagocytosis in the gastropod, *Helix pomatia.* They concluded that a single molecular mechanism mediated both phagocytosis and agglutination, a suggestion that has been supported by a number of other studies (Ratcliffe et al., 1985).

Alternative Explanations for Phagocytic Recognition in Invertebrates

While lectin-mediated opsonization remains a potentially important mechanism for phagocytic recognition in invertebrates, it cannot explain all of the available data. Inert particles such as India ink, carmine, polystyrene, latex beads, and glass fragments have been employed as targets for phagocytosis in numerous studies (Coombe et al., 1984). These materials do not display surface carbohydrate moieties, and so their recognition cannot be explained by the binding of carbohydrate-specific lectins or by any other mechanism based on recognition of non-self biological determinants. Two alternative recognition systems may account for these observations. First, phagocytosis of inert particles in the absence of opsonization may depend on the general surface properties of the ingested material, including surface tension or hydrophobicity (Capo et al., 1979). Alternatively, generalized phagocytic recognition may require "self recognition" (Fig. 2–2). A number of independent models have suggested that "self" identity is clonally expressed in all metazoans by discrete cellular antigens (H-ag) (Cohen, 1983; Kolb, 1977; Lafferty and Talmage, 1976; Lafferty and Woolnough, 1977; Mäkelä et al., 1976). According to self recognition theories for phagocytic activation, recognition of H-ag by clonally identical ("self") cells actively suppresses defensive reactivity, whereas failure to recognize "self" determinants allows phagocytosis to proceed. Thus, only particles that do not bear self H-ag promote phagocytosis. The degree of selectivity evident in invertebrate phagocytosis can be explained either by a limited diversity of recognition specificities or through secondary enhancement, possibly mediated by lectins (Coombe et al., 1984).

Although the above description of the physical events in phagocytosis appears simple, it is less easy to account for the biochemical evolution of self recognition. Any proposed model must account for the simultaneous development of complementary antigen/receptor interactions within a single individual. Rothenberg (1978) resolved this problem by proposing that a single H-ag gene determines the molecular specificity of three distinct macromolecules, so that clonal self recognition occurs through complementary protein/carbohydrate interactions. More parsimonious molecular explanations for self recognition that do not rely on complex biochemical pathways are possible, however. Mäkelä et al. (1976) suggested that self recognition involves direct interaction

(a) Interaction of identical cells

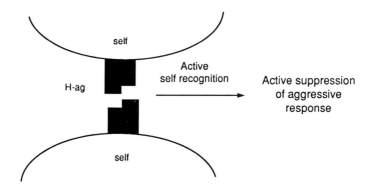

(b) Interaction of heterogeneous cells

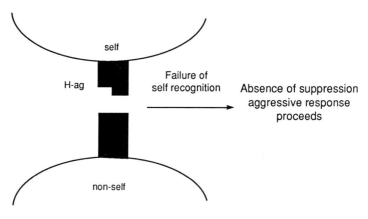

Figure 2–2.   Schematic representation of self recognition models for primitive phagocytic recognition. H-ag represents the putative self marker.

between the H-ag's of different cells. Each molecule effectively acts as both receptor and ligand.

Self Recognition and the Problem of Autoreactivity

The above self recognition mechanisms reasonably account for the phagocytosis of inert particles in the absence of opsonization. More importantly, they resolve a major conceptual inconsistency implicit in lectin-mediated recognition models. Carbohydrate binding receptors in invertebrates generally have broad specificity. Parrinello and Canicatti (1982) showed that hemagglutinins from the tunicate *Ascidia malaca* have combining sites complementary to generalized D-galacto configurations. Thus, a single receptor species has the capacity to bind a range of different polysaccharides. There are also indications that somatic variation may give rise to multiple agglutinin binding specificities within a single individual, further expanding the range of susceptible carbohydrates. Such broad specificities expressed by a restricted number of lectin binding sites are likely to incorporate self-reactivity. It is doubtful that mechanisms exist for the selective depletion of self-reactive lectins, or to restrict the range of carbohydrate moieties expressed on the surface of "self" cells. Indeed, lectin-like molecules have recently been identified on the surface of autologous molluscan coelomocytes (Vasta and Marchalonis, 1985). It has been suggested that these lectins act as integral cell surface receptors, and this could account for those instances in which phagocytosis proceeds in the absence of humoral factors. Alternatively, these authors may have identified autoreactive multivalent humoral lectins bound to the surface of "self" cells. Certainly, the bound molecules were analogous, if not identical, to serum opsonins.

The potential for autoreactivity, which is predicted by the broad specificities of invertebrate lectins, represents a powerful selective stimulus for the development of self recognition mechanisms. Coombe et al. (1984) proposed a two-signal system for recognition, which may have developed to preclude autodestruction. They suggested that phagocytosis is initiated by opsonization but that it proceeds only in the absence of stimulation from anti-self H-ag receptors.

EXAMPLES OF SELF RECOGNITION–NON-FUSION REACTIONS

Diversification of Self Recognition Systems

Mechanisms based on an obligatory requirement for self recognition are extremely attractive, particularly in any consideration of the molecular and functional evolution of immunological reactivity. The putative H-ag's involved in invertebrate self recognition may correspond to the primordial Ig domains described above in the discussion of molecular homologies and cellular recognition. If this is the case, the development of specific immunological recognition would come about through gradual diversification of these molecules and their related effector and cellular functions. This form of diversification corresponds to the classical evolutionary scheme proposed initially by Hildemann and Reddy (1973) (Fig. 2–3). However, there is still little experimental evidence to support such evolutionary transitions. Indeed, data supporting the existence of self recognition mechanisms have come only from systems in which alloreactivity has allowed definitive comparison with autologous controls.

Non-Fusion Reactions in Tunicates

The allogeneic mechanisms that have been used to demonstrate self recognition are typified by the non-fusion reactions (NFR) of colonial tunicates. NFRs were first described by Bancroft (1903), who found that colonies of *Botryllus schlosseri* from different clonal origins do not fuse when apposed. In contrast, fragments from a single colony rapidly coalesce. Similar systems of allogeneic discrimination were identified in a number of colonial and solitary tunicate species (Fuke, 1980; Mukai and Watanabe, 1974). For instance, hemocytes from the solitary tunicate *Halocynthia roretzi* undergo rapid in vitro lysis in both allogeneic and xenogeneic combinations (Fuke and Nakamura, 1985).

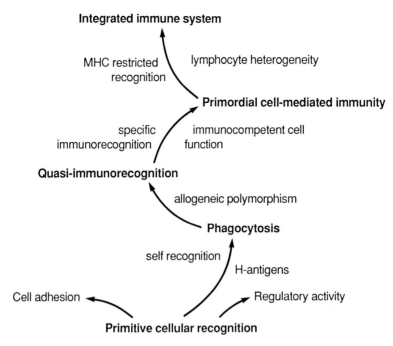

Integrated immune system

MHC restricted recognition   lymphocyte heterogeneity

Primordial cell-mediated immunity

specific immunorecognition   immunocompetent cell function

Quasi-immunorecognition

allogeneic polymorphism

Phagocytosis

self recognition   H-antigens

Cell adhesion ◀   ▶ Regulatory activity

Primitive cellular recognition

Figure 2–3.   Hypothetical evolutionary divergence of immunological reactivity from primitive mechanisms of cellular recognition.

Genetics of Non-Fusion Reactions

The dependence of these allogeneic interactions on self recognition has been inferred from genetic analyses. In *Botryllus* species, NFRs are controlled by a single highly polymorphic gene locus that encodes up to several hundred alleles (Oka and Watanabe, 1957; Scofield et al., 1982a,b). For fusion to occur, colonies must share at least one allele at the fusion (F) locus. This pattern was initially demonstrated by testing for fusion between F_1 sibs and parental generations (Oka and Watanabe, 1957) (Fig. 2–4). F_1 individuals were consistently found to fuse with their parents and 75% of their siblings, suggesting that only a single compatible allele was required for fusion.

Burnet (1971, 1974) argued that this genetic pattern fulfills the predictions of self recognition hypotheses. He attributed NFR to the recognition of discrete cell surface proteins equivalent to the putative self markers involved in phagocytic activation (Burnet, 1974). Receptors specific to the individual's self genotype were proposed to regulate aggressive responses. Positive recognition of one or both "self" allelic determinants encoded by an individual's F locus would prevent reactivity. Genetic dominance applies so that a single shared allele yields sufficient stimulation to inhibit responsiveness.

Relationship of Non-Fusion Reactions to Phagocytosis

Burnet's explanation of the genetic basis of NFR corresponds closely with the expectations of putative phagocytic self recognition mechanisms. NFR may therefore represent a direct evolutionary extension of the recognition mechanisms that operate in primitive

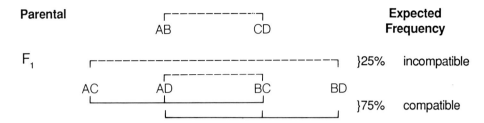

Figure 2–4.   Patterns of allogeneic reactivity in F_1 and parental generations of the colonial urochordate, *Botryllus.* Dotted lines represent incompatible combinations; compatible individuals are linked by solid lines. These results conform with the expectations of self recognition hypotheses.

phagocytic systems (Cohen, 1983; Crichton and Lafferty, 1975; Kolb, 1977). Hence, the high degree of allogeneic specificity typical of NFRs could reflect increased polymorphism of self recognition markers and their complementary receptors. Certainly, selective stimuli capable of driving such diversification are evident among colonial invertebrate species. In the colonial tunicate *Botryllus,* colonies represent clones of individuals (zooids) enclosed in a common tunic. Overlapping generations of clonally budded progeny (oozoids) are actively integrated into the colony by vascular anastomosis and tunic fusion. The selective advantage of this colonial integration would be maintained only if the incorporation of non-clonemates were prevented (Buss and Green, 1985). Allodiscriminative reactions, such as NFR, would thus be required to prevent the fusion of genetically incompatible individuals while still allowing the integration of clonal progeny.

SPECIFIC IMMUNORECOGNITION

Adaptive Immunity in Vertebrates

The invertebrate recognition mechanisms, which I have described above, are clearly distinct from those that predominate among vertebrates. This disparity is best seen in the capacity for specific adaptive reactivity. Specific anamnesis, or memory, is characteristic of both humoral and cell-mediated immune responses in all vertebrate taxa (Hildemann, 1984). For instance, specific antibody production and adaptive transplant rejection were both detected in the agnathan *Eptatretus stoutii* (Gilbertson et al., 1986). Even in this very primitive vertebrate, anamnesis is mediated by discrete immunocompetent cells, presumably equivalent to the lymphocytes of higher vertebrates. Moreover, lymphocyte heterogeneity was demonstrated among osteichthyeans and possibly agnathians, suggesting divergence of cellular and humoral functions (Gilbertson et al., 1986).

Definitions of Immunorecognition

The adaptive immune responses that are characteristic of vertebrates do not only represent experimentally convenient expressions of immunity. They embody the fundamental parameters of clonal selection and so provide the basis for a generally acceptable definition of "specific immunorecognition," or "immunocompetence." According to this definition, development of specific immunological memory requires at a minimum the presence of immunocompetent cells that clonally express specific anti-foreign receptors. Recognition leads to selective clonal proliferation and results in adaptive effector responses, which are specific for the original sensitizing determinant. These basic functions are unlikely to be affected by secondary adaptation or expansion of this mechanism (Hildemann, 1974, 1984), and so should act as principal criteria for the identification of primordial "immunorecognition" systems among primitive animals.

Immunocompetence and Primitive Recognition Systems

On the basis of this definition, neither invertebrate phagocytic recognition nor allogeneic NFR meets the criteria for primordial immunocompetence. Invertebrate phagocytosis is usually not enhanced by preimmunization. Apparent cases of phagocytic memory can most often be explained by nonspecific factors. Van der Knaap et al. (1983) attributed induced clearance in the pond snail, *Lymnaea stagnalis,* to increased activation of amoebocytes after primary bacterial challenge. The specificity of the response did not change after activation, thus excluding specific anamnesis. The failure of phagocytic systems to show adaptive immunorecognition is also consistent with studies of invertebrate humoral recognition molecules. Although adaptive production of lectins and bacteriolytic factors has been observed in invertebrates, most studies have suggested that levels of humoral recognition molecules are decreased after immunization. Moreover, the lack of specificity of lectins and bacteriolysins limits their potential as mediators in true anamnestic responses.

Similarly, putative self recognition mechanisms of the type attributed to tunicate NFR are incompatible with specific immunorecognition. By definition, self recognition should be insensitive to varied forms of foreignness because only self determinants are recognized (Burnet, 1974). In the absence of additional "educational" mechanisms, these systems cannot yield the adaptive specificity required of true immunorecognition. The failure to detect specific memory during NFR supports this conclusion.

Molecular Relationships of Opsonic Mechanisms

The functional disparity between primitive invertebrate defensive reactions and adaptive vertebrate immunity is also supported by available molecular data. Attempts to attribute invertebrate opsonization to vertebrate-like Ig systems have been inconclusive. Marchalonis and Warr (1978) described the isolation of a protein from tunicates with specificity for DNP and erythrocyte determinants similar to those of primitive vertebrate Ig. Similarly, Vasta et al. (1984) found that monoclonal antibodies specific for TEPC-15 idiotypes cross-reacted with a hemagglutinin from the horseshoe crab, *Limulus polyphemus.* While the arthropod hemagglutinin did not show significant overall homology to Ig (TEPC-15), short stretches of peptides (8–10 amino acids) were similar in the lectin, the TEPC-15 myeloma antibody, and vertebrate C-reactive protein (Vasta and Marchalonis, 1985; Vasta et al., 1984). The authors concluded that the observed serological cross-reactivity reflected either convergent evolution or conservation of primitive combining sites. Delmotte et al. (1986) also reported the purification of a TNP-binding protein from the sea star, *Asterias rubens,* that induced lysis of erythrocytes in the presence of serum complement. This molecule was composed of tetramerically arranged 30-kD subunits and was suggested to have an immunoglobulin-like structure.

Although these studies suggest that invertebrate antibody homologues may exist, the weight of evidence indicates that Ig-like molecules are not generally involved in invertebrate phagocytic opsonization (Ratcliffe et al., 1985). Most often, molecular analyses of putative invertebrate opsonins have failed to reveal substantial similarity to Ig (Vasta and Marchalonis, 1985).

Specific Immunorecognition in Invertebrates

The lack of specific anamnestic responses and molecular homology with Ig-like molecules in invertebrate phagocytosis and NFR has been interpreted by a number of authors to suggest that immunorecognition has developed only among vertebrates (Cunningham, 1978). This implies either that the selective stimuli for the development of specific reactivity operate exclusively among vertebrates or that phylogenetically primitive animals are of insufficient complexity to maintain such systems. It has been suggested that unique vertebrate characteristics—including longevity, cellular complexity, homeostasis, and limited fecundity—increase the requirement for immunological surveillance against pathogens or neoplasms and so have acted to promote and maintain specific immunorecognition (Cunningham, 1978).

Such arguments, however, are limited by their anthropocentrism. Many of the factors proposed as contributing to the development of specific recognition mechanisms are not common to all vertebrates, nor are they absent from all invertebrates. The cellular complexity and homeostasis of many invertebrates cannot be questioned. The develop-

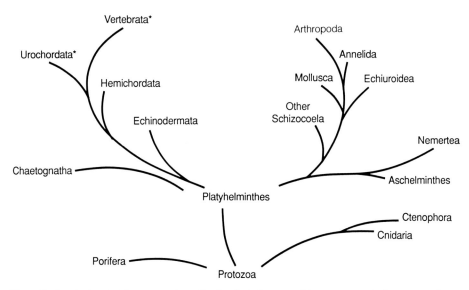

Figure 2–5.   A classical interpretation of animal phylogeny with metazoan phyla representing a progressive but divergent radiation from protozoan ancestors. The deuterostome line produced the urochordates and vertebrates, as indicated by asterisks.

ment of neoplasia has been documented in invertebrates, as has their substantial load of pathogenetic organisms (Sparks, 1985). Indeed, theories that postulate the de novo development of immunological reactivity among vertebrates ignore a large body of evidence which suggests that invertebrates possess defense mechanisms, distinct from phagocytosis and NFR, that fulfill the prerequisites for immunocompetence.

These primordial immune responses have most often been characterized using tissue transplant rejection systems. The majority of invertebrate phyla tested so far express adaptive histocompatibility reactions having a specific alloimmune memory component. For instance, sponges (Bigger et al., 1982), corals (Hildemann, et al., 1980), sea stars (Karp and Hildemann, 1976), and tunicates (Raftos et al., 1987a) all show enhancement of second-set graft rejection that is generally specific to the immunizing tissue type. These studies clearly indicate a broad phylogenetic distribution of adaptive rejection reactions. However, the following discussion, which summarizes the evidence that these reactions fulfill the criteria of primordial immunity, is based on studies in a single system, namely, graft rejection in the solitary tunicate *Styela plicata*. The choice of this organism is based on phylogenetic considerations. Both traditional and revised views of the molecular phylogeny of the animal kingdom suggest that tunicates are the closest extant ancestors of vertebrates (Field et al., 1989) (Fig. 2–5). Hence, they have the greatest chance of exhibiting immunological characteristics directly comparable to those of vertebrates.

HISTOCOMPATIBILITY IN *STYELA PLICATA*

Allograft Rejection in *S. Plicata*

Given the close phylogenetic relationship between tunicates and vertebrates, it is not surprising that recent analyses of histocompatibility in the solitary tunicate *S. plicata* have identified a cell-mediated immune system with functional similarities to that of higher chordates (Kingsley et al., 1989; Raftos, 1991; Raftos and Briscoe, 1990; Raftos et al., 1987a,b, 1988). Tissue transplantation experiments have shown that approximately 75% of first-set allografts taken from the integumentary tunic layer of *S. plicata* are actively rejected within 40 days of grafting, whereas the majority of first-set autografts remain fully viable (Raftos et al., 1987a). The kinetics of graft elimination are similar to those for transplant rejection in primitive vertebrates and presumably reflect comparable antigenic disparities (Raftos et al., 1987a). Moreover, population genetic analyses indicate that histocompatibility antigens in *S. plicata* are controlled by a single gene region (Raftos and Briscoe, 1990). In keeping with this observation, preimmunization experiments have implied the existence of cellular determinants that initiate specific immunorecognition (Raftos, 1991) such that differences at a single histocompatibility allele are sufficient to promote graft rejection in allogeneic combinations (Raftos et al., 1988).

Cellular Basis of Allograft Rejection

The specific patterns of immunorecognition exhibited by *S. plicata* are consistent with expression of anti-allogeneic receptors limited to discrete populations of immunocompetent cells. The existence of such receptor activity is strongly supported by the finding that specific effector functions are restricted to a single hemocyte subpopulation. Lymphocyte-like cells (LLCs) are the only cell type that shows prolonged accumulation around and within incompatible transplants prior to rejection. These hemocytes appear to be responsible for the cytotoxic destruction of allogeneic graft cells, a process known to initiate graft degeneration (Raftos et al., 1987b; Raftos, 1991; unpublished data). Similarly, cell populations that have been enriched for presensitized LLCs have the capacity to induce specific rejection reactions when adoptively transferred to naive individuals (Raftos, unpublished data).

Specific Alloimmune Memory and LLC Proliferation

Analysis of secondary graft rejection indicates that LLCs involved in this phase of allograft rejection are derived by proliferation of sensitized cells. LLCs penetrate second-set grafts in *S. plicata* more rapidly than in first-set combinations (Raftos et al., 1987b). Similar accelerated cell accumulation is not seen with third-party grafts, confirming the specificity of the LLC response. Enhanced LLC activity also correlates with increased rapidity of second-set graft rejection, directly indicating the involvement of these cells in rejection processes. As with LLC responses, transplant rejection shows precise allogeneic specificity characterized by the failure of most recipients of third-party grafts to mount rapid (memory-derived) rejection responses following preimmunization with allogeneic hemocytes or tunic grafts (Raftos et al., 1987a; Raftos, 1991).

An adaptive role for LLCs in generating such specific responses to histocompatibility antigens is supported by the demonstration of LLC proliferation following allogeneic immunization. Enhanced proliferative activity is seen in LLCs from allografted individuals as compared to autologous controls (D. A. Raftos, unpublished data). Blastogenesis is localized in discrete "lymphoid" tissues of the body wall and, again, exhibits a specific adaptive potential. LLC proliferation is enhanced by second-set but not by third-party immunizations.

Homology of Tunicate and Vertebrate Cell-Mediated Immunity

These functional characteristics of histocompatibility reactions in *S. plicata* fulfill the definitive criteria for primordial immunity. Immunocompetent cells, which apparently bear anti-allogeneic receptors, proliferate on antigen exposure to yield adaptive responses that are specific to the presensitizing tissue type. Such close resemblances to vertebrate cell-mediated immune reactions are most easily explained by evolutionary homology. The limited disparity with vertebrate reactions, which is seen in some aspects of tunicate recognition, is clearly outweighed by the similarities and can be explained on the basis of specific adaptation. For instance, a number of authors have identified ultrastructural differences between tunicate LLCs and vertebrate lymphocytes, even though the general morphologies of these two cell types are similar (Ermak, 1982; Warr et al., 1977). Despite some arguments to the contrary, ultrastructural variations may be explained by adaptation to the particular requirements of individual taxa, and so they need not preclude evolutionary contiguity per se.

DEVELOPMENT OF SPECIFIC IMMUNORECOGNITION

Origins of Primordial Immunorecognition

Given its functional similarities to vertebrate systems, it is likely that adaptive graft rejection in tunicates exemplifies an evolutionary state intermediate between primitive nonspecific recognition and integrated immunity. I have postulated that primordial immunorecognition of the type evident in *S. plicata* arose from the primitive systems of self recognition described in the discussion above. According to this hypothesis, anti-self receptors, which originally evolved to regulate phagocytosis, subsequently diversified to mediate specific anti-allogeneic recognition.

The occurrence of this process of diversification does not, however, imply that self recognition and specific immunorecognition are mutually exclusive in primitive systems. Indeed, theoretical considerations favor the retention of both mechanisms (Cohen, 1983). Self recognition may have been maintained as an essential first step to prevent the development of autoreactivity in histocompatibility reactions. Organisms such as *S. plicata* probably do not have the capacity to selectively deplete autoreactive clones from immunocompetent cell populations (Cohen, 1983). There is no evidence of clonal depletion in studies of tunicate hemopoietic tissues, which lack the defined primary lymphoid organs of higher chordates (Ermak, 1977). This lack of capacity for somatic selection contrasts with the demonstrated ability to generate specific anti-allogeneic responses seen in *S. plicata.* Any system of precise allodiscrimination would be predicted to lead to the generation of anti-self reactivity (Cohen, 1983). Accordingly, a number of authors have proposed "dual receptor" systems in which adaptive histocompatibility responses, mediated by specific anti-allogeneic receptors, could proceed only in the absence of self interaction mediated by distinct anti-self receptors (Cohen, 1983; Lafferty and Woolnough, 1977). The existence of such dual reactivity is suggested by studies of the in vitro responses of hemocytes from the solitary tunicate *H. roretzi.* Allogeneic cell lysis in this species conforms with the predictions of self recognition models (Fuke and Nakamura, 1985; Fuke and Numakunai, 1982) and so supports the existence of self as well as specific immunorecognition in solitary tunicate species.

Selective Basis of Evolutionary Diversification

The coexistence of self and allogeneic recognition mechanisms in invertebrates may reflect a transitional evolutionary state in which the identification of one or the other system depends on the analytic procedures employed and the life history of the species tested. In colonial tunicates, for example, fusion between individuals is an integral part of the life cycle, and so it may be actively promoted. Systems that act to prevent incorporation of incompatible individuals recruited by the above mechanism would likely require the generalized cellular reactivity and speed characteristic of nonspecific self recognition (NFR). Solitary tunicate species, on the other hand, do not reproduce asexually, and so they are unlikely to require mechanisms for the active integration of budded progeny or genetically compatible adults. However, solitary tunicates are sessile and usually aggregative animals, and thus they may come into contact with incompatible conspecifics. It follows that systems that inhibit passive fusion in aggregates of solitary individuals need not be as rapid as the self-restricted NFR of colonial forms, which prevent active incorporation. The slower rate of allogeneic reactions in solitary species may have allowed gradual diversification of the anti-self interactions involved in self-restriction, ultimately giving rise to specific anti-allogeneic receptors that depend on a process of proliferative selection for effector functions.

The evolutionary transition from self to specific recognition may have been promoted by the capacity of specific immunorecognition systems to develop adaptive responses. Repeated allogeneic challenges are more likely to occur in close aggregations of sessile organisms, and alloimmune memory would be of considerable selective value in this situation. Alternatively, this change may have been driven by the potential for enhanced discriminatory capacity that is inherent in systems of specific immunorecognition. The greater discriminatory capability offered by specific recognition mechanisms is clearly seen in the widely disparate levels of histocompatibility polymorphism evident in colonial and solitary tunicates. In the colonial tunicate genus, *Botryllus,* NFRs are controlled by 80 to several hundred alleles at a single fusion locus (Fuke and Nakamura, 1985), while population genetic analysis of *S. plicata* has revealed only five discrete tissue haplotypes. Despite the substantial difference in polymorphism, the histocompatibility systems of *Botryllus* and *S. plicata* yield similar frequencies of reciprocal compatibility between individuals. Such a marked disparity between the level of polymorphism and discriminatory efficacy probably reflects the ability of specific immunorecognition mechanisms to identify single haplotype differences. The ability to react against single disparate histocompatibility alleles would provide a proportionally greater selective benefit than that afforded by self recognition, where a single shared allele yields compatibility.

Functions of Primordial Immunity

These arguments over the evolutionary basis for specific recognition are predicated on the assumption that the major physiological function of adaptive immunity in invertebrates is allorecognition rather than pathogen or neoplasm elimination. In *S. plicata,* this assumption has been validated by studies of conspecific fusion. Histocompatibility, as determined by tunic graft rejection, correlates strongly with the prevention of fusion between individuals in their native environment. Tunicates that have undergone fusion with closely opposed neighbors can be found only at extremely low frequencies in natural populations of *S. plicata.* It has been demonstrated that these fused pairs always share histocompatibility tissue types, suggesting a direct anti-allogeneic basis for histocompatibility reactions (Kingsley et al., 1989).

EVOLUTION OF VERTEBRATE IMMUNITY

Critical Distinctions Between Vertebrates and Invertebrates

The apparent predilection of primordial immunity for allorecognition may account for the continued diversification of immunological mechanisms seen among vertebrates. Vertebrates have little physiological need for allorecognition. They do not exhibit sessile aggregative behavior and so are not in danger of conspecific fusion. Nor can the potential for allogeneic challenge, which arises in utero among amniotes, account for the maintenance of complex discriminative systems in oviparous vertebrates. Furthermore, the role of immune surveillance against neoplasia as the predominant function of vertebrate histocompatibility systems has been progressively discounted. The evolution of vertebrates may therefore have eliminated the original requirement for allodiscrimination as the basis for immunorecognition. However, evolution is a conservative process, and it is unlikely that specific recognition responses developed among invertebrates would be abandoned and replaced by a comparable mechanism of integrated immunity. A more parsimonious explanation is that invertebrate allorecognition underwent adaptation to fulfill the requirements for pathogen surveillance that predominate among vertebrates.

The Radiation of Immunorecognition

The changes required to bring about this expansion of immunorecognition functions are easily envisaged (Lafferty and Talmage, 1976; Lafferty and Woolnough, 1977; Mäkelä et al., 1976) (Fig. 2–6). Anti-allogeneic receptors may have diversified, so that their repertoire of specificities included not only alloantigens but also autologous histocompatibility antigens complexed with viral or parasitic determinants. Hence, the capacity to destroy infected somatic cells would be superimposed upon preexisting effector mechanisms for allogeneic cytotoxicity. Subsequent specialization of effector cell functions may then have given rise to the heterogenous activities that are typical of vertebrate lymphocytes, including T-lineage helper, suppressor, and cytotoxic cells, as well as B cells.

Molecular Phylogenies

This postulated development of integrated immune functions from primitive systems of allorecognition is supported by the observed molecular homologies between MHC antigens, lymphocyte receptors, and immunoglobulins. Primitive self recognition was presumably based on complementarity between identical cell surface (H-ag) determinants (Mäkelä et al., 1976; Rothenberg, 1978). Divergence of these primitive H-ag genes during the development of specific immunorecognition systems may have given rise to distinct but complementary antigens and receptors. Mäkelä et al. (1976) proposed that this functional divergence occurred as a sequential process. They suggested that complementary self interactions gradually showed increased affinity toward altered-self components. These altered-self antigens may initially have represented allogeneic H-ag, thus providing the potential for adaptive histocompatibility reactions. Continued diversification of receptors for alloantigens could then have led to expression of receptors with specificity for H-ag modified by exogenous determinants in a manner analogous to MHC restriction of vertebrate T cell antigen receptors (Mäkelä et al., 1976; Schwartz, 1985).

MHC Restriction as a Relic of Primordial Immunorecognition

The development of mechanisms for recognition of pathogenic organisms from allodiscriminative processes also provides an evolutionary explanation for the role of histocompatibility antigens in vertebrate immunity. Most vertebrate T cell immune responses require that exogenous antigens be associated with self MHC (histocompatibility) deter-

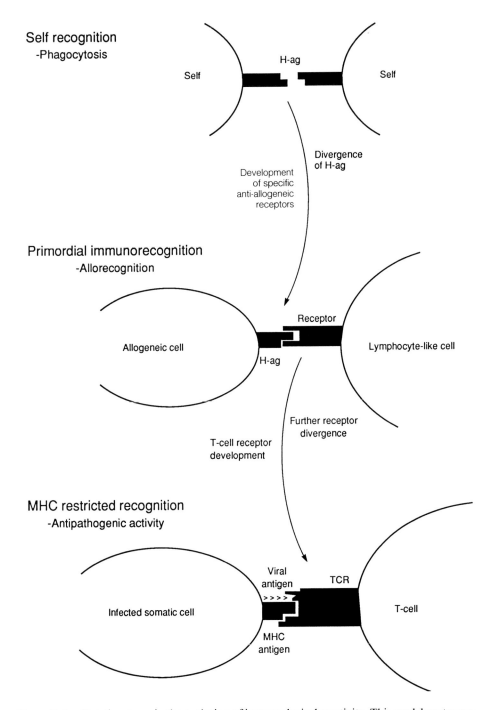

Figure 2–6. Putative stages in the evolution of immunological reactivity. This model portrays a gradual alteration from self recognition to MHC-restricted reactivity with primordial immune responses characterized by invertebrate adaptive histocompatibility as an intermediate stage.

minants prior to recognition, so-called MHC restriction (Schwartz, 1985; Zinkernagel and Doherty, 1979). To explain the requirement for MHC restriction of anti-pathogen responses, various hypotheses have been proposed, including that MHC antigens act to localize recognition responses on cell surfaces (Langman, 1978), to guide antigen to the most appropriate effector mechanism (Cunningham and Lafferty, 1977), to selectively prevent autoreactivity (Cohen, 1983), and to regulate interactions between immunocompetent cells during ontogeny. These complex arguments would be effectively circumvented if histocompatibility antigens had a primordial allodiscriminative function. According to the evolutionary model presented above, histocompatibility antigens were the original targets for immunorecognition. MHC-restricted reactivity could have arisen from anti-allogeneic receptors that were coincidentally cross-reactive with self antigens which became complexed with exogenous determinants during somatic infection. This evolutionary origin may have shaped recognition processes to the extent that the MHC component of recognition is retained in advanced mechanisms largely as a relic of the system's evolutionary past.

Final Comment

The phylogenetic pathways for the development of immunological recognition that are described in this chapter remain almost entirely conjectural. There is no doubt that invertebrates possess sophisticated mechanisms of adaptive immunity. However, the relationship of these mechanisms to the integrated immune systems of vertebrates is largely unresolved. It should also be stressed that conjectural models are prone to anthropocentrism. Invertebrates are themselves complex and specialized organisms that should not be thought of merely as physiological stepping-stones to vertebrates. Viewed in this light, the potential for convergent evolution should not be underestimated. It may eventually turn out that the defensive systems of different phyla are unrelated but have developed convergently, reflecting the similar physiological requirements for recognition in diverse organisms. We will have to await definitive molecular evidence that can discriminate between analogy and homology. More importantly, we must not be disappointed if the results of such analysis eventually show evolutionary convergence. The capability to develop unrelated systems of immunorecognition, which nonetheless show close functional similarities, would be a splendid illustration of the power inherent in adaptive evolution.

References

Bancroft, F. W. (1903). Variation and fusion of colonies in compound ascidians. *Proc. Calif. Acad. Sci. 3*:137–186.

Bigger, C. H., Jokiel, P. L., Hildemann, W. H., and Johnston, I. S. (1982). Characterization of alloimmune memory in sponges. *J. Immunol. 129*:1570–1572.

Burnet, F. M. (1971). "Self recognition" in colonial marine forms and flowering plants in relation to the evolution of immunity. *Nature 232*:21.

Burnet, F. M. (1974). Invertebrate precursors to immune response. In *Contemporary Topics in Immunobiology,* vol. 4 (Cooper, E. L., ed.). New York and London: Plenum Press. pp. 13–24.

Buss, L. W., and Green, D. R. (1985). Histocompatibility in vertebrates: The relic hypothesis. *Dev. Comp. Immunol. 9*:191–201.

Capo, C., Bongrand, P., Benoliel, A., and Depieds, R. (1979). Non-specific recognition in phago-cytosis: Ingestion of aldehyde-treated erythrocytes by rat peritoneal macrophages. *Immu-nology* 36:501–508.

Cohen, N. (1983). Some thoughts about the possible early phylogeny of MHC restriction. *Dev. Comp. Immunol.* 7:775–780.

Coombe, D. R., Ey, P. L., and Jenkin, C. R. (1984). Self/non-self recognition in invertebrates. *Quart. Rev. Biol.* 59:231–255.

Crichton, R., and Lafferty, K. J. (1975). The discriminatory capacity of phagocytic cells in the chiton *(Liolophura gaimardi). Adv. Exp. Med. Biol.* 64:89–98.

Cunningham, A. J. (1978). A comparison of immune strategies of vertebrates and invertebrates. *Dev. Comp. Immunol.* 2:243–252.

Cunningham, A. J., and Lafferty, K. J. (1977). A simple, conservative explanation of the H-2 restric-tion of interactions between lymphocytes. *Scand. J. Immunol.* 6:1–6.

Delmotte, F., Brillouet, C., Leclerc, M., Luquet, G., and Kader, J.-C. (1986). Purification of an anti-body-like protein from the sea star *Asterias rubens* (L.). *Eur. J. Immunol.* 16:1325–1330.

Ermak, T. H. (1976). The hematogenic tissues of tunicates. In *The Phylogeny of Thymus and Bone-Marrow-Bursa Cells* (Wright R. K., and Cooper E. L., eds.). Amsterdam: Elsevier/North Holland. pp. 45–55.

Ermak, T. H. (1982). The renewing cell populations of ascidians. *Am. Zool.* 22:795–805.

Field, K. G., Olsen, G. J., Lane, D. J., Giovannoni, S. J., Ghiselin, M. T., Raff, E. C., Pace, N. R., and Raff, R. A. (1989). Molecular phylogeny of the animal kingdom. *Science* 239:748–753.

Fuke, M. T. (1980). "Contact reactions" between xenogeneic and allogeneic coelomic cells of soli-tary ascidians. *Biol. Bull.* 158:304–315.

Fuke, M. T., and Nakamura, I. (1985). Pattern of cellular alloreactivity of the solitary ascidian, *Hal-ocynthia roretzi,* in relation to genetic control. *Biol. Bull.* 169:631–637.

Fuke, M. T., and Numakunai, T. (1982). Allogenic cellular reactions between intra-specific variants of the solitary ascidian, *Halocynthia roretzi. Dev. Comp. Immunol.* 6:253–261.

Gilbertson, P., Wotherspoon, J., and Raison, R. L. (1986). Evolutionary development of lympho-cyte heterogeneity: Leucocyte sub-populations in the Pacific hagfish. *Dev. Comp. Immunol.* 10:1–10.

Hildemann, W. H. (1974). Some new concepts in immunological phylogeny. *Nature* 250:116–120.

Hildemann, W. H. (1984). A question of memory. *Dev. Comp. Immunol.* 8:747–756.

Hildemann, W. H., and Reddy, A. L. (1973). Phylogeny of immune responsiveness: Marine inver-tebrates. *Fed. Proc.* 32:2188–2194.

Hildemann, W. H., Jokiel, P. L., Bigger, C. H., and Johnston, I. S. (1980). Allogeneic polymorphism and alloimmune memory in the coral *Montipora verrucosa. Transplantation* 30:297–301.

Hood, L., Kronenberg, M., and Hunkapiller, T. (1985). T-cell antigen receptors and the immuno-globulin supergene family. *Cell* 40:225–229.

Hunkapiller, T., and Hood, L. (1986). The growing immunoglobulin gene superfamily. *Nature* 323:15–16.

Karp, R. D., and Hildemann, W. H. (1976). Specific allograft reactivity in the sea star *Dermasterias imbricata. Transplantation* 22:434–439.

Kolb, H. (1977). On the phylogenetic origins of the immune system—a hypothesis. *Dev. Comp. Immunol.* 1:193–206.

Kingsley, E., Briscoe, D. A., and Raftos, D. A. (1989). The role of histocompatibility in fusion between conspecifics of *Styela plicata* (Urochordata, Ascidiacea). *Biol. Bull.* 176:282–289.

Lafferty, K. J., and Talmage, D. W. (1976). Theory of allogenic reactivity and its relevance to trans-plantation biology. *Transplant. Proc.* 8:305.

Lafferty, K. J., and Woolnough, J. (1977). The origin and mechanism of the allograft reaction. *Immunol. Rev.* 35:231–262.

Langman, R. E. (1978). Cell-mediated immunity and the major histocompatibility complex. *Rev. Physiol. Biochem. Pharmacol. 81*:1–37.

Mäkelä, O., Koskimies, S., and Karjalainen, K. (1976). Possible evolution of acquired immunity from self recognition structures. *Scand. J. Immunol. 5*:305–310.

Mansour, M. H., DeLange, R., and Cooper, E. L. (1985). Isolation, purification, and amino acid composition of the tunicate hemocyte Thy-1 homolog. *J. Biol. Chem. 260*:2681–2686.

Marchalonis, J. J., and Warr, G. W. (1978). Phylogenetic origins of immune recognition: Naturally occurring DNP-binding molecules in chordate sera and hemolymph. *Dev. Comp. Immunol. 2*:443–460.

Matsunaga, T., and Mori, N. (1988). The origin of the immune system. The possibility that immunoglobulin superfamily molecules and cell adhesion molecules of chicken and slime mould are all related. *Scand. J. Immunol. 25*:485–495.

Mukai, H., and Watanabe, H. (1974). On the occurrence of colony specificity in some compound ascidians. *Biol. Bull. 147*:411–421.

Ohno, S. (1970). *Evolution by Gene Duplication.* Heidelberg: Springer-Verlag.

Oka, H., and Watanabe, H. (1957). Colony specificity in compound ascidians as tested by fusion experiments (a preliminary report). *Proc. Jap. Acad. 33*:657.

Parrinello, N., and Canicatti, C. (1982). Carbohydrate binding specificity and purification by biospecific affinity chromatography of *Ascidia malaca* haemagglutinins. *Dev. Comp. Immunol. 6*:53–64.

Prowse, R. H., and Tait, N. N. (1969). In vitro phagocytosis by amoebocytes from the haemolymph of *Helix aspersa:* I. Evidence for opsonic factors in serum. *Immunology 17*:437–443.

Raftos, D. A. (1991). Cellular restriction of histocompatibility responses in the solitary urochordate, *Styela plicata. Dev. Comp. Immunol. 15*:93–98.

Raftos, D. A., and Briscoe, D. A. (1990). Genetic basis of allograft rejection in the solitary urochordate, *Styela plicata. J. Hered. 81*:96–100.

Raftos, D. A., Briscoe, D. A., and Tait, N. N. (1988). Mode of recognition of allogeneic tissue in the solitary urochordate, *Styela plicata. Transplantation 45*:1113–1126.

Raftos, D. A., Tait, N. N., and Briscoe, D. A. (1987a). Allograft rejection and alloimmune memory in the solitary urochordate, *Styela plicata. Dev. Comp. Immunol. 11*:343–351.

Raftos, D. A., Tait, N. N., and Briscoe, D. A. (1987b). Cellular basis of allograft rejection in the solitary urochordate, *Styela plicata. Dev. Comp. Immunol. 11*:713–726.

Ratcliffe, N. A., Rowley, A. F., Fitzgerald, S. W., and Rhodes, C. P. (1985). Invertebrate immunity: Basic concepts and recent advances. *Int. Rev. Cytol. 97*:184–350.

Renwrantz, L., Schancke, W., Harm, H., Erl, H., Liebsch, H., and Gercken, J. (1981). Discriminative ability and function of the immunobiological recognition system of the snail, *Helix pomatia. J. Comp. Physiol. 141*:477–488.

Roch, P. G., and Cooper, E. L. (1983). A β_2-microglobulin-like molecule on earthworm *(L. terrestris)* leukocyte membranes. *Dev. Comp. Immunol. 7*:633–636.

Rothenberg, B. E. (1978). The self recognition concept: An active function for the molecules of the major histocompatibility complex based on the complementary interaction of protein and carbohydrate. *Dev. Comp. Immunol. 2*:23–38.

Schwartz, R. H. (1985). T-lymphocyte recognition of antigen in association with gene products of the major histocompatibility complex. *Annu. Rev. Immunol. 3*:237–261.

Scofield, V. L., Schlumpberger, J. M., and Weissman, I. L. (1982a). Colony specificity in the colonial tunicate *Botryllus* and the origins of vertebrate immunity. *Am. Zool. 22*:783–794.

Scofield, V. L., Schlumpberger, J. M., West, L. A., and Weissman, I. L. (1982b). Protochordate allorecognition is controlled by an MHC-like gene system. *Nature 295*:499–502.

Shalev, A., Greenberg, A. H., Logdberg, L., and Bjorck, L. (1981). β_2-microglobulin-like molecules in lower vertebrates and invertebrates. *J. Immunol. 127*:1186–1191.

Shalev, A., Pla, M., Ginsburger-Vogel, T., Echalier, G., Logdberg, L., Bjorck, L., Colombani, J., and Segal, S. (1983). Evidence for β_2 microglobulin-like and H-2-like antigenic determinants in *Drosophila. J. Immunol. 130*:297–302.

Sparks, A. K. (1985). *Synopsis of Invertebrate Pathology (Exclusive of Insects).* Amsterdam: Elsevier.

van der Knaap, W.P.W., Boots, A.M.H., van Asselt, L. A., and Sminia, T. (1983). Specificity and memory in increased defense reactions against bacteria in the pond snail, *Lymnaea stagnalis. Dev. Comp. Immunol. 7*:435–443.

Vasta, G. R., and Marchalonis, J. J. (1985). Humoral and cell associated lectins from invertebrates and lower chordates. *Dev. Comp. Immunol. 9*:531–539.

Vasta, G. R., Marchalonis, J. J., and Kohler, H. (1984). Invertebrate recognition protein crossreacts with an immunoglobulin idiotype. *J. Exp. Med. 159*:1270–1276.

Warr, G. W., Decker, J. M., Mandel, D. D., DeLuca, D., Hudson, R., and Marchalonis, J. J. (1977). Lymphocyte-like cells of the tunicate *Pyura stolonifera:* Binding of lectins, morphological and functional studies. *Aust. J. Exp. Biol. Med. Sci. 55*:151–164.

Williams, A. F., and Gagnon, J. (1982). Neuronal cell Thy-1 glycoprotein: Homology with immunoglobulin. *Science 216*:696–703.

Zinkernagel, R. M., and Doherty, P. C. (1979). MHC restricted cytotoxic T-cells: Studies on the biological role of polymorphic major transplantation antigens determining T-cell restriction specificity, function and responsiveness. *Adv. Immunol. 27*:51–177.

3

Ontogeny and Phylogeny of MHC-like Molecules

JIM KAUFMAN

The major histocompatibility complex (MHC) is an extensive genetic region that encodes a large number of different proteins and which determines a variety of biological functions. The hallmark of the MHC, for which it was first discovered and characterized, is the polymorphic cell surface heterodimeric glycoproteins that constitute the classical MHC transplantation antigens. These are divided into two multigene families, which encode the class I and class II molecules, and which are related in structure, function, and presumably evolution. These molecules are recognized by certain thymus-derived (T) lymphocytes via their clonally variable T cell receptors (TCR) to bring about tissue graft rejection and a variety of other phenomena related to T cell recognition. This chapter will consider only these classical MHC transplantation antigens and their relatives, which include molecules with different functions, structural features, and chromosomal locations. The large number of unrelated genes found in the MHC has been reviewed elsewhere (Guillemot et al., 1988, 1989; Klein, 1986; Trowsdale et al., 1991).

T LYMPHOCYTES AND RECOGNITION OF POLYMORPHIC MHC TRANSPLANTATION ANTIGENS

The classical transplantation MHC molecules are interesting because they are recognized by T lymphocytes. These T cells are an important part of the immune system, in both resistance to disease and involvement in autoimmune disease. While B cells and T cells share many properties, there are four phenomena specific for T cell recognition that perplexed immunologists for many years: MHC restriction, self/non-self recognition, immune response gene effects, and the high proportion of the immune response that is directed to MHC alloantigens. These are now beginning to be understood at the cellular and molecular levels.

1. *MHC restriction.* Both T cells and B cells recognize antigen with those properties considered to be basic to the immune system: diversity, specificity, and memory. In addition, they both express antigen receptors that are very similar in structure: integral membrane glycoprotein heterodimers with extracellular immunoglobulin variable (V) and constant (C) domains. These antigen receptors also have very similar mechanisms for achieving V region diversity: multiple members of a multigene family, germline rear-

rangement of V, D (diversity), and J (joining) gene segments, N and P nucleotide addition at the junctions, and combinatorial diversity of chain association. (The only exception is that B cells also have mechanisms for somatic hypermutation after antigen contact, which is lacking in T cells presumably to avoid breaking self tolerance of T cells in the periphery, as discussed below). However, unlike B cells, T cells steadfastly refuse to recognize antigen alone; instead they must see it along with a particular MHC molecule on the surface of another cell. This phenomenon of MHC restriction of T cell recognition, or antigen presentation by MHC molecules, was discovered by Zinkernagel and Doherty in 1974 (Schwartz, 1985; Zinkernagel and Doherty, 1979).

2. *Self and non-self recognition.* T lymphocytes can distinguish between self and nonself, being tolerant to most self antigens. This is a property that they acquire, at least in part, in the thymus and is dependent on the particular MHC molecules present there. This process of MHC-dependent thymic education was suggested by Jerne in the early 1970s (von Boehmer et al., 1989). In contrast, it is relatively easy to detect or elicit lowaffinity antibodies to autoantigens (though there may be more similarities between T cell and B cell maturation and tolerance than has been previously suspected).

3. *Immune response genes.* The system of T cell–MHC recognition of antigen is not perfect. It breaks down under a number of situations, for example, the lack of a particular T cell receptor V region or the similarity of self antigen to an exogenous antigen. The beststudied cause of such breakdowns is that a particular MHC molecule can present many but not all antigens. This "hole in the response repertoire" was discovered and mapped to the MHC by Benacerraf and by McDevitt and Sela in the early 1970s, and was christened the immune response (or Ir) gene defect (Schwartz, 1986). After the discovery of MHC restriction, it was quickly realized (Zinkernagel and Doherty, 1979) that an animal with only one MHC molecule (e.g., a homozygote with one functional MHC locus) would fail to respond to some pathogens due to this Ir gene phenomenon, and that the survival of heterozygotes would be a good selection for multiple alleles in the MHC, giving rise to polymorphism.

4. *Alloantigen response.* The MHC was originally discovered as the predominant locus responsible for tissue rejection, and this is mirrored in a variety of other in vivo and in vitro assays, including the graft vs. host reaction (GvH) and the mixed lymphocyte reaction (MLR). These reactions to MHC molecules are much stronger than the reactions to any other single allo- or xenoantigen (with the exception of the products of the *mls* locus and certain bacterial enterotoxin "superantigens") and can involve up to 10% of T lymphocytes. The strong response requires only differences in the MHC; in fact, only a few amino acids in a single MHC molecule, as in the mouse *bm* mutants, are sufficient to elicit a response (Klein 1986; Nathenson et al. 1986). In contrast, the number of B cells directed to MHC molecules is not overwhelmingly greater than the number directed to a variety of other alloantigens or xenoantigens. However, there is some preferential production of alloantibodies to MHC alloantigens, and evidence for the preferential recognition of polymorphic chicken MHC B-G antigens by some B cells is discussed below.

Molecular Basis of T Cell Recognition of MHC Antigens

The molecular basis of these four perplexing T cell recognition phenomena is the following: MHC molecules bind particular peptides that are derived from antigen by proteolytic processing, and these MHC/peptide complexes are recognized by the TCR (Bjorkman and Parham, 1990; Bjorkman et al., 1987a,b; Davis and Bjorkman, 1988). This explains many of the mysteries that were associated with T cell recognition of antigen, such as why T cells do not recognize native antigen, why other cells are required to process and present antigen, why antibodies to MHC molecules but not to antigen block T cell recognition, how the same MHC molecules can present antigens that vary enormously in size, why antibodies were never found to so-called minor transplantation antigens, and so on.

The cell biology resulting from these molecular phenomena is complex and depends in part on the particular class of MHC transplantation antigen (Fig. 3–1). In general, MHC class I molecules bind peptides derived from endogenously synthesized molecules,

Cytolysis: generally CD8$^+$ T cells recognize class I + processed antigen

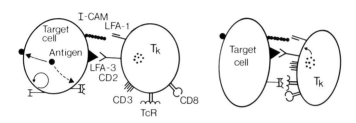

Regulation: generally CD4$^+$ T cells recognize class II + processed antigen

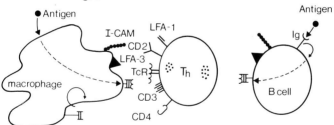

Figure 3–1. Cell–cell interaction in T cell recognition of antigen and MHC molecules. Endogenous antigen is synthesized inside a cell as shown for the "target cell," while exogenous antigen can be taken up either nonspecifically, as in the macrophage (and other antigen presenting cells), or specifically via cell surface Ig, as in the B lymphocyte. The antigens are proteolysed into fragments, probably in different intracellular compartments, and some of them bind to MHC molecules as "processed antigen" during biosynthesis, during recycling, or at some other point. The T cells first interact with another cell in an antigen-independent manner via accessory molecules such as LFA-1 and CD2 (which bind I-CAM and LFA-3, respectively). The subsequent antigen-dependent recognition of MHC molecule and processed antigen is via the αβ TCR/CD3 complex. In addition, CD4 binds to class II and CD8 binds to class I, perhaps as part of the signaling process via the associated cytoplasmic *lck* kinase. Finally, the T cells release molecules from intracellular vesicles (cytotoxic factors from cytotoxic T cells and various lymphokines or inflammatory agents from helper T cells) before dissociating for the next round of interactions.

and they are recognized by cytotoxic T cells via the TCR composed of α and β chains and by the accessory molecule CD8. On the other hand, MHC class II molecules generally bind peptides derived from exogenously acquired antigen, and they are recognized by T cells, which are generally but not always involved in positive regulation ("help"), via the $\alpha\beta$ TCR and the accessory molecule CD4. Evidently, the T cell first binds to the other cell via nonspecific accessory molecules like CD2 and LFA-1, whose ligands are LFA-3 and I-CAM, respectively. This is followed by interaction of the TCR/CD3 complex and the CD8 or CD4 molecules with the peptide/MHC class I or class II molecule complex. Signals delivered by some of these cell surface molecules (perhaps via CD4 and CD8 associated with the *lck* tyrosine kinase inside T cells) lead to the activation of these cells and the release of the contents of intracellular vesicles. These vesicles may contain cytotoxic components (like perforins and granzymes) in cytotoxic T cells, or a variety of lymphokines and inflammatory signals (such as interleukins 2 through 7, γ interferon, etc.) in helper T cells, which may also mediate delayed-type hypersensitivity (DTH) and other reactions.

Much has been learned in the last few years about the cell biology of antigen processing, leading to the view that endogenous and exogenous antigens are degraded in different intracellular locations and that the class I and class II molecules meet their peptides in different intracellular compartments (Brodsky and Guagliardi, 1991; Koch et al., 1989; Long 1989; Robertson 1991). There is an enormous cytoplasmic complex called the proteosome that degrades cytoplasmic proteins into peptides; it is thought that these peptides are fed into the endoplasmic reticulum by peptide transporters and then may be bound by newly synthesized class I molecules. Both the proteosome and the peptide transporter include polymorphic components that are encoded in the MHC. In contrast, newly synthesized class II molecules in the endoplasmic reticulum are bound to an "invariant chain" that both blocks the peptide binding site and directs the class II molecules to the endosomal compartment. In the acidic endosomes, the invariant chain falls off and is degraded, while the class II molecules bind peptides that have been produced there from endocytosed proteins. There is still much to be learned about the nature of the processing events themselves, including which enzymes are actually responsible, whether these enzymes are the same in every cell and in every individual, and whether there is trimming after binding.

Only some peptides are bound to any particular MHC molecule, and the features of the peptide that are crucial for binding are becoming clear. Initially, studies with chemically synthesized peptides led to suggestions that the correct peptides either have a propensity to form amphipathic helices or that they have a particular five amino acid motif for binding. However, a different view has developed from analysis of the three-dimensional structures of human class I molecules at high resolution and analysis of the amino acid sequences of natural peptides eluted from MHC molecules (Germain, 1991; Madden et al., 1991; Saper et al., 1991). The structures of HLA-A2, Aw68, and B27 show that there are distinct subsites in the peptide binding site. The subsites at either end are composed of invariant residues, while the ones in between vary in size, shape, and charge. The peptides bound to the HLA-B27 molecule are all nonamers with the same irregular extended conformation, in which the invariant main chain atoms at the N- and C-termini (residues 1 and 9) interact with the invariant subsites at either end. The specificity of binding is due to the side chains of particular amino acids in between (residues 2, 3, 7, and 9) that are buried in the polymorphic subsites. This is a general property because the natural peptides eluted from many different class I molecules are all octamers or nonamers with several nearly invariant positions, which are different for each MHC molecule. In contrast, the peptides eluted from class II molecules are longer and include several peptides that are identical except for different lengths at the C-terminal ends. The ragged ends suggest that long peptides bound to class II molecules are trimmed back with no precise binding to the peptide C-termini. No invariant positions were found for the peptides naturally bound to class II molecules; on the other hand, synthetic peptides with only two specific amino acids bound stably to class II molecules, presumably via particular subsites.

As noted above, the structure and mechanisms for generation of diversity of the TCR give no clues as to how the T cell can distinguish self and non-self. Nor do the MHC molecules themselves distinguish between self and non-self peptides. In fact, education in the thymus is at the cell population level, involving two steps of selection for those T cells that bind self MHC molecules, but do not recognize self peptides bound to those self MHC molecules (Fig. 3–2) (von Boehmer et al., 1989). Before thymic education, the immature $\alpha\beta$ thymocytes apparently express all possible combinations of rearranged TCR variable regions (i.e., an unselected repertoire) as well as bearing both CD4 and CD8 molecules. These "double-positive" cells could presumably bind to all kinds of MHC

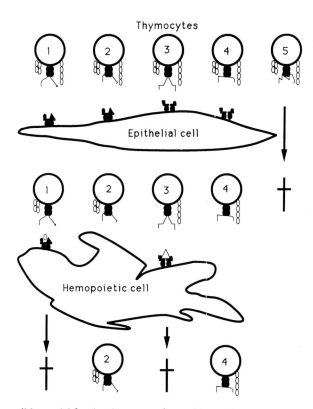

Figure 3–2. A possible model for development of $\alpha\beta$ TCR-bearing T cells in the thymus. Early $\alpha\beta$ TCR-bearing thymocytes have both CD4 and CD8. Those thymocytes that fail to bind thymic epithelial cells simply die, whereas those that bind MHC molecules (though it is not yet clear whether the MHC molecules bear peptides) on the thymic epithelial cells with their TCR are positively selected. The thymocytes that bind class I lose their CD4 molecule, while those binding class II lose their CD8 molecule. Those surviving thymocytes that bind self peptide/MHC molecule complexes on cells of hemopoietic origin are tolerized, and the survivors of this negative selection leave the thymus as educated T lymphocytes.

molecules as well as to some normal antigens, and they would not be tolerant of self anti-gens. In one step, thymocytes that bind the self MHC molecules on certain epithelial cells in the thymic cortex are positively selected. Cells that bind self class I molecules stop expressing CD4, while cells that bind self class II molecules stop expressing CD8, to give so-called "single-positive" thymocytes. The large majority of cells, perhaps 85%, fail to bind and die in the cortex as "double-positive" small cortical thymocytes. In another step, thymocytes that recognize self peptides bound to self MHC molecules on cells derived from the hemopoietic lineage (dendritic cells, macrophages, or perhaps even other thy-mocytes in the cortex, cortical-medulary junction, or the medulla) die. While the order of these two steps is not clear, the survivors of both selection steps can presumably only recognize non-self peptides bound to self MHC in the periphery of the animal. In exper-imental situations like tissue grafts, however, a particular mature T cell may cross-react with a particular non-self MHC molecule bound to a particular self or non-self peptide; this happens frequently enough to explain the high level of reactivity to MHC alloanti-gens.

Thus the thymus selects the useful, ignores the useless, and destroys the dangerous (von Boehmer et al., 1989)—but the detailed molecular mechanisms of positive and neg-ative selection remain to be clarified. For instance, it is not clear whether MHC molecules on the epithelial cells in the positive selection steps bear peptides. If they do not, then it is not clear how they avoid acquiring them; if they do, then it is not clear how a TCR bias for self MHC molecules bound to these particular peptides (perhaps the same self peptides used for deletion in the negative selection step) is avoided. Perhaps peptides are not required under all circumstances, or perhaps special innocuous "filler peptides" are pro-vided in these epithelial cells. Another problem is how tolerance is generated for proteins that are not obviously produced in the thymus. One possibility is that all proteins are in fact produced at extremely low levels in the thymic stromal cells. But it is generally thought that specific binding in the negative selection step is the same as that which occurs in antigen presentation and T cell activation in the periphery, and that only the conse-quences are different, due to the state of the T cell. Thus, "physiological amounts" of antigen would be required for tolerance induction. A second possibility is that protein antigens from all over the body are brought into the thymus by macrophages and other antigen presenting cells (APCs). Finally, there may be additional tolerance-inducing mechanisms (like T cell–mediated suppression, VETO cells, etc.) in the periphery.

In addition to $\alpha\beta$ T cells, other T cells bear TCR composed of γ and δ chains (Blue-stone and Matis, 1989). They may be a much more heterogeneous population of cells than the $\alpha\beta$ T cells, but at least some of them require the thymus for maturation and have very complex ontogeny and tissue distribution. Their biological functions or the nature of the antigens they recognize are not clear, but different $\gamma\delta$ T cells are stimulated in vitro by heat shock proteins, mycobacterial proteins, nonclassical class I molecules like Qa molecules (either allogeneic or self in association with an exogeneous antigen), or CD1 molecules.

Structure of the Molecules Involved in T Cell Recognition

Most cell surface molecules involved in T cell recognition (like many other cell surface proteins with unrelated functions) are members of the immunoglobulin (Ig) gene super-family (Fig. 3–3) (Williams and Barclay, 1988). As mentioned above, both the B cell antigen receptor (antibody) and the TCR consist of two chains, each with an N-terminal V domain followed by C domain(s), a hydrophobic transmembrane region, and a hydrophilic cytoplasmic tail. The three-dimensional structures of a number of antibodies have been determined by X-ray crystallographic techniques. The antibody V domains of the H (heavy) and L (light) chains interact via the five-strand faces and together form the antigen binding site from three complementarity determining regions (CDRs) of high nucleotide and amino acid variability located in some of the bends. Some of the antibody C domains interact via the four-strand faces in a well-defined way (Fig. 3–4). Presumably the α and β chains of the TCR interact in a similar manner.

The CD4 and CD8 molecules also have N-terminal V-like domains, although without the clonal variability found in antibodies and TCR. The V-like domain of the CD8 molecule is followed by a presumed extended region of protein, a transmembrane region, and a cytoplasmic tail. On the cell surface, CD8 is a homodimer or a heterodimer of closely related chains. The V-like domain of the CD4 molecule is followed by a long sequence that is thought to be organized into three regions derived from Ig-like domains, and on the cell surface CD4 is apparently a monomer. The N-terminal CD4 V-like domain binds class II molecules as well as gp120 of the human immunodeficiency virus (HIV).

Both CD2 and the molecule to which it binds, LFA-3, have extracellular regions composed of two Ig-like domains (V- and C-like) followed by a hydrophobic transmembrane tail (or an inositol phopholipid tail in some forms of LFA-3). These two molecules are apparently derived by gene duplication from a common ancestor that originally mediated "like–like" interactions between cells. LFA-1 is a member of the integrin gene superfamily (like collagen receptor, laminin receptor, MAC-1, and many other adhesion molecules), but it binds to a member of the Ig gene superfamily called I-CAM (immunoglobulin cell adhesion molecule), which is also the receptor for the rhinovirus responsible for the common cold.

The MHC molecules are integral membrane glycoprotein heterodimers, with extracellular regions composed of two membrane-proximal Ig C–like domains and two membrane-distal non-Ig–like regions. The two glycoprotein chains of the class II heterodimer each consist of an N-terminal non-Ig region (β1 with an intrachain disulfide bond and α1 without) followed by the Ig C–like domain (β2 and α2), a small hydrophobic transmembrane region, and a small hydrophilic cytoplasmic tail. The class I heterodimer consists of a small Ig C–like protein called β_2-microglobulin (β_2-m) and a large transmembrane glycoprotein called the α chain. The class I α chain consists of two N-terminal non-Ig regions (α2 with an intrachain disulfide bond and α1 without) followed by an Ig C–like domain (α3), a small hydrophobic transmembrane region, and a rather long cytoplasmic tail with several different phosphylation sites. These different regions of the MHC molecules are generally encoded by different exons (see Fig. 3–8), with the long cytoplasmic region of the class I α chain encoded by several exons.

The three-dimensional structure of the extracellular region of several human class I molecules have been determined by X-ray crystallographic techniques (Figure 3–5) (Bjorkman and Parham, 1990; Bjorkman et al., 1987a,b; Madden et al., 1991; Saper et al., 1991). The α3 domain and β_2-m are associated through four-strand faces, but in a different manner than Ig C domains, with α3 in very limited contact with the rest of the molecule. Analyses of mutant class I molecules show that residues in the α3 domain are crucial for interaction with the CD8 coreceptor (Bjorkman and Parham, 1990; Lawlor et al., 1990; Salter et al., 1990); two protruberant surface residues probably constitute most

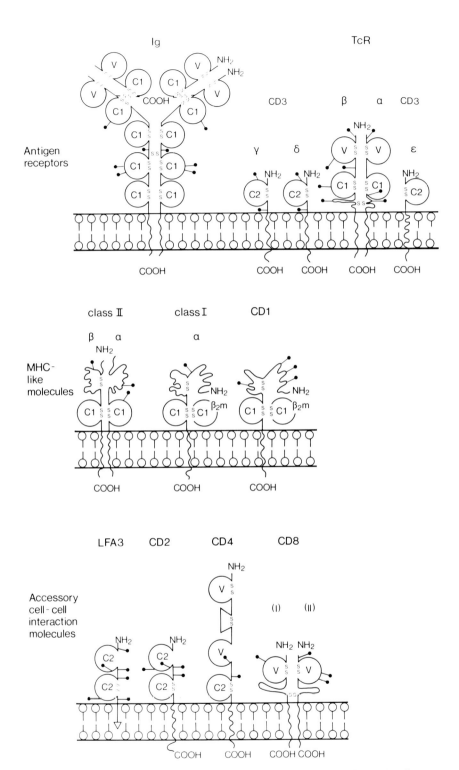

Figure 3–3.   Domain organization of molecules involved in T cell recognition of antigen and MHC molecules (modified from Williams and Barclay, 1988). Many cell surface proteins are

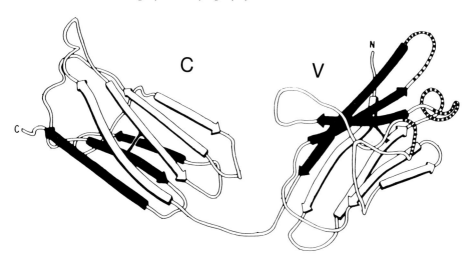

Figure 3–4.   Model of an Ig light chain (modified from Schiffer et al., 1973). In this ribbon diagram, the β strands of the four-strand faces are represented by flat white arrows, the β strands of the three-strand faces are represented by flat black arrows, and the invariant disulfide bonds are represented by black lines. Two other long sequence stretches of the variable (V) domain are often considered to be β strands, making the three-strand face into a "five-strand face." The three bends functioning as complementarity determining regions (CDRs) at the N-terminal ends of the V domain are indicated by hatched lines. Not shown are the domains of the other chain, which are located roughly symmetrically (but with a different angle between them) above the domains shown with the two constant (C) domains interacting via the four-strand faces and the two V domains interacting via the "five-strand faces."

of the binding site. Above the two Ig-like domains, and mostly in contact with β_2-m, is a symmetrical domain composed of α1 and α2, which together form a β-sheet platform of eight anti-parallel β strands that support two roughly parallel broken α helices. The groove formed by the two α helices and the β sheet contains most of the polymorphic residues of the MHC molecule and binds the antigenic peptide. This groove is closed at either end by subsites composed of invariant residues; in between are distinct subsites

---

composed in part of one or more extracellular domains with amino acid sequence homology to Ig variable (V) or constant (C1, C2) regions. These include the antigen receptors of B cells (Ig) and of T cells (TCR $\alpha\beta$ or TCR $\gamma\delta$ in a noncovalent complex with CD3 $\gamma\delta\epsilon$), shown on the top line; the antigen-binding "restriction elements" class II $\alpha\beta$ and class I α–β_2-m (as well as many homologous molecules such as CD1), shown on the second line; and the antigen-independent accessory molecules such as LFA-3 (which interacts with CD2), CD4 (which interacts with class II), and CD8 (which interacts with class I), shown on the third line. Not shown is the accessory molecule I-CAM, composed of multiple V-like regions, which interacts with LFA-1 (itself a member of the integrin family). Many of these molecules are glycoproteins (but not Ig light chain, CD3 ϵ and β_2-m proteins), which bear hydrophobic transmembrane segments (but note the Ig light chain and β_2-m without any, and LFA-3 with an inositol phospholipid). Most of the Ig-like domains have intradomain disulfide bonds (S–S, but note LFA-3, CD2, and CD4). The MHC-like molecules also bear membrane-distal extracellular domains with quite a different structure: a platform of β strands under a pair of roughly parallel α helices. (From Kaufman, 1988.)

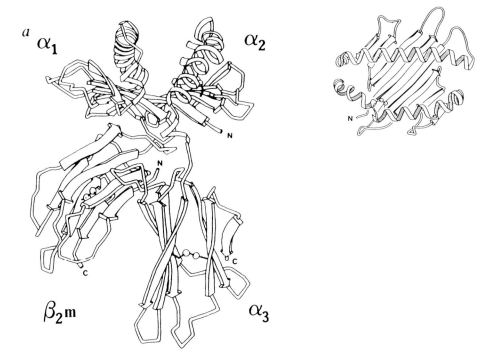

Figure 3–5.   Model of the human class I molecule HLA-A2 from the side *(left)* and the $\alpha 1/\alpha 2$ domain from the top *(right)* (from Bjorkman et al., 1987a,b). In this ribbon diagram, the β strands are represented by flat arrows, the α helices by coils of ribbon, and the disulfide bonds by connected balls. (N, amino-terminus; C, carboxy-terminus.)

whose size, shape, and charge depend on the polymorphic residues, so that each class I molecule has a unique binding site for peptides of eight or nine amino acids. Modeling studies suggest that the TCR CDR1 and CDR2 bind the two α helices, while the CDR3 (which in TCR is by far the most variable region) binds the peptide (Davis and Bjorkman, 1988). Modeling studies also suggest that the class II non-Ig–like regions form a very similar structure, although the groove may permit binding of longer peptides; there is no indication of how the Ig-like domains associate or where the CD4 binding site is located (Brown et al., 1988).

    Many of the key features of mammalian class I molecules are also found in the class I molecules of the chicken (Guillemot et al., 1989; Kaufman et al., 1990). Nearly all of the residues that differ between the two B-F alleles examined are found in the peptide-binding site, mostly in those positions known to be polymorphic in mammals. Those residues in the peptide-binding site that are invariant among mammals are also identical in the B-F molecule. The β_2-m is also highly conserved among birds. However, the homologous domains have not evolved at the same rate. The $\alpha 1$ domain is much more diverged between mammals and chicken than the $\alpha 2$ domain, even though they are thought to be involved in the same function. Similarly, the $\alpha 3$ domain is much more diverged between mammals and the chicken than is β_2-m; the two residues involved in CD8 binding are among the few residues conserved on the surface of the $\alpha 3$ domain. It may be that the β_2-m is under greater conservation pressure because it makes more interdomain contacts than $\alpha3$. There is also evidence that β_2-m may have other functions besides binding to the class I α chain (Brinckerhoff et al., 1989; Dargemont et al., 1989), for example, as a chemotactic factor for thymic stem cells.

Biosynthesis of MHC Molecules

The biosynthesis of MHC molecules is like that of many membrane proteins, involving cotranslational insertion in the lipid bilayer of the endoplasmic reticulum and acquisition of N-linked glycans, followed by transport through the endoplasmic reticulum and Golgi apparatus with maturation of the carbohydrate moieties and eventual expression on the plasma membrane. Many distinctive features of their biosynthesis, exist, however, evidently due in part to the peculiar requirements of antigen presentation (Fig. 3–6) (Kaufman et al., 1984; Koch et al., 1989; Long, 1989). There is also evidence in some cells for recycling of mature class I and class II molecules that have bound antibodies.

    The biosynthesis of class II molecules is dominated by interaction with the so-called invariant chain (also called Ii, γ, p33, p31, etc.) whose sequence has no homology with MHC molecules (Germain, 1991; Germain and Hendrix, 1991; Kaufman et al., 1984; Koch et al., 1989; Long, 1989). The invariant chain is encoded on a different chromo-

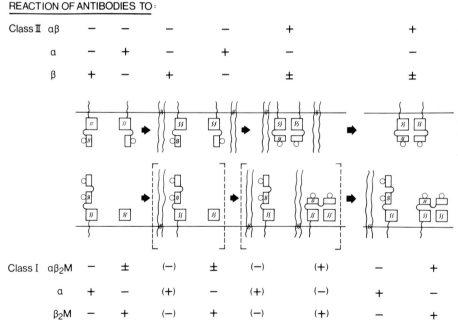

Figure 3–6. A model of the biosynthesis of class I and class II molecules based on the apparent biochemical similarities and possible functional similarity of the class II–associated invariant chain found in all vertebrates studied and the class I α chain–associated red blood cell molecule found on the surface of erythrocytes in reptiles and in *Xenopus.* The reactivities of antisera to the heterodimer and separated chains are indicated, with hypothetical reactivities in parentheses. The hypothetical biosynthetic intermediates of class I and the red blood cell molecule are enclosed in dotted boxes. (From Kaufman et al., 1990, 1991a.)

some than are the MHC genes (chromosome 5 in man, 18 in mouse), but its transcription is regulated in concert with the class II genes. The invariant chain is a transmembrane glycoprotein with the N-terminus in the cytoplasm; it forms noncovalently-linked dimers, and a small proportion is found on the cell surface as a large proteoglycan. Upon synthesis into the endoplasmic reticulum, the two class II chains associate separately with the invariant chain before association with each other into a complex that is sensitive to denaturing detergents like sodium dodecyl sulfate (SDS). This complex moves to the Golgi apparatus and, mediated by signals in the invariant chain, to endosomal vesicles. In the acidic endosomal vesicles, the invariant chain falls off and is degraded; the class II heterodimer may bind antigenic peptides, assume an SDS-resistant compact conformation, and move to the cell surface. The separated chains, the SDS-sensitive complex, and the SDS-resistant complex can all be distinguished by different antisera, with the most important changes evident in the $\alpha 1$ domain. Both class II chains, but not the invariant chain, are required for correct folding and transport out of the endoplasmic reticulum to the cell surface. However, in the absence of the invariant chain, the carbohydrate maturation is affected and certain exogenous antigens are not presented, presumably because the class II molecule bypasses the endosomal compartment. Other molecules are probably also important for this process; for instance, some mutant cell lines bear class II molecules without peptides on the surface, but the mutation is not in the invariant chain.

Class I molecules have no apparent requirement for a molecule equivalent to the class II–associated invariant chain, although class I α chains in the mouse endoplasmic reticulum associate with an 88 kD protein and class I α chains on *Xenopus* and reptile erythrocytes associate with a disulfide-linked dimer of 45 kD glycoprotein chains (Degen and Williams, 1991; Du Pasquier et al., 1989; Kaufman et al., 1990). Some minutes after synthesis into the endoplasmic reticulum, the class I α chain and β_2-m associate in a heterodimer; they travel to the Golgi apparatus and the plasma membrane via the default pathway, without the lengthy detour seen for class II molecules. As with class II molecules, marked antigenic changes occur upon assembly, particularly in the α chain. Unassociated β_2-m is secreted from the cell, while virtually all class I α chains require β_2-m for correct folding and export from the endoplasmic reticulum, the exception being H-2D^b. Certain mutant cells lack genes responsible for the presence of peptides in the endoplasmic reticulum; they produce class I heterodimers without peptide that travel to the cell surface but are unstable at normal body temperature.

Nature and Origin of Polymorphism in MHC Antigens

Most genes in an animal have at most a few common alleles because without either a high mutation rate or some form of selection new alleles appear slowly and are not fixed in the population, while old alleles are lost by drift until only one remains. The MHC transplantation antigens have apparently by far the highest number of common alleles among mammalian molecules, with over 200 common alleles at some loci (Klein 1986). The basis of this extraordinary polymorphism has been the subject of debate for many years, and the relative importance of creation of new alleles by mutation, recombination, and gene conversion versus selection of alleles for functional or structural features is still not clear.

Most immunologists would agree that the high polymorphism must be the result of selection for pathogen recognition and disease resistance. This view is based on two points (Zinkernagel and Doherty, 1979). First, graft rejection is not physiological in vertebrates;

second, a species with a single MHC molecule (like a homozygous single gene) would be susceptible to any pathogen that produced no peptides that bound to the MHC molecule (an Ir, immune response, gene defect). The existence of multiple MHC molecules would give an individual more chance to bind peptides and thus respond to the pathogen. The advantage of individuals with multiple MHC alleles in disease resistance would select for heterozygotes, either by overdominance (heterozygote advantage) or by frequency-dependent selection (rare allele advantage). Thus, at the population level, MHC polymorphism would ensure that some individuals would always survive a new pathogen. It should be pointed out that a large number of different but monomorphic MHC genes would probably be a good alternative to polymorphism for binding as many non-self peptides as possible. However, for each different MHC molecule in an animal, a set of T cells is tolerized due to binding of self peptides, and with too many MHC genes, there would be no T cells left (Vidovic and Matzinger, 1988). Thus, a few polymorphic genes would apparently provide an optimal solution for pathogen resistance, as seen for *Xenopus* polyploid animals (Du Pasquier et al., 1989).

It has been suggested that pathogen resistance is not a sufficient selective pressure to explain the high level of polymorphism observed in some species, based on two points (Klein, 1986, 1987). The first point argues that while there are many examples of MHC associations with autoimmune disease, there are few examples of MHC-determined resistance to infectious diseases in real populations. The best example is in chickens, where the *B21* haplotype confers resistance to Marek's disease virus; however, it is not clear whether this resistance is due to polymorphic class I or class II genes, or to one of the many other unrelated genes that are located in this region of the chicken MHC (Crone and Simonsen, 1987; Guillemot et al., 1989). A second example has recently been reported, in which HLA-Bw53 and a DRw13/DQw1 haplotype are associated with resistance to endemic malaria in Africa (Hill et al., 1991). A selection strong enough to maintain 100 common alleles should be observable at the population level, although it has been argued that the use of modern medicines and the overlapping defenses in the immune system would usually obscure such MHC-determined resistance. The second point is that there are a number of examples of populations and species with oligomorphic or even monomorphic MHCs. There is no competition based on MHC between members of a monomorphic population, but enormous fluctuations in population size due to epidemics should be observable.

These interpretations have led to two suggestions. The first is the trans-species hypothesis (Klein, 1986, 1987), which proposes that only a few alleles are sufficient to protect a population, with any particular allele being selectively neutral and produced by simple point mutations over evolutionary time. Hence the level of polymorphism in a given species (or population) would depend on the number of alleles that are inherited during speciation (or isolation) and on the age of the species. In fact, whole MHC alleles are shared between humans and chimpanzees, and some features of rodent polymorphism are found in many mouse species as well as in rats; however, reassortment of these features (by recombination or gene conversion) is also important (Lawlor et al., 1990). In fact, such events have clearly taken place in strains of laboratory mice during this century (Nathenson et al., 1986). Also, the polymorphism in MHC molecules is clearly not neutral. The nucleotide changes that lead to amino acid changes (replacement or nonsynonymous mutations) between alleles in mouse and humans are virtually limited to the codons for amino acids postulated to bind the peptide antigen (Hughes and Nei, 1988, 1989), implying that there is selection for diversity in the peptide antigen binding groove and against diversity elsewhere in the MHC molecules. Despite these difficulties with the trans-species hypothesis, the major point, that only a few alleles are sufficient to protect the species against pathogens, may still be true.

The other suggestion to explain the high level of polymorphism possible (but not necessary) in the MHC proposes that functions of the MHC other than pathogen defense exist in vertebrates (Flaherty, 1987). The best example is the kin recognition, mating selection, and abortion induction system based on odors, which has been ascribed to the rodent MHC (Boyse et al., 1987). Detailed studies over many years have shown that the major genetic factor in olfactory discrimination among laboratory strains of mice is the MHC, and that the difference could be as little as three amino acids in a single MHC molecule. The behavioral effects included male and female mating preferences and female abortion of preimplantation embryos in favor of mating with a new male. These behaviors were usually in favor of an allogeneic over a syngeneic mate, although they are overlaid by a complicated and unexplained hierarchy between allogeneic mates or between syngeneic mates. Urine is a good source of the important odors; soluble and fragmented MHC molecules are present in rodent blood and urine, but it is not clear whether the MHC molecules themselves, some organic substances selected by the MHC molecules, or some complex of the two are being sensed. Whatever the mechanism, half-wild mouse populations in barn enclosures mate preferentially with MHC-incompatible partners and overwhelmingly produce MHC heterozygotes over several generations (Potts et al., 1991).

One difficulty with the rodent olfactory model as originally proposed is that the only selective force invoked was an ill-defined advantage of overall heterozygosity, for which MHC polymorphism was only a very specific marker for kinship. Another problem is that some animals with high levels of MHC polymorphism (like birds) simply do not have good olfactory systems. One possible hypothesis to overcome these objections (Kaufman, et al., 1990, 1991a) is that the level of MHC polymorphism is increased in birds (and other vertebrates) by mating preferences based on pathogen levels. In birds, pathogen (thus far ectoparasite) load determines visible secondary sexual characteristics and subsequent mate choice during lekking. Thus the males chosen by the females would be the healthiest (and thus display the most desirable secondary sexual characteristics) due to some MHC molecules one year, and a different set of MHC molecules the next year, depending on the particular pathogens that appear. This model would both link mating selection directly to the proposed disease association of MHC molecules (allowing weak selection to be magnified by mating preference) and generalize the phenomenon to any mechanism to detect heterozygous advantage.

A second proposed driving force for polymorphism in rodents stems from the observation that MHC heterozygosity can lead to a thicker placenta and more successful pregnancy due to increased production of growth factors (Green and Wegmann, 1987). A third phenomenon that might select for polymorphism is direct binding of MHC molecules to membrane receptors and enzymes to affect activity or affinity (Edidin, 1983).

GENES OF THE MHC TRANSPLANTATION ANTIGENS AND THEIR RELATIVES

Genes of the Mammalian Polymorphic MHC Transplantation Antigens

The genes of the polymorphic chains of the mammalian MHC transplantation antigens are all found in the MHC, but many other related and unrelated genes are also located in the MHC, and some related genes are found on other chromosomes (Fischer-Lindahl et al., 1991; Guillemot et al., 1988; Klein, 1986; Stroynowski, 1990; Trowsdale et al., 1991). The chromosomal regions corresponding to the MHC are known in quite some molecular detail for humans (*HLA* on chromosome 6), mouse (*H-2* on chromosome 17), rat (*RT-1* on chromosome 14), and chicken (*B* on the 16th microchromosome in size along with the nucleolar organizer region) (Fig. 3–7). The organization of the mouse MHC is nearly identical to that of the rat MHC (separated for perhaps 8 million years), is very similar to that of the human MHC (separated for perhaps 90 million years), and is quite different from that of the chicken MHC (separated for as much as 250 million years). Describing these regions in detail has been complicated by the fact that there has been expansion and contraction of the gene families, intergenic exchange, degeneration of genes to pseudogenes, as well as insertions, deletions, and rearrangement of large pieces of DNA. The mouse MHC has been separated by recombination into regions *(K, A, E, S, Qa, and Tla)* encoding different kind of molecules and, in some cases, having different levels of overall polymorphism. These recombinational events did not occur randomly along the chromosome, but are mostly located at short "recombinational hot spots." It is not really clear whether the hot spots give rise to the definable regions, or if differences in the regions give rise to the hot spots. One possibility is that the initial recombinational break occurs almost randomly along the DNA, and then the chiasma rolls to a hot spot where it resolves, due to mismatch between the two chromosomes in the polymorphic regions.

The mammalian class II regions (*HLA-D, H-2I, RT-1B,* and *D*) are the most conserved in organization, with homologous genes organized in similar fashions, but with different numbers of expressed genes. Four isotypic subregions in the human *D* region—called *DP, DO/DN, DQ,* and *DR* (arranged centromeric to telomeric)—roughly correspond to the $A\beta3$ pseudogene, the $A\beta2$ gene, the $A\alpha/A\beta$ parts of the *A* subregion, and the *E* subregion in the mouse *I* region. The *DR* subregion has one invariant *DRα (DRA)* gene and three *DRβ (DRB1–3)* genes and pseudogenes, which encode one or two polymorphic DR antigen(s), depending on the haplotype. The *E* subregion in the mouse contains the $E\beta2$ gene (which is not known to be transcribed) and the genes that encode the I-E antigen: the nearly invariant $E\alpha$ gene (which in many mouse strains is a pseudogene) and a polymorphic $E\beta$ gene that spans the *A/E* subregion boundary. The polymorphic $E\beta1$ domain in the polymorphic *A* subregion is separated by the hot spot from the rest of the genes in the conserved *E* subregion. The *DQ* subregion contains the rather polymorphic *DQα (DQA1)* and *DQβ (DQB1)* genes, which encode the DQ antigen and a pair of closely related genes, *DXα* and *DXβ* (now called *DQA2* and *DQB2*, not known to be transcribed), as well as a *DQB3* gene fragment. The *DO/DN* region contains the *DOβ (DOB)* and *DZα (DNA)* genes, which are transcribed but may not be translated. The *DP* subregion contains the somewhat polymorphic *DPα (DPA1)* and *DPβ (DPB1)* genes, which encode the DP antigen, and a pair of closely related pseudogenes, *SXα* and *SXβ* (now called *DPA2* and *DPB2*). The *A* subregion of the mouse contains the polymorphic $A\alpha$ and $A\beta$ genes (which encode the I-A antigen homologous to the DQ antigen), the $A\beta2$ gene (which is homologous to *DOβ* and is transcribed but may not be translated), and the $A\beta3$ pseudogene (homologous to *DPβ*). Thus the number of class II molecules can range from one or two in mice to three or four in humans, but, as discussed below, the levels of expression (and the regulation) vary among these isotypes. Despite these differences, all of these mammalian class II molecules expressed at the cell surface are apparently used for antigen presentation to $\alpha\beta$ T cells. These class II isotypes may be very ancient because separate class II β chains with N-terminal sequences homologous to DR/E and DQ/A molecules

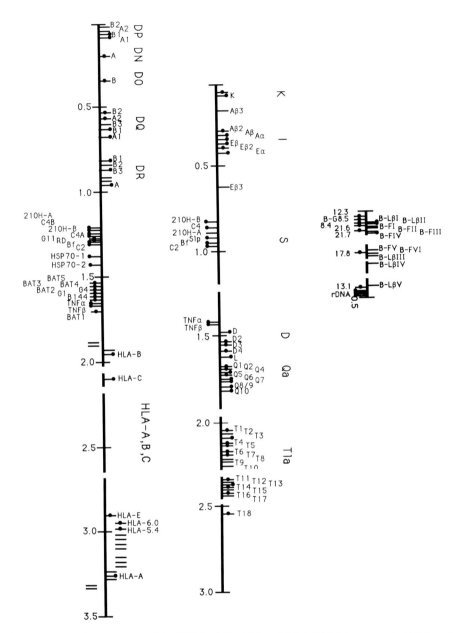

Figure 3–7. Molecular maps of the MHC of CB chicken *(right)*, BALB/c mouse *(middle)*, and human *(left)*. All of the maps are to the same scale, with the distances determined by molecular cloning (contiguous heavy lines) or by pulse field or field inversion gel electrophoresis (line breaks). Genes that encode known protein products are indicated by a thin line with a circle on top; pseudogenes that cannot be transcribed are indicated by a thin line; all others (transcribed but not known to be translated, intact but not known to be transcribed, or status unknown) are indicated by a thin line with a circle in the middle. MHC-like class I and class II genes are above the line, and MHC-encoded unrelated genes are below the line. (From Guillemot et al., 1988, 1989; for a more recent map of the human MHC, see Trowsdale et al., 1991.)

have been isolated from a salamander called the axolotl. In addition, quite distantly related class II genes have recently been found (*DMA* and *DMB* in the *DO/DN* region of humans; *Ma, Mb1* and *Mb2* in the *Aβ2* region of mice); they are transcribed in B cells (Cho et al., 1991; Kelly et al., 1991).

In a number of senses, the mammalian class I genes seem to have been even more plastic in evolution than the class II genes. In addition to the polymorphic MHC transplantation antigens, there are 20–40 other related "nonclassical" class I α chain genes in man and mouse. Most of the class I genes are located telomeric to the class II region, but some are located centromeric or even on other chromosomes. While many of these "nonclassical" genes may be pseudogenes, quite a number are expressed or at least transcribed, and these have minimal if any polymorphism, quite restricted and unexpected tissue distributions, a range of different structural features, and unknown or different functions compared to the polymorphic, and ubiquitous transmembrane transplantation antigens involved in antigen presentation. Even the common subunit of class I molecules, β_2-m, is encoded by a nonpolymorphic gene on a chromosome other than the MHC (chromosome 15 in humans, 2 in mouse). In addition, multiplication and diversification events have mostly taken place after separation of the lineages, so that many of the class I gene subfamilies may not have homologues between species.

All three polymorphic MHC transplantation antigen genes of humans (*HLA-A, B,* and *C*) are located telomeric to the class III region, whereas those of the mouse are split (with *H-2D* and *L* telomeric to the *S* region as in humans, but *H-2K* centromeric to the *I* region). The sequences of the human genes are more similar among themselves than they are to the mouse genes, and vice versa. This situation could be the result of concerted evolution, in which genes within a species are homogenized for sequence changes (including simple drift) by mechanisms like gene conversion. But it is more likely that the multiple genes in man and mouse are the result of independent duplication events, as presumably occurred in the *H-2^d* haplotype of BALB/c and DBA/2 mice. Most mouse strains have only the *K* and *D* genes, but a gene multiplication event occurred in the *d* haplotype, giving rise to five closely linked *D* genes, of which only *D* and *L* are expressed. In fact, the mouse transplantation antigen genes are all closely related to members of the "nonclassical" *Qa* gene subfamily described below, as though one subfamily was the ancestor of the other.

Genes of the Nonclassical MHC-Like Class I Molecules

Some 20 nonpolymorphic human genes and pseudogenes are homologous to *HLA-A, B,* and *C*, mostly mapping between *B* and *A* (Fig. 3–7). (Guillemot et al., 1988; Klein, 1986). Many of these have been sequenced, and this has shown that they are not especially related to the "nonclassical" mouse class I genes. Only a few, like the *HLA-E* gene, are transcribed in the tissues examined, and it is not clear whether they are translated. In contrast, many of the homologous "nonclassical" genes in the mouse MHC are expressed, but with unexpected tissue distributions. There are 7–10 *Qa* region genes, depending on the gene family expansion and contraction events in particular haplotypes located immediately telomeric to the *D* genes. These genes encode a number of antigens, including the Q10 product, which is secreted into the bloodstream by the liver, and the Q6–Q9 products, which constitute the oligomorphic Qa2 antigens, transmembrane, lipid-linked or released by proteolyens from the cell surface. Telomeric to the Q genes are

15–25 *Tla* region genes and pseudogenes, which are more distantly related to the *K*, *D*, and *Qa* region genes. Some have been shown to be transcribed in some of the tissues examined, but only a few are known to be expressed on the cell surface (like the *T13* gene product on thymocytes and activated T cells). However, it is possible that expression of these presumed pseudogenes is confined to some very rare cell types (like stem cells), particularly because there are intriguing regulatory phenomena in tumor cells. Telomeric to the *Tla* region is the recently-described *M* region containing eight class I genes or pseudogenes, including the *Hmt (M3)* gene expressed on many cell types. The Hmt molecule has many sequence differences compared to the classical class I molecules, but it binds β_2-m and presents at least one mitochondrial peptide as a target for killer T cells. The Hmt molecule may be involved in the presentation of bacterial proteins because the binding site will accept only N-formylated peptides (Fischer-Lindahl et al., 1991). Some of the other MHC encoded "nonclassical" class I genes may present antigens to $\gamma\delta$ T cells (though the scattered experiments have not ruled out cross-reaction); while others, for example those with highly restricted tissue distributions, could have important functions in development. On the other hand, they may have no biological function and may simply be the result of unchecked gene expansion, given the apparent lack of similar genes in humans or in other rodents.

Other more distantly related "nonclassical" class I genes map outside the MHC. For instance, five *CD1* genes on human chromosome 11 encode three monomorphic CD1 antigens. These bind β_2-m poorly and are found on cortical thymocytes, dendritic cells, and a few other cell types. Two mouse *CD1* genes also map outside the MHC; these are transcribed in thymus, liver, and spleen and are apparently translated. Their function is unknown, but there is some evidence that they are recognized by T cells (including the mysterious $\gamma\delta$ T cells) in patients with systemic lupus erythematosus (SLE). Another presumably monomorphic class I gene encodes a β_2-m binding molecule in the intestinal epithelium of the neonatal rat (Simister and Mostov, 1989). This class I heterodimer is an Fc receptor that binds maternal antibody from the mother's milk and transports it from the intestine to the blood. Because maternal antibody protects the suckling rat until the neonatal immune system matures, the maintenance of this functional gene would clearly be under very intense selection.

Parenthetically, there is a very interesting example of a class I gene acquired by a virus, which opens the possibility of horizontal transmission of MHC molecules between individuals, perhaps even between species. This class I–like gene is found in human cytomegalovirus (CMV), and its product is thought to bind β_2-m and to be involved in viral infectivity (Beck and Barrell, 1988).

Genes of the Chicken MHC

Organization of the chicken MHC genes is apparently quite different from that of mammals (Fig. 3–7) (Crone and Simonsen, 1987; Guillemot et al., 1989; Kaufman et al., 1990, 1991b). This may be related to the fact that birds (and some reptiles) have reduced many of their chromosomes to microchromosomes. The chicken MHC, called the *B* complex, is located on an 8 million base pair (Mbp) microchromosome, of which 6 Mbp code for ribosomal RNA in the nucleolar organizer. The rest of the chromosome, including the centromere, has been separated by recombination into two regions. The *B-F/B-L* region of the *B12* haplotype contains five class II (*B-L*) β genes, six reported class I (*B-F*) α genes, a *B-G* gene, and numerous unrelated genes. These genes are very small compared to those in mammals, with 50–150-bp introns and very short intragenic regions. It appears as though all these genes are somewhat intermingled, rather than being located in discrete subregions. The *B-G* region encodes another multigene family of polymorphic cell surface proteins, the B-G antigens.

One invariant class II α and two polymorphic class II β chains have been identified in the chicken, reminiscent of the mammalian DR/E antigens. Two subfamilies of class II β genes are based on hybridization and nucleotide sequences, although the four class II β genes sequenced, while quite homologous to mammalian class II β genes, do not correspond to mammalian isotypes. The class II α gene has not yet been cloned. If the class II α chain gene were located on one of the cosmids already isolated, it should have been detected, so it is possible that the gene maps outside the MHC.

One β_2-m and one or more polymorphic class I α chain glycoproteins have been isolated from erythrocytes. The β_2-m protein has been fully sequenced, with one position evidently having two allelic variants, and Southern blots identify a single β_2-m gene that maps outside the MHC, as in mammals. The multiple class I molecules isolated from erythrocytes apparently differ only at their C-termini, and are presumably derived by alternative splicing, as two cDNA clones for chicken class I α chains differ only in a 33-bp stretch corresponding to exon 7 of the cytoplasmic region. This major erythrocyte (and leukocyte) class I molecule is most closely related to the mammalian class I transplantation antigens, rather than to the nonclassical class I molecules. All the fragments that strongly cross-hybridize with class I α chain cDNAs map to the *B-F/B-L* region of the MHC, but other more weakly hybridizing bands map outside this region and perhaps even outside the MHC. Thus there may be only one major class I molecule (B-F) expressed on erythrocytes and leukocytes, along with other minor class I molecules with different tissue distributions like the mammalian "nonclassical" class I molecules.

Although many genes in the chicken MHC are unrelated to class I and class II genes, they are not located in a special region. The avian counterparts to the mammalian class III genes (for example, C4, C2, Bf, TNF, etc.) have not been found on the MHC microchromosome.

An extremely interesting, and thus far unique, aspect of the chicken MHC is the presence of a third multigene family of highly polymorphic cell surface antigens, called the B-G molecules (Kaufman et al., 1991b). The functions and the detailed structure of the B-G molecules are not known, but the similarities with class I and class II molecules, both at the level of genes (the presence of a multigene family with extensive polymorphism within the MHC) and at the level of gene products (expression of dimeric molecules with extracellular regions of comparable sizes on a variety of blood cell types), are intriguing.

The B-G molecules on erythrocytes are disulfide-linked dimers of integral membrane proteins that lack N-linked and sizeable O-linked carbohydrates. Each chicken *B* haplotype examined shows a completely different pattern of some 5–6 bands on SDS-gel electrophoresis, ranging in size from 35 to 55 kD (with fainter smaller bands). The size differences are apparently due to differences in the length of the C-terminal cytoplasmic tail. The sequence of this region as determined from the isolated cDNAs consists of multiple small repeats, and it is easy to see how alternative splicing, recombination, or deletions could account for the variations in size of the B-G molecule. Preliminary sequencing of the B-G gene in the *B-F* region shows a membrane proximal V-like Ig domain, but none of the many genes in the *B-G* region has yet been characterized. There is as yet nothing known about the molecules recognized by monoclonal antibodies to B-G molecules on leukocytes or on stromal cells of the bursa.

A number of phenomena may give hints about the function of the B-G molecules or the *B-G* chromosomal region. First, and most compelling, they are highly polymorphic and map to the MHC. Second, so-called "natural antibodies" to the products of the *B-G* region are found in the serum of unimmunized allogeneic chickens, as well as alligators, mice, rats, and humans. Third, B-G molecules have a striking "adjuvant effect" in the production of alloantibodies to other erythrocyte surface antigens such as B-F when expressed in the same erythrocyte or liposome, apparently due to the presence of the natural antibodies. Fourth, the *B-G* region apparently has effects on various in vivo phenomena, including claims for histoincompatibility, GvH-associated hemolytic anemia, and perhaps resistance to fowl cholera. Finally, some monoclonal antibodies to erythrocyte B-G molecules stain thrombocytes (a blood cell with properties like a platelet and a macrophage), leukocytes, and various organs, including the bursa, thymus, and intestine.

Despite these hints, the biological function of B-G molecules is not known. It is not clear whether B-G polymorphism reflects selection based on some property of the molecule, or a genetic phenomenon such as genetic hitchhiking. Alternatively, it may reflect a property of the DNA, (for example, a gene conversion target or a multiple small repeat duplication/deletion). It would be very exciting if the B-G molecules had some immunological function like antigen presentation to T cells. Another possibility is that B-G molecules are involved in the maturation of B cells, based on the intricate staining patterns of various monoclonal antibodies to B-G on stromal cells in the bursa, the "natural antibody" phenomenon, and the notion that perhaps the selection process in the thymus has a homologue in the bursa.

EXPRESSION OF MHC MOLECULES DURING ONTOGENY

The tissue distribution of a particular MHC molecule may depend on the species, age, and immunological state of the animal (Crone and Simonsen, 1987; Du Pasquier et al., 1989; Flajnik et al., 1987; Klein, 1986). Some aspects of the tissue distribution can be interpreted in the context of the biological functions of the molecules. Other aspects are at present inexplicable and may actually have no biological significance. This may be due to accidental possession of a series of transcriptional regulatory elements that are activated in certain cells at certain stages, or to evolutionary vestiges of expression patterns that were once meaningful.

In adult mammals, the polymorphic class I transplantation molecules are considered to be ubiquitous. This would accord with their presumed function in protection against virus infection and neoplasia. In fact, the level of expression varies tremendously between cell types, from essentially none on erythrocytes to extremely high on cells of the immune system. These levels can be affected by various stimuli; in particular, α and β interferons, often produced in response to virus infection, stimulate the cell surface expression of class I molecules or many different cells. In contrast, the "nonclassical" class I molecules each have very distinctive tissue distributions. The different Qa antigens are found variously on cells of the immune system, especially thymocyte and T cell subsets, activated T and B cells, natural killer (NK), cells and various tumors, but also on fibroblasts and some hemopoietic stem cells. The exception is the Q10 gene product,

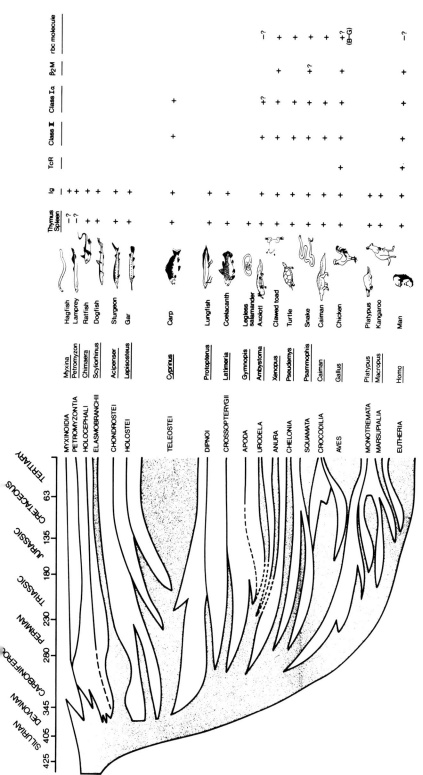

Figure 3–9. An evolutionary tree of the vertebrates with the known features of the immune system for representative extant species. The width of the branches are related to the number of genera described in the fossil record or found in extant animals. (Modified from McFarland et al., 1979.) Demonstration of lymphoid organs (spleen and thymus), antigen receptors (immunoglobulin and T cell receptor), MHC-like molecules (class II heterodimer, class I α chain, β_2-m), and red blood cell homodimers (like B-G or those associated with class I α chains) are indicated for each species as referenced in the text. (+ means presence; − means absence, ? indicates uncertainty; blank indicates unknown.) (From Kaufman et al., 1991a.)

Evidence for MHC and MHC-like Molecules in Vertebrates

Phylogeny of Vertebrates

The vertebrates have had a complex evolutionary history since the appearance of the first jawless fish as much as 500 million years ago (Fig. 3–9) (McFarland et al., 1979; Young, 1981). It is important to realize that living vertebrates are only a dim and distorted reflection of this story, and that many previously dominant forms and important transitional forms are either extinct or are represented by a few extant species (which have themselves continued to evolve from their ancestral forms). Because fossils yield little information about cells and molecules, the evolutionary history of the MHC must be inferred from living forms, and it may therefore be quite incomplete. It is also important to realize that each living species is a separate evolutionary lineage, subject to the selection pressures and historical accidents of that lineage. Thus, while there will be similarities between closely related animals, there may also be differences due to accident or selection, and there may be similarities between quite distantly related animals also based on accident or selection. Both the incompleteness and the heterogeneity of the vertebrates obscure the story of MHC and MHC-like molecules.

Mammals, birds, and reptiles are all amniotes and are related to each other in a complicated evolutionary tree. There are three groups of mammals: the oviparous (egg-laying) monotremes of the subclass Prototheria (represented now by the duckbill platypus and three species of spiny anteater), the viviparous, pouched marsupials of the subclass Metatheria (represented by 250 species of opossums and kangaroos, etc.), and the placental mammals of the subclass Eutheria (around 4,000 species—including all the familiar rodents, shrews, and rabbits; hoofed animals and elephants; carnivores and primates; and whales, bats, manatees, sloths, and others). The birds, or class Aves, consist of feathered, toothless, egg-laying animals (perhaps 9,000 species, ranging from primitive ratite birds like the ostrich and gallinaceous forms like the chicken to very advanced songbirds), which may have all descended from a single ancestral species. In contrast, the recent reptiles consist of four orders of animals that are quite widely separated from each other in evolution: turtles and tortoises (Chelonia, with roughly 250 species); snakes and lizards (Sauria and Squamata, with some 6,000 species); crocodiles and alligators (Crocodilia, with 22 species); and Tuatara (the single species of the order Rhynchocephalia). Mammals may have diverged from the common ancestor only shortly before the various reptile branches diverged from each other (or perhaps even after the turtles diverged from the rest of the amniotes). Birds derived from archaeosaurs (including dinosaurs), which are most closely related to the crocodilians, and thus birds are more closely related to one branch of the reptiles than the various branches of the reptiles are to each other. Hence, the similarities between the avian and mammalian MHC (and immune system in general) will either be found in all of the reptiles as well, if they have not been lost by secondary evolution, or will have arisen by convergent evolution (perhaps due to endothermy). The connection between the ancestral reptile and the amphibians is still obscure.

There are three orders of living amphibians: legless salamanders (Gymnophiona or Apoda, represented by some 150 species); salamanders and newts (Urodela or Caudata, with some 350 species); and frogs and toads (Anura or Salienta, with some 3,500 species). The most striking aspect of amphibian development is the metamorphosis between the juvenile and adult stages, which can involve enormous changes in anatomy, size, physiology, and behavior, selected by lives in two different environments. The changes in the immune system at metamorphosis may still be reflected in some aspects of the ontogeny of the amniotes. The amphibians are thought to have arisen from ancestors similar to lung or lobe-finned fish, although the fossil record for true frogs and salamanders does not connect well with the really primitive amphibians. Indeed, some evidence indicates that the amphibians are a polyphyletic group, with the anurans and urodeles having separate fish ancestors.

The living fish are divided into three great groups: the jawless fish of the subphylum or superclass Agnatha (represented now by the order Cyclostomata, with some 30 species of lampreys and hagfish); the cartilaginous fish, or class Chondrichthyes, including the Holocephalans (represented by 25 species of rat fish, rabbit fish, and elephant fish) and the Elasmobranchs (represented by some 600 species of sharks, rays, skates, and the like); and the bony fish, or class Osteichthyes, including the Chondrosteans (represented by a few species like the sturgeon), the Holosteans (represented by a few species like the gar), the Teleosteans (including some 20,000 species ranging from salmon, trout, and catfish to perch and sunfish), and the Dipnoi and Crossopterygians (represented by six species of lungfish and one lobe-finned fish). The jawless fish appear first in the fossil record, followed by the bony armored fish with jaws (Placoderms, perhaps most similar to Holocephalans like the ratfish, or *Chimaera*). The armored bony fish presumably lost their armor to become the true bony fish and lost the process of ossification to become cartilaginous fish.

Immune Systems of Living Vertebrates

All vertebrates have serum Ig, although many details of the antibody response differ (Cohen and Sigel, 1982; Du Pasquier, 1989; Kaufman et al., 1991a). Mammals, birds, and anuran amphibians show antibody diversity, affinity maturation, and the switch in isotypes, all of which seem to be less obvious or lacking in urodele amphibians and the fish. The organization of Ig genes is quite similar in those mammals, reptiles, amphibians, and bony fish that have been studied. Birds have an unusual Ig gene organization, which may be related to the presence of a bursa and the existence of self-renewing surface Ig–bearing B cells in the bloodstream, or to the presence of microchromosomes. Cartilaginous fish also have an unusual Ig gene structure, which shares many features of organization and sequence with TCR genes. However, all of these sequences correspond to Ig genes that are closely related to mammalian antibody genes, and thus all vertebrates should have B cells (except possibly for jawless fish, which have not yet been studied).

All vertebrates display allograft rejection, though the speed and strength of the reaction varies (Cohen and Sigel, 1982; Du Pasquier, 1989; Kaufman et al., 1991a). Mammals, birds, anuran amphibians, and bony fish generally show acute graft rejection, while most reptiles, urodele amphibians, cartilaginous fish, and jawless fish have been reported to have only chronic graft rejection. A similar hierarchy is generally found for other T cell phenomena, including mixed lymphocyte reaction (MLR), GvH, DTH, and T cell help for antibody production. It is not clear whether the molecular and cellular bases for the apparently poor T cell function in some of these animals are the same. All vertebrates except the jawless fish have a thymus, which presumably means that they possess some kind of T cell and therefore that they express MHC-like molecules.

However, demonstration of an MHC and MHC-like molecules in nonmammalian vertebrates has been beset by difficulties in identifying suitable animal models and in developing functional assays and reagents. An MHC, as defined by a single major genetic locus that determines graft rejection and other T cell recognition phenomena, has been defined only in certain mammals, the chicken, and *Xenopus*. Some or all of the elements of class I and class II molecules have also been found in many mammals, birds, reptiles, and amphibians, with the limitations discussed above. However, no MHC has been defined in the bony, cartilaginous, or jawless fish. We have not identified molecules with the expected structural features of class I or class II antigens in fish, but recently class II β-like and class I α-like genes have been reported in a bony fish, the carp (Hashimoto et al., 1990).

Important events in MHC evolution almost certainly occurred in the ancestors of these fish lineages. Bony fish clearly have a thymus and show acute graft rejection, a strong MLR, and T cell help for antibody production. Their Ig genes are much like those of mammals. Cartilaginous fish have a thymus, but they display chronic graft rejection and have no other evidence for T cell function. In addition, their Ig genes have much in common with TCR genes. Jawless fish have no defined thymus, show chronic graft rejection, and have no other evidence of T cell function, except for a reported "hagfish MLR." The responding cell in this allogeneic proliferation assay binds a monoclonal antibody that is directed to hagfish serum Ig, meaning either that Ig and the T cells share a nonclonal epitope, that B cells respond in this MLR, or that T cells secrete their TCR (Raison et al., 1987). The overall impression is that the main stream of evolution leading to the mammalian immune system is present in bony fish, that cartilaginous fish split from this path shortly after B cells and T cells diverged, and that the jawless fish may not even have separated lineages.

Ancestors of the MHC and MHC-like Molecules in Invertebrates

Despite the difficulties in defining the MHC of many vertebrates, there has been much speculation that primordial MHC molecules were involved in invertebrate allorecognition (Burnet, 1971; Cohen and Sigel, 1982). Allorecognition phenomena with some similarities to the graft rejection phenomena of vertebrates have been described in a variety of invertebrate species, although they differ in many details with those of vertebrates and also among themselves. These allorecognition phenomena are particularly well documented in colonial marine invertebrates, ranging from metazoans as primitive as sponges (Phylum Porifera) to the tunicates (Subphylum Urochordata), which are thought to be closest to the ancestors of vertebrates. In the tunicate *Botryllus* (for which the genetics is relatively well established), a single highly polymorphic locus determines rejection/fusion events, but unlike most vertebrate allorecognition, single-haplotype differences allow fusion. The same highly polymorphic locus is also responsible for compatibility/incompatibility in fertilization. In addition to this locus, there are evidently other unlinked polymorphic loci that determine survival of chimeras after fusion, apparently in response to selection pressures involving somatic and germ cell parasitism (Weissman et al., 1990). In the hydroid *Hydractinia* (Phylum Cnidaria or Coelenterata), in which the genetics are not well understood, the observation of initial fusion followed by separation has been given a theoretical basis, involving the advantages of large colonies versus the dangers of somatic cell parasitism by unsequestered germlines (Shenk, 1991). It has also been suggested, so far without evidence, that the primary function of polymorphic allorecognition systems in invertebrates, as in vertebrates, is to recognize pathogens.

However, allorecognition phenomena in vertebrates and invertebrates could have evolved independently, with the invertebrate systems based on molecular families unrelated to the MHC (Kaufman, 1988; Kaufman et al., 1990). As an example of such a superficially similar system, the fertilization incompatibility system of some plants is determined by the highly polymorphic S locus, which evidently encodes proteins with no similarity to MHC molecules (Ebert et al., 1989). Moreover, there is some evidence to indicate that vertebrate allorecognition depends not on direct recognition of the polymorphic residues in the MHC molecules (which mostly point into the groove rather than up toward the TCR), but to the peptides that are bound (Davis and Bjorkman, 1988). If the invertebrate allorecognition systems function in a similar manner, then there is no need for a polymorphic molecule because the polymorphisms of the bound peptides (like minor antigens) should be sufficient for allorecognition. Finally, alternative origins of vertebrate MHC molecules have been proposed. Other functions of MHC molecules, such as their involvement in cell–cell interactions, peptide binding, or intracellular transport, might have been more important than polymorphism for the function of ancestral MHC molecules (Kaufman, 1988; Kaufman et al., 1990).

It is easy to imagine that some monomorphic cell–cell interaction molecule was adopted by the fledgling immune system during evolution as the target for T lymphocyte recognition, followed by selection for polymorphism. The most striking possibility is that a multigene family of these monomorphic molecules was (and still might be) involved in recognition events during differentiation and development. These recognition events could involve positive signals, such as those involved in homing for cell movement or induction; or negative signals, like those in tissue restructuring (Flajnik et al., 1987; Williams and Barclay, 1988). Thus far, candidate molecules like Qa, TL, or CD1 have not been shown to be involved in developmental interactions, although polymorphic MHC molecules are clearly involved in positive and negative selection of T cells during their development in the thymus (von Boehmer et al., 1989).

The non-Ig domains of many MHC molecules bind peptides in the groove formed by the two α helices and the β sheet, which could be their most ancient property. Even single-celled organisms possess chemoreceptors, mating factor receptors, and aggregation factor receptors, although thus far such molecules have not been shown to have any similarity to MHC molecules. Another possibility is that MHC-like molecules arose as regulators by binding to receptors and enzymes on the cell surface or in intracellular vesicles (Edidin, 1983).

Important events in the functions of polymorphic MHC molecules take place in intracellular vesicles (Germain, 1991; Koch et al., 1989; Long, 1989), and it is not hard to imagine that ancestral MHC molecules were involved in regulation, transport, or other events inside the cell. The primordial MHC molecule was suggested to have been involved in transepithelial transport (Klein, 1986). A class I–like molecule indeed functions as an Fc receptor for transport of maternal Ig across the intestinal epithelium of suckling rats (Simister and Mostov, 1989), though this seems likely to have appeared after the evolution of the MHC. An even more intriguing possibility is that MHC molecules arose from intracellular heat shock proteins and chaparonins, which also bind peptides (Flajnik et al., 1991).

Acknowledgments

Due to the small number of references allowed in this article, I was forced to use review articles to cover important points, so I would like to apologize to the many people whose work is mentioned without proper citation. I would like to thank my friends and colleagues who helped me over the years to begin to understand this large body of knowledge; the staff of the BII for help with computers, artwork, and photography; Susan Carson and Jan Salomonsen for critical reading; and the Basel Institute for Immunology, founded and supported by Hoffman-LaRoche SA, for letting me work in my corner, more-or-less undisturbed, for the last eight years. Finally, I would like to thank Ed Cooper, who has taught me a lesson that I will never forget.

References

Beck, S., and Barrell, B. G. (1988). Human cytomegalovirus encodes a glycoprotein homologous to MHC class I antigens. *Nature 331*:269–272.

Bjorkman, P., and Parham, P. (1990). Structure, function, and diversity of class I major histocompatibility complex molecules. *Annu. Rev. Biochem. 59*:253–288.

Bjorkman, P. J., Saper, M. A., Samraoui, B., Bennet, W. S., Strominger, J. L., and Wiley, D. J. (1987a). Structure of the human class I histocompatibility antigen HLA-A2. *Nature 329*:506–512.

Bjorkman, P. J., Saper, M. A., Samraoui, B., Bennet, W. S., Strominger, J. L., and Wiley, D. J. (1987b). The foreign antigen binding site and T cell recognition of class I histocompatibility antigens. *Nature 329*:512–518.

Bluestone, J. A., and Matis, L. A. (1989). TCRγδ cells—minor redundant T cell subset or specialized immune system component? *J. Immunol. 142*:1785–1788.

Boyse, E. A., Beauchamp, G. K., and Yamazaki, K. (1987). The genetics of body scent. *Trends Genet. 3*:97–102.

Brinckerhoff, C. E., Mitchell, T. I., Karmilowicz, M. J., Kluve-Beckerman, B., and Benson, M. D. (1989). Autocrine induction of collagenase by serum amyloid A-like and β_2-microglobulin-like proteins. *Science 243*:655–657.

Brodsky, F., and Guagliardi, L. (1991). The cell biology of antigen processing and presentation. *Annu. Rev. Immunol. 9*:707–744.

Brown, J. H., Jardetzky, T., Saper, M. A., Samraoui, B., Bjorkman, P. J., and Wiley, D. J. (1988). A hypothetical model of the foreign antigen binding site of class II histocompatibility molecules. *Nature 332*:845–850.

Burnet, F. M. (1971). "Self-recognition" in colonial marine forms and flowering plants in relation to the evolution of immunity. *Nature 232*:230–235.

Cho, S., Attaya, M., and Monaco, J. (1991). New class II-like genes in the murine MHC. *Nature 353*:573–576.

Cohen, N., and Sigel, M. M., eds. (1982). *The Reticuloendothelial System: Vol. 3. Phylogeny and Ontogeny.* New York: Plenum Press, pp. 423–459.

Crone, M., and Simonsen, M. (1987). Avian major histocompatibility complex. In *Avian Immunology: Basis and Practise,* vol. 2 (Toivnanen A., and Toivanen P., eds.). Boca Raton, Fla.: CRC Press, pp. 25–42.

Dargemont, C., Dunon, D., Deugnier, M. A., Denoyelle, M., Girault, J. M., Godeau, F., Thiery, J. P., and Imhof, B. A. (1989). Thymotaxin, a chemotactic peptide, is identical to β_2-microglobulin. *Science 246*:803–806.

Davis, M. M., and Bjorkman, P. J. (1988). T-cell antigen receptor genes and T-cell recognition. *Nature 334*:395–402.

Degen, E., and Williams, D. 1991. Participation of a novel 88 kD protein in the biogenesis of murine class I histocompatibility molecules. *J. Cell Biol. 112*:1099–1115.

Du Pasquier, L. (1989). Evolution of the immune system. In *Fundamental Immunology,* 2nd ed. (Paul, W. E., ed.). New York: Raven Press.

Du Pasquier, L., Schwager, J., and Flajnik, M. (1989). The immune system of *Xenopus. Annu. Rev. Immunol. 7*:251–275.

Ebert, P. R., Anderson, M. A., Bernatzky, R., Altshuler, M., and Clarke, A. E. (1989). Genetic polymorphism of self-incompatibility in flowering plants. *Cell 56*:255–262.

Edidin, M. (1983). MHC antigens and nonimmune functions. *Immunol. Today* 4:269–270.

Fischer-Lindahl, K., Hermel, E., Loveland, B., and Wang, C.-R. (1991). Maternally transmitted antigen of mice: A model transplantation antigen. *Annu. Rev. Immunol.* 9:351–372.

Flaherty, L. (1987). MHC complex polymorphism: A nonimmune theory for selection. *Human Immunol.* 21:3–14.

Flajnik, M., Canel, C., Kramer, J., and Kasahara, M. (1991). Which came first, MHC class I or class II? *Immunogenetics* 33:295–300.

Flajnik, M. F., Hsu, E., Kaufman, J. F., and Du Pasquier, L. (1987). Changes in the immune system during metamorphosis of *Xenopus*. *Immunol. Today* 8:58–64.

Germain, R. (1991). Antigen presentation: The second class story. *Nature* 353:605–607.

Germain, R., and Hendrix, L. (1991). MHC class II structure, occupancy and surface expression determined by post-endoplasmic reticulum antigen binding. *Nature* 353:134–139.

Green, D. R., and Wegmann, T. G. (1987). Beyond the immune system: The immunotrophic role of T cells in organ generation and regeneration. *Prog. Immunol.* 6:1100.

Guillemot, F., Auffray, C., Orr, H. T., and Strominger, J. L. (1988). MHC antigen genes. In *Molecular Immunology* (Hames, B. D., and Glover, D. M., eds.). Oxford: IRL Press, pp. 81–144.

Guillemot, F., Kaufman, J., Skjoedt, K., and Auffray, C. (1989). The major histocompatibility complex of the chicken. *Trends Genet.* 57:300–304.

Hashimoto, K., Nakanishi, T., and Kurosawa, Y. (1990). Isolation of carp genes encoding major histocompatibility complex antigens. *Proc. Natl. Acad. Sci. USA* 87:6863–6867.

Hedley, M. L., Drake, B. L., Head, J. R., Tucker, P. W., and Forman, J. (1989). Differential expression of the class I MHC genes in the embryo and placenta during midgestational development in the mouse. *J. Immunol.* 142:4046–4053.

Hill, A., Allsopp, C., Kwiatkowski, D., Anstey, N., Twumasi, P., Rowe, P., Bennet, S., Brewster, D., McMichael, A., and Greenwood, B. (1991). Common West African HLA antigens are associated with protection from severe malaria. *Nature* 352:595–600.

Hughes, A. L., and Nei, M. (1988). Pattern of nucleotide substitution at major histocompability complex class I loci reveals overdominant selection. *Nature* 335:167–170.

Hughes, A. L., and Nei, M. (1989). Nucleotide substitution at major histocompatibility complex class II loci: Evidence for overdominant selection. *Proc. Natl. Acad. Sci. USA.* 86:958–962.

Kaufman, J. (1988). Vertebrates and the evolution of the MHC class I and class II molecules. *Verh. Dtsch. Zool. Ges.* 81:131–144.

Kaufman, J. F., Auffray, C., Korman, A. J., Shackelford, D., and Strominger, J. L. (1984). The class II molecules of the human and murine major histocompatibility complex. *Cell* 36:1–13.

Kaufman, J., Flajnik, M., and Du Pasquier, L. (1991a). The MHC molecules of ectothermic vertebrates. In *Phylogenesis of Immune Function* (Cohen, N., and Warr, G., eds.). Boca Raton, Fla.: CRC Press, pp. 125–150.

Kaufman, J., Skjødt, K., and Salomonsen, J. (1990). The MHC molecules of nonmammalian vertebrates. *Immunol. Rev.* 113:83–117.

Kaufman, J., Skjødt, K., and Salomonsen, J. (1991b). The B-G multigene family of the chicken major histocompatibility complex. *Crit. Rev. Immunol.* 11:113–143.

Kelly, A., Monaco, J., Cho, S., and Trowsdale, J. (1991). A new human HLA class II–related locus, DM. *Nature* 353:571–573.

Klein, J. (1986). *Natural History of the Major Histocompatiblity Complex.* New York: Wiley.

Klein, J. (1987). Origin of major histocompatibility complex polymorphism: The trans-species hypothesis. *Human Immunol.* 19:155–162.

Koch, N., Lipp, J., Pessara, U., Schenck, K., Wraight, C., and Dobberstein, B. (1989). MHC class II invariant chains in antigen processing and presentation. *Trends Biochem. Sci.* 14:383–386.

Lawlor, D., Zemmour, J., Ennis, P., and Parham, P. (1990). Evolution of class-I MHC genes and proteins: from natural selection to thymic selection. *Annu. Rev. Immunol.* 8:23–64.

Long, E. (1989). Intracellular traffic and antigen processing. *Immunol. Today* 10:232–234.

Madden, D., Gorga, J., Strominger, J., and Wiley, D. (1991). The structure of HLA-B27 reveals nonamer self-peptides in an extended conformation. *Nature 353*:321–325.

McFarland, W. N., Paugh, F. H., Cade, T. J., and Heisen, J. B. (1979). *Vertebrate Life.* New York: Macmillan.

Nathenson, S. G., Geliebter, J., Pfaffenbach, G. M., and Zeff, R. A. (1986). Murine major histocompatibility complex class I mutants: Molecular analysis and structure-function implications. *Annu. Rev. Immunol. 4*:471–502.

Potts, W., Manning, C., and Wakeland, E. (1991). Mating patterns in seminatural populations of mice influenced by MHC genotype. *Nature 352*:619–621.

Raison, R. L., Gilbertson, P., and Wotherspoon, J. (1987). Cellular requirements for mixed lymphocyte reactivity in the cyclostome, *Eptatretus stoutii. Immunol. Cell Biol. 65*:183–188.

Robertson, M. (1991). Proteasomes in the pathway. *Nature 353*:300–301.

Salter, R., Benjamin, R., Wesley, P., Buxton, S., Garret, T., Clayberger, C., Krensky, A, Norment, A., Littman, D., and Parham, P. 1990. A binding site for the T-cell co-receptor CD8 on the a3 domain of HLA-A2. *Nature 345*:41–46.

Saper, M., Bjorkman, P., and Wiley, D. (1991). Refined structure of the human histocompatibility antigen HLA-A2 at 2.6 Å resolution. *J. Mol. Biol. 219*:277–319.

Schiffer, M., Girling, R. L., Ely, K. R., and Edmundson, A. B. (1973). Structure of a λ-type Bence-Jones protein at 3.5-Å resolution. *Biochemistry 12*:4620–4631.

Schwartz, R. H. (1985). T lymphocyte recognition of antigen in association with gene products of the MHC. *Annu. Rev. Immunol. 3*:237–261.

Schwartz, R. H. (1986). Immune response (Ir) genes of the murine major histocompatibility complex. *Adv. Immunol. 38*:31–202.

Shenk, M. (1991). Allorecognition in the colonial marine hydroid *Hydractinia* (Cnidaria/Hydrozoa). *Am. Zool. 31*:549–557.

Simister, N. E., and Mostov, K. E. (1989). An Fc receptor structurally related to MHC class I antigens. *Nature 337*:184–187.

Stroynowski, I. (1990). Molecules related to class-I major histocompatibility complex antigens. *Annu. Rev. Immunol. 8*:501–530.

Trowsdale, J., Ragoussis, J., and Campbell, R. (1991). Map of the human MHC. *Immunol. Today 12*:443–446.

Vidovic, D., and Matzinger, P. (1988). Unresponsiveness to a foreign antigen can be caused by self-tolerance. *Nature 336*:222–225.

von Boehmer, H., Teh, H. S., and Kisielow, P. (1989). The thymus selects the useful, neglects the useless and destroys the harmful. *Immunol. Today 10*:57–61.

Weissman, I., Saito, Y., and Rinkevich, B. (1990). Allorecognition histocompatibility in a protochordate species: Is the relationship to MHC somatic or structural? *Immunol. Rev. 113*:227–241.

Williams, A. F., and Barclay, A. N. (1988). The immunoglobulin superfamily—domains for cell surface recognition. *Annu. Rev. Immunol. 6*:381–405.

Young, J. 1981. *The Life of the Vertebrates.* Oxford: Clarendon Press.

Zinkernagel, R. M., and Doherty, P. C. (1979). MHC-restricted cytotoxic T cells: Studies on the biological role of polymorphic major transplantation antigens determining T cell restriction specificity, function and responsiveness. *Adv. Immunol. 27*:52–177.

4

Early Development of the Immunoglobulin Gene Superfamily

MOHAMED H. MANSOUR,

EDWIN L. COOPER

The immune system is among the first complex eukaryotic systems to be investigated at the cellular, molecular, and genetic levels. Its major cell types, the lymphocytes, are freely wandering cells that undergo programmed migrations during development, that can be triggered to differentiate and divide by hormonal, mitogenic and antigenic stimuli, and that interact with one another and with other cell types to form a finely balanced cellular network that mediates and regulates the immune response (Jerne, 1974). The cellular basis for these interactions depends, in part, on the presence of two categories of lymphocytes: T lymphocytes (thymus-derived) and B lymphocytes (bursal or bone marrow–derived). Both cell types, along with antigen processing and presenting cells, mediate almost all aspects of immune responses to a virtually limitless number of foreign antigens.

Both lymphocyte types behave in a similar manner when confronted by antigen in that they are stimulated to divide and differentiate into highly specialized regulatory and effector cells. T cells can directly mediate cellular effector responses, or they may secrete various regulatory lymphokines that are involved in effector responses. B lymphocytes synthesize and export specific antibody molecules responsible for humoral responses (Cathou and Dorrington, 1974). As lymphocytes differentiate, they express an array of cell surface molecules, encoded by structural genes, which reflect the cells' functions. The differentiation of T and B lymphocytes is thus manifested by the coordinated expression of distinct combinations of cell surface molecules, each of which plays a specified role in the acquisition of immunocompetence.

In this chapter, we review the molecular structure of certain cell surface molecules that are involved in antigen recognition, as well as other functions affected by T and B lymphocytes. In spite of their apparent heterogeneity, all these molecules are made up of a number of homologous units that correspond in three-dimensional structure to a single immunoglobulin (Ig) domain. Current views suggest that this domain may have evolved to serve as the basic structural and functional unit of recognition and has been employed, during evolution, to generate a large number of different molecules that participate not only in the immune response but also in aspects of recognition within the nervous system.

EVOLUTIONARY ASPECTS OF THE Ig SUPERFAMILY

A wide variety of cell surface molecules participate in vertebrate immune responses and are members of the Ig superfamily. The evolution of the immune system, as a complex phenotype, may have therefore proceeded within the limits of this superfamily of molecules. Their common evolutionary and structural denominator is a highly conserved, yet very versatile, three-dimensional structure that incorporates two important features. First, a highly conserved Ig fold may accommodate many different sequences, and hence can evolve efficiently to generate diverse structures, for example, V domains in Ig and T cell receptors. Second, although involved in generating receptor diversity, the ability of such a domain structure to interact with structurally related sequences is highly conserved; thus, its duplication may generate new sets of molecules that are already capable of interacting, including T cell receptor–MHC polypeptides; poly Ig receptor–IgA and IgM; and CD8–MHC class I polypeptide interactions.

Based on the divergence of sequences among Ig-related domains in the superfamily and consistent with the fact that complex molecules like Ig and MHC are only found in vertebrates, the evolutionary starting point of the superfamily (the primordial domain) may have thus evolved before the emergence of vertebrates. The gene encoding the primordial domain may have encoded a cell surface protein with an Ig fold having both V- (variable) and C- (constant) domain characteristics. The tendency for contemporary domains to interact suggests that this primitive molecule may have been a homodimer or interacted with similar units expressed on different cell types. Partial duplication and independent divergence within a network of these primitive molecules may have led to more specialized V-like and C-like domains, and this probably was an early critical event in the evolution of the superfamily. This would make it possible to generate proteins with separate regions for recognition and for membrane-anchoring or effector functions. At some point, early in the evolution of the Ig superfamily, genes encoding Thy-1, β_2-m, MRC-OX2, and the poly-Ig receptor may have begun to diverge from one another, and these may have been followed by MHC-encoded class I and II polypeptides. The Thy-1 gene may have diverged first as it encodes a single domain that is intermediate between the V and C domains.

Acquisition of the ability to rearrange DNA was a second critical event in the evolution of the Ig superfamily. The development of the rearrangement mechanism presented new opportunities for molecules like Ig and the T cell receptor and its accessory molecules to diversify through extensive duplication of gene segments, through combinatorial joining of these gene segments, and through somatic mutation arising from joining imprecisions. The nonrearranging CD4 and CD8 molecules, being more closely related to the V_κ and V_λ domains than any other structure in the superfamily, may have diverged from the V domains before acquisition of the rearrangement mechanism, or they may have lost the ability to rearrange and subsequently acquired flexible, hinge-like segments to serve as static components in T cell receptor–mediated responses.

This complex hierarchy for organizing information suggests that early Ig-related domains were involved in recognition in general, rather than in functions strictly linked to immunity. Molecules corresponding to a single domain were first expressed at cell surfaces at an early stage in the evolution of multicellular organisms, and these were capable of interacting with identical receptors on other cells to mediate simple cell–cell interactions. Both arms of this recognition system then duplicated and diverged to provide more sophisticated recognition systems to control the specificity of cell interactions during tissue formation. Immunorecognition molecules could have evolved from these recognition systems. This idea fits well with the fact that two of the simplest structures in the Ig super-family (Thy-1 and MRC-OX2) seem to be more neuronal than lymphoid in their expression, and they may have functions that are not associated with immune responses. It remains to be determined whether other complex systems, like the nervous system in eukaryotes, may have also employed evolutionary strategies similar to the Ig superfamily to evolve and elaborate its complex function.

THE Ig FAMILY

General Structural Features of Ig Molecules

Immunoglobulin molecules comprise a family of glycoproteins with the same basic molecular architecture, but which exhibit an array of different functional properties. Thus, they bind specifically with antigen, activate the complement system, mediate several cytotropic reactions, and act as receptors for antigen on B lymphocyte membranes (Cathou and Dorrington, 1974). Although these different biological activities are reflections of extensive structural variations, all Ig molecules share an overall basic monomeric structure that consists of two light (L) and two heavy (H) polypeptide chains linked together by noncovalent interactions and interchain disulfide bridges (Fig. 4–1). Depending on the Ig class, this monomeric unit may form polymeric structures that consist of two or more basic units (Singer and Doolittle, 1966). Consistent with its functional specialization, limited proteolysis of the basic Ig unit results in the production of three functionally distinct fragments, two Fab's and one Fc (Fig. 4–1). The antigen binding site is located on the Fab portion of the molecule, whereas other effector functions are carried out by the Fc portion. Between each Fab and Fc lies a hinge region of approximately 25 amino acids, which is characteristically rich in proline residues and contains the inter-heavy chain disulfide bridges. The number of bridges and the length of the hinge region

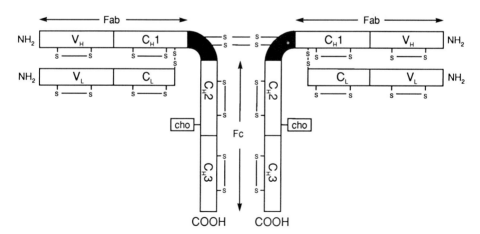

Figure 4–1. Schematic presentation of polypeptide chains in the basic Ig structure. The light (L) chains are divided into two homology regions: V_L and C_L. The heavy (H) chains are divided into four homology regions: V_H, C_H1, C_H2, and C_H3. C_H1 and C_H2 are joined by a "hinge" region, indicated by the solid area. Fab and Fc refer to fragments produced by enzymatic cleavage. Glycosylation sites (Cho), as well as interchain and intrachain disulfide bonds (S–S), are indicated. (Modified from Edelman et al., 1969.)

vary between different Ig classes and subclasses within a species and among different species (Hill et al., 1966).

All light chains are on average 217 amino acids in length and have a molecular mass of approximately 22.5 kD (Nisonoff et al., 1975). They are subdivided into two structurally and antigenically distinguishable types, κ and λ chains. A given Ig molecule contains two identical chains of either type, and their relative proportion among all Ig molecules is species-dependent. Each type may be further classified into various subtypes based on serologically-detectable differences in amino acid sequence. However, sequence homologies between κ chains of different species are in general much greater than those between κ and λ chains within a given species, which suggests that the separation of the two types was an early event during the evolution of Ig molecules.

Each light chain is structurally divisible into two approximately equal halves. The amino-terminal half contributes to the conformation of the antigen combining site and hence helps to determine the antigenic specificity and structural variability of the Ig molecule. The carboxyl-terminal half, on the other hand, is constant or shared by different light chains of the same type, and it provides the half-cysteine residue necessary for interaction with the adjacent heavy chain (Fig. 4–1) (Edelman et al., 1969).

All heavy chains are on average 450–576 amino acid residues in length with a range of molecular mass of 50–70 kD, depending on the Ig class (Nisonoff et al., 1975). On the basis of serological and structural differences, five classes of heavy chains have been found in humans and are designated γ, μ, α, δ, and ε. The class of the heavy chain determines the class of the Ig molecule, which is thus divided into five classes: IgG, IgM, IgA, IgD, and IgE. Within a given heavy chain class, structural variations account for its further subdivision into subclasses, each sharing similar serological and physicochemical properties.

As in light chains, a stretch of about 110 amino acid residues at the amino-terminal end of all heavy chains shows typical structural variability, unique to each Ig, which contributes to the conformation of the antigen combining site. A constant segment lies at the carboxyl-terminal end of this variable segment and shows the least variability within heavy chains of the same subclass. The constant region typically provides different sites for N- and O-linked oligosaccharide side chains, and it mediates the various secondary or effector functions of Ig's (Cathou and Dorrington, 1974; Nisonoff et al., 1975).

The Domain Hypothesis

The divisibility of Ig light and heavy polypeptide chains into a number of structurally homologous segments, two in light chains and four in heavy chains, was first recognized over 20 years ago (see Edelman et al., 1969). This type of segmentation was first proposed for IgG by Singer and Doolittle (1966) and was supported by the significant internal homology observed among sequences in Fc fragments and Bence-Jones proteins (Hill et al., 1966). Together, these pioneering observations helped in formulating the hypothesis of a common ancestral gene, coding for about 110 amino acids, which served as a building block from which all Ig polypeptides subsequently evolved by duplication and translocation. Subsequent determination of the complete primary structure of IgG (Edelman, 1970; Edelman al., 1969) revealed the striking arrangement of both light and heavy chains into linearly connected variable (V) and constant (C) regions within which there are significant homologies.

Two conformationally similar homologous regions were defined in the light chain

(V_L, C_L) and four in the heavy chain (V_H, C_H1, C_H2, C_H3). Each consists of about 110 residues and is predicted to be folded into a compact domain (Edelman, 1970; Edelman and Gall, 1969). Each domain is stabilized by a single intrachain disulfide loop of a roughly similar size and is linked to its neighboring domains by less compactly folded stretches of the polypeptide chain (Fig. 4–1). Subsequent comparison of amino acid sequences of other classes of Ig in different species showed that all of their chains are tandemly arranged into basically similar domains, confirming the universality of the domain structure among all Ig molecules.

In striking agreement with the domain hypothesis, high resolution X-ray crystallography of Ig fragments revealed that the V_L, V_H, C_L, and C_H1 regions share a similar three-dimensional configuration and a common pattern of folding that is characteristic for all Ig polypeptide chains (the Ig fold) (Fig. 4–2). Within the Ig fold, the sequence is folded into twisted, stacked anti-parallel β strands forming two β-pleated sheets that surround an inner volume which is tightly packed with hydrophobic side chains. These two β sheets are covalently linked by a centrally placed, highly conserved disulfide bridge that spans about 65 amino acid residues and which is aligned approximately perpendicular to the

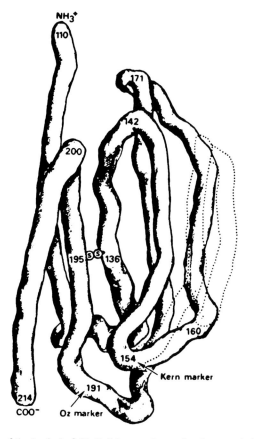

Figure 4–2. Diagram of the basic Ig fold. Solid trace shows the characteristic folding of the polypeptide chain in the constant subunits (C_L and C_H). Numbers designate light (λ) chain residues beginning at NH_3^+, which corresponds to residue 110 for the light chain. Broken lines indicate the additional loop of polypeptide chain characteristic of V_L and V_H subunits. (Reproduced from Amzel and Poljak, 1979.)

plane of the sheets. Both V and C domains share a common core structure, but they differ in the middle of the domain, where V domains have an extra loop of sequence and other structural variations in the antigen binding crevice (Amzel and Poljak, 1979).

At the gene level, each domain is encoded by a separate exon interspersed with introns, which demonstrate a correlation between the domain structure of the protein and the exon/intron organization of their corresponding genes. Accordingly, this relationship confirms that contemporary V and C domains probably evolved from a primordial gene that coded for a polypeptide of about 110 amino acids and corresponds to a single Ig domain. Partial duplication and divergence of this gene leading to V- and C-like domains was probably an early event in the evolution of the Ig family of molecules. This duplication would make it possible to generate polypeptides with separate regions for recognition (V domains) and for membrane anchorage and/or effector functions (C domains). The gradual acquisition of the structural and functional characteristics unique for each domain type must have occurred in positions appropriate for their functions, but which were not deleterious to the general structural integrity of the common fold.

Conservation of the Structural Configuration of Ig Domains

One major advantage of knowing the X-ray crystallographic structure of the Ig fold is that it allows accurate alignment of available amino acid sequences of different domains in various Ig classes and species, which permits us to assess the general structural features and factors that determine each type of domain. According to the crystallographic model, a typical C domain is made up of one β sheet formed of strands A, B, D, and E that is packed face-to-face with another β sheet formed of strands C, F, and G (Fig. 4–3). These linearly arranged strands are joined by connecting loops, which constitute bends, helices, and other structures of various lengths (b1–6), and the whole structure is stabilized by a conserved disulfide bridge between strands B and F (Amzel and Poljak, 1979). When comparisons of different C-domain sequences were made in relation to the crystallographic model (Beale and Feinstein, 1976), the following conclusions were reached.

1. The only invariant residues are the two cysteines that form the intradomain disulfide bridge, as well as a shielding tryptophan residue that occurs 14 residues carboxyl-terminal to the first invariant cysteine. Together with a highly conserved hydrophobic residue next to the cysteine of strand B, these four residues define a "pin" structure that limits the relative movement of the two β sheets. The spatial relationship of the side chains of the "pin" residues is almost identical in all C domains examined.

2. The closest sequence homologies occur in segments corresponding to β-sheet strands, which collectively define a domain "core." In these segments, hydrophobic and non-polar residues tend to be conserved and occur at alternating positions so that their side chains fill the space between the two faces of the domain and make important contributions to its hydrophobic nature. Although deletions and insertions within the segments forming the domain faces tend to be limited, so as to maintain the hydrophobic nature of the domain interior, the side-chain volumes of residues occupying inward-pointing positions in the "core" vary substantially. These variations are compensated by shifts in the relative position of β sheets, by lateral insertions of residues outside the domain "core," and by changes in the orientation of side chains.

3. Other segments of relatively poor homology constitute the domain "periphery," which is mainly defined by the connecting loops between the β strands. These may vary

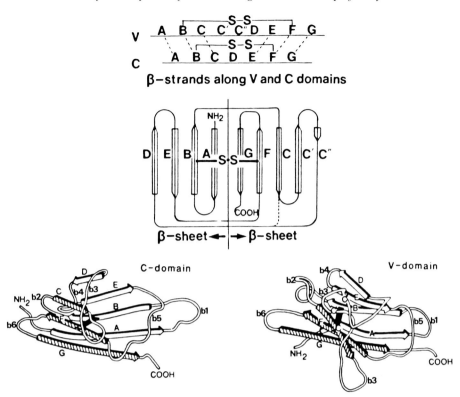

Figure 4–3. β strands in Ig V and C domains. The top shows, as letters, the regularly spaced patches of sequence-forming β strands along the V and C domains and the position of the conserved disulfide (–S–S–) bond; the dotted lines show β strands in which sequence homology is clear. The middle shows the folding pattern of V and C domains with two β sheets laid out and separated by the line down the middle. For a C domain, the β strand continues directly to D (see dotted line), whereas a V domain contains an extra loop of sequence, as shown by C′ and C″. (From Williams, 1984, in turn modified from Amzel and Poljak, 1979.) Shown below are C and V domains drawn to fit the crystallographic model of an Ig light chain. (Modified from Beale and Feinstein, 1976; Edmundson et al., 1975.)

in length, sequence, and conformation; nevertheless, their role in regulating the basic domain folding is not entirely passive.

In all C domains, the principal driving force behind changes in sequence is the maintenance of a closely packed interior. Changes in the size of interior residues may potentially produce extensive structural adjustments so that the net effect is the maintenance of the overall packing and the three-dimensional conformation of the whole domain. A more pronounced feature in all C domains is the high conservation of complementarity regions that allow interdomain interactions. Residues involved in this type of association are always located along the four-stranded β sheet (Beale and Feinstein, 1976), which constitutes part of the highly conserved "core" of all C domains.

Similar structural comparisons for the V-domain sequences of different Ig classes in different species were also made with the X-ray crystallographic model serving as a basis for alignment (Beale, 1984, 1985). All V domains examined, although bearing essentially no homology to C-domain sequences, show a three-dimensional structure similar to that of the C domains. As in the C domains, the polypeptide is folded into a four-stranded β sheet made of strands A, B, D, and E (Fig. 4–3) facing a three-stranded β sheet made of strands C, F, and G. These strands are connected by bends (b1–6) of varying lengths and shapes, and the whole structure is stabilized by a centrally placed disulfide bridge between strands B and F. The segments that form each face are extensively hydrogen-bonded into the β-pleated sheets, and the space between the two faces is filled with hydrophobic amino acid side chains. One of these is a tryptophan residue that together with the two cysteines of the intrachain bridge make up three invariant residues that correlate exactly with the "pin" structure of the C domains. A major difference between C and V domains, however, lies in the existence of an additional loop of sequence in the V domains that has no equivalent in the C domains. This loop (designated C' and C″ in Fig. 4–3) does not lie in either β sheet, and, thus, the sandwich nature of the V domains is less apparent, and the two faces of the domain are more distorted than those of the C domains. A connecting bend (b3′) in this additional loop shows, together with bends (b2) and (b6), the highest frequency of deletions and insertions among different V domains. The spatial proximity of these segments, however, is highly maintained, and together they constitute the three hypervariable regions that determine the conformation of the antigen combining site in all V domains.

A comparison of the sequences of the C and V domains, when aligned to fit the three-dimensional structure, confirms earlier views that sequence homologies have been obscured by divergent evolution (Edelman, 1970; Edelman and Gall, 1969). In particular, the interactions of the three- and four-stranded anti-parallel β sheets, which constitute the domain "core," have changed during the evolution of the two domains. In V-domain dimers, the three-stranded β sheets face each other across a solvent channel in which hapten-like molecules can be bound. The corresponding sheets in C-domain dimers supply the external surfaces and such binding therefore cannot occur. Conversely, the four-stranded β sheet furnishes the noninteracting external surface of the V domains, whereas it allows interdomain interactions across a solvent-free zone and helps maintain the compact arrangement of C domains. Sequence patterns in parts of the β sheets can be correlated with these differences (Poljak et al., 1973; Segal et al., 1974). In V domains, the alternating pattern of polar and apolar residues is broken in the three-stranded β sheet by the substitution of aromatic for aliphatic polar residues in positions important for conservation of the general architecture of the antigen combining site. Similarly, the alternating sequence pattern in the four-stranded β sheet of C domains is interrupted by the substitution of hydrophobic for polar residues at the conserved sites of complementarity between C-domain interfaces. The four-stranded β sheet in V domains is marked by an invariant buried glutamine residue associated with a β bulge, and the alternating polar constituents are mainly seryl and threonyl residues, whereas the corresponding polar residues on C-domain surfaces are more diversified.

Analyses of the structure of Ig domains lend strong support to the concept of a common evolutionary origin of both types of domains. Clearly, structural changes have occurred during the course of evolution which have resulted in specialized functions for the two types of domains. Two significant aspects of the variability in sequences are as follows: First, the antigen combining site, which involves the end of the β sheets and mainly the loops between their strands, is subject to extensive insertion, deletion, and substitution of amino acids. This is related to the versatility of the structure in creating binding sites for a wide variety of antigens. Here different sequences produce *different* configurations without affecting the overall three-dimensional structure. Second, the residues packed in the "core" structure show extensive substitution. Only three residues are invariant: the cysteines of the disulfide bridge between the sheets and a tryptophan that packs against it. Here, different sequences produce *similar* secondary and tertiary structures that provide a scaffolding on which new binding sites may be built and mainly allow an important aspect of domain function, namely interdomain interactions.

CONSERVATION OF STRUCTURAL CONFORMATION AMONG C- AND V-LIKE DOMAINS

Although comparisons of the amino acid sequences of the Ig superfamily extracellular domains provide tentative support for their structural homology, such sequence homology would be more meaningful if it correlated with the maintenance of certain, presumably essential, features of three-dimensional structures. X-ray crystallographic data are not yet available for any of the other members of the superfamily. However, sequences of C- and V-like domains were compared and conserved features were correlated with properties of the high-resolution crystallographic structures of Ig fragments (Beale, 1984, 1985).

Among all C-like domains in the superfamily, the polypeptide chains are apparently folded into globular domains closely related in three-dimensional structure to Ig C domains. The only invariant residues, however, are the two cysteines that form the intradomain disulfide bridge and, in most sequences, the tryptophan that packs against it. Together, these residues constitute a "pin" structure identical to that found in all Ig domains. A prediction of β-strand formation in all sequences is apparent, especially around the invariant cysteines in strands B and F (Fig. 4–3). Obviously, all sequences seem to conserve segments in which alternating patterns of polar and apolar residues clearly correspond to positions on the crystallographic model that lie in the strands of the β-pleated sheets.

Among these segments are six positions that contain residues with hydrophobic side chains and eight that contain residues with uncharged and generally hydrophobic side chains. In the crystallographic model these positions correspond to parts of the domain core, and their side chains always point to the domain interior and make essential contributions to its hydrophobic nature. Although all C-like domain sequences seem to be closely related to each other and to Ig C domains, they show characteristic divergences from each other, so that any further conservation depends on the type of the C-like domain. This is partially dependent on conserving the basic structure of the C-like domain, while allowing the development of contact residues essential for the interactions of each particular domain. Among these are the conserved modification in β-strand G of class I α_3 domains and class II α_2 and β_2 domains that mediate interdomain contacts similar to the Ig C_H3/C_H3 type, and the modified type of contact seen in class I $\alpha_3-\beta_2$-m (Beale, 1984). These differences seem to be highly conserved and contribute to the fact that a particular type of domain shows less divergence between different species than different types of domain show within the same species.

Similar analyses also apply to all V-like domains in the superfamily (Beale, 1985). All basic features of Ig V domains are maintained, including the domain "pin" structure, the domain "core" with the inner volume packed with hydrophobic side chains, the extra loop of sequence in the middle of the domain that creates a four- and five-stranded β-sheet arrangement, as well as characteristic V-like sequences that are involved in salt bridges and the formation of β bulges. However, the group of V-like domains seems to be, in general, under less constraint than the C-like domains, due to the fact that they show appreciable divergence from each other and from the Ig V domains. Except for the V-like domains of the T cell receptors, which are predicted to be closely related to the three-dimensional structure of Ig V domains, V-like domains in the poly-Ig receptor and in MRC-OX2, CD4, and CD8 are expected to have undergone appreciable displacement in their β sheets and strands to accommodate the divergent features specific to each type of domain. These differences may be involved in optimizing the biological roles of each particular domain, however.

The degree of homology observed between Ig's and members of the superfamily at the primary, secondary, and tertiary structural levels, as well as at the DNA level, strongly supports the hypothesis that these superfamilies of molecules share a common ancestral gene and that all evolved by divergent evolutionary processes from a single primordial domain sequence. Among all contemporary domains in the superfamily, one candidate for approximating the primordial structure is a sequence synthesized as a single domain that shares some features with the C and V domains rather than exactly fitting one type. Both β_2-m and Thy-1 fit one or the other or both criteria, and they are discussed below as candidates for the primordial domain from which all members of the Ig superfamily may have evolved.

β_2-Microglobulin: A Free C-Like Domain

β_2-m is a small polypeptide (99 amino acids, molecular mass of 11.8 kD) found in serum, as well as other biological fluids, and on the surface of almost all mammalian cells (Parnes and Seidman, 1982; Poulik, 1975). Although it is not an integral membrane protein, β_2-m is found in close but noncovalent association with several membrane proteins. In mice, all of the identified cell surface molecules associated with β_2-m are encoded on chromosome 17 within the MHC, and they include both class I molecules and a series of differentiation antigens (thymus leukemia antigen, Qa-1 and Qa-2) that are present on restricted classes of lymphocytes and other blood cells. In contrast, β_2-m is not genetically linked to MHC, and its gene is located on chromosome 2. Most of the polypeptide is encoded by a single uninterrupted exon flanked by intervening introns, and, in this regard, its genetic organization is similar to the domains of all members of the Ig superfamily (Ploegh et al., 1979).

β_2-m plays a critical role in ensuring efficient expression of MHC class I molecules on the cell membrane. Through its association with MHC class I molecules, β_2-m is suggested to act either by rescuing the relatively insoluble α chains, which otherwise would aggregate in the cytoplasm, or by generating a specific conformational determinant necessary for intracellular transport. Consistent with this idea is the fact that the class I α chain becomes alloantigenic only after association with β_2-m (Hyafil and Strominger, 1979). Thus the α_3 domain–β_2-m interaction generates a quaternary structure that allows the proper expression of alloantigenic sites on the α_1 and α_2 domains. These observations are extended by the finding that human-mouse hybrid cells express hybrid MHC–β_2-m molecules on the cell surface and by the ability of MHC-associated β_2-m to be spontaneously exchanged with exogenous β_2-m in an equilibrium reaction. Collectively, the data prompt the speculation that evolutionary constraint has limited diversification of interacting sites on β_2-m and β_2-m binding molecules.

The primary structures of β_2-m from mouse, rabbit, man, and guinea pig are similar, although a number of amino acid substitutions occur throughout the molecule. Position 6 from the amino-terminal has a different amino acid in all four β_2-m molecules, and there are 14 positions in which there are three different amino acid residues. The degree of sequence homology varies between 68% and 75% among the four β_2-m molecules, which emphasizes the strong degree of structural similarity across wide species differences. All β_2-m sequences show striking homology to Ig C domains and can be aligned and characterized, in terms of three-dimensional structure, with the crystallographic model of Ig fragments (Beale and Feinstein, 1976).

The β_2-m molecule incorporates the following features of C domains. (1) The two invariant cysteines are located in their exact expected positions and form the domain disulfide bridge. (2) The invariant tryptophan of all Ig domains is replaced by leucine or methionine, but these hydrophobic residues could play the role of tryptophan in shielding the disulfide bridge. (3) The absence of tryptophan is further compensated for by conserved hydrophobic chains found at various points along the sequence. (4) The polypeptide is folded into four- and three-stranded β sheets with apparent alternation of polar-apolar residues in a manner characteristic of Ig C domains. (5) Contact sites are located along the four-stranded β sheet in positions analogous to C_H3/C_H3-type contacts, and it is this surface of the molecule that shows minimal divergence among sequences of different species.

Interpretations of such comparisons of tertiary structure suggest that the role of β_2-m is critical enough to warrant a high level of conservation in evolution, particularly of the key residues through which the molecule mediates its functions. It is noteworthy that β_2-m is one of only two domains in the superfamily that naturally exists as an individual domain, corresponding exactly to the C-domain type. Its function is simple, yet crucial, and it has basically developed to interact with a similar highly conserved C-like sequence in the MHC Class I α_3 domain. This has led to the suggestion that β_2-m may approximate the structure of the common domain ancestor of the Ig superfamily. This notion is strengthened by the detection of β_2-m homologues in species as primitive as earthworms (Roch et al., 1983) and insects (Shalev et al., 1983). However, these ideas should ultimately be tested by sequence studies of the invertebrate β_2-m homologue.

Thy-1 Glycoprotein: A Free V-Like Domain

The Situation in Mammals

The Thy-1 glycoprotein is a differentiation marker of unknown function expressed predominantly on thymocytes, T cells, brain tissue, and some other cell types. It was first identified as the θ alloantigen of mouse brain and thymocytes (Reif and Allen, 1964), and structural homologues have subsequently been identified in various animal species. In all species, Thy-1 expression on brain cells is conserved, as contrasted with its expression on different cell types which is subject to species variation. The molecule is expressed as a single glycoprotein of 18 kD, of which one-third is carbohydrate (Williams and Gagnon, 1982). The protein moieties of rat and murine Thy-1 molecules have been sequenced and consist of 111 and 112 amino acid residues, respectively.

Studies of cDNA clones in rodents and humans suggest a similar organization of the Thy-1 gene in both species (Seki et al., 1985). The majority of the polypeptide is encoded by a single uninterrupted exon surrounded by intervening introns, in a manner analogous to all Ig-related domains in the superfamily. A major discrepancy between the cDNA and protein sequence data concerns the mechanism by which Thy-1 is anchored in the lipid bilayer. The protein sequence data suggest that a hydrophobic non-protein glycophospholipid segment covalently attached to the carboxyl-terminal end of the polypeptide is used, whereas cDNA data predict an extra 20 residue–long hydrophobic segment, which is not accounted for in the protein sequence data. Both sets of data, however, agree on the strong structural similarities of Thy-1 in rodents and humans. Apart from three patches of sequence, Thy-1 molecules of these species have all features in common. Despite that the structure of the molecule is highly conserved in rodents and humans, its expression on T cells differs dramatically in these species. This reflects fundamental differences in the way the Thy-1 gene is regulated during evolution.

Among the members of the Ig superfamily, Thy-1 is exceptional in that it is synthesized and expressed as a free V-like domain that does not associate either with itself or with any other polypeptide. On the basis of sequence identities with V_L and V_H chains, putative β strands C' and C'' (characteristic of V domains) were identified along a five-stranded β sheet similar to a V-domain fold (Cohen et al., 1981). This assessment is supported by the location of the conserved disulfide bridge, which spans 65 amino acid residues. When computer analysis of sequence alignment between the poly-Ig receptor and Ig sequences is carried out, it places Thy-1 between V_κ and V_H domains in significance of fit. Thy-1, however, lacks the shielding tryptophan residue, which is replaced by a leucine residue as is also the case in the C-like β_2-m domain. At variance from nearly all V domains, Thy-1 does not have the buried glutamine residue and thus lacks the β bulges characteristic of V domains. In addition, Thy-1 also fits well with a number of Ig C domains in terms of sequence identities. This is particularly strengthened by the location of an extra disulfide bridge within the Thy-1 sequence, at a site comparable to that seen in C_H1 domains. Thus, Thy-1 bears remarkable structural similarities to both V- and C-type domains. Of particular interest is that these comparisons suggest the evolution of a C-type sequence from a Thy-1-like sequence by a deletion in the middle of the domain.

Comparative Aspects of Thy-1 Distribution

The structural characteristics ascribed to Thy-1 strongly argue that this molecule is a better candidate for a primordial domain sequence than is any other contemporary Ig domain. A search for a Thy-1 homologue in an invertebrate species may establish whether this molecule is more primitive in evolution than is any other member of the Ig superfamily, which is not found in species more primitive than vertebrates. Recently, optic and central nervous tissue from the squid was used to look for a putative Thy-1 antigen (Williams and Gagnon, 1982; Williams et al., 1988). Solubilization in deoxycholate and fractionation by lentil lectin affinity chromatography and gel filtration yielded abundant small glycoproteins. Material similar to Thy-1 was found, and it consisted of two glycoproteins that were subsequently purified using monoclonal antibody affinity columns. The glycoproteins yielded sequences of 84 residues for Sgp-1 and 92 residues for Sgp-2, which were then analyzed for similarities to Thy-1 and other Ig-related sequences. Although Sgp-1 was similar at >3 standard deviation units away from mean standard scores using the ALIGN program, the sequence patterns were not typical of Ig-related domains, leaving the relationship of Sgp-1 to the Ig superfamily open to question; similarly, Sgp-2 showed no relationship to Ig genes. However, there were similarities to Ly-6 including the presence of ten Cys residues in each sequence.

Within the protochordates an Ig-related molecule has been sought mainly using serological cross-reactivity. The tunicate Thy-1 homologue is a 27–30 kD glycoprotein, with a protein moiety of about 12 kD (Mansour and Cooper, 1984; Mansour et al., 1985, 1987). Full sequence data of this molecule are not yet available, but the molecule bears some structural resemblance to rodent Thy-1 as assessed by amino acid composition, peptide mapping, and location of the intrachain disulfide bridge, as well as by limited sequence analysis of two tryptic peptides. The full sequence data of this molecule and others should be informative in determining the structural elements that strengthen their likeness to a primordial domain ancestor of the Ig superfamily (Cooper and Mansour, 1989; El Deeb et al., 1988).

OTHER MEMBERS OF THE Ig SUPERFAMILY WHOSE PHYLOGENY IS LESS CLEAR

Members of the Ig Superfamily and Their Relationship to MHC and Receptor Structures

On the basis of primary amino acid sequence and gene organization studies, all molecules with known structures that are involved in antigen recognition by vertebrate T and B lymphocytes, as well as other cell surface glycoproteins, constitute a structurally related superfamily of molecules (Williams, 1984, 1985). The argument that all the molecules shown in Fig. 4–4 are structurally related is based on their sharing sequence domains of

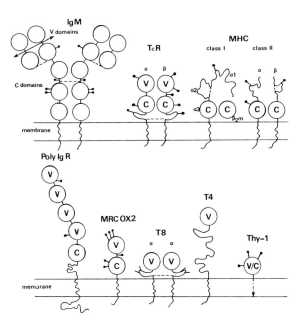

Figure 4–4.   Schematic models for molecules in the Ig superfamily. Circles marked V are like Ig V domains, and those marked C are like Ig C domains. Intrachain disulfide bonds are represented by broken lines, N-linked glycans by (|) and O-linked glycans by (|). (Modified from Clark et al., 1985.)

110 amino acid residues, each corresponding to Ig, V, and C domains. The superfamily is therefore named after its prototypical member, the Igs. The degree of amino acid sequence homology between domain-like structures ranges from about 15%–40% across the different members within the superfamily. As for Ig, each domain-like structure is usually encoded by a separate exon, which demonstrates the striking correlation between the distinct structural features of these proteins and the exon/intron organization of their corresponding genes. Different molecules corresponding to the known members of the superfamily contain different numbers of domain-like structures and other sequences, giving rise to structures with molecular mass ranging from about 12 kD (Thy-1 or β_2-m) to 95 kD (poly Ig receptor).

Considering the distribution of V- or C-like domain types among the different polypeptides, one or the other and/or both types of domains can be included in each member. The major histocompatibility complex (MHC) class I encoded molecule is a 45-kD polymorphic integral membrane protein that is associated noncovalently with the 12-kD non-MHC–encoded, monomorphic or dimorphic β_2-m (Cushly and Owen, 1983). The class I molecule is found on the surface membrane of virtually all nucleated cells and acts as a restricting element for cytotoxic T cells—hence, it is a key molecule in self recognition. Structural analyses of classical class I molecules, including proteolytic digestions and amino acid and nucleotide sequence analyses, suggest that the polypeptide is divided into three extracellular domains, each of about 90 amino acid residues, and designated α_1, α_2, and α_3; a hydrophobic transmembrane region of about 20 amino acids; and a hydrophilic cytoplasmic region of 30–40 amino acids.

Sequence heterogeneity among different class I molecules resides in the amino-terminal α_1 and α_2 domains, where the putative alloantigeneic sites are located. By contrast, the carboxyl-terminal α_3 domain and β_2-m are both conserved and display sequence homology to each other and to Ig C domains. A similar analysis applies to the MHC-encoded class II molecules that are expressed on some lymphocytes, macrophages, and endothelial cells and which act as restricting elements for helper T cells. Thus, these molecules are key structures for cell–cell cooperation in the initiation of an immune response (Kaufman et al., 1984). The class II molecule is a heterodimer made up of two noncovalently-associated integral membrane proteins, termed the α and β chains, 34 kD and 28 kD, respectively. The extracellular regions of both chains are divisible into two extracellular domains. The amino-terminal domains, α_1 and β_1, show extensive sequence variability among the different class II molecules examined, whereas the carboxyl-terminal domains, α_2 and β_2, are conserved and show sequence homology to each other and to Ig C domains.

The T cell antigen receptor, from species as far apart in evolution as mouse and man, is a cell surface bound heterodimeric glycoprotein composed of a 34-kDα chain and a 32-kD β chain, which are linked covalently by an interchain disulfide bridge, as well as by noncovalent hydrogen bonding. Two different classes of T cell–specific cDNA clones, corresponding to the α and β components, have been isolated (Hedrick et al., 1984; Hood et al., 1985). Amino acid sequence analyses of the two chains and nucleotide sequence data for both clones show that each chain can be divided into an amino-terminal V-like domain and a carboxyl-terminal C-like domain, joined by a hinge-like region. An additional class of T cell–specific cDNA clone has been identified and designated γ, but it does not encode any of the T cell receptor components that have been identified by monoclonal antibodies and, as yet, its function is unknown (Hood et al., 1985).

Despite their sequence homologies and similarities in gene organization and rearrangement, the T cell receptor differs fundamentally from Ig molecules in its mechanisms of antigen recognition. Ig molecules bind readily to antigens in the absence of MHC gene products, whereas T cells can only recognize antigens in conjunction with molecules encoded by MHC class I and II antigens. Correlations between functional T cell subclasses and the class of MHC-encoded molecule that is recognized in conjunction with antigens suggest that antigen recognition by cytotoxic T cells tends to occur in association with class I molecules, whereas that of helper T cells tends to occur in conjunction with class II molecules. The molecular basis for this correlation has not yet been elucidated, however, and it even seems that, at least for the type of β chain expressed, there is no correlation with either the T cell subclass or with recognition of class I or class II molecules (Hood et al., 1985).

Another characteristic feature of the functional T cell receptor that distinguishes it from Ig is that the T cell receptor alone appears to be insufficient for stable antigen recognition. The molecule is part of a macromolecular complex that includes the three invariant polypeptides of the CD3 antigen, and other additional T cell surface-associated molecules that are required to form a stable interaction among the T cell receptor, the appropriate class of MHC-encoded molecules, and the antigen. These additional T cell surface structures include the CD8 (human equivalent to the murine Lyt-2) and the CD4 (human equivalent to the murine L3T4) molecules, which may act as accessory receptor molecules for the nonpolymorphic regions of MHC class I and II molecules, respectively (Littman et al., 1985).

Interestingly, cloning and sequence analyses of CD8 and CD4 cDNA demonstrate that these two polypeptides are also structurally homologous to Ig. The CD8 molecule is a cell surface homodimeric glycoprotein formed of two covalently linked α and $\dot{\alpha}$ chains, each with a molecular mass of 34 kD. Each chain can be divided into an amino-terminal domain followed by a hinge-like segment, a hydrophobic transmembrane region, and a cytoplasmic tail. The amino-terminal domain of the polypeptide chains shows significant sequence homologies to both Ig and T cell receptor V domains (Littman et al., 1985). The CD4 molecule, on the other hand, exists as a single 55-kD polypeptide that does not appear to form a covalently-linked dimer, either in pair-wise association with itself or with another cell surface molecule. The overall structure of this molecule consists of an amino-terminal V-like domain sequence, a joining (J)-like region, a stretch of 374 amino acid residues, a membrane spanning region and a highly charged cytoplasmic tail. The amino-terminal domain sequence shows particularly strong homologies to the V domains of Ig light chains (Maddon et al., 1985).

Additional cell surface proteins of known structure that are considered to be members of the Ig superfamily include the receptor for transepithelial transport of IgA and IgM (poly-Ig receptor) and two cell surface differentiation antigens (rat MRC-OX2 antigen and the Thy-1 glycoprotein). The poly-Ig receptor is expressed on the basal surface of epithelial cells in the gut and acts by binding polymorphic IgA and IgM. The receptor-Ig molecule complex is then internalized and transported across the cytoplasm in vesicles that are exocytosed at the apical cell surface to the gut lumen. In this process, the receptor is proteolytically cleaved to release the Ig molecule with part of the receptor attached to it as a "secretory component." Structural analyses of the poly-Ig receptor reveal that it is a 95-kD glycoprotein made up of a single polypeptide chain, which can be divided into an extracellular poly-Ig binding segment, a membrane-spanning hydrophobic segment, and an unusually long cytoplasmic tail (Mostov et al., 1984). Starting at the amino-terminus, the extracellular poly-Ig binding segment comprises six domains—four that show clear homology to each other and to Ig V domains, followed by an Ig C-like domain sequence and a membrane-proximal segment that is followed by the membrane-spanning portion of the molecule.

When compared to the poly-Ig receptor, the rat MRC-OX2 is a relatively newer member of the superfamily, which is smaller in size and exhibits a wider tissue distribution. The molecule is a single glycoprotein of 41–47 kD that is found on neurons, thymocytes, B cells, follicular dendritic cells, activated T cells, and all endothelial cells. There is no evidence of its function and no correlation between the different cell types that express it, except that it may play a role in recognition events. The polypeptide consists of two extracellular domains, a transmembrane segment and a cytoplasmic tail. The amino-terminal domain sequence is similar to Ig V domains, whereas the carboxyl-terminal sequence is more like Ig C domains (Clark et al., 1985). Thus, the overall structure is similar to an Ig light chain or the T cell receptor α or β chains.

The Ig domain hypothesis states that the evolutionary starting point for complex Ig molecules was a single domain, which gave rise to a structure similar to an Ig light chain from which all Ig polypeptides in turn evolved. This concept may be extended to include all the molecules in the Ig superfamily. The primary structural relatedness of all molecules, as shown in Figure 4–4, provides evidence that all are evolutionarily related and that they diverged from a common "primordial domain" ancestor (Williams, 1984, 1985).

According to Williams (1987), the superfamily of molecules with Ig-like domains has recently been extended by the addition of new members—largely on the basis of sequence homology. The concept of the Ig superfamily as a set of structures involved in basic cell surface recognition events has been greatly strengthened by the identification of Ig-related structures present on neural and hemopoietic but not lymphoid tissues. For example, these include the neural cell adhesion molecule (N-CAM), myelin-associated glycoprotein (MAG), the major glycoprotein of peripheral myelin (P_0), the platelet-derived growth factor receptor (PDGFR), the colony stimulating factor-1 receptor (CSF1R), the mouse macrophage Fc receptor (FcR), and the carcino-embryonic antigen (CEA). CEA has a flamboyant structure with seven Ig-related domains. T cell antigens also have Ig-like features, which have recently been recognized with the reports of sequences for CD1, CD2, the CD3 ϵ chain, and the CD8 chain II, along with further analysis of the Ig characteristics of the CD4 antigen and the CD3 α and δ chains. CD5 antigen may also be Ig-related, but this observation is contentious.

With respect to the Ig superfamily, certain structures apparently break generalizations about molecules that are at some stage associated with cell surfaces, with secreted or degraded molecules as possible soluble forms. For example, the serum glycoprotein α 1B-gp has a five-domain structure, and the link protein of basement membranes has a V-like domain at the amino-terminus of a chain of 334 amino acids. Alpha 1B-gp may yet be found as a cell surface form, but the link protein may not be expressed as an integral plasma membrane protein as its function is to bind together the hyaluronic and proteoglycan components of the basement membrane matrix.

T Cell Receptor Molecules

In a recent review, Strominger (1989) brought us up to date with T cell receptors. T cell receptors (TCR) are antigen-recognizing elements on effector cells. Two isotypes have been discovered, TCR $\gamma\delta$ and TCR $\alpha\beta$, which appear in that order during ontogeny. Maturation of prothymocytes, which later colonize the thymic rudiment at defined gestational stages, occurs mainly within the thymus; there is also evidence for extrathymic maturation. At the molecular level, maturation includes the rearrangement and expression of TCR genes. Determination of the mechanisms, cell lineages, and subsequent thymic selection that results in self tolerance is one of the central problems in developmental immunology and is important for understanding autoimmune diseases.

Our understanding of immune recognition began very early with the simple discoveries that transplant rejection and immune responsiveness were both controlled by polymorphic cell surface molecules. A single genetic region of all vertebrates, the major histocompatibility complex (MHC), which is located on chromosome 6 in humans and on chromosome 17 in mice, encoded two classes of cell surface molecules. We are now reasonably certain that the primary role of these molecules, which are the class I and class II MHC antigens, is the presentation of foreign peptides derived from foreign antigens to T lymphocytes, which are the effector cells of the immune system. In this manner, the rejection of transplanted cells, tissues, and organs is a by-product of this primary role. Because these molecules are polymorphic, they play an essential role in immune recognition. This ensures that a large number of foreign peptides will be recognized by most species, particularly in mammals and to some extent in amphibians, and that sufficient population diversity will be a by-product. Thus the species as a whole will escape potentially catastrophic population variations due to detrimental environmental pathogens such as viruses. Relevant information in ectothermic vertebrates is not so clear, and among invertebrates it is essentially spurious.

An enormous amount of information now exists on the structure and function of these two classes of molecules and about the genes that encode them. T lymphocytes recognize foreign antigens in an MHC-restricted fashion. The effector T lymphocytes from one individual recognize a foreign antigen, which may be presented by cells of another individual only if the two individuals share at least one allelic MHC antigen. There was extensive debate as to whether the effector cell possessed two receptors, one for foreign antigen and one for the MHC restricting element, or only a single receptor that recognized both. It seems clear now that a single receptor recognizes the complex of foreign peptide with MHC antigen.

Despite the enormous progress that revealed the organization and structure of B cell genes, for several years the nature of the TCR was contentious. What pushed the search for this molecule was the realization that a great deal of diversity was required to ensure that numerous foreign antigens would be recognized appropriately in association with many different MHC molecules. It seems only natural that these would, of necessity, be generated by mechanisms analogous to those that generate diversity of immunoglobulins. Recently, a molecule was identified, first as a heterodimeric protein on the surfaces of murine and human T cells that was identified by clone-specific monoclonal antibodies and subsequently as the α and β genes that encode the two chains of the protein. Based on earlier predictions, these genes are encoded as V, J, and C segments of the α gene on chromosome 14 in both humans and mice, and as V, D, J, and C segments of the β gene on chromosome 7 in humans and on chromosome 6 in mice. The diversity of the α and β genes is generated by selection of different segments for joining and by the insertion of random nucleotides (called N segments) between the rearranging VJ and VDJ elements. Somatic mutation, now known to be a crucial mechanism in generating diversity of Ig genes, apparently plays no role in generating the diversity of TCR chains. We now refer to this as the $\alpha\beta$ T cell receptor, or TCR $\alpha\beta$.

Later work on the TCR $\alpha\beta$ led to the discovery of a third rearranging T cell gene also made up of V, J, and C segments, first in mice and then in humans, called TCR γ. During ontogeny, TCR γ gene rearrangements and most, if not all, of the process of T cell maturation occur in the thymus. Moreover, TCR γ gene rearrangements precede TCR α and β rearrangements and TCR αβ expression. TCR γ expression in the mouse thymus reaches relatively high levels early in fetal development and then declines before birth. A second T cell receptor species, TCR γδ, and the gene segments that encode the fourth chain of the rearranging T cell receptor family, TCR δ, were subsequently discovered. This second receptor is found on only 1%–10% of peripheral blood T cells in humans or mice, but these T cells represent 30% of the total in chickens. The function of this second T cell receptor remains unclear.

T cell repertoire development within the thymus involves rearranging of TCR α, β, γ, and δ genes, which may be accompanied by the formation of TCR γδ and TCR αβ receptors. This process includes the selection for export to the periphery of T cells that recognize foreign antigens in the context of MHC molecules (positive selection). However, these are not self-reactive (negative selection), at least in the case of cells bearing TCR αβ.

CONCLUSIONS: SIGNIFICANCE OF THE Ig SUPERFAMILY FOR THE IMMUNE SYSTEM AND THE CNS: RECOGNITION FUNCTIONS AND EVOLUTION OF AN IMMUNE SYSTEM

Fasciclin II

According to a recent review by Harrelson and Goodman (1988), insects possess two members of the Ig superfamily: fasciclin II and amalgam. Using oligonucleotide probes based on protein microsequence data, they isolated fasciclin II cDNA clones from a grasshopper embryo cDNA library. The deduced amino acid sequence generates a mature protein with a molecular size of about 97 kD; the purified protein runs as a single band on reducing SDS-polyacrylamide gels with an apparent molecular size of 95 kD (glycosylated) or 87 kD (deglycosylated). The deduced amino acid sequence of fasciclin II predicts a hydrophobic signal sequence of 22 amino acids, a mature extracellular region of 742 amino acids, a hydrophobic transmembrane domain of 25 amino acids, and a cytoplasmic region of 108 amino acids.

Among the different subclasses of the Ig superfamily, fasciclin II is most highly related to the vertebrate neural adhesion molecules with C2-type Ig domains, including N-CAM; myelin-associated glycoprotein (MAG), a glial adhesion molecule; and L1, an axonal adhesion molecule (also called NILE in mammals and G4 and N-CAM in the chicken). In addition to multiple Ig repeats, two of these vertebrate neural cell adhesion molecules, N-CAM and L1, also contain Fn-type III repeats near their transmembrane domains, and the two Fn-type III repeats in fasciclin II have significant amino acid homology with these domains.

Fasciclin II and this group of Ig superfamily-related vertebrate neural cell adhesion molecules are most alike in the third Ig-like domain. The third Ig-like domain of fasciclin II has its greatest similarity to the third Ig-like domain of MAG (Dayoff ALIGN score 11.1), less to the L1 third domain (ALIGN score 9.15), and least to the N-CAM third domain (ALIGN score 8.77). The fibronectin-related domains in fasciclin II bear greater resemblance to those of L1 than to those of N-CAM. Yet the greatest overall similarity occurs between fasciclin II and N-CAM.

In addition to fasciclin II, there is another insect protein that contains Ig-related domains, and it is known as amalgam. The amalgam gene in *Drosophila* encodes a protein that contains three Ig domains, and the Ig domains of fasciclin II and the amalgam protein are highly related. Although fasciclin II is transiently expressed on a subset of axon fascicles in the developing grasshopper central nervous system (CNS), antibodies against the amalgam protein reveal that it is expressed on the surface of all axons in the developing *Drosophila* CNS and is similar in some respects to N-CAM.

Because insects have at least two different members of the Ig superfamily expressed on either a subset (fasciclin II) or all (amalgam) of the axon pathways in the developing CNS, at least three interesting hypotheses can be advanced. First, just as their closest vertebrate relatives are all N-CAMs, fasciclin II and amalgam are likely to be N-CAMs in insects. Second, because fasciclin II is expressed only on a restricted subset of axon pathways in the developing insect CNS, and primarily around two of the 30 neuroblast families during neurogenesis, perhaps many other Ig-like molecules exist in insects and are used for specific adhesion and recognition during neuronal development. Third, the discovery of fasciclin II and amalgam shows that molecules with repeated Ig-like domains clearly evolved long ago, before the phyletic split of the arthropod and chordate lines, as previously suggested by Williams (1987). The expression of Ig-like molecules during neuronal development in insects supports the notion that (1) a gene coding for Ig-like domains evolved first as a cell adhesion molecule; (2) this gene, through duplication and divergence early in evolution, expanded into a gene family that functioned at least in part as neural cell adhesion and recognition molecules; and (3) only later in chordate evolution were some of these genes selected for special functions in the immune system.

Axon Guidance

According to Dodd and Jessell (1988), many specialized functions of the vertebrate nervous system depend on the intricate network of neuronal connections that are generated during development. To begin this developmental program, there is the projection of axons onto their targets by means of diverse and changing environments. How accurately axons select pathways, and the extent to which neuronal connections form, results from the precise guidance of axons to their targets; these two points are contentious.

In vitro studies have demonstrated that soon after neuronal differentiation, the axons of many central and peripheral neurons, for example, retinal and ciliary ganglion neurons, use a combination of integrins and N-cadherin to mediate their extension onto cell surface and extracellular matrix substrates. The ability of axons of the same neurons to extend on laminin substrates decreases dramatically later in development, and their extension is no longer inhibited by antibodies to integrins. Some integrins can interact with adhesion molecules that belong to the Ig gene superfamily. Integrins that are expressed on axons at later times could therefore interact with other ECM proteins or with axonal glycoproteins such as N-CAM, contactin, and L1 that possess Ig-like domains.

Evolution of an Immune System

The Ig superfamily probably originated from molecules that first evolved to mediate interactions between cells; the vertebrate immune system developed out of this set of

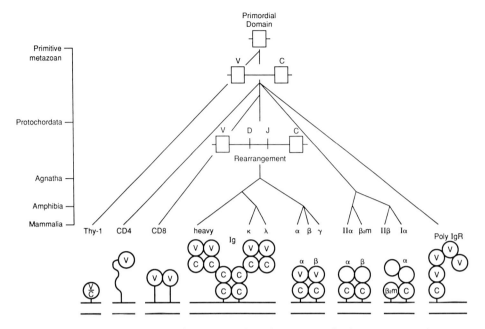

Figure 4–5.   Hypothetical scheme for the evolution of the Ig superfamily. The scheme is constructed based on current data regarding Thy-1 antigens, and by assuming that evolutionary relatedness correlates with the degree of sequence similarity among the members. (Scheme modified from Hood et al., 1985.)

structures (Fig. 4–5). Extensive diversification in the superfamily may have occurred as primitive sensory systems developed; Thy-1 and P_0, the only Ig-related molecules known to exist alone as single-domain structures, are major molecules in membranes of neurons and glial cells. The MRC-OX2 antigen, which has a simple two-domain structure, and N-CAM are expressed on neuronal cells; glial cells express N-CAM and MAG. Genetic linkage brings together structures involved in neural and immune development, the genes for Thy-1, N-CAM, and the CD3 ϵ, γ, and δ chains are all located on the q23 band of chromosome 11. This linkage may identify a chromosomal segment where extensive early duplication of Ig superfamily genes occurred.

Within the context of Williams's hypotheses, Ig superfamily evolution probably stems from the development of heterophilic recognition between related molecules (Williams, 1987). A heterophilic recognition system that underwent extensive diversification would allow sophisticated signaling in cell interactions. To pose one question, how can recognition-mediating interactions within an organism be turned outward to produce an immune system? According to one view, this represents a derivation from a cytotoxicity system that mediated programmed cell death. For example, in neural differentiation in *Caenorhabditis elegans* many cells die in an orderly fashion; usually cells differentiate into a cell type that dies and is phagocytosed, but in some cases one cell appears to kill another before phagocytosis. These very early cytotoxic cells could provide a precursor for the vertebrate immune system. Furthermore, integration of the functional and structural aspects could occur if the specificity of a natural killer cell for programmed cell death was mediated by Ig-related molecules. Finally, assuming that this specificity could have been modified to incorporate a determinant of a common virus or parasite, then a recognition system would develop which possessed features that are similar to those of T lymphocyte antigen receptors. Because natural cell cytotoxicity is gaining favor as a strong link between invertebrate and vertebrate cell-mediated responses, these views seem to be not far-fetched, at least for now.

Acknowledgment

The preparation of this chapter was partially supported by NSF Research grant DCB 85-19848.

References

Amzel, L. M., and Poljak, R. J. (1979). Three-dimensional structure of immunoglobulins. *Annu. Rev. Biochem. 48*:961–997.

Beale, D. (1984). A comparison of the amino acid sequences of different classes of immunoglobulin and histocompatibility C-domains from different mammalian species: Correlations between conservancy and three-dimensional structure. *Comp. Biochem. Physiol. 77B*:399–412.

Beale, D. (1985). A comparison of the amino acid sequences of the extracellular domains of the immunoglobulin superfamily: Possible correlations between conservancy and conformation. *Comp. Biochem. Physiol. 80B*:181–194.

Beale, D., and Feinstein, A. (1976). Structure and function of the constant regions of immunoglobulins. *Quart. Rev. Biophys. 9*:135–180.

Cathou, R. E., and Dorrington, K. J. (1974). Structure and function of immunoglobulins. In *Biological Macromolecules—Subunits in Biological Systems* (Timasheff, S. N., and Fasman, G. D., eds.). New York: Marcel Dekker.

Clark, M. J., Gagnon, J., Williams, A. F., and Barclay, A. N. (1985). MRC-OX2 antigen: A lymphoid/neuronal membrane glycoprotein with a structure like a single immunoglobulin light chain. *EMBO J., 4*:113–118.

Cohen, F. E., Novotny, J., Sternberg, M.J.E., Campbell, D. G., and Williams, A. F. (1981). Analysis of structural similarities between brain Thy-1 antigen and immunoglobulin domains: Evidence for an evolutionary relationship and a hypothesis for its functional significance. *Biochem. J. 195*:31–40.

Cooper, E. L., and Mansour, M. H. (1989). Distribution of Thy-1 in invertebrates and ectothermic vertebrates. In *Cell Surface Antigen Thy-1: Immunology, Neurology, and Therapeutic Applications.* (Reif, A. E., Schlesinger, M., eds.). New York: Marcel Dekker, pp. 197–219.

Cushly, W., and Owen, M. J. (1983). Structural and genetic similarities between immunoglobulins and class I histocompatibility antigens. *Immunol. Today 4*:88–93.

Dodd, J., and Jessell, T. M. (1988). Axon guidance and the patterning of neuronal projections in vertebrates. *Science 242*:692–708.

Edelman, G. M. (1970). The covalent structure of human γ G immunoglobulin: XI. Functional implications. *Biochemistry 9*:3197–3205.

Edelman, G. M., Cunningham, B. A., Gall, W. E., Gottlieb, P. D., Rutishauser, U., and Waxdal, M. J. (1969). The covalent structure of an entire γ G immunoglobulin molecule. *Proc. Natl. Acad. Sci. USA 63*:78–85.

Edelman, G. M., and Gall, W. E. (1969). The antibody problem. *Annu. Rev. Biochem. 38*:415–466.

Edmundson, A. B., Ely, K. R., Abola, E. E., Schiller, M., and Panagiotopoulos, N. (1975). Rotational allomerism and divergent evolution of domains in immunoglobulin light chains. *Biochemistry 14*:3953–3961.

El Deeb, S., Saad, A. H., and Zapata, A. (1988). Detection of Thy-1[+] cells in the developing thymus of the lizard *Chalcides ocellatus. Thymus* 12:3–9.

Harrelson, A. L., and Goodman, C. S. (1988). Growth cone guidance in insects: Fasciclin II is a member of the immunoglobulin superfamily. *Science 242*:700–708.

Hedrick, S. M., Cohen, D. I., Nielsen, E. A., and Davis, M. M. (1984). Isolation of a cDNA clone encoding T cell specific membrane-associated proteins. *Nature 308*:149–153.

Hill, R. L., DeLaney, R., Fellows, R. E., and Lebovitz, H. E. (1966). The evolutionary origins of the immunoglobulins. *Proc. Nat. Acad. Sci. USA 56*:1762–1769.

Hood, L., Kronenberg, M., and Hunkapiller, T. (1985). T cell antigen receptors and the immuno-globulin supergene family. *Cell 40*:225–229.

Hyafil, F., and Strominger, J. L. (1979). Dissociation and exchange of the β_2-microglobulin subunit of HLA-A and HLA-B antigens. *Proc. Natl. Acad. Sci. USA 76*:5834–5838.

Jerne, N. K. (1974). Toward a network theory of the immune system. *Ann. Immunol. (Institut Pasteur) 125C*:373–384.

Kaufman, J. F., Auffray, C., Schamboeck, A., Korman, A. J., Schackelford, D. A., and Strominger, J. L. (1984). The class II molecules of the human and murine major histocompatibility complex. *Cell 36*:1–13.

Littman, D. R., Thomas, Y., Maddon, P. J., Chess, L., and Axel, R. (1985). The isolation and sequence of the gene encoding T8: A molecule defining classes of T lymphocytes. *Cell 40*:237–246.

Maddon, P. J., Littman, D. R., Godfrey, M., Maddon, D. E., Chess, L., and Axel, R. (1985). The isolation and nucleotide sequence of a cDNA encoding the T cell surface protein T4: A new member of the immunoglobulin gene family. *Cell 42*:93–104.

Mansour, M. H., and Cooper, E. L. (1984). Serological and partial molecular characterization of a Thy-1 homolog in tunicates. *Eur. J. Immunol. 14*:1031–1039.

Mansour, M. H., DeLange, R., and Cooper, E. L. (1985). Isolation, purification and amino acid composition of the tunicate hemocyte Thy-1 homolog. *J. Biol. Chem. 260*:2681–2686.

Mansour, M. H., Negm, H. E., and Cooper, E. L. (1987). Tunicate Thy-1: An invertebrate member of the Ig superfamily. *Dev. Comp. Immunol. 11*:3–36.

Mostov, K. E., Friedlander, M., and Blobel, G. (1984). The receptor for transepithelial transport of IgA and IgM contains multiple immunoglobulin-like domains. *Nature 308*:37–43.

Nisonoff, A., Hopper, J. E., and Spring, S. B. (1975). *The Antibody Molecule.* New York: Academic Press.

Parnes, J. R., and Seidman, J. G. (1982). Structure of wild-type and mutant mouse β_2-microglobulin genes. *Cell 29*:661–669.

Ploegh, H. L., Cannon, L. E., and Strominger, J. L. (1979). Cell-free translation of mRNAs for the heavy and light chains of HLA-A and HLA-B antigens. *Proc. Natl. Acad. Sci. USA 76*:2273–2277.

Poljak, R. J., Amzel, L. M., Avey, H. P., Chen, B. L., Phizacherly, R. P., and Saul, F. (1973). Three-dimensional structure of the Fab' fragment of a human immunoglobulin at 2.8 Å resolution. *Proc. Natl. Acad. Sci. USA 70*:3305–3310.

Poulik, M. D. (1975). β_2-microglobulin. In *The Plasma Proteins.* New York: Academic Press, pp. 433–454.

Reif, A. E., and Allen, J.M.V. (1964). The AKR thymic antigen and its distribution in leukemias and nervous tissues. *J. Exp. Med. 120*:413–433.

Roch, P., Cooper, E. L., and Eskinazi, D. P. (1983). Serological evidences for a membrane structure related to human β_2-microglobulin expressed by certain earthworm leukocytes. *Eur. J. Immunol. 13*:1037–1042.

Segal, D. M., Padlan, E. A., Cohen, G. H., Rudikoff, S., Potter, M., and Davies, D. R. (1974). The three-dimensional structure of a phosphorylcholine-binding mouse immunoglobulin Fab and the nature of the antigen-binding site. *Proc. Natl. Acad. Sci. USA 71*:4298–4302.

Seki, T., Spurr, N., Obata, F., Goyert, S., Goodfellow, P., and Silver, J. (1985). The human Thy-1 gene: Structure and chromosomal location. *Proc. Natl. Acad. Sci. USA 82*:6657–6661.

Shalev, A., Pla, M., Ginsburger-Vogel, T., Echalier, G., Lögdberg, L., Björck, L., Colombani, J., and Segal, S. (1983). Evidence for β_2-microglobulin-like and H-2-like antigenic determinants in *Drosophila. J. Immunol. 130*:297–302.

Singer, S. J., and Doolittle, R. F. (1966). Antibody active sites and immunoglobulin molecules. *Science 153*:13–25.

Strominger, J. L. (1989). Developmental biology of T cell receptors. *Science 244*:943–950.

Williams, A. F. (1984). The immunoglobulin superfamily takes shape. *Nature, 308*:12–13.

Williams, A. F. (1985). Immunoglobulin-related domains for cell surface recognition. *Nature 314*:579–580.

Williams, A. F. (1987). A year in the life of the immunoglobulin superfamily. *Immunol. Today 8*:298–303.

Williams, A. F., and Gagnon, J. (1982). Neuronal cell Thy-1 glycoproteins: Homology with immunoglobulin. *Science 216*:696–703.

Williams, A. F., Tse, A..G.-D., and Gagnon, J. (1988). Squid glycoproteins with structural similarities to Thy-1 and Ly-6 antigens. *Immunogenetics 27*:265–272.

5

Evolutionary Origins
of Immunoglobulin Genes

GARY W. LITMAN, CHRIS T. AMEMIYA,

KRISTIN R. HINDS-FREY,

RONDA T. LITMAN, FUMIO KOKUBU,

DAIJIRO SUZUKI,

MICHAEL J. SHAMBLOTT,

FIONA A. HARDING, ROBERT N. HAIRE

As is apparent from the preceding discussions in this text, the immunoglobulins are unique among multigene families. In addition to representing the most complex gene system described to date, the immunoglobulin (Ig) and T cell receptor (TCR) genes undergo somatic reorganization in individual cells committed to immunologic recognition. In mammals, these rearrangements take place over extended chromosomal distances by a mechanism involving sequence-specific recombination (Tonegawa, 1983). The Ig heavy chain variable region (V_H) genes, as defined in mice and humans, are the most complex among variable (V) gene systems in terms of both the number and relative diversity of individual segmental components, and they may occupy chromosomal regions exceeding those of the light chain variable region (V_L) and TCR variable region (V_T) genes. The primary focus of this chapter will be on the evolutionary development of V genes, as they account for primary antibody function. Recently, considerable insight also has been gained into the nature of both heavy chain (C_H) and light chain (C_L) constant region gene structure in several nonmammalian species.

BACKGROUND

Evolutionary Inferences from Studies of Mammalian Ig Genes

Essentially, there are two different approaches for characterizing an evolutionary process at the DNA level: (1) Genes or gene segments can be compared within a single species, e.g. mouse α and β globin genes, or (2) comparisons can be made between homologous genes found in different species, e.g., human vs. mouse β globin genes. Comparisons can be made between species that are separated either by relatively short or by extended phylogenetic periods. Although this chapter focuses on Ig genes that are found in species that diverged early in evolution, studies in mouse and man have led to several important evolutionary conclusions:

1. The antigen binding receptors represent a closely related, extended gene family that may have originated from a single ancestral gene (Tonegawa, 1983);
2. A unique recombination signal sequence (RSS) invaded (or arose through recombination with) the ancestral gene, thereby segmenting the coding region and introducing obligatory somatic recombination (Hood et al., 1985; Sakano et al., 1979);
3. Individual V_H and V_L components, diversity (D) and heavy and light chain joining (J_H, J_L) segments as well as C_H genes (and their individual exons) expanded by gene duplication;
4. Individual Ig gene families were translocated to different chromosomes (Tonegawa, 1983);
5. The gene families encoding individual segmental elements, e.g., V_H, kappa variable region (V_κ) have undergone dramatic contraction and expansion during relatively brief evolutionary periods;
6. Unique families of flanking, repetitive DNA sequences may play a direct role in their evolution as a multigene family (Straubinger et al., 1984).

Given the unusual nature of this gene system and the complex control mechanisms that govern its specific expression, we have sought to understand its evolutionary diversification by examining antibody gene structure and function in species that occupy critical and distant phylogenetic positions. Our attention has been focused primarily on the V_H system and has addressed a series of questions:

1. Could a paucity of V_H genes explain the apparent restricted nature of the lower vertebrate immune response?
2. Did this highly complex system evolve early or late in vertebrate development?
3. What is the origin of the sequence-mediated recombination system?
4. Is imprecision an intrinsic property of this system, or is it a more recent, adaptive acquisition?
5. How stable is a multigene family whose members are not necessarily selected for during the reproductively active portion of an individual life span?
6. Is the evolution of a multigene system that undergoes somatic change equivalent to systems that do not?

A direct means for addressing questions such as these involves analysis of corresponding gene structure and organization in the modern forms of species representing major departure points in the evolutionary radiation of vertebrates. Before discussing these approaches (and findings), it is instructive to briefly consider the overall evolution of the immune system as well as its humoral mediators.

General Comments on the Evolution of Immunity

The nature of cellular and humoral immunity and physicochemical as well as biological properties of lower vertebrate immunoglobulins are well known and were reviewed recently (Litman, 1984); the statements that follow are intended to provide a broad overview of this subject.

Invertebrates, as phylogenetically distant as the metazoans, possess the capacity to reject allografts (Hildemann, 1974; Hildemann et al., 1979). In one particularly well-documented study with an echinoderm, the process was accompanied by a marked temporal acceleration of second- and third-set rejections (Karp and Hildemann, 1976). Autoreactivity and self tolerance were both documented in *Hydractinia echinata,* a colonial marine hydroid (Buss et al., 1985), and in some insects unequivocal evidence for the induction of highly specific humoral immune responses was offered (Rheins and Karp, 1982). The relationship of these macromolecules to the naturally occurring agglutinins (lectins) and lysins typically found in invertebrates and some of the early vertebrates is unclear. Notable cross-reactions between a naturally occurring agglutinin (lectin), C-reactive protein (an acute phase reactant), and the murine T-15 idiotype, a conformational determinant of a specific murine antibody to phosphorylcholine, were all described (Vasta et al., 1984).

At the cellular level, major histocompatibility complex (MHC)-like function and genetic polymorphism were described in a tunicate (protochordate) (Scofield et al., 1982), although the nature of the specific mediators is not known as yet. Ig-like molecules are seen first in the two surviving groups of jawless vertebrates, the lampreys and hagfishes. The structure of the hagfish Ig is now well characterized and appears to be an Ig-type heterodimer (Kobayashi et al., 1985; Varner et al., 1991). The structural-functional instability of both hagfish and lamprey immunoglobulins, however, is in marked contrast to the mammalian Ig-like, stable characteristics of antibodies found in the elasmobranchs, the earliest of the jawed vertebrates (Litman et al., 1982a). These antibodies possess both heavy and light chains and closely resemble mammalian Ig at the level of protein organization (Marchalonis and Edelman, 1966).

Multiple Ig classes or subclasses were described in advanced teleosts, dispelling earlier notions that class divergence did not occur until the level of the *Sarcopterygii* (Lobb and Clem, 1983a,b). Recent evidence suggests that Ig heavy chain class diversification, either through gene duplication or via alternative RNA splicing pathways, may have occurred even earlier, at the level of elasmobranchs (Kobayashi et al., 1984). While the progressive phylogenetic development of antibody and Ig complexity is well documented, until recently there has been little structural information, either at the protein or gene levels, from which to evaluate the mechanisms involved. Alternative processes and mechanisms are less easy to recognize and, in this regard, there may be considerable bias in some of the approaches and interpretations that we and others have used in characterizing the phylogenetic development of immunity.

Methods to Detect Genes in Distantly Related Species

Several different approaches can be used to identify specific gene products; among them are the following.

1. mRNA can be copied into a complementary DNA (cDNA) and cloned in a drug-resistant plasmid or bacteriophage. Specific genes can be identified by hybridization with synthetic oligodeoxynucleotides that complement a region of known protein structure, by subtractive hybridization that enriches a single species of specific messenger RNA

(mRNA), or by a differential hybridization assay that uses in vitro translation of mRNA to identify specific and related protein products.

2. Using expression vectors, cDNAs form fusion products with a vector-encoded polypeptide, and specific products can be recognized by reactivity with an antibody.

3. Gene sequences can be detected by cross-hybridization with a cDNA or genomic sequence derived from a related member of a gene family within the same species (homologous cross-hybridization) or between different species (heterologous cross-hybridization). The latter approach is invaluable when relatively little protein structural information is available (for probe construction) and when selective enrichment of mRNA cannot be achieved efficiently. Notably, both of these complications apply to most lower vertebrate systems where neither plasmacytomas (or even stable lymphoid cell lines) nor extensive amino acid sequence information is available. When a suitable probe is available, heterologous cross-hybridization can identify cDNA or genomic sequences efficiently. The latter provide information on both gene structure and organization, including the nature of intervening, flanking, and regulatory sequences. It generally is conceded that heterologous cross-hybridization can detect regions exhibiting up to 30% mismatch in Southern blotting and up to ~40% mismatch in plaque screening (Jeffreys, 1981). These estimates, primarily based on analyses of globin genes, are consistent with our observations to date for immunoglobulin genes. The recent introduction of technology based on the polymerase chain reaction permits new probing strategies that potentially are more powerful than preexisting methods (Gould et al., 1989).

Ig GENES IN NONMAMMALIAN SPECIES

Reptilian Ig V_H Genes

Caiman V_H Genes

Initially, we felt that it was advantageous (and necessary) to adopt a systematic (successive) approach that would use a gene sequence characterized at one level of evolutionary development as a probe to identify a related sequence in a species occupying a more distant level (Litman et al., 1983). A murine V_H cDNA probe, S107V (Early et al., 1980a), complementing a member of the T15 family of phosphorylcholine binding antibodies was successfully used to isolate several different reptilian Ig V_H genes from a genomic DNA bacteriophage λ library (Litman et al., 1985b). Comparison of the coding segments of reptilian *(Caiman crocodylus)* and mammalian sequences (including the probe sequence) revealed striking organizational homology, as well as 65%–70% nucleotide sequence identity (Litman et al., 1983, 1985b). In addition, the reptilian and mammalian sequences typically have been found to be 60%–65% identical at the amino acid level (Fig. 5–1).

The reptilian sequence can be organized into mammalian-like framework regions (FR) and complementarity determining regions (CDR). FR2, which contains a functionally critical tryptophan residue common to the binding site of all V regions (Ohno et al., 1985), exhibits the highest degree of nucleotide identity with the murine sequence. As would be predicted for Ig genes selected in this fashion, there is little homology in CDR segments. When FR segments of the reptilian and prototype mammalian FR segments

Figure 5–1. Alignment of three coding DNA segments and three predicted amino acid sequences for V_H genes from *Caiman* C3, mouse S107, and human H11 (Rechavi et al., 1982); sequences are organized into framework regions (FR) and complementarity determining regions (CDR), as described (Litman et al., 1983). The "common bases" lines show uppercase letters for bases common to all three DNA sequences and lowercase letters for bases common to two of three sequences. The "common acids" lines show the first letter of the three-letter amino acid abbreviation as uppercase for acids common to all three sequences (e.g., Ala) and as lowercase (e.g., ala) for acids common to two of the three predicted peptide sequences.

```
                    627         640           650           660           670           680           690           700          710      716
caiman C3 DNA:      CAGGTGCAGCTGGTGGAGTCCGGAGGAGGATGGAGGAAACTGGAAACTCTTTGCCCTCTCTGCAAAGCTCGGGGTTCACTTCGGT
                    216 220        230           240           250           260           270           280           290         300 305
murine S107 DNA:    GAGGTGAAGCTGGTGGAGTCTGGAGGAGGCTTGGTACAGCCTGGGGGTTCTCTGAGACTCTCCTGTCAACTTCTGGTTCACTTCAGT
                    1         10            20            30            40            50            60            70          80
human H11 DNA:      GAGGTGCAGCTGGTGGAGTCCGGGGGAGGCTTAGTTCAGCCTGGGGGGTCCCTGAGACTCTCCTGTGCAGCCTCTGGATTCACCTTCAGT
COMMON bases:       gAGGTgcAGCTGGTGGAGTCcGGaGGAGGcggcTggt,cAgcCTGGggg,TCtcTGaGaCTCTCCTGtcAgcCTcCGgtTCACtTCaGT

gene segment (Ig region):    <--- FR1 segment (first framework region) ----------------------->

                    1        5           10           15           20          25          30
caiman Ig C3:       GlnValGlnLeuValGluSerGlyGlyGlyValArgLysProGlyAsnLeuArgLeuSerCysLysAlaSerGlyPheThrPheGly
                    1        5           10           15           20          25          30
murine Ig S107:     GluValLysLeuValGluSerGlyGlyGlyLeuValGlnProGlyGlySerLeuArgLeuSerCysAlaThrSerGlyPheThrPheSer
                    1        5           10           15           20          25          30
human Ig H11:       GluValGlnLeuValGluSerGlyGlyGlyLeuValGlnProGlyGlySerLeuArgLeuSerCysAlaAlaSerGlyPheThrPheSer
COMMON acids:       gluValglnLeuValGluSerGlyGlyGlyleuvalglnProGlyglySerLeuArgLeuSerCysalaalaSerGlyPheThrPheSer

                    717          730
C3 DNA:             GGCTACGGCATGTTC
                    306 310       320
S107 DNA:           GATTTCTACATGGAG
                    91          100
H11 DNA:            AGCTACTGGATGCAC
COMMON:             ggctactacATG,ac

<CDR1 segment >
                    31           35
Ig C3:              GlyTyrGlyMetPhe
                    31           35
Ig S107:            AspPheTyrMetGlu
                    31           35
Ig H11:             SerTyrTrpMetHis
COMMON:             tyr    Met

                    740          750          760          770            780              790            800            810            820 824
C3 DNA:             TGGGTCCGCCAGGCTCCTGGGAAGGGGCTGGACTGG GTG GCT     ACAATTAATA     CTGATGATCCAGC CAGTGGTACTCCCCCGGCGTTCAGGGG
                    330          340          350          360            370              380            390            400            410      419
S107 DNA:           TGGGTCCGCCAGCCTCCAGGGAAGGAAGGGAGTGGATTGTTCA     AGTAGAAACAAAGC CTAATGATTATACAAC AGTACAGTCCATCTGTGAAGGGT
                    110          120          130          140            150              160            170            180            190      198
H11 DNA:            TGGGTCCGCCAGCTCCAGGGAAGGGGCTGGAGTGG GTC TCA     CGTATTAATA     GTGATGGGAGTAGCACA CGTACCCGGACTCCGTGAAGGGC
COMMON:             TGGGTCCGCCAGgcTCcaGGGAAGGgGCTGGAg TGg gTG gCa   agtAttAAtA     ctgATGa,t,tagc Cag ,GTAC,c,gc,tccGTgaAGGG,

<--- FR2 segment (second framework region) -->    <CDR2 segment (2nd complementarity-determining region)>

                    40          45          50              55           60               65 66
Ig C3:              TrpValArgGlnAlaProGlyLysGlyLeuAspTrp Val Ala    ThrIleAsnT   hrAspGlySerSer GlnTrpTyrSerProAlaValGlnGly
                    40          45          50              55           60               65
Ig S107:            TrpValArgGlnProProGlyLysArgLeuGluTrpIleAlaIle   SerArgAsnLysAlaAsnAspTyrThrThrG luTyrSerAlaSerValLysGly
                    40          45          50              55           60               65 66
Ig H11:             TrpValArgGlnAlaProGlyLysGlyLeuValTrp Val Ser    ArgIleAsnS  erAspGlySerSerThrT hrTyrAlaAspSerValLysGly
COMMON:             TrpValArgGlnalaProGlyLysglyLeu  Trp val  ala    IleAsn      aspGlyserserthr Tyrser  serValgInGly

                    825          840          850          860           870            880             890          900         910        920
C3 DNA:             AAATTCACCATCTCCAGAGGACCAACTCCCAGAAACATGCTACCTGCAAATGAGCAGCCTGAGGACACAGCCACGTATTACTGGCGCAGA
                    420          430          440          450           460            470             480          490         500        510 515
S107 DNA:           CGGTTCATCGTCTCCAGAGACACTTCCCAAAGCACCCTACCTTCAGTGAATGCCCTGAGACTGAGGACACTGCCATTTATTACTGTGCAAGA
                    199          210          220          230           240            250             260          270         280        290 294
H11 DNA:            CGATTCACCATCTCCAGAGACAACGCCAAGAACAACTGTATCTGCAAATGAACAGTCTGAGACCGAGGACACGGCGTGTATTACTGTGCAAGA
COMMON:             cgaTTCAcCaTCTCCAGaGaCaacteCCAgAaacaTgcTaCCTgcAaATGAagcaGCTgAgGaCgGaCgaGGACAC,GCcatgTATTACTGcGcAaGA

<--- FR3 segment (third framework region) ----------------------->

                    67           70           75           80             85              90           95          98
Ig C3:              LysPheThrIleSerArgGlyAsnSerGlnAsnMetLeuTyrLeuGlnMetSerSerLeuThrProGluAspThrAlaThrTyrTyrCysAlaArg
                    69 70         75           80             85              90           95          100
Ig S107:            ArgPheIleValSerArgAspThrSerGlnSerThrLeuTyrLeuGlnMetAsnAlaLeuArgAlaGluAspThrAlaIleTyrTyrCysAlaArg
                    67           70           75           80             85              90           95          98
Ig H11:             ArgPheThrIleSerArgAspAsnAlaLysAsnAsnThrLeuTyrLeuGlnMetAsnSerLeuArgAlaGluAspThrAlaValTyrTyrCysAlaArg
COMMON:             argPheThrIleSerArgaspasnserglnasn     LeuTyrLeuGlnMetasnserLeuargalaGluAspThrAla    TyrTyrCysAlaArg
```

were compared, it was apparent that some segments of reptilian genes are more related to certain corresponding segments of mammalian genes than they are to each other (Litman et al., 1983, 1985b) (Fig. 5–2). In a certain sense, these data contradict the notion of an evolutionary molecular clock; however, it must be emphasized that members of extensively diversified and collectively grouped gene families, rather than vertically related genetic pairs, are being compared. Because hybridization conditions were permissive and relatively few genes actually were studied, clock partners may exist but they are not being identified.

In addition to emphasizing a need for caution in interpreting nucleotide sequence differences between individual members of complex gene families, these studies suggest that the chances for identifying homologues in phylogenetically distant species may increase as gene families expand and undergo different types of limited diversification. Such approaches may become particularly important in investigations (in progress) with the even more distant jawless vertebrates, which lack a spleen and thymus but nevertheless mount a humoral immune response.

Unusual *Caiman* RSSs

In addition to the apparent homology in amino acid sequence between the reptilian and higher vertebrate V_H genes, their organizational homology is similar (Litman et al., 1983, 1985b). In most reptilian genes, a mammalian-like V_H RSS is located 3′ of the coding region; however, one *Caiman* gene, E1, possesses two striking differences involving this structure. A RSS is located within the intervening sequence (IVS) that divides the leader region (LR) (Fig. 5–3); the presence of near-perfect direct repeats and an inverted repeat

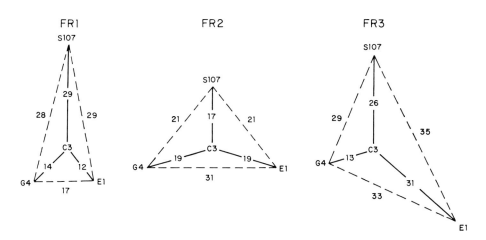

Figure 5–2. Spanning trees defining the sequence distances between framework region (FR) segments of *Caiman* genes C3, E1, and G4 (Litman et al., 1985b), and of mouse S107 (Early et al., 1980a), a murine V_H sequence. Each distance (edge length) is the number of base changes per 100 bases (% bc/b) between two segments (vertices). For FR1, *Caiman* C3 is farthest from murine S107, but for FR2, C3 is closest to S107. For FR3, C3 is closer to murine S107 than to *Caiman* E1. Only a comparison of the FR1 segments distinguishes the reptilian and mammalian genes.

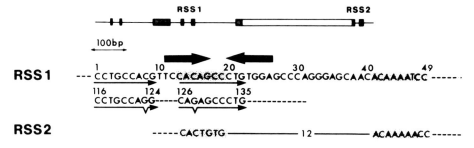

Figure 5–3.   Chromosomal organization of the *Caiman* V$_H$ gene E1. The essential features (5′ →
3′) include a hyperconserved upstream octamer and putative promoter for RNA polymerase II
transcription (solid bars), interrupted leader (striped bars), mature coding (open bars), and RSSs
(hatched bars) (RSS1–2). Scale (↔) is 100 bp. The DNA sequences of the 7-mer and 9-mer
components of RSS1–RSS2 are highlighted. The complete nucleotide sequence of a 135-bp
segment from the leader region IVS has been aligned vertically—i.e., 1–9 vs. 116–124; 13–22 vs.
126–135—to illustrate direct repeats that differ at only one position, shown by line dip. The
thicker arrows define an inverted repeat that includes the heptamer component of RSS1.

suggest that it was transposed into this site either directly or through an intermediate
product of a recombination event (Litman et al., 1985b). This finding offered the first
evidence that the RSS may be mobile, an important event in the evolution of the Ig gene
superfamily (Litman et al., 1985b). The second unique characteristic involves the 12- vs.
23-base pair (bp) length of the spacer within the 3′ RSS. This configuration, typical of
V$_\kappa$ and V$_T$ but not V$_H$, would permit a direct V$_H$ ↔ J$_H$ joining, bypassing the heavy chain
diversity (D$_H$) segmental element. In an extensively diversified V$_H$ gene family, this mechanism may provide an important exception to V-D-J combinatorial diversity (Litman et
al., 1985b).

Chelydra V$_H$ Gene Complexity

Of the reptilians examined to date, it appears as if *Chelydra serpentina,* a species that has
sustained little morphologic change over an extended period of evolutionary development, possesses the most extensive family of individual hybridizing components. When
approximately two genomic equivalents of a λ recombinant DNA library were screened
with a murine V$_H$ probe, it was possible to identify 576 hybridizing individual phage
clones (Litman et al., 1984). Although many of these may be repeat isolates or may represent overlaps, a large number were found, by restriction mapping and Southern blotting, to contain multiple hybridizing regions, thus increasing the overall number of individual genes. It is not unreasonable to assume that under the stringency conditions
employed, several hundred individual hybridizing components are related to the particular mouse probe employed. Presumably, other heterologous and, more importantly,
homologous probes would be able to define additional components. While it is tempting
to speculate that this species has sustained an enormous accumulation of V$_H$ genes, we
recognize that a large percentage of the homologous regions may represent nonfunctional
pseudogenes in various stages of inactivation or perhaps, reactivation. Recently, we
found that *Chelydra* V$_H$ genes may be present on at least four different chromosomes
(unpublished).

V_H and V_L Genes in an Avian

Both V_H and V_L genes (and IgM-type constant region [C_μ], have been characterized in an avian species. The poly(A)-containing mRNA from the spleens of immunized chickens was size enriched and used to construct a cDNA library. Differential screening (light chain–enriched vs. liver mRNA) was used to isolate light chain–containing clones (Reynaud et al., 1983). Comparison of the chicken light chain constant region with both human and mouse λ constant sequences showed it to be 61% homologous at the amino acid level. The chicken variable region sequence is more related (11%–21%) to certain human sequences than to a mouse lambda variable region ($V_\lambda 1$) sequence, similar to the results reported with *Caiman* V_H genes (Litman et al., 1983, 1985b). Recent studies of the genomic arrangement pattern of the chicken *(Gallus domesticus)* λ genes has shown the single lambda constant region gene to be linked downstream from a lambda joining (J_λ) component (Reynaud et al., 1985). A single $V_\lambda 1$ gene is linked closely to the J_λ segment. Gene conversion-like events, involving closely adjacent, nonfunctional V_λ segments (pseudogenes) account for antibody diversity in this species (Reynaud et al., 1987). A homologous probe spanning the V_H-5' $C_\mu 2$ was shown to detect major differences between Ig V_H arrangements in bursal vs. erythrocyte DNA (Weill et al., 1986). Somatic mechanisms similar to those that diversify the V_λ locus appear to account for the generation of antibody diversity at the heavy chain locus. Recent studies have shown that a single V_H gene is flanked by multiple pseudogenes that lack RSSs. This gene is linked closely to several D segments and to a single J_H, which are immediately upstream of the C_H (Reynaud et al., 1989). The apparent coevolution of these loci is of considerable biological significance.

Amphibian Ig Genes

The value of amphibian models in understanding the ontogenetic development of the B cell repertoire has prompted several laboratories, including our own, to isolate Ig genes from these species. Unlike the other cases described in this chapter, we were able to recover only a single V_H^+ genomic clone by screening several equivalents of a *Xenopus laevis* (clawed toad) genomic DNA λ bacteriophage library with the murine S107 V_H probe, and we failed to recover positive clones under low stringency conditions using other murine and reptilian V_H probes. The overall organization of the *Xenopus* gene and structure of the RSS closely resemble other vertebrate V_H genes. Reptilian and most elasmobranch V_H genes are more related than the *Xenopus* gene to mammalian prototypes. When a V_H-specific probe derived from this clone was used to screen a genomic DNA library, a large number of different V_H^+ clones were detected, suggesting that *Xenopus* possesses a complex V_H system. The recent description of a *Xenopus* germline V_H gene is entirely consistent with these observations (Yamawaki-Kataoka and Honjo, 1987). Using an expression library-antibody approach, other investigators were able to recover full and near-full copy length cDNAs from a *Xenopus laevis* spleen library (Schwager et al., 1988b). Nucleotide sequence analysis indicates the presence of a typical signal peptide as well as V_H, J_H, and C_H coding segments. Additional nucleotide sequences at the V_H-J_H boundary can be accounted for by D and/or N region-type sequence segments, similar to previous observations with the more phylogenetically distant elasmobranchs.

Multiple V_H segments have been identified in individual isolates from λ genomic libraries, implying relatively close linkage for some members of this multigene family (Schwager et al., 1988a,b). Both sequencing studies and extensive Southern blot analyses show the presence of three V_H families. Multiple (>4) J_H segments also have been detected using both blot hybridization and sequencing approaches. It appears that the J_H gene segments are located ~ 8 kilobase (kb) from the C_H (μ-type) gene; however, their linkage relationship to V_H genes is not known at present (Schwager et al., 1988a). Recently, we showed that at least ten different V_H families are present in *Xenopus laevis;* these differ more from one another than do V_H families in higher vertebrates. Furthermore, extensive D-region CDR and J_H diversity are equivalent to or exceed that observed in higher vertebrates (Haire et al., 1990). The basis for restricted antibody diversity in *Xenopus* does not appear to involve less complex rearranging antibody genes.

Teleost V_H Genes

Our laboratory has isolated V_H genes from *Elops* (ladyfish), a more basal (less derived) teleost species, using heterologous hybridization. Southern blot analyses of restriction endonuclease–digested *Elops* genomic DNA using homologous probes are consistent with an extensive V_H gene family. DNA sequence analyses confirmed the complexity of the V_H gene families and showed the presence of significant numbers of pseudogenes. Certain V_H^+-λ clones have been shown to contain 4–5 V_H genes in 17-kb insert regions. This result is consistent with a mammalian vs. elasmobranch-type linkage arrangement. The presence of only a single C_H gene, linkage of J_H-C_H and absence of V_H-J_H (or C_H) linkage also are consistent with a mammalian-type gene organization (Amemiya and Litman, 1990). Preliminary results of field inversion gel electrophoretic analyses suggest that V_H genes are contained in several, relatively large chromosomal segments. In addition, a V_H gene was recovered from a *Carassius auratus* (goldfish) genomic DNA λ library using heterologous cross-hybridization with the murine S107V probe (Wilson et al., 1988/ 1989).

Elasmobranch Ig Genes

Heterodontus V_H Genes

Our efforts focused next on *Heterodontus,* the earliest vertebrate species in which typical (non-labile) disulfide stabilized heterodimeric structures are found (Litman et al., 1982a). The immune response in the species is characterized by a significant repertoire of hapten-specific immune responses that lack interindividual variation, a characteristic of the anti-hapten response of even inbred mammals. Furthermore, with prolonged immunization, antibody neither changes fine specificity nor undergoes affinity maturation, both characteristics of the secondary immune response (Mäkelä and Litman, 1980; Litman et al., 1982b). We screened an elasmobranch genomic DNA library using the murine S107V probe and recovered a large number of V_H genes that were organized similarly to the mammalian V_H genes. Specifically, a segment that would encode a 19 amino acid LR peptide is interrupted by a 128-bp IVS, longer than the majority of mammalian-reptilian V_H IVSs. This region is in the same continuous reading frame as a 97 amino acid mature coding region and is followed by a characteristic Ig V_H RSS.

The structure of the HXIA V_H gene is compared at the amino acid and nucleotide levels to S107, the murine probe, in Figure 5–4. The coding segments (including LR, not

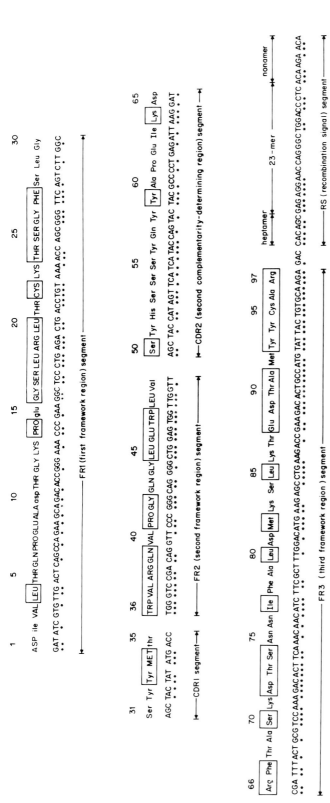

Figure 5–4. The complete nucleotide and predicted amino acid sequence of the *Heterodontus* (horned shark) V_H gene HX1A compared with mammalian (mouse) S107 and with portions of the experimentally determined amino acid sequence of pooled heavy chains. Nucleotide identities are indicated by an asterisk; amino acid identities are enclosed in open bars. This comparison eliminates the insertions and deletions needed to optimize homology between the shark sequence and the longer prototype (mouse) sequence (Litman et al., 1985a). Amino acids indicated in uppercase letters (NNN) represent positions where the determined amino acid sequence corresponds to the predicted sequence; lowercase letters (nnn) correspond to predicted amino acids that differ from determined sequences (note that the predicted amino acid sequence varies from the sequences of pooled immunoglobulins, suggesting that this is not an abundantly expressed gene); an uppercase letter followed by two lowercase letters (Nnn) represents positions that have not been determined by amino acid sequencing. The amino acid sequence of positions 1–27 has been reported (Litman et al., 1982a); positions 34–48 were determined by sequencing an interior peptide, V_H-CN5. (Litman et al., 1985a.)

shown) of the shark and mouse V_H genes exhibit 60% nucleotide identity, similar to values of 56% for shark and *Caiman,* and 63% for *Caiman* and mouse comparisons (Litman et al., 1985a). The pattern of localized nucleotide and amino acid identities between the *Heterodontus* and murine sequence(s) is consistent with the presence of FR and CDR gene segments. The six polypeptide regions encoded by these gene segments (LR, FR1, CDR1, FR2, CDR2, and FR3) exhibit an average of 40% common amino acid residues between shark and mouse, which is only slightly lower than the values of 44% between shark and *Caiman* and 50% between *Caiman* and mouse. The FR1 segment of shark shares about one-fifth fewer nucleotides with either the *Caiman* or mouse than they share with each other. The FR2 segments of shark, *Caiman,* and mouse exhibit approximately the same level of nucleotide identity. The nucleotide sequence of shark FR3 is more related to the mouse than to the *Caiman,* although the latter species occupies an intermediate phylogenetic position. Similar correlations are also seen at the amino acid level, suggesting that considerable selective pressure for maintaining Ig V_H gene structure has been exerted over an extended period of evolutionary time. In addition, the significant difference between the FR1 segments of the elasmobranch and the higher vertebrates is consistent with different rates of evolution-selection that influence the individual segments of these genes. Regardless of this FR1 deviation, it is likely that the Ig V_H genes in the majority of species occupying phylogenetic positions between sharks and man can be identified by the heterologous cross-hybridization approach, using an appropriately selected probe.

Hyperconservation of the V_H RSS

The 3' noncoding segment has a 3-nucleotide spacer segment (that potentially could encode a single amino acid residue) followed by a typical V_H RSS that contains a conserved heptamer, a variable 23-nucleotide spacer, and a conserved nonamer (Figs. 5–4 and 5–5). The RSSs of HXIA, human, and murine V_H, V_K, V_λ, and V_T (Chien et al., 1984) and reptilian V_H genes (Litman et al., 1985b), are closely related. The lack of phylogenetic variation in this segment throughout the radiations of the jawed vertebrates suggests that it may have originated at an early point in evolution and may be associated with other developmental processes that involve DNA reorganization, such as is suggested for C_H switch sequences.

A 40-nucleotide segment that spans the 3' coding sequence of HXIA and includes the heptamer of the RSS is highly conserved between the shark and mouse genes (Fig. 5–4). The 15-nucleotide segment encoding Tyr-Tyr-Cys-Ala-Arg (residues 93–97) is identical in a large number of murine and human V_H segments (Litman et al., 1985a), even though the amino acid residues encoded possess two- to six-fold codon degeneracy. This lack of silent substitutions may be related to codon utilization frequencies, involvement of a specific gene segment in mRNA secondary structure(s), or other selective mechanisms. For example, the five-nucleotide segment that encodes the conformationally critical second cysteine codon (C-T-G-T-G) is complemented inversely by the first five nucleotides of the recombination signal heptamer segment (C-A-C-A-G) and, based on preliminary calculations, could form a thermodynamically stable hairpin loop. Although the signal segment undergoes recombination and is excised before RNA transcription, the 40-nucleotide segment may be stabilized against genetic change by involvement in a transient secondary structure or interaction with an enzyme(s) associated with the recom-

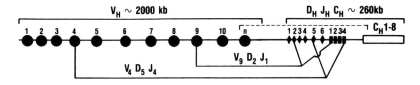

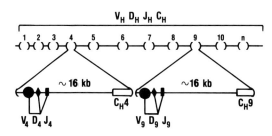

Figure 5–5. **(A)** Schematic depiction (number and position of segments not based on absolute physical map data) of the mammalian (mouse) V_H locus illustrating the linear relationships of V_H, D_H, J_H, and C_H components. In the mouse and presumably other mammals, ~ 100–1,000 V_H segments may occupy $\sim 2,000$ kb; the D_H, J_H, and C_H segments occupy ~ 260 kb (Wood and Tonegawa, 1983), and the physical distance between the two regions may be as little as 100 kb (human) (Berman et al., 1988). In both the mouse and human, DQ52 is located within the J_H cluster (not illustrated). Two combinatorial joinings are illustrated, each involving $>2,000$ kb of DNA. **(B)** In the elasmobranch (shark), V_H, D_H, and J_H segments occur within ~ 1.3 kb, and with the complete C_H segment they occupy ~ 15 kb. Recently a second D segment was detected between V_H and J_H (Kokubu et al., 1988b). Both 5' and 3' recombination elements contain 12-bp spacers. Thus the 12/23 joining rule would be obeyed because the 5' D segment contains 12- and 22-bp spacers (Hinds and Litman, 1986) and both V_H and J_H contain 23-bp spacers. The two joinings are shown as being noncombinatorial, which we suggest may be a consequence of this specific type of gene arrangement, as described.

bination process (Litman et al., 1985a). This region has been shown to be the site of a unique type of a secondary replacement gene rearrangement (Kleinfield et al., 1986; Reth et al., 1986).

V_H Gene Complexity

Resolution of the complete nucleotide sequence of the HXIA gene permitted the synthesis of a CDR2-specific oligonucleotide probe which established that HXIA (or a closely related gene) is selectively expressed in the spleen (Litman et al., 1985a). In addition, it was possible to identify and subclone a sequence segment that includes the complete coding and only a small segment of the leader-IVS regions. This probe, HXIA V_H, could be used in Southern blot analyses of restriction endonuclease–digested *Heterodontus* DNA, as the hybridization analyses (and interpretations) would not be confounded by non-immunoglobulin, e.g., repeat sequence, homologies arising in flanking regions. An autoradiographic exposure of a genomic Southern blot–hybridization analysis using this probe indicates that a number of different V_H-related sequences are present, similar to previous reports for higher vertebrates including reptilian (Litman et al., 1984) and avian (Weill et al., 1986) DNAs. Although allelic differences and homologous pseudogenes could account for some of the complexity, it is likely that this phylogenetically early vertebrate possesses an extensive V_H multigene family. Furthermore, the persistence of a number of components after high-stringency washing suggests that the HXIA gene may belong to a 6–7 member subset of closely related genes, which presumably could be identified and selectively recovered from a genomic library using higher stringency hybridization-wash procedures than we employ normally. A number of other elasmobranchs also exhibit complex hybridization patterns with this V_H probe (Litman, unpublished observations).

When the HXIA probe was used to screen a *Heterodontus* genomic library (1.5 genome equivalents) under moderately stringent conditions, 366 positive clones were recovered (Kokubu et al., 1988b). Experiments presently in progress are examining the cross-hybridization characteristics of a large number (>100) of these different V_H genes with different mouse V_H subgroup-specific probes. It appears, thus far, that S107V hybridizes most strongly to the greatest number of isolates, although several genes hybridize preferentially with probes representing other families, e.g. V_H 7183. Perhaps the T15-like genes possess a "basic" structure that can be altered functionally through minimal variation, as suggested by a recent report describing a single amino acid substitution that changed the antigen binding specificity of S107 from phosphocholine to DNA (Giusti et al., 1987).

Other evidence for functional gene complexity has been obtained through sequencing large numbers of different genomic isolates as well as cDNAs that were recovered by screening a spleen cDNA library with the HXIA probe (Litman and Hinds, 1987). Nucleotide sequence differences are localized primarily in CDR1 and CDR2, although FR segments also show differences at multiple positions. While there is no evidence that any of the isolates represent productively translated mRNAs, it is apparent that extensive variability is concentrated primarily in CDR segments. Taken together, these results suggest that the multigene nature, subset distributions, and nucleotide substitution patterns of V_H genes in phylogenetically primitive vertebrates closely resemble those seen in higher vertebrates.

Segmental Joining and Chromosomal Organization of *Heterodontus* V_H Genes

As indicated above, the detection of a typical immunoglobulin RSS immediately 3′ to the coding region of the HXIA germline gene suggests that elasmobranch V_H genes are capable of undergoing segmental joining. In order to characterize the joining mechanism and to determine the relative complexity of the individual gene segments involved, a *Heterodontus* spleen cDNA library was screened with the HXIA V_H probe. A large number of positively hybridizing clones were detected, and one of these was shown to contain V_H, J_H, and a μ-like C_H region and to possess a "D" (or "N")-like segment (Hinds and Litman, 1986). Specific probes then were used to isolate these segments in their germline configuration. V_H, D_H, and J_H segments were shown to possess mammalian-like 3′ and/or 5′ RSSs; however, individual V_H, D_H, J_H, and C_H segments were found to be closely linked (~ 10 kb, linkage distance from V_H-C_H1) and arrayed in multiple clusters (Fig. 5–5). The linkage relationships between clusters are not known, although preliminary results of field inversion electrophoretic analyses suggest that most genes are flanked by identical restriction sites—that is, there are few distinguishable loci. If recombination does not occur between clusters, combinatorial diversity may be limited, which perhaps explains the restricted nature of the *Heterodontus* immune response alluded to above. Although combinatorial diversity may not be as important in generating antibody diversity, gene segmentation may provide some degree of junctional diversity that affects the combining site directly or through network-type regulation by anti-idiotype antibody. The phylogenetic distribution of this unusual linkage pattern extends to include at least one other elasmobranch *(Raja erinacea)* (Harding et al., 1990a). The close linkage between DQ52 and J_H in both humans and mice is reminiscent of this relationship (Ravetch et al., 1981; Sakano et al., 1981) as is the close linkage of segmental elements in certain TCR gene loci.

It appears that the organization of the individual germline components that comprise the V_H gene, rather than their sequences, has changed during evolution, raising new questions: (1) Did a common ancestral form possess a single segmented V_H gene cluster that underwent unit duplication (as in the shark) or segmental duplication (as in mammals)? (2) Is this linkage pattern unique to *Heterodontus,* or is it found in other ectothermic vertebrates? (3) If the organization is not unique to this species, how did the transition(s) from the multicluster arrangement to the contemporary (mammalian) pattern take place? Genetic regulation of the clusters is probably novel because current models for selective expression and allelic exclusion of V_H genes assume the involvement of a single, extended linkage group. The possible role(s) of enhancer sequences and their relationships to 5′ promoter components is intriguing as these functions are associated with the mammalian-type gene organization. In this regard it is interesting to note that *Heterodontus* V_H genes lack the ubiquitous octamer (ATGCAAAT) component of the promoter (Falkner and Zachau, 1984; Parslow et al., 1984) that is found upstream of all higher vertebrate V_H genes thus far examined (Kokubu et al., 1988b; Litman et al., 1985b).

Variation in Elasmobranch Gene Organization

In the genomic configuration (gonad DNA libraries), approximately 50% of the *Heterodontus* genes are arranged as V_H-D_{H1}-D_{H2}-J_H-C_H. The majority of other genes appear to be $V_H D_H$-J_H or $V_H D_H J_H$-joined (Kokubu et al., 1988b). The cluster-type arrangement patterns are not found in species above the phylogenetic level of the cartilaginous fish (Amemiya and Litman, 1990), and they resemble TCRs in terms of the close linkage of rearranging segments, the presence of two D segments in a single cluster, and, as indicated above, the absence of regulatory octamer (Kokubu et al., 1988b). The VD-and VDJ-joined genes could reflect the structure(s) of primordial genes before integration of the complete recombination mechanism(s), or they may represent a unique type of germline joining that may be facilitated by the very close proximity (~300–350 nucleotides) of segmental elements. Because identical genes have been recovered from both liver and gonad DNA of unrelated animals, it is unlikely that the VD- and VDJ-joined genes derive from typical somatic rearrangement. Germline-joined gene segments have been detected in *Raja,* another elasmobranch (Harding et al., 1990). Joined genes may provide predetermined specificities, i.e., not rely on chance somatic rearrangement, or they may be selectively used during different developmental stages. Alternatively, they may be inactive and used in gene conversion or secondary recombination.

The relative order of gene segments within a mammalian-type, single extended cluster may be a significant factor in selective gene utilization in higher vertebrates (Yanco-poulos and Alt, 1986; Yancopoulos et al., 1984), whereas entirely different schemes may be employed in lower vertebrate species. Variations in the arrangement and possible join-ing of segments within individual clusters may be a factor in the regulation and selective expression of the *Heterodontus* Ig genes. In addition, relationships between V_H and V_L gene organization (and expression) may be of considerable regulatory importance.

Several additional forms of gene organization have been detected. A single example exists of a V_H-D_1-D_2-J_H-C- type gene in which the RSSs flanking the D_1 segment are 12/12 vs. 12/22 for the other genes. In the atypical gene, adherence to the 12/23 joining rule is accomplished by a 22/12 RSS spacer arrangement flanking the D_2 segment. In this gene, both D_1 and D_2 can recombine or D_1 can recombine with V_H and J_H independently. It appears that the gene arose by chromosomal inversion of the sequence segment between D_1 and D_2. We have suggested that such inversions may have been major factors in the evolution of Ig gene loci (Kokubu et al., 1988b). A few examples of other gene arrange-ments, including the linkage of multiple V_H pseudogenes (and apparent absence of D_H and J_H segments) in ~ 15 kb and the linkage of V_H and J_H (D_H is absent) segments, were recently detected (Hinds-Frey and Litman, unpublished observation).

Ig Constant Regions and Constant Region Gene Heterogeneity

Avian immunoglobulin constant regions of the IgM (secretory) type have been charac-terized (Dahan et al., 1983). Comparisons of the sequences of individual domains to mammalian immunoglobulins reveal varying gradients of increasing $5' \rightarrow 3'$ sequence homology, which are similar to those observed in amino acid sequence comparisons of different mammalian IgM heavy chains (Kehry et al., 1979). A second class of avian Ig heavy chain gene was also characterized at the DNA sequence level (Parvari et al., 1988). This gene consists of four exons, which distinguishes it from the IgG's that contain three constant-region exons. Overall, this gene is most related to mammalian C sequences; one interpretation of these findings is that the avian gene was ancestral to both mammalian IgG and IgE.

The IgM-type gene of *Xenopus* was fully characterized (Schwager et al., 1988b). *Xen-opus* C_H (μ-type) domains vary in the extent to which they resemble the domains of other immunoglobulins and members of the Ig gene superfamily, suggesting independent evo-lutionary histories for these domains. Our laboratory also identified non-C_μ clones from a spleen cDNA library encoding two other C_H isotype cDNAs—IgX and IgY. The nucle-otide and amino acid homologies of IgX to *Xenopus* IgM are $\sim 30\%$ overall (Haire et al., 1989). Significant regions of similarity are present in C_H4 and in the $3'$ secretory region. IgX is more homologous to *Heterodontus* IgM than to *Xenopus* IgM. The nucleotide sequence of IgY also has been resolved, and it differs markedly from the sequence of IgX and IgM (Haire et al., 1989).

The nucleotide structure of a portion of a teleost Ig heavy chain gene was inferred from the sequences of partial copy length cDNAs (Ghaffari and Lobb, 1989). One of the six clones that were sequenced differs at a single position in C_H3, suggesting that multiple isotypes may be present or that a somatic mutation occurred, accounting for the loss of a glycosylation site. The unique arrangement of cysteines found in the heavy chain may account for the unusual organization of the Ig tetramer in this species. Our laboratory resolved the complete nucleotide sequence of a teleost *(Elops)* Ig constant region (Ame-miya and Litman, 1990).

Heterodontus immunoglobulin C_H originally was identified by selecting $V_H J_H^+$ clones from a spleen cDNA library (Hinds and Litman, 1986). Using moderate stringency hybridization conditions and a homologous C_H-specific probe, 186 clones were identified by screening 1.5 equivalents of a genomic DNA λ library. Both restriction mapping and DNA sequencing showed that ~80% of these are unique (Kokubu et al., 1987). The complete nucleotide sequence through the transmembrane exons of one of these has been determined (Kokubu et al., 1988a), and its organization was compared to the corresponding genomic sequence of mouse IgM (Fig. 5-6; unpublished observations). Although exon length is equivalent to the higher vertebrate counterpart, the *Heterodontus* introns are appreciably longer. Two polyadenylation signal sequences are located more proximal to the C_H4-secretory exon than is the single polyadenylation signal sequence in mouse; conversely, the transmembrane polyadenylation signal is more distal in *Heterodontus.* These *cis* relationships may be important in the differential expression of secretory vs. membrane Ig in this species. An extremely high degree of nucleotide sequence identity exists between the transmembrane sequences in *Heterodontus* and higher vertebrate immunoglobulins. Probes derived from these sequence segments may prove useful in identifying Ig genes in more phylogenetically distant species.

The DNA sequences of the first C_H exons of several different C_H genes have been compared, and differences primarily are localized to 5′ positions. Extended regions of nucleotide and predicted amino acid sequence identity are apparent at 3′ positions. The five sequences lack frameshift or premature termination codons, and at least three of the five genes (or closely related genes) were shown to be transcriptionally active using Northern blot analyses with gene-specific oligodeoxynucleotide probes. Quantitative gene titration data are consistent with ~100–200 different C_H genes, as suggested by the library screening analyses. Based on oligodeoxynucleotide probe hybridization and restriction fragment length analyses, it was shown that ~90% of these genes belong to two major C_H families (Kokubu et al., 1987). In this representative early vertebrate, it appears that (V_H-D_H-J_H) segments are associated with individual C_H genes that differ at the predicted protein level. This variation, which occurs in all exons (Kokubu et al., 1988a), could be of considerable functional significance, and it is possible that specific V regions may associate

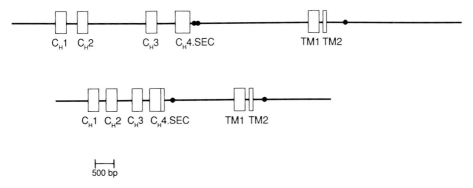

Figure 5–6.   Genomic organization of (**A**) *Heterodontus* (horned shark) and (**B**) *Mus* (mouse) IgM-type immunoglobulin heavy chains (Early et al., 1980b; Kawakami et al., 1980). Constant region exons C_H1–C_H4, the secretory (SEC) portion of C_H4, and the transmembrane (TM) exons are indicated. The polyadenylation signal sequences are shown by dots and scale is designated below. (Kokubu et al., 1988a.)

selectively with C_H regions that possess unique biological functions or even that certain substitutions in C_H regions influence antigen combining site function directly.

Light Chain Genes

Heterodontus Ig light chain genes have been identified by antibody screening of a spleen cDNA library. The structural organization and predicted amino acid sequences of the light chain constant region are most related to lambda light chains of higher vertebrates (Shamblott and Litman, 1989a). At the genomic level, V_L, J_L, and C_L segments are closely linked in multiple ~ 2.7-kb clusters, similar to the heavy chain genes (Shamblott and Litman, 1989b). V_L and J_L segments are separated by less than 300 nucleotides. It appears that in this species, V_L and V_H gene loci may be co-evolving. When the entire nucleotide sequence database is searched, the sequences of light chain genes are shown to be most homologous, overall, to mammalian β T cell receptors (Shamblott and Litman, 1989a). The extensive regions of shared nucleotide sequence identity may define essential features of the putative common ancestor of the contemporary TCR and Ig gene families.

SUMMARY

From this discussion, it is apparent that an Ig gene system resembling that seen in higher vertebrates arose early in evolution. When and how it diverged (or emerged) from the TCR system is unclear; however, there is little question that these gene families are closely related in structural and functional terms. Even at distant levels of evolutionary development, the V_H gene family is complex and appears to undergo a form of somatic rearrangement reminiscent of that seen in higher vertebrates. It is difficult to reconcile the genetic complexity of lower vertebrate Ig with the apparent restrictions in antibody diversity noted in these species. A wide range of somatic effects, unusual cell population kinetics, and different cell–cell interactions, nevertheless, could explain most biological observations.

The Ig gene systems of two lower vertebrate species (*Heterodontus* and *Gallus*) are not merely mouse-human equivalents. The V_H gene family in *Heterodontus* reflects one of the most extreme organizational changes that has occurred in any genetic system characterized to date. The multiple, close linkage "cluster" pattern necessitates the evolutionary maintenance and regulation of a series of individual segmental elements. A major biological issue arising from these studies, which suggests the presence of multiple C_H genes, involves the possiblity that they encode different biological functions. The recent finding of VD- and VDJ-joined germline genes is without precedent in any other immune system; the functional consequences of this arrangement pattern are not understood. How this extended gene family is coordinately regulated is of considerable interest, and it is expected that these studies, as well as parallel investigations now being initiated in species representing more distant periods of vertebrate evolutionary development, will continue to provide unique insight into the genetic basis for specific immunologic recognition.

Acknowledgment

The editorial assistance of Barbara Pryor is appreciated. This work was supported by grants from the National Institutes of Health, AI-23338 and GM-38656. C. T. Amemiya is a Sloan Foundation fellow; F. A. Harding is supported by an N.I.H. training grant, 5-26882, to the University of Rochester.

References

Amemiya, C. T., Haire, R. N., and Litman, G. W. (1989). Nucleotide sequence of a cDNA encoding a third distinct *Xenopus* immunoglobulin heavy chain isotype. *Nucleic Acids Res. 17*:5388.

Amemiya, C. T., and Litman, G. W. (1990). The complete nucleotide sequence of an immunoglobulin heavy chain gene and analysis of immunoglobulin gene organization in a primitive teleost species. *Proc. Natl. Acad. Sci. USA 87*:811–815.

Berman, J. E., Mellis, S. J., Pollock, R., Smith, C. L., Suh, H., Heinke, B., Kowal, C., Surti, U., Chess, L., Cantor, C. R., and Alt, F. W. (1988). Content and organization of the human Ig V_H locus: Definition of three new V_H families and linkage to the Ig C_H locus. *EMBO J. 7*:727–738.

Buss, L. W., Moore, J. L., and Green, D. R. (1985). Autoreactivity and self-tolerance in an invertebrate. *Nature 313*:400–402.

Chien, Y.-h., Gascoigne, N.R.J., Kavaler, J., Lee, N. E., and Davis, M. M. (1984). Somatic recombination in a murine T-cell receptor gene. *Nature 309*:322–326.

Dahan, A., Reynaud, C. A., and Weill, J. C. (1983). Nucleotide sequence of the constant region of a chicken μ heavy chain immunoglobulin mRNA. *Nucleic Acids Res. 11*:5381–5389.

Early, P., Huang, H., Davis, M., Calame, K., and Hood, L. (1980a). An immunoglobulin heavy chain variable region gene is generated from three segments of DNA: V_H, D and J_H. *Cell 19*:981–992.

Early, P., Rogers, J., Davis, M., Calame, K., Bond, M., Wall, R., and Hood, L. (1980b). Two mRNAs can be produced from a single immunoglobulin μ gene by alternative RNA processing pathways. *Cell 20*:313–319.

Falkner, F. G., and Zachau, H. G. (1984). Correct transcription of an immunoglobulin κ gene requires an upstream fragment containing conserved sequence elements. *Nature 310*:71–74.

Ghaffari, S. H., and Lobb, C. J. (1989). Cloning and sequence analysis of channel catfish heavy chain cDNA indicate phylogenetic diversity within the IgM immunoglobulin family. *J. Immunol. 142*:1356–1365.

Giusti, A., Chien, N., Zack, D., Shin, S.-U., and Scharff, M. (1987). Somatic diversification of S107 from an antiphosphocholine to an anti-DNA autoantibody is due to a single base change in its heavy chain variable region. *Proc. Natl. Acad. Sci. USA 84*:2926–2930.

Gould, S. J., Subramani, S., and Scheffler, I. E. (1989). Use of the DNA polymerase chain reaction for homology probing: Isolation of partial cDNA or genomic clones encoding the iron-sulfur protein of succcinate dehydrogenase from several species. *Proc. Natl. Acad. Sci. USA 86*:1934–1938.

Haire, R. N., Amemiya, C. T., Suzuki, D., and Litman, G. W. (1990). A high degree of immunoglobulin V_H gene complexity in a lower vertebrate: *Xenopus laevis. J. Exp. Med. 171*:1721–1737.

Haire, R. N., Ohta, Y., Litman, R. T., Amemiya, C. T., and Litman, G. W. (1991). The genomic organization of immunoglobulin V_H genes in *Xenopus laevis* shows evidence for interspersion of families. *Nucleic Acids Res. 19*:3061–3066.

Haire, R., Shamblott, M. J., Amemiya, C. T., and Litman, G. W. (1989). A second *Xenopus* immunoglobulin heavy chain constant region isotype gene. *Nucleic Acids Res. 17*:1776.

Harding, F. A., Cohen, N., and Litman, G. W. (1990). Immunoglobulin heavy chain gene organization and complexity in the skate, *Raja erinacea. Nucleic Acids Res. 18*:1015–1020.

Hildemann, W. H. (1974). Some new concepts in immunological phylogeny. *Nature 250*:116–120.

Hildemann, W. H., Johnson, I. S., and Jokiel, P. L. (1979). Immunocompetence in the lowest metazoan phylum: Transplantation immunity in sponges. *Science 204*:420–422.

Hinds, K. R., and Litman, G. W. (1986). Major reorganization of immunoglobulin V_H segmental elements during vertebrate evolution. *Nature 320*:546–549.

Hood, L., Kronenberg, M., and Hunkapiller, T. (1985). T cell antigen receptors and the immuno-globulin supergene family. *Cell 40*:225–229.

Jeffreys, A. J. (1981). Recent studies of gene evolution using recombinant DNA. In *Genetic Engineering,* vol. 2. (Williamson, R., ed.). London: Academic Press, pp. 1–48.

Karp, R. D., and Hildemann, W. H. (1976). Specific allograft reactivity in the sea star *Dermasterias imbricata. Transplantation 22*:434–439.

Kawakami, T., Takahashi, N., and Honjo, T. (1980). Complete nucleotide sequence of mouse immunoglobulin μ gene and comparison with other immunoglobulin heavy chain genes. *Nucleic Acids Res. 8*:3933–3945.

Kehry, M., Sibley, C., Fuhrman, J., Schilling, J., and Hood, L. E. (1979). Amino acid sequence of a mouse immunoglobulin μ chain. *Proc. Natl. Acad. Sci. USA 76*:2932–2936.

Kleinfield, R., Hardy, R., Tarlinton, D., Dangl, J., Herzenberg, L., and Weigert, M. (1986). Recombination between an expressed immunoglobulin heavy-chain gene and a germline variable gene segment in a Ly-1$^+$ B-cell lymphoma. *Nature 322*:843–846.

Kobayashi, K., Tomonaga, S., and Hagiwara, K. (1985). Isolation and characterization of immunoglobulin of hagfish, *Eptatretus burgeri,* a primitive vertebrate. *Mol. Immunol. 22*:1091–1097.

Kobayashi, K., Tomonaga, S., and Kajii, T. (1984). A second class of immunoglobulin other than IgM present in the serum of a cartilaginous fish, the skate, *Raja kenojei:* Isolation and characterization. *Mol. Immunol. 21*:397–404.

Kokubu, F., Hinds, K., Litman, R., Shamblott, M. J., and Litman, G. W. (1987). Extensive families of constant region genes in a phylogenetically primitive vertebrate indicate an additional level of immunoglobulin complexity. *Proc. Natl. Acad. Sci. USA 84*:5868–5872.

Kokubu, F., Hinds, K., Litman, R., Shamblott, M. J., and Litman, G. W. (1988a). Complete structure and organization of immunoglobulin heavy chain constant region genes in a phylogenetically primitive vertebrate. *EMBO J. 7*:1979–1988.

Kokubu, F., Litman, R., Shamblott, M. J., Hinds, K., and Litman, G. W. (1988b). Diverse organization of immunoglobulin V_H gene loci in a primitive vertebrate. *EMBO J. 7*:3413–3422.

Litman, G. W. (1984). Phylogeny of immunoglobulins. In *Molecular Immunology* (Atassi, M. Z., Van Oss, C. J., and Absolom, D. R., eds.). New York: Marcel Dekker, pp. 215–230.

Litman, G. W., Berger, L., Murphy, K., Litman, R., Hinds, K., and Erickson, B. W. (1985a). Immunoglobulin V_H gene structure and diversity in *Heterodontus,* a phylogenetically primitive shark. *Proc. Natl. Acad. Sci. USA 82*:2082–2086.

Litman, G. W., Berger, L., Murphy, K., Litman, R., Hinds, K., Jahn, C. L., and Erickson, B. W. (1983). Complete nucleotide sequence of an immunoglobulin V_H gene homologue from *Caiman,* a phylogenetically ancient reptile. *Nature 303*:349–352.

Litman, G. W., Chisholm, R., and Erickson, B. W. (1984). Structure and organization of immunoglobulin genes in phylogenetically diverse species: Studies at the DNA level. *Dev. Comp. Immunol. 3*:131–138.

Litman, G. W., and Erickson, B. W. (1985). A molecular genetic approach to understanding the evolution of immunoglobulin gene structure and diversity. *Am. Zool. 25*:713–726.

Litman, G. W., Erickson, B. W., Lederman, L., and Mäkelä, O. (1982a). Antibody response in *Heterodontus. Mol. Cell. Biochem. 45*:49–57.

Litman, G. W., and Hinds, K. R. (1987). The Ig V_H system of *Heterodontus,* a phylogenetically primitive elasmobranch. In *Evolution and Vertebrate Immunity: The Antigen-Receptor and MHC Gene Families* (Kelsoe, G., and Schulze, D., eds.). Austin: University of Texas Press, pp. 35–51.

Litman, G. W., Murphy, K., Berger, L., Litman, R. T., Hinds, K. R., and Erickson, B. W. (1985b). Complete nucleotide sequences of three V_H genes in *Caiman,* a phylogenetically ancient reptile: Evolutionary diversification in coding segments and variation in the structure and organization of recombination elements. *Proc. Natl. Acad. Sci. USA 82*:844–848.

Litman, G. W., Stolen, J., Sarvas, H. O., and Mäkelä, O. (1982b). The range and fine specificity of the anti-hapten immune response: Phylogenetic studies. *J. Immunogenet. 9*:465–474.

Lobb, C. J., and Clem, L. W. (1983a). Distinctive subpopulations of catfish serum antibody and immunoglobulin. *Mol. Immunol.* 20:811–818.

Lobb, C. J., and Clem, L. W. (1983b). Phylogeny of immunoglobulin structure and function: XXI. Secretory immunoglobulins in the bile of the marine teleost *Archosargus probatocephalus. Mol. Immunol. 18*:615–619.

Mäkelä, O., and Litman, G. W. (1980). Lack of heterogeneity in anti-hapten antibodies of a phylogenetically primitive shark. *Nature 287*:639–640.

Marchalonis, J., and Edelman, G. M. (1966). Polypeptide chains of immunoglobulins from the smooth dogfish *(Mustelus canis). Science 154*:1567–1568.

Ohno, S., Mori, N., and Matsunaga, T. (1985). Antigen-binding specificities of antibodies are primarily determined by seven residues of V_H. *Proc. Natl. Acad. Sci. USA 82*:2945–2949.

Parslow, T. G., Blair, D. L., Murphy, W. J., and Granner, D. K. (1984). Structure of the 5′ ends of immunoglobulin genes: A novel conserved sequence. *Proc. Natl. Acad. Sci. USA 81*:2650–2654.

Parvari, R., Avivi, A., Lentner, F., Ziv, E., Tel-Or, S., Burstein, Y., and Schechter, I. (1988). Chicken immunoglobulin γ-heavy chains: Limited V_H gene repertoire, combinatorial diversification by D gene segments and evolution of the heavy chain locus. *EMBO J. 7*:739–744.

Ravetch, J. V., Siebenlist, U., Korsmeyer, S., Waldmann, T., and Leder, P. (1981). Structure of the human immunoglobulin μ locus: Characterization of embryonic and rearranged J and D genes. *Cell 27*:583–591.

Rechavi, G., Bienz, B., Ram, D., Ben-Neriah, Y., Cohen, J. B., Zakut, R., and Givol, D. (1982). Organization and evolution of immunoglobulin V_H gene subgroups. *Proc. Natl. Acad. Sci. USA 79*:4405–4409.

Reth, M., Gehrmann, P., Petrac, E., and Wiese, P. (1986). A novel V_H to $V_H DJ_H$ joining mechanism in heavy-chain negative (null) pre-B cells results in heavy-chain production. *Nature 322*:840–842.

Reynaud, C.-A., Anquez, V., Dahan, A., and Weill, J.-C. (1985). A single rearrangement event generates most of the chicken immunoglobulin light chain diversity. *Cell 40*:283–291.

Reynaud, C. A., Anquez, V., Grimal, H., and Weill, J.-C. (1987). A hyperconservation mechanism generates the chicken light chain preimmune repertoire. *Cell 48*:379–388.

Reynaud, C-A., Dahan, A., Anquez, V., and Weill, J-C. (1989). Somatic hyperconversion diversifies the single V_H gene of the chicken with a high incidence in the D region. Cell 59:171–183.

Reynaud, C.-A., Dahan, A., and Weill, J.-C. (1983). Complete sequence of a chicken λ light chain immunoglobulin derived from the nucleotide sequence of its mRNA. *Proc. Natl. Acad. Sci. USA 80*:4099–4103.

Rheins, L. A., and Karp, R. D. (1982). An inducible humoral factor in the American cockroach *(Periplaneta americana):* Precipitin activity that is sensitive to a proteolytic enzyme. *J. Invert. Pathol. 40*:190–196.

Sakano, H., Huppi, K., Heinrich, G., and Tonegawa, S. (1979). Sequences at the somatic recombination sites of immunoglobulin light-chain genes. *Nature 280*:288–294.

Sakano, H., Kurosawa, Y., Weigert, M., and Tonegawa, S. (1981). Identification and nucleotide sequence of a diversity DNA segment (D) of immunoglobulin heavy-chain genes. *Nature 290*:562–565.

Schwager, J., Grossberger, D., and Du Pasquier, L. (1988a). Organization and rearrangement of immunoglobulin M genes in the amphibian *Xenopus. EMBO J. 7*:2409–2415.

Schwager, J., Mikoryak, C. A., Steiner, L. A. (1988b). Amino acid sequence of heavy chain from *Xenopus laevis* IgM deduced from cDNA sequence: Implications for evolution of immunoglobulin domains. *Proc. Natl. Acad. Sci. USA 85*:2245–2249.

Scofield, V. L., Schlumpberger, J. M., West, L. A., and Weissman, I. L. (1982). Protochordate allorecognition is controlled by a MHC-like gene system. *Nature 295*:499–502.

Shamblott, M. J., and Litman, G. W. (1989a). Complete nucleotide sequence of primitive vertebrate immunoglobulin light chain genes. *Proc. Natl. Acad. Sci. USA 86*:4684–4688.

Shamblott, M. J., and Litman, G. W. (1989b). Genomic organization and sequences of immunoglobulin light chain genes in a primitive vertebrate suggest coevolution of immunoglobulin gene organization. *EMBO J.* 8:3733–3739.

Straubinger, B., Pech, M., Muhlebach, K., Jaenichen, H.-R., Bauer, H.-G., and Zachau, H. G. (1984). Molecular footprints of human immunoglobulin gene evolution: A new sequence family. *Nucleic Acids Res.* 12:5265–5275.

Tonegawa, S. (1983). Somatic generation of antibody diversity. *Nature 302*:575–581.

Varner, J., Neame, P., and Litman, G. W. (1991). A serum heterodimer from hagfish *(Eptatretus stoutii)* exhibits structural similarity and partial sequence identity with immunoglobulin. *Proc. Natl. Acad. Sci. USA 88*:1746–1750.

Vasta, G. R., Marchalonis, J. J., and Kohler, H. (1984). Invertebrate recognition protein cross-reacts with an immunoglobulin idiotype. *J. Exp. Med. 159*:1270–1276.

Weill, J. C., Reynaud, C. A., Lassila, O., and Pink, J.R.L. (1986). Rearrangement of chicken immunoglobulin genes is not an ongoing process in the embryonic bursa of Fabricius. *Proc. Natl. Acad. Sci. USA 83*:3336–3340.

Wilson, M. R., Middleton, D., and Warr, G. W. (1988/1989). Immunoglobulin heavy chain variable region gene evolution: Structure and family relationships of two genes and a pseudogene in a teleost fish. *Proc. Natl. Acad. Sci. USA 85*:1566–1570. *Proc. Natl. Acad. Sci. USA 86*:3276 (correction).

Wood, C., and Tonegawa, S. (1983). Diversity and joining segments of mouse immunoglobulin heavy chain genes are closely linked and in the same orientation: Implications for the joining mechanism. *Proc. Natl. Acad. Sci. USA 80*:3030–3034.

Yamawaki-Kataoka, Y., and Honjo, T. (1987). Nucleotide sequences of variable region segments of the immunoglobulin heavy-chain of *Xenopus laevis. Nucleic Acids Res. 15*:5888.

Yancopoulos, G. D., and Alt, F. W. (1986). Regulation of the assembly and expression of variable-region genes. *Annu. Rev. Immunol. 4*:339–368.

Yancopoulos, G. D., Desiderio, S. V., Paskind, M., Kearney, J. F., Baltimore, D., and Alt, F. W. (1984). Preferential utilization of the most J_H-proximal V_Y segments in pre-B-cell lines. *Nature 311*:727–733.

6

Organization and Assembly
of Immunoglobulin Genes

STEPHEN V. DESIDERIO

BACKGROUND AND HISTORICAL PERSPECTIVES

Vertebrates are able to mount specific antibody responses to an enormous variety of antigens, including those that have never been encountered before. In this sense the humoral immune response is said to be adaptive. When an individual has already encountered a particular antigen, a second exposure often elicits an antibody response that is more rapid, and of higher titer, than the original response. In this respect the humoral response is said to be anamnestic—that is, to possess a memory.

The adaptive and anamnestic features of the humoral immune response were elegantly explained at the cellular level by the clonal selection theory of Macfarlane Burnet (1958). According to the theory, an organism gives rise to a large number of B lymphocytes, each bearing immunoglobulin (Ig) of a single specificity. When an antigen is encountered, it binds to Ig on the surfaces of only those B lymphocytes that express appropriate specificities. The B cells that bind antigen are induced to proliferate and to differentiate into cells that secrete Ig of the same or similar specificity as the parental cell. In the years since the clonal selection theory was first proposed, a large body of experimental evidence has accrued in its support.

The clonal selection theory requires (1) that diversification of the antibody population precede the initial exposure to antigen, (2) that a given B lymphocyte express Ig of one and only one antigenic specificity, and (3) that the antibody bound at the cell surface be of the same specificity as the antibody that is later secreted in response to antigen. The theory therefore places unusual constraints on Ig structure and expression. How these constraints are met can best be understood by examining the molecular genetic basis of antibody diversity.

Although antibodies were recognized by von Behring and Ehrlich before the turn of the century, it was not until the development of techniques for the analysis of macromolecules that the molecular properties of antibodies could be understood. In the 1930s Tiselius and Kabat demonstrated that antibodies belong to an electrophoretically heterogeneous class of serum proteins, the gamma globulins (Tiselius and Kabat, 1939). With the advent of analytical ultracentrifugation, it became clear that immunoglobulins were heterogeneous with respect to their sedimentation coefficients as well. Two major classes were first recognized: the 7S gamma globulins (about 150 kD) and the 19S macroglobulins (about 900 kD). Ultimately, seven distinct classes of immunoglobulin were found in the mouse: IgM, IgD, IgG1, IgG2a, IgG2b, IgE, and IgA. The bulk of the 7S class is now known to consist of IgG, while the 19S class consists of IgM. With the discovery that myelomas secrete homogeneous immunoglobulin, it became possible, by about 1970, to determine the complete amino acid sequences of immunoglobulin heavy and light chains (Edelman et al., 1969). From these and other studies (Porter, 1967), a general picture of immunoglobulin structure emerged. Immunoglobulins contain two identical heavy chains (50–70 kD, depending on the class) and two identical light chains (25 kD). These are held together by interchain disulfide bonds. The heavy and light chains exhibit periodicity in their amino acid sequences and can be divided into a series of homologous domains, each about 110 amino acids in length. Each domain contains an intrachain disulfide linkage, and X-ray crystallography has revealed that the various domains assume a similar tertiary structure (the Ig fold) (Amzel and Poljak, 1979).

There are two types of light chain, called kappa and lambda. Either of these can be found in any class of immunoglobulin, but only one type of light chain is found in any one molecule. Immunoglobulin molecules of different classes are distinguished by the type of heavy chain found in the molecule. There are, in the mouse, seven distinct heavy chain isotypes (μ, δ, $\gamma 1$, $\gamma 2a$, $\gamma 2b$, ϵ, and α), corresponding to the seven Ig classes. For example, an IgG1 molecule always contains two $\gamma 1$ chains and may contain either two κ or two λ chains.

When the amino acid sequences of light chains from different sources were compared, a remarkable pattern emerged. Whereas the carboxy-terminal halves of the chains (residues 108–214) were nearly identical for a given type of light chain, the amino-terminal residues (1–107) showed considerable sequence variability. Consequently, these regions were designated constant (C) and variable (V), respectively. Similarly, heavy chains were found to consist of V and C regions; in heavy chains, the V region varies in length from about 107 to about 132 amino acids.

Closer examination of the amino acid sequences of heavy and light chains revealed that variability in amino acid sequence is not constant across the V region (Wu and Kabat, 1970). The V regions of heavy and light chains were each found to contain three hypervariable regions, now called CDR1 (for complementarity determining region), CDR2, and CDR3. Interspersed between the hypervariable regions are less variable framework regions (FR). By several methods, including the crystallographic analysis of antigen-antibody complexes (Amit et al., 1986; Sheriff et al., 1987), the hypervariable regions have been shown to mediate almost all specific contacts with antigen.

The structure of immunoglobulins posed several genetic problems, the most obvious being (1) how the genome encodes such a large number (10^7–10^8) of variable regions and (2) how different variable regions come to be associated with the same constant region. As early as 1965, Dreyer and Bennett suggested one solution—that V and C regions are encoded by separate gene segments. While the hypothesis had no precedent, it was supported by genetic evidence available at the time. One genetic argument is the following. Within a species, individuals may express variant forms of a particular class of Ig chain. These variants, or allotypes, generally result from single amino acid substitutions. Most allotypes are inherited in simple Mendelian fashion. Three allotypes of the human κ chain, Km(1), Km(1,2), and Km(3), result from amino acid substitutions at residues 153 and 191 within the C region. Any one of these allotypes can be found in association with a number of different V regions, despite the fact that the allotypes correspond to a single genetic locus. Such observations suggested that V and C regions are encoded by separate loci.

A direct test of the Dreyer-Bennett hypothesis awaited the development of ways to study higher eukaryotic genomes, in particular the discovery of restriction enzymes and their application to the physical analysis of DNA. In 1976, Hozumi and Tonegawa obtained the first evidence for somatic rearrangement of Ig genes. Genomic DNA from murine embryos or from a κ-producing plasmacytoma was digested with a restriction endonuclease, fractionated, and hybridized to radiolabeled κ chain mRNA. The results suggested that κ chains are encoded by separate DNA segments in the embryo and that these are brought together during lymphocyte differentiation. The advent of methods for the molecular cloning of DNA soon made it possible, in large part through work performed in the laboratories of S. Tonegawa, P. Leder, and L. Hood, to describe the organization and reorganization of Ig genes in detail.

DEVELOPMENTAL SIGNIFICANCE

Immunoglobulin gene rearrangement is an essential feature of B cell differentiation. Two recombinational mechanisms participate in the construction of Ig genes. The first generates DNA sequences that encode the V regions of heavy and light chains and occurs via a developmentally ordered series of rearrangements. The regulation of this process underlies the phenomena of allelic and isotypic exclusion (the expression, by a given B lymphocyte, of only one allelic form of an Ig chain and of only one light chain isotype, respectively). The second type of rearrangement is responsible for class switching (a change in the class of Ig produced by a given B cell lineage). An increasing body of evidence suggests that class switching, like other aspects of B cell development, can be modulated by specific lymphokines. Thus, both types of Ig gene rearrangement are subject to developmental control.

STRUCTURE AND ASSEMBLY OF Ig GENES

V Regions Encoded by Multiple Germline Gene Segments

The κ, λ, and heavy chains of Ig genes are encoded at three genetic loci; in the mouse, these loci are found on chromosomes 6, 16, and 12, respectively. (The corresponding loci in humans are on chromosomes 2, 22, and 14.) For each type of Ig chain, the variable region is encoded in the germline by multiple DNA segments. During B cell differentiation, these segments are brought together by site-specific recombination to form a single transcriptional unit (Tonegawa, 1983).

The DNA sequences that specify the variable regions of κ chains reside in the germline as two separate DNA segments: V_κ and J_κ (Sakano et al., 1979; Seidman and Leder, 1978; Seidman et al., 1979). The joining of a V_κ segment to a J_κ segment creates a continuous stretch of DNA that encodes the variable domain of the κ protein (Fig. 6–1). The coding sequence of each V_κ segment is interrupted by an intron. The first exon of the V_κ segment encodes most of the κ chain's leader sequence (amino acid residues −20 to −5); the second block of coding sequence encodes amino acid residues −4 to 95. The remainder of the variable region (residues 96–108) is encoded by the J_κ segment. There are five J_κ segments in the mouse, of which one ($J_\kappa 3$) is inactive. The murine J_κ segments are clustered within about 1.4 kb of DNA; this cluster lies about 2.5 kb to the 5′ side of a single exon, C_κ, that encodes the constant region of the κ chain.

A.

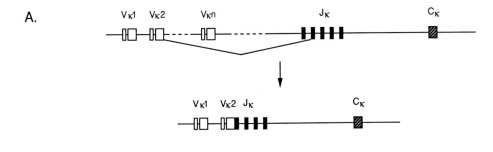

B.

Figure 6–1.   Organization and assembly of Ig light chain genes. (**A**) The murine κ locus: V gene segments are designated by open boxes, J segments by filled boxes, and the C segment by a hatched box. Joining of a V segment to J2 is indicated. (**B**) The murine λ locus. V, J, and C segments are designated as in A. Drawings are schematic.

Estimates of the number of V_κ segments in the mouse genome range from about 100 to about 300 (D'Hoostelaere et al., 1988). The V_κ segments lie to the 5′ side of the J_κ cluster; their exact distance from the J_κ segments is unknown because they have not yet been physically linked to the J_κ locus. Likewise, the absolute orientations of the V_κ segments, relative to the J_κ cluster, have not been determined, although it seems that germline V_κ segments may lie in either orientation.

The variable regions of λ chains (Fig. 6–1B), like those of kappa chains, are encoded in two separate germline segments, V_λ and J_λ. The organization of the lambda locus (Blomberg et al., 1981), however, differs from that of the kappa and resembles the locus encoding the β chain of the T cell antigen receptor. There are only two V_λ segments in the mouse; these resemble V_κ segments. While there is only one C_κ gene in the mouse, there are four C_λ genes, each linked, at its 5′ side, to a single J_λ segment. The V_λ, J_λ, and C_λ elements occur in two linked clusters, each containing a V_λ segment and two J_λ-C_λ pairs (Fig. 6–1). One of the J_λ segments ($J_\lambda 4$) has suffered a deletion in its RNA splicing signal and is presumably inactive. The overall organization of the lambda locus in the human is similar, but the number of tandem V_λ-J_λ-C_λ groups is somewhat larger.

The variable regions of heavy chains are encoded in three separate, germline DNA segments: V_H, D, and J_H (Early et al., 1980a,b; Sakano et al., 1980). During B cell differentiation these segments are joined to form a V_H-D-J_H unit (Fig. 6–2). Thus, two recombination events are required for assembly of the variable exons of heavy chains: D-to-J_H and V_H-to-DJ_H. Like the V segments of κ and λ chains, the V_H segments are interrupted by an intron that separates an exon encoding most of the leader peptide sequence from the remainder of the coding block. These two coding portions of the V_H segment specify the amino-terminal portion of the heavy chain through amino acid residue 101. The remainder of the V domain, comprising roughly 20–25 amino acids, is encoded by the D and J_H segments.

The murine heavy chain locus contains a cluster of four J_H segments that lies about

6.5 kb 5′ to the first heavy chain constant gene segment, C_μ (Gough and Bernard, 1981). The J_H segments contain from 15 to 17 codons. In the mouse, about 12 D segments were identified (Kurosawa and Tonegawa, 1982); they are variable in sequence and in length (8–18 codons) and fall into three families on the basis of sequence. Two of these families lie within a 60-kb region that is separated from the J_H cluster by about 20 kb. The third family has a single member that lies 0.7 kb to the 5′ side of the J_H cluster.

Estimates of the number of V_H segments in the mouse genome vary from 100 to greater than 1,000 (Brodeur et al., 1984; Livant et al., 1986). About ten murine V_H families were identified on the basis of DNA sequence homology; members of a given family are clustered, and families show little overlap. The known V_H families of the mouse occupy 2 to 3 centimorgans (cM) on chromosome 12. In the human, five families of V_H segments were identified (Berman et al., 1988); members of these families are interspersed throughout the V_H locus. The known human V_H segments all reside within about 2,500 kb of DNA; the most J_H-proximal of these segments lies 77 kb to the 5′ side of the J_H cluster (Berman et al., 1988; Schroeder et al., 1988). In both mouse and man, the V_H segments appear to be oriented in a single direction, so that V_H-D-J_H joining invariably results in deletion of the intervening DNA.

Ig Gene Segments Joined During B Cell Differentiation by Site-Specific Recombination

All Ig gene segments are accompanied by characteristic DNA sequence elements that lie at or near the sites of recombination (Early et al., 1980a; Max et al., 1979; Sakano et al., 1979, 1980; Seidman et al., 1979). These recombinational signals consist of three elements: a heptamer sequence, a spacer region, and a nonamer sequence (Fig. 6–3). The heptamer and nonamer sequences are conserved, both among gene families and among vertebrate species. The spacer regions are less conserved with respect to sequence, but their lengths fall into two classes of 12 and 23 base pairs (bp). Recombination generally occurs only between a gene segment that carries a 12-bp spacer and a segment that carries a 23-bp spacer (Early et al., 1980a). For example, the D segments of heavy chain genes are flanked on either end by conserved heptamer and nonamer elements, separated by a 12-bp spacer (abbreviated 7-12-9). The J_H and V_H segments, which recombine with the 3′

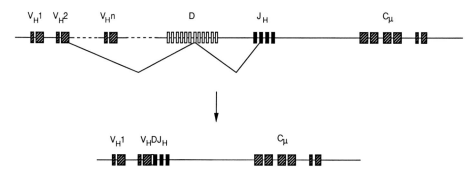

Figure 6–2.   Assembly of a heavy chain gene. V and C_μ segments are designated by hatched boxes, D segments by open boxes, and J segments by filled boxes. A V_H-D-J_H joining is indicated. Joining of D to J_H precedes joining of V_H to DJ_H in developing B cells (see text).

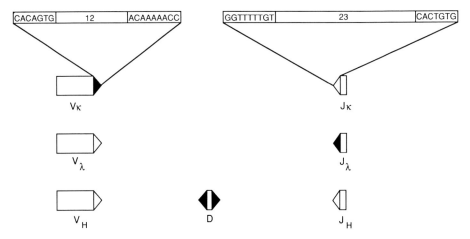

Figure 6–3.   Consensus recombinational signal sequences at the flanks of Ig gene segments. Gene segments are indicated by open boxes. Signal sequences that contain 12-bp spacer regions are designated by filled triangles; sequences that contain 23-bp spacer regions are designated by open triangles. Consensus DNA sequences for conserved heptamer and nonamer elements are shown indicated in the top line. Only sequences of the upper DNA strands are shown.

and 5′ ends of the D segment, respectively, carry heptamer and nonamer elements that are separated by 23-bp spacer elements (7-23-9). Recombination between segments carrying spacers of similar length (e.g., between two J_H segments) is generally not observed. T cell receptor gene segments, which are joined during T cell differentiation, are accompanied by similar sequence elements.

The heptamer-spacer-nonamer sequences are sufficient to mediate joining: model substrates that carry 7-12-9 and 7-23-9 signals, but little or no coding sequence, are observed to rearrange efficiently, and in a manner characteristic of normal rearrangement, after introduction into B-progenitor cells (Akira et al., 1987; Hesse et al., 1987). The recombinational signal sequences of Ig and T cell receptor (TCR) gene segments are functionally equivalent; Ig and TCR gene rearrangement appear to be catalyzed by the same or similar mechanisms (Yancopoulos et al., 1986). The heptamer sequence, in the absence of a nonamer element, can itself mediate rearrangement, but such reactions are apparently rare (Kleinfield et al., 1986; Reth et al., 1986). In experiments with model substrates, deletion of the nonamer sequence results in a large diminution in the frequency of rearrangement (Li et al., 1989). Rearrangement is compatible with nucleotide substitutions within the spacer sequences, although changes within spacers may have some effect on the frequency of rearrangement (Akira et al., 1987).

A large number of mutant joining sequences were tested for their ability to support rearrangement (Hesse et al., 1989). No mutant signal sequence was more efficient than the consensus nonamer (GGTTTTTGT) or heptamer (CACTGTG); the heptamer sequence was more sensitive than the nonamer to base substitution. As expected, changes in spacer length of more than 1 bp greatly impaired rearrangement. Restoration of homology between altered signal sequences did not restore activity, supporting the idea that the heptamer and nonamer sequences serve primarily as recognition sites for components of the recombinational machinery. A specific DNA binding protein whose recognition sequence precisely coincides with the conserved nonamer was identified in extracts of lymphoid cell nuclei (Halligan and Desiderio, 1987) and purified to apparent homogeneity (Li et al., 1989). A protein that binds DNA fragments that contain a heptamer sequence was also identified (Aguilera et al., 1987), but the DNA sequence requirements for recognition by this protein have not been determined precisely. While one or both of these proteins may participate in recombination, this remains to be shown.

A third protein that binds a subset of recombinational signal sequences at a site overlapping the heptamer was identified and its gene molecularly cloned (Matsunami et al., 1989). The predicted protein sequence exhibits limited homology to the integrase family of prokaryotic, site-specific recombinases. The gene that encodes this protein is expressed in non-lymphoid and lymphoid cells, and within the lymphoid lineages its expression is not correlated with recombination activity. A role for this protein, if any, in the assembly of Ig genes is therefore not clear.

Imprecise Joining of Ig Coding Sequences

The mechanism of Ig gene rearrangement is apparently different from all other known recombinational mechanisms. Recombination between gene segments produces two new junctions: a junction between coding sequences (a coding joint) and a junction between the sequences that previously flanked the gene segments (a non-coding joint). Joining is often accompanied by loss of a variable number of nucleotides from the coding sequences at the site of recombination. The joint that is formed between the flanking sequences, however, is almost always a precise head-to-head fusion of the heptamer elements, without loss or addition of nucleotides. Thus, recombination between Ig gene segments is asymmetric (Lewis et al., 1985).

At the recombinant joints between heavy chain segments, insertion of novel nucleotides is often observed. These insertions, which have been called "N regions," are highly variable with respect to length and sequence and do not appear to be encoded in the germline (Alt and Baltimore, 1982; Desiderio et al., 1984). It is likely that they are synthesized de novo during recombination, before the joining of the coding segments. It has been proposed that the insertion of N regions is accomplished by the enzyme terminal deoxynucleotidyl transferase (TdT), which catalyzes the polymerization of deoxyribonucleotides in the absence of a template and is expressed transiently in immature B cells at the time of heavy chain rearrangement. The appearance of N regions is positively correlated with expression of TdT (Blackwell et al., 1986; Desiderio et al., 1984; Landau et al., 1987), and it is likely that TdT catalyzes their insertion. Even in the absence of TdT, very short insertions (typically 1–2 bp) are sometimes observed at recombinant joints in which no nucleotides have been lost from one or both of the participating coding sequences. The inserts, termed P regions, are the inverse complements of the intact coding sequences that flank them. This suggests that flanking, germline-encoded sequences serve as the template for P element insertion (LaFaille et al., 1989; McCormack et al., 1989).

The imprecision in the joining of coding regions suggests that most products of rearrangement may be nonfunctional because coding regions have been joined out of frame. This is borne out by examination of rearrangements that have occurred during propagation of B cells in culture, in the apparent absence of positive selection for functional products. In these cases, the majority of rearrangements are found to be nonfunctional (Reth et al., 1985). In experiments with exogenous substrates for rearrangement, the length distribution of nucleotide loss from coding joints is rather broad and incompatible with mechanisms in which the reading frame is preserved (Lieber et al., 1988a).

Unusual Rearrangements in Normal and Mutant Cells: Implications for the Mechanism of Recombination

While the usual products of Ig gene rearrangement are a coding joint and a non-coding joint, these are not the only products observed. Experiments with model substrates for recombination revealed two additional, site-specific reactions. One of these reactions called "signal sequence replacement" or "hybrid junction" formation, results in the replacement of the recombinational signal sequences of one gene segment by the signal sequences of another (Lewis et al., 1988; Morzycka-Wroblewska et al., 1988). In one series of experiments (Morzycka-Wroblewska et al., 1988) the 7-23-9 sequences of an Ig V_H gene segment were observed to recombine with the 5' end of a DJ_H element; joining was accompanied by a loss of nucleotides from the D coding sequence, but not from the heptamer sequence. In other words, the reaction shows the same asymmetry, with respect to imprecision of joining, as is observed in normal rearrangement. In these experiments, signal sequence replacement was observed to occur at about one-tenth the frequency of normal rearrangement.

The signal sequence replacement reaction has important implications for the generation of antibody diversity because it allows the interchangeability of signal sequences. A J_H segment (signal sequence 7-23-9), for example, might acquire the signal sequence of a D segment (signal sequence 7-12-9), and thereby the ability to join directly to V_H. An apparent signal sequence replacement product has been isolated from B cells (Stenzel-Poore and Rittenberg, 1987), and functional heavy chain genes that lack D coding sequences may owe their existence to this mechanism.

A second unusual reaction, apparently rare, is the formation of what has been called an "open-shut" joint (Lewis et al., 1988). Products of this reaction are indistinguishable from unrearranged gene segments, except for the loss of nucleotides from the coding sequence at its boundary with the recombinational signals. The common features of classical rearrangement, signal sequence replacement, and open-shut joint formation suggest a model for recombination in which DNA segments carrying 7-12-9 and 7-23-9 signal sequences are paired and then cleaved at the heptamer coding sequence junctions, to yield an intermediate in which four DNA ends (two coding ends and two flanking ends) are held in proximity (Fig. 6–4). This intermediate may be resolved in more than one way. Commonly, the two coding sequences and the two flanking sequences are joined (Fig. 6–4a). Less frequently, the coding sequences of one segment may become joined to the flanking sequences of the other, resulting in signal sequence replacement (Fig. 6–4b). In some instances, after transient scission of the DNA, a coding sequence may be rejoined to its own signal sequence (Fig. 6–4c); if loss of nucleotides from the coding sequence precedes the joining step, the product is detectable as an "open-shut" joint. Whether the formation of open-shut joints requires an interaction between 7-23-9 and 7-12-9 signals is not resolved.

Genes that Function in Ig Gene Assembly

With the possible exception of N region insertion, which is likely catalyzed by TdT, the enzymology of Ig gene assembly is as yet undefined. As is clear from the preceding discussion, Ig gene rearrangement involves multiple enzymatic steps, and it therefore seems likely that the recombinational apparatus contains multiple components. A major

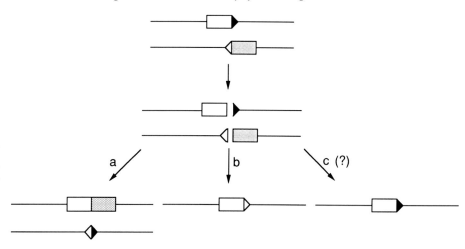

Figure 6–4.   Alternative resolution of an intermediate in Ig gene rearrangement. Open and shaded boxes represent different Ig coding segments; filled and open triangles represent recombinational signal sequences bearing 12- and 23-bp spacer regions. For explanation, see text.

advance was made with the observation by Baltimore and coworkers that the products of two genes—RAG-1 and RAG-2—were necessary and sufficient to activate Ig gene rearrangement in cells that ordinarily do not express recombinational activity, such as NIHY–3T3 fibroblasts (Oettinger et al., 1990; Schatz et al., 1989). RAG-1 and RAG-2 are adjacent genes that reside on mouse chromosome 2. The murine RAG-1 and RAG-2 genes encode polypeptides of 1040 and 527 amino acids, respectively, whose sequences are highly conserved among vertebrates. In the mouse, RAG-1 and RAG-2 are coexpressed preferentially in immature lymphoid cells; in cell lines, expression of these genes is correlated with recombinational activity. In immature B lymphoid cell lines, agents that increase intracellular cAMP lead to a tenfold increase in recombinational activity and increased expression of RAG-1 and RAG-2 (Menetski and Gellert, 1990).

RAG-1 and RAG-2 transcripts are not confined, however, to cells undergoing antigen receptor gene rearrangement. Among normal thymocytes, RAG-1 and RAG-2 are expressed not only in cells undergoing T cell receptor (TCR) gene assembly but also in a subset of immature cells (the TCR$^+$ CD4$^+$ CD8$^+$ subset) that have already rearranged their receptor genes; in these thymocytes, RAG-1 and RAG-2 expression can be terminated by TCR crosslinking (Turka et al., 1991). In chicken, expression of RAG-2, but not RAG-1, was observed in bursal B cells undergoing Ig diversification by gene conversion (Carlson et al., 1991). Expression of RAG-2 was not, however, required for gene conversion in a chicken B progenitor cell line (Takeda et al., 1992). RAG-1 transcripts were found at low levels outside the immune system—notably, in the central nervous system (Chun et al., 1991). Inactivation of the RAG-1 (Mombaerts et al., 1992) or RAG-2 (Shinkai et al., 1992) genes in mice resulted in the complete absence of functional B and T cells but no apparent defects outside of the immune system. The functions of RAG-1 or RAG-2, if any, in cells other than those undergoing antigen receptor rearrangement have not been determined.

It is not yet known how RAG-1 and RAG-2 activate rearrangement. One possibility is that they represent the sole lymphoid-specific components of the recombinase. A second possibility, not exclusive of the first, is that they act indirectly by inducing expression of recombinase components. The predicted amino acid sequences of RAG-1 and RAG-2 provide few clues to function, as they lack extensive homology to known recombinational enzymes or DNA-binding proteins. RAG-1 does, however, contain a cysteine-rich sequence motif that is conserved among several proteins that function in DNA repair or transcription (Freemont et al., 1991). Elsewhere in its sequence RAG-1 exhibits modest homology to a motif conserved among topoisomerases (Wang et al., 1990). The RAG-1 protein has an apparent molecular size of about 120 kD, similar to that predicted by its sequence. RAG-2, in contrast, has an apparent molecular size of about 70 kD, somewhat larger than its predicted 58 kD; this suggests post-translational modification. Both proteins reside mainly in the nucleus and both are phosphorylated, primarily on serine (Lin and Desiderio, unpublished data). The latter observation suggests that the activities of the RAG-1 and RAG-2 proteins may be modulated by specific protein kinases.

From the foregoing it is clear that the generation of functional Ig and T cell receptor molecules depends on an intact recombinational apparatus; mutations that inactivate a component of the apparatus would be expected to result in a combined T and B cell immunodeficiency. Mice that are homozygous for the *scid* (severe combined immunodeficiency) mutation lack functional B and T cells and exhibit a specific defect in the joining of Ig gene segments. In B-progenitor cells from *scid* mice, Ig gene segments are recognized and cleaved, but appropriate joining of coding segments to each other does not occur (Lieber et al., 1988b). The *scid* mutation does not affect lymphoid cells alone; non-lymphoid and lymphoid cells from homozygous *scid* mice exhibit increased sensitivity to agents that cause double-strand breaks in DNA, but not to agents that cause single-strand breaks (Fulop and Phillips, 1990; Hendrickson et al., 1991). The defects in DNA repair and Ig gene rearrangement are cotransmitted as autosomal recessive traits, indicating that they are caused by identical or closely linked mutations (Fulop and Phillips, 1990). The simplest interpretation of these results is that the *scid* mutation affects a widely expressed protein that functions in DNA repair. This protein is clearly distinct from the products of RAG-1 or RAG-2, because the *scid* mutation maps to a different genetic locus, on chromosome 16.

Ig Gene Rearrangement and Generation of Diversity

Except for somatic hypermutation, which occurs after the assembly of Ig genes (reviewed by Perlmutter et al., 1984), the antibody repertoire is established by the time rearrangement is complete. Rearrangement acts on families of gene segments, selecting a single member of each family for incorporation into an antibody gene. The number of antibody molecules that can be generated by a combinatorial process alone may be estimated as follows. If we consider that there are about 100 V_H segments, 12 D segments, and 4 J_H segments in the mouse genome, and if we assume that all members of a family have an equal chance of participating in rearrangement, then about $100 \times 12 \times 4 = 4{,}800$ different heavy chains can be produced. By a similar argument, the mouse genome can encode about $100 \times 4 = 400$ different κ chains. If we further assume that any heavy chain can associate with any light chain, the number of different antibody molecules that can be generated by a combinatorial process is about $4{,}800 \times 400 = 1.9 \times 10^6$.

An additional source of antibody diversity is the loss of nucleotides from the coding junctions of heavy and light chain genes, and the insertion of N regions at the coding junctions of heavy chain genes. The V_H-D, D-J_H, and V_L-J_L junctions lie in the CDR3-

encoding regions of heavy and light chain genes. Crystallographic studies of specific antibody-lysozyme complexes show that the CDR3 regions of heavy and light chains form a large part of the surface that contacts antigen (Amit et al., 1986); junctional diversity contributes to the structural diversity of this part of the antigen binding site. As a conservative estimate, let us assume that junctional diversity increases the number of possible heavy and light chains by a factor of three. Then the number of different Ig molecules that can be generated by antibody gene rearrangement is increased to about $14,400 \times 1,200 = 1.7 \times 10^7$, well within the range of estimates for the size of the antibody repertoire (10^6–10^9).

Coupling of Rearrangement and Expression of Ig Genes

All Ig V segments carry 5′ sequence elements that direct initiation of transcription (Falkner and Zachau, 1984). These include an AT-rich region about 25 bp upstream of the transcription initiation ("cap") site, called a TATA box, and a conserved 8-bp sequence 50–80 bp upstream of the cap site (the "octanucleotide sequence"). Transcription from a V promoter, however, occurs only at a low level or not at all when it is unrearranged. After Ig gene assembly, the steady-state level of transcription from the rearranged V segment promoter increases greatly (Mather and Perry, 1981). In a B cell or plasma cell, only rearranged promoters are activated; unrearranged V segments remain silent.

The activation of transcription following rearrangement, and its specificity for rearranged genes, is explained, at least in part, by the presence of transcriptional enhancer elements within the J-C introns of heavy and κ chain genes (Banerji et al., 1983; Gillies et al., 1983), and in the regions 3′ of the C_H cluster (Pettersson et al., 1990), C_κ (Meyer et al., 1989), $C_\lambda 1$ and $C_\lambda 4$ (Hagman et al., 1990). Transcriptional enhancers activate transcription in *cis;* they can act over distances of several thousand base pairs, and in an orientation-independent fashion. It is now clear that enhancers contain multiple binding sites for proteins that probably interact with other proteins at the promoter to activate transcription. The intronic enhancers of heavy and κ chain genes lie between the J segment clusters and the C gene segments, and they are the best studied of the enhancers associated with Ig genes. The activities of both these enhancers are specific to lymphoid cells. Removal of the intronic enhancer from a heavy chain gene impairs its transcription by at least two orders of magnitude (Grosschedl and Baltimore, 1985). Furthermore, an active intronic enhancer is required not only for the establishment of Ig transcription but also for its continued expression (Grosschedl and Marx, 1988). The enhancer-dependence of Ig transcription provides a means by which rearrangement and transcriptional activation are coupled. Before rearrangement, V promoters are separated from the Ig enhancers by distances of roughly 100 kb or more. Rearrangement brings a V promoter within 1 to 2.5 kb of an intronic enhancer, thereby activating transcription.

Immunoglobulin genes are expressed specifically in B-lymphoid cells. While it is true that Ig genes are assembled only in lymphoid cells, their tissue-specific expression cannot be explained by tissue-specific rearrangement alone. This has been demonstrated by experiments in which rearranged κ or heavy chain genes have been introduced into transgenic mice. Under such circumstances, an assembled Ig gene is present in all tissues of the mouse, but is expressed only in lymphoid tissues (Grosschedl et al., 1984; Storb et al., 1984).

The molecular basis for tissue-specific Ig gene expression has been examined at several levels and is now beginning to be understood. A detailed summary of this area is beyond the scope of this chapter; Ig gene transcription has been reviewed recently (Staudt and Lenardo, 1991). Immunoglobulin promoters and enhancers act synergistically to drive transcription (Garcia et al., 1986), suggesting that proteins that bind these regions interact. Footprinting in vivo of DNA within the heavy chain enhancer identified several putative protein binding sites (Ephrussi et al., 1985). One of these was found to coincide with a sequence identical to the conserved octanucleotide element found upstream of V segments. The other sites, designated μE motifs (μE1 through μE5) are distinct from the octanucleotide sequence but homologous to each other. Mutations in the E motifs were observed to decrease, but not to abolish, enhancer activity; mutations in the octanucleotide had a more profound negative effect (Lenardo et al., 1987). Distinct proteins that bind to at least four of these sites in vitro (μE1, μE3, μE5 and the octanucleotide) have been identified (Peterson et al., 1986; Singh et al., 1986). The proteins known to bind to μE1, μE3 and μE5 are ubiquitous. At least three different proteins that recognize the octanucleotide have been identified: two or more are ubiquitous (Fletcher et al., 1987; Singh et al., 1986; Sturm et al., 1987); one is preferentially expressed in early development (Rosner et al., 1990); another is B cell–specific (Staudt et al., 1986). The B cell–specific protein was purified and was shown to activate transcription of a κ gene (which carries an upstream octanucleotide sequence) in vitro (Scheidereit et al., 1987). Complementary DNA clones encoding ubiquitous and lymphoid-specific octanucleotide binding proteins have been obtained; expression of the lymphoid-specific cDNA in a non-lymphoid cell line was shown to activate transcription of a B cell–specific promoter in trans (Mueller et al., 1988).

The intronic κ enhancer functions in mature B cells and plasma cells, in distinction to the intronic μ enhancer, which functions at all stages of B lymphoid ontogeny. The intronic κ enhancer contains several E motifs; two of these bind proteins that also recognize E motifs in the intronic μ enhancer. In addition, the intronic κ enhancer contains a binding site for a factor called NF-κB (Sen and Baltimore, 1986b). The presence of the NF-κB binding site is necessary for activity of this enhancer. NF-κB is constitutively active in mature B-cells. It is also present in inactive form in other cell types, including pre-B-cells, T-cells, and fibroblastoid cells, in which its activity can be induced post-translationally (Sen and Baltimore, 1986a). Induction of κ chain expression by lipopolysaccharide (LPS) in the pre-B cell line 70Z/3 correlates with the activation of NF-κB (Sen and Baltimore, 1986a); mutations in the NF-κB binding site abolish enhancer activity and inducibility by LPS (Lenardo et al., 1987).

An outline of the mechanism through which NF-κB is activated in response to extracellular stimuli is now understood (Ghosh et al., 1990; Nolan et al., 1991 and references therein). NF-κB is a heterodimer consisting of 50- and 65-kD subunits; in the inactive form of the protein, these are found in the cytoplasm bound to an inhibitor, IκB (Baeuerle and Baltimore, 1988). Phosphorylation of IκB by cellular kinases, including protein kinase C, results in its inactivation and its dissociation from NF-κB (Ghosh and Baltimore, 1990); active NF-κB, free of the inhibitor, then migrates to the cell nucleus and activates transcription (Baeuerle and Baltimore, 1988).

In summary, the promoters and enhancers of heavy and light chain genes contain binding sites for several proteins, some of which have been found to activate transcription in vitro. At least two of these binding sites—the octanucleotide and the NF-κB site—are recognized by positive regulatory proteins whose expression or activity is restricted to B cells. The specificity of Ig transcription derives in part from the B cell–specific expression or activation of these DNA binding proteins. These proteins, however, are not sufficient to fully account for the B lymphoid specificity of Ig regulatory elements, and it seems likely that negative regulatory mechanisms—expression of transcriptional silencer proteins in non-B cells, for example—also contribute to the specificity of Ig gene expression.

HEAVY CHAIN ISOTYPE SWITCHING

IgM$^+$ B Cells and Cells that Express Other Ig Classes

The process of V_H-D-J_H joining creates heavy chain genes in which a variable coding region lies to the 5′ side of a C_μ gene segment (see Fig. 6–2). A cell that carries a functionally rearranged heavy chain gene in this configuration is able to express an Ig/μ chain; if the cell also expresses a functional light chain, IgM is assembled and exposed at the cell surface. Subsequent differentiation of cells that express surface IgM may yield progeny that express other Ig classes (class switching; reviewed by Honjo, 1983). For example, an IgM-expressing cell may ultimately give rise to a cell that expresses IgG, IgA, or IgE. An Ig molecule expressed after class switching differs from the progenitor IgM molecule only in the constant region of the heavy chain; the light chains, and the variable region of the heavy chain, are identical to those of the original IgM molecule.

Class Switching Accompanied by a Second Type of DNA Rearrangement

In the mouse, the constant regions of heavy chain genes are specified by a tandem array of eight gene segments: C_μ, C_δ, $C_\lambda 3$, $C_\lambda 1$, $C_\lambda 2b$, $C_\lambda 2a$, C_ϵ, and C_α. These are clustered in a 200-kb region that extends from the 3′ side of the J_H cluster (Shimuzu et al., 1982) (Fig. 6–5). In the human, at least nine constant gene segments are present in a similar array. Unlike the constant segments of light chain genes, which consist of a single exon, the

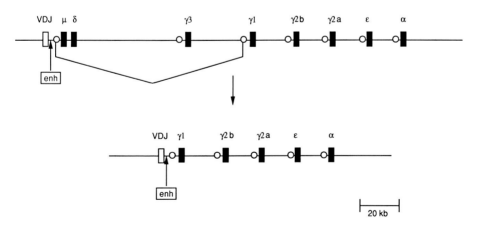

Figure 6–5.   Switch recombination at the heavy chain locus. Constant gene segments are indicated by filled boxes; the product of V_H-D-J_H joining is indicated by an open box. Switch regions are symbolized by open circles. The position of the heavy chain enhancer (enh) is indicated. The product of switch recombination between μ and γ 1 is diagrammed.

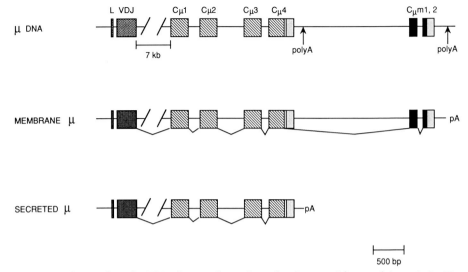

Figure 6–6. Generation of mRNA for membrane-bound and secreted forms of the μ chain. *Top line:* Structure of a functional μ gene. DNA encoding variable and constant domains are indicated by shaded and hatched boxes, respectively. Regions that encode the carboxyl ends of secreted and membrane-bound μ are indicated by open and filled boxes, respectively. Regions that are transcribed but not translated are indicated by hatched boxes. Polyadenylation sites are marked with arrows. *Middle line:* Generation of mRNA for the membrane-bound μ chain. The splicing pattern is indicated by kinked lines. The polyadenylated 3′ end of the RNA is labeled (pA). *Bottom line:* Generation of mRNA for the secreted μ chain. For explanation, see text.

heavy chain constant segments contain multiple exons, corresponding to different functional domains in the protein (Honjo, 1983). For example, the C_μ segment contains six exons. Four exons encode the first, second, third, and fourth globular domains (including the carboxyl-terminus of the secreted chain); two encode the distinctive carboxyl end of the membrane form of μ (Fig. 6–6).

The expression of heavy chains other than μ or δ is generally associated with the translocation of a $V_H DJ_H$ coding unit to a new site about 1–4 kb upstream of another C_H gene segment (Fig. 6–5). (Expression of the δ chain occurs via differential processing of RNA, which will be discussed briefly below.) This process, called switch recombination, occurs within extensive regions of repetitive DNA (S or switch regions) that terminate about 1 to 2 kb upstream of all C_H segments except for C_δ (Kataoka et al., 1981). The S_μ region lies between the J_H cluster and the C_μ segment, to the 3′ side of the heavy chain enhancer. Recombination between the S_μ region and the S region of a downstream C_H segment translocates the enhancer region together with the $V_H DJ_H$ element, thereby forming a new heavy chain transcriptional unit without altering the variable coding sequence. Recombination between two switch regions is accompanied by deletion of the intervening DNA (Honjo and Kataoka, 1978). Successive switch recombination is known to occur. By this route, for example, a cell that expresses IgG may give rise to progeny that express IgA.

Switch regions vary in length from about 1 kb (S_ϵ) to 10 kb ($S_\gamma 1$). The repetitive units of which they are composed also vary with respect to sequence and length. The $S_\gamma 1$, $S_\gamma 2a$, and $S_\gamma 2b$ regions, for example, contain repeat units of 49 bp (Kayaoka et al.,

1981), while the S_α repeat unit is 80 bp long (Davis et al., 1980); the precise sequences of the repeat units diverge both within a given switch region and among different switch regions. Nonetheless, switch regions do share the short sequences YAGGTTG, GAGCT, and GGGGT, which are imperfectly repeated within a switch region.

Recombination between switch regions can occur at a large number of sites within the participating repeat units, and it does not exhibit the exquisite site-specificity of V_H-D-J_H and V_L-J_L joining; switch recombination is clearly mechanistically distinct from the formation of V coding regions. It is possible that homology between pairs of switch regions plays a role in recombination. Experiments in which an exogenous substrate for switch recombination was introduced into lymphoid and non-lymphoid cell lines have shown that S_μ and S_γ2b are regions sufficient to mediate recombination in B cells but not in fibroblasts (Ott et al., 1987). An enzyme has been identified and purified from mammalian cells that catalyzes strand-transfer between DNA molecules (Hsieh et al., 1986) in a reaction that requires only very short regions of DNA homology. It has been suggested that this enzyme participates in switch recombination. Its relevance to class switching remains to be shown, however, because the reaction is not specific to switch recombination sequences, nor is the enzyme specific to lymphoid cells.

Class Switching Modulated by External Signals

The expression of specific classes of immunoglobulin can be induced in mitogen-activated B cells by T cell–derived lymphokines. For example, interleukin 4 (IL-4) induces the expression of IgG1 and IgE as it suppresses the expression of IgM, IgG3, IgG2a, and IgG2b (Snapper and Paul, 1987). The commitment to switch to a particular class of Ig can be correlated with transcription of the appropriate heavy chain constant segment before recombination. For example, IL-4, which induces IgG1 expression, induces the expression of germline γ1 transcripts within several hours of culture (Berton et al., 1989). Conversely, the ability of IL-4 to suppress germline transcription of other C_H segments (e.g., C_λ2b) correlates positively with its ability to suppress switching to those classes of immunoglobulin (Lutzker et al., 1988).

It was proposed that the propensity of a locus to undergo recombination is affected by its accessibility to the enzyme(s) that catalyze recombination (Yancopoulos and Alt, 1985). If this is so, then the effects of interleukins on germline transcription of C_H gene segments may be coupled to their participation in switch recombination. It should be underscored that these two events need not be causally linked; it is possible that transcriptional activation and switch recombination are parallel consequences of a proximal event that would simultaneously permit interaction of the DNA with transcriptional and recombinational factors.

DIFFERENTIAL RNA PROCESSING AND CONTROL OF Ig EXPRESSION

While a detailed discussion of the role of RNA processing in the developmental regulation of Ig gene expression is beyond the scope of this chapter, two mechanisms with important consequences for developmental immunology deserve mention. The first mechanism underlies the developmental switch from expression of surface to secreted immunoglobulin. The second accounts for the simultaneous expression of IgM and IgD on the surfaces of antigen-responsive B cells. These mechanisms have been reviewed (Blattner and Tucker, 1984; Rogers and Wall, 1984).

Switching from Membrane-Bound to Secreted Forms of Ig

Exposure to antigen triggers B cell mitogenesis, and, subsequently, some progeny cells differentiate to plasma cells. Among the many phenotypic changes that occur during this process is a switch from synthesis of a membrane-bound form of heavy chain to synthesis of a secretory form of the same chain. The membrane-bound form of μ is slightly larger than the secreted form. The carboxyl end of the membrane-bound form contains an acidic region, followed by a 26–amino acid uncharged region and a 3–amino acid basic region; this motif is characteristic of many transmembrane proteins. The sequence of the secreted form differs only in this region.

These two forms of μ are specified by distinct processed forms of mRNA (Early et al., 1980b; Rogers et al., 1980). The C_μ gene segment possesses two exons that specify the carboxyl end of the membrane form. These exons are spliced to the $C_\mu 4$ exon to generate an mRNA that encodes the membrane form. If the membrane exons are not spliced to the $C_\mu 4$ exon, the mRNA specifies the secreted form (Fig. 6–6). This mechanism is not unique to μ chain expression; the other heavy chains also exhibit membrane and secreted forms, and the C_H segments that encode them carry membrane-specific exons.

Coexpression of IgM with Other Classes of Ig

At a developmental stage after the assembly of heavy and light chain genes, but before encountering antigen, most B cells simultaneously express IgM and IgD on their surfaces. On any given cell, these IgM and IgD molecules have identical antigenic specificities (i.e., identical V regions). We have already seen that class switching is generally the result of recombination between S regions; because C_μ is deleted by this process, switch recombination cannot account for coexpression of IgM and IgD. In fact, the C_δ gene segment is not preceded by a typical switch recombination site. Coexpression of μ and δ chains may occur by differential processing of a long primary transcript that initiates at the V_H cap site and contains μ and δ exons. B cells that transiently coexpress IgM with a class other than IgD have also been observed; these may represent cells that have become committed to switch to a particular class but that have not yet undergone switch recombination. Coexpression in these instances may also involve differential RNA processing, although the primary transcripts would need to be at least 70 kb long. Evidence for the existence of such transcripts is indirect, and more exotic mechanisms, such as the covalent joining of two RNA species by trans-splicing (Sutton and Boothroyd, 1986), have been invoked.

REGULATION OF Ig GENE REARRANGEMENT AND ITS IMPLICATIONS FOR DEVELOPMENTAL IMMUNOLOGY

Allelic and Isotypic Exclusion

The genetic mechanisms described above explain two requirements of the clonal selection theory: the generation of antibody diversity before antigenic exposure, and the synthesis of membrane and secreted antibody with identical antigenic specificities. We now address the remaining requirement: that a given B cell express antibody of only one specificity.

A given B cell expresses only one of its two heavy chain alleles (Cebra et al., 1966; Pernis et al., 1965). Thus, B cells from mice that are heterozygous for a heavy chain allotype express one or the other allotype, but not both. This property is termed "allelic exclusion." Furthermore, B cells express either a κ or λ chain, but not both (Pernis and Chiappino, 1964); this property is called "isotypic exclusion." Taken together, these phenomena satisfy the clonal selection theory with respect to expression of unique antibody specificities by B cells.

Historically, at least four models have been proposed to account for these phenomena (reviewed by Alt et al., 1987): (1) selective transcription of one allele; (2) selective rearrangement of one allele; (3) a stochastic model, based on a high rate of aberrant, unregulated rearrangement; and (4) a regulated model in which functional rearrangement suppresses further rearrangement at the same locus. The first two models were excluded by the observation that many B cells possess two rearranged alleles (one functional, one nonfunctional), and that in these cases both alleles are usually transcribed. The second two models have proven more difficult to distinguish, but the available data favor a regulated model.

Rearrangement Ordered During B Cell Ontogeny

At about day 12 of murine development, B cell precursors that express cytoplasmic μ (cμ) chain, but not light chain ("pre-B cells"), appear in the mouse fetal liver (Levitt and Cooper, 1980). These cells precede the appearance of surface IgM$^+$ B cells in fetal liver by several days. Analysis of rearrangements in normal and immortalized cells of the B lineage reveals an underlying developmental order in the assembly of heavy and light chain genes (see Alt et al., 1987, for review).

The earliest rearrangements are D-to-J_H joinings; V_H-to-DJ_H rearrangement generally begins only after D-to-J_H recombination has occurred on both chromosomes, and only rarely have V_H-to-D joinings been observed. DJ_H transcripts are expressed from loci bearing DJ_H rearrangements (Reth and Alt, 1984); DJ_H transcription is coordinated with D-to-J_H rearrangement by approximation of enhancer-dependent D promoter elements to the Ig heavy chain enhancer (Alessandrini and Desiderio, 1991). A subset of DJ_H transcripts encode a truncated form of μ protein, termed D_μ (Reth and Alt, 1984), which apparently functions in the selection of D segment reading frame. A D segment may be joined to a J_H segment in one of three reading frames—RF1, RF2, or RF3—but in functional Ig genes, a bias toward RF1 is observed. Genes that encode the D_μ protein are assembled exclusively by joining of D to J_H in RF2. Cells that express D_μ protein on their surfaces are apparently prevented from further maturation (Gu et al., 1991). As a result, mature B cells that bear DJ_H rearrangements in RF2 are rarely observed. The bias against use of RF3 in functional Ig heavy chain genes is explained in large part by the fact that most D segments carry termination codons in that reading frame. Light chain gene rearrangement occurs after the productive assembly of a heavy chain gene. From the arrangement of light chain genes in κ- or λ-producing B cells, it has been inferred that rearrangement of κ genes precedes rearrangement of λ genes. If rearrangement at either κ allele is productive, the λ genes retain their germline configuration. Lambda rearrangements are generally observed in cells that carry nonproductive rearrangements or deletions at both κ alleles.

Models for the Developmental Regulation of Rearrangement

These observations have led to the proposal that the expression of a functional μ chain serves (1) to inactivate further heavy chain gene rearrangement and (2) to activate κ chain gene rearrangement (Alt et al., 1981). Several lines of evidence support this hypothesis. First, in experiments with a cell line that undergoes V_H-to-DJ_H joining in culture, κ rearrangement was activated only in subclones that had undergone productive heavy chain gene rearrangement (Reth et al., 1986b). Second, expression of an exogenous μ gene in μ-deficient derivatives of this cell line was positively correlated with activation of a κ rearrangement (Reth et al., 1987). Activation was specific to expression of the membrane form of μ; expression of the secreted form had no effect on κ rearrangement. Third, in mice that carry a transgene for the membrane form of human μ, B cells that express the human gene do not express endogenous murine IgM (Nussenzweig et al., 1987). This result suggests that expression of the membrane form of μ establishes allelic exclusion, perhaps at the level of heavy chain rearrangement.

Similarly, it has been proposed that isotypic exclusion is mediated by expression of intact immunoglobulin. In this model, the association of a functional κ or λ chain with an intracellular μ chain suppresses further rearrangement. The model is supported by experiments in which endogenous κ chain rearrangement was suppressed by expression of intact Ig consisting of an endogenous μ chain and a transgenic κ chain (Ritchie et al., 1984). Subsequent experiments indicated that this effect is mediated by Ig molecules that contain the membrane form of μ, but not the secreted form (Manz et al., 1988).

Because Ig gene segments are rearranged by a common apparatus and share similar recombinational signal sequences, it is unlikely that locus-specific regulation of rearrangement is effected at the substrate level. The assembly of heavy and light chain genes is mediated by a common recombinational apparatus that is active in immature B cells but not in mature B cells or plasma cells (Desiderio and Wolff, 1988; Lieber et al., 1987); as we have seen, coexpression of RAG-1 and RAG-2 is correlated with recombinational activity. While the suppression of recombination following productive light chain rearrangement may involve termination of RAG-1 and RAG-2 expression, locus-specific regulation of recombination requires a more selective process. Several lines of evidence support the hypothesis that rearrangement of a given gene segment is facilitated by transcription at a nearby site (Blackwell et al., 1986; Van Ness et al., 1981). First, rearrangement of exogenous, integrated substrates in pre-B-cells was positively correlated with expression of a linked genetic marker (Blackwell et al., 1986). Second, rearrangement of the κ locus in pre-B-cell lines could be induced by treatment with lipopolysaccharide (LPS); under these conditions, activation of rearrangement was accompanied by induction of transcripts initiating within the J_κ-C_κ intron (Schlissel and Baltimore, 1989). Interpretation of this experiment, however, was complicated by the fact that LPS affects multiple intracellular processes. Third, in transgenic mice carrying an artificial substrate for D-to-J rearrangement, joining was observed in lymphoid cells only when these segments were linked to the Ig heavy chain enhancer (Ferrier et al., 1990). The results of this more direct experiment suggested that transcription and rearrangement share common elements of control.

References

Aguilera, R., Akira, S., Okazaki, K., and Sakano, H. (1987). A pre-B nuclear protein that specifically interacts with the immunoglobulin V-J recombination sequences. *Cell 51*:909–917.

Akira, S., Okazaki, K., and Sakano, H. (1987). Two pairs of recombinational signals are sufficient to cause immunoglobulin V-(D)-J joining. *Science 238*:1134–1138.

Alessandrini, A., and Desiderio, S. V. (1991). Coordination of immunoglobulin DJ_H transcription and D-to-J_H rearrangement by promoter-enhancer approximation. *Mol. Cell. Biol. 11*:2096–2107.

Alt, F. W., and Baltimore, D. (1982). Joining of immunoglobulin heavy chain gene segments: Implications from a chromosome with evidence of three D-J_H fusions. *Proc. Natl. Acad. Sci. USA 79*:4118–4122.

Alt, F. W., Blackwell, T. K., and Yancopoulos, G. D. (1987). Development of the primary antibody repertoire. *Science 238*:1079–1087.

Alt, F. W., Rosenberg, N., Enea, E., Siden, E., and Baltimore, D. (1981). Organization and reorganization of immunoglobulin genes in A-MuLV-transformed cells: Rearrangement of heavy but not light chain genes. *Cell 27*:381–390.

Amit, A. G., Mariuzza, R. A., Phillips, S.E.V., and Poljak, R. J. (1986). Three-dimensional structure of an antigen-antibody complex at 2.8 Å resolution. *Science 233*:747–753.

Amzel, L. M., and Poljak, R. (1979). Three dimensional structure of immunoglobulins. *Annu. Rev. Biochem. 48*:961–997.

Baeuerle, P. A., and Baltimore, D. (1988). Activation of DNA-binding activity in an apparently cytoplasmic precursor of the NF-KB transcription factor. *Cell 53*:211–217.

Banerji, J., Olson, L., and Schaffner, W. (1983). A lymphocyte-specific enhancer is located downstream of the joining regions in immunoglobulin heavy chain genes. *Cell 33*:729–740.

Berman, J. E., Mellis, S. J., Pollock, R., Smith, C. L., Suh, H., Heinke, B., Kowal, C., Surte, U., Chess, L., Cantor, C. R., and Alt, F. W. (1988). Content and organization of the human Ig V_H locus: Definition of three new V_H families and linkage to the Ig C_H locus. *EMBO J. 7*:727–738.

Berton, M. T., Uhr, J. W., and Vitetta, E. S. (1989). Synthesis of germline gamma-1 transcripts in resting B cells: Induction by interleukin-4 and inhibition by interferon-gamma. *Proc. Natl. Acad. Sci. USA 86*:2829–2833.

Blackwell, T. K., Moore, M. W., Yancopoulos, G. D., Suh, H., Lutzker, S., Selsing, E., and Alt, F. W. (1986). Recombination between immunoglobulin variable region gene segments is enhanced by transcription. *Nature 24*:585–589.

Blattner, F. R., and Tucker, P. W. (1984). The molecular biology of immunoglobulin D. *Nature 307*:417–422.

Blomberg, B., Traunecker, A., Eisen, H., and Tonegawa, S. (1981). Organization of four mouse lambda light chain immunoglobulin genes. *Proc. Natl. Acad. Sci. USA 78*:3765–3769.

Brodeur, P. H., Thompson, M. A., and Riblet, R. (1984). The content and organization of mouse Igh-V families. *UCLA Symp. Mol. Cell. Biol.* New Series *18*:445. New York: Academic Press.

Burnet, M. (1958). *The Clonal Selection Theory of Acquired Immunity.* Nashville: Vanderbilt University Press.

Cebra, J. J., Colberg, J. E., and Dray, S. (1966). Rabbit lymphoid cells differentiated with respect to alpha-, gamma- and mu-heavy polypeptide chains and to allotypic markers Aa1 and Aa2. *J. Exp. Med. 123*:547–558.

Chun, J.J.M., Schatz, D. G., Oettinger, M. A., Jaenisch, R., and Baltimore, D. (1991). The recombination activating gene-1 (RAG-1) transcript is present in the murine central nervous system. *Cell 64*:189–200.

Davis, M. M., Kim, S. K., and Hood, L. (1980). DNA sequences mediating class switching in alpha-immunoglobulins. *Science 209*:1360–1365.

Desiderio, S. V., and Wolff, K. R. (1988). Rearrangement of exogenous immunoglobulin V_H and

DJ$_H$ gene segments after retroviral transduction into immature lymphoid cell lines. *J. Exp. Med. 167*:372–389.

Desiderio, S. V., Yancopoulos, G. D., Paskind, M., Thomas, E., Landau, N., Buss, M., Alt, F., and Baltimore, D. (1984). Insertion of N regions into heavy-chain genes is correlated with expression of terminal deoxytransferase in B cells. *Nature 311*:752–755.

D' Hoostelaere, L. A., Huppi, K., Mock, B., Mallett, C., and Potter, M. (1988). The IgK L chain allelic groups among the IgK haplotypes and IgK crossover populations suggest a gene order. *J. Immunol. 141*:652–661.

Dreyer, W. J., and Bennett, J. C. (1965). The molecular basis of antibody formation: A paradox. *Proc. Natl. Acad. Sci. USA 54*:864–869.

Early, P., Huang, H., Davis, M., Calame, K., and Hood, L. (1980a). An immunoglobulin heavy chain variable region is generated from three segments of DNA: V$_H$, D and J$_H$. *Cell 19*:981–992.

Early, P., Rogers, S., Davis, M., Calame, K., Bond, M., Wall, R., and Hood, L. (1980b). Two mRNAs can be produced from a single immunoglobulin mu gene by alternative RNA processing pathways. *Cell 20*:313–319.

Edelman, G. M., Cunningham, B. A., Gall, W. E., Gottlieb, P. D., Rutishauser, U., and Waxdal, M. J. (1969). The covalent structure of an entire gamma-G immunoglobulin molecule. *Proc. Natl. Acad. Sci. USA 63*:78–85.

Ephrussi, A., Church, G. M., Tonegawa, S., and Gilbert, W. (1985). B-lineage-specific interactions of an immunoglobulin enhancer with cellular factors in vivo. *Science 227*:134–140.

Falkner, F. G., and Zachau, H. G. (1984). Correct transcription of an immunoglobulin-kappa gene requires an upstream fragment containing conserved sequence elements. *Nature 310*:71–74.

Ferrier, P., Krippl, B., Blackwell, T. K., Furley, A.J.W., Suh, H., Winoto, A., Cook, W. D., Hood, L., Costantini, F., and Alt, F. W. (1990). Separate elements control DJ and VDJ rearrangement in a transgenic recombination substrate. *EMBO J. 9*:117–125.

Fletcher, C., Heintz, N. H., and Roeder, R. G. (1987). Purification and characterization of OTF-1, a transcription factor regulating cell cycle expression of a human histone H2b gene. *Cell 51*:773–740.

Freemont, P. S., Hanson, I. M., and Trowsdale, J. (1991). A novel cysteine-rich sequence motif. *Cell 64*:483–484.

Fulop, G. M., and Phillips, R. A. (1990). The *scid* mutation in mice causes a general defect in DNA repair. *Nature 347*:479–482.

Garcia, J. V., Stafford, J., and Queen, C. (1986). Synergism between immunoglobulin enhancers and promoters. *Nature 322*:383–385.

Ghosh, S., and Baltimore, D. (1990). Activation in vitro of NF-κB by phosphorylation of its inhibitor, IκB. *Nature 344*:678–682.

Ghosh, S., Gifford, A. M., Riviere, L. R., Tempst, P., Nolan, G. P., and Baltimore, D. (1990). Cloning of the p50 DNA binding subunit of NF-κB: homology to *rel* and *dorsal. Cell 62*:1019–1029.

Gillies, S. D., Morrison, S. L., Oi, V. T., and Tonegawa, S. (1983). A tissue-specific transcription enhancer element is located in the major intron of a rearranged immunoglobulin heavy chain gene. *Cell 33*:717–728.

Gough, N., and Bernard, O. (1981). Sequences of the joining region genes for immunoglobulin heavy chains and their role in generation of antibody diversity. *Proc. Natl. Acad. Sci. USA 78*:509–513.

Grosschedl, R., and Baltimore, D. (1985). Cell-type specificity of immunoglobulin gene expression is regulated by at least three DNA sequence elements. *Cell 41*:885–897.

Grosschedl, R., and Marx, M. (1988). Stable propagation of the active transcriptional state of an immunoglobulin gene requires continuous enhancer function. *Cell 55*:645–654.

Grosschedl, R., Weaver, D., Baltimore, D., and Costantini, F. (1984). Introduction of a mu immunoglobulin gene into the mouse germ line: Specific expression in lymphoid cells and synthesis of functional antibody. *Cell 38*:647–658.

Gu, H., Kitamura, D., and Rajewsky, K. (1991). B cell development regulated by gene rearrangement: Arrest of maturation by membrane-bound D_μ protein and selection of D_H element
reading frames. *Cell* 65:47–54.

Hagman, J., Rudin, C. M., Haasch, D., Chaplin, D., Storb, U. (1990). A novel enhancer in the
immunoglobulin λ locus is duplicated and functionally independent of NF-KB. *Genes
Develop.* 4:978–992.

Halligan, B., and Desiderio, S. V. (1987). Identification of a DNA-binding protein that recognizes
the nonamer recombinational signal sequence of immunoglobulin genes. *Proc. Natl. Acad.
Sci. USA* 84:7019–7023.

Hendrickson, E. A., Qin, X.-Q., Bump, E. A., Schatz, D. G., Oettinger, M., and Weaver, D. T.
(1991). A link between double-strand break-related repair and V(D)J recombination: The
scid mutation. *Proc. Natl. Acad. Sci. USA* 88:4061–4065.

Hesse, J., Lieber, M., Gellert, M., and Mizuuchi, K. (1987). Extrachromosomal DNA substrates in
pre-B cells undergo inversion or deletion at immunoglobulin V(D)J joining signals. *Cell*
49:775–783.

Hesse, J. E., Lieber, M. R., Mizuuchi, K., and Gellert, M. (1989). V(D)J recombination: A functional definition of the joining signals. *Genes Devel.* 3:1053–1061.

Honjo, T. (1983). Immunoglobulin genes. *Annu. Rev. Immunol.* 1:499–528.

Honjo, T., and Kataoka, T. (1978). Organization of immunoglobulin heavy chain genes and allelic
deletion model. *Proc. Natl. Acad. Sci. USA* 75:2140–2144.

Hozumi, N., and Tonegawa, S. (1976). Evidence for somatic rearrangement of immunoglobulin
genes coding for variable and constant regions. *Proc. Natl. Acad. Sci. USA* 73:3628–3632.

Hsieh, P., Meyn, M. S., and Camerini-Otero, R. D. (1986). Partial purification and characterization
of a recombinase from human cells. *Cell* 44:885–894.

Kataoka, T., Miyata, T., and Honjo, T. (1981). Repetitive sequences in class-switch recombination
regions of immunoglobulin heavy chain genes. *Cell* 23:357–368.

Kleinfield, R., Hardy, R. R., Tarlinton, D., Dangh, J., Herzenberg, L. A., and Weigert, M. (1986).
Recombination between an expressed immunoglobulin heavy-chain gene and a germline
variable gene segment in a Lyl$^+$ B-cell lymphoma. *Nature* 322:843–846.

Kurosawa, Y., and Tonegawa, S. (1982). Organization, structure and assembly of immunoglobulin
heavy-chain diversity DNA segments. *J. Exp. Med.* 155:201–218.

LaFaille, J. J., DeCloux, A., Bonneville, M., Takagaki, Y., and Tonegawa, S. (1989). Junctional
sequences of T cell receptor γδ genes: implications for γδ T cell lineages and for a novel intermediate of V-(D)-J joining. *Cell* 59:859–870.

Landau, N., Schatz, D. G., Rosa, M., and Baltimore, D. (1987). Increased frequency of N-region
insertion in a murine pre-B-cell line infected with a terminal deoxynucleotidyl transferase
retroviral expression vector. *Mol. Cell. Biol.* 7:3237–3243.

Lenardo, M., Pierce, J. W., and Baltimore, D. (1987). Protein-binding sites in Ig gene enhancers
determine transcriptional activity and inducibility. *Science* 236:1573–1577.

Levitt, D., and Cooper, M. D. (1980). Mouse pre-B cells synthesize and secrete mu heavy chains but
not light chains. *Cell* 19:617–625.

Lewis, S. A., Gifford, A., and Baltimore, D. (1985). DNA elements are asymmetrically joined during
the site-specific recombination of kappa immunoglobulin genes. *Science* 228:677–685.

Lewis, S. M., Hesse, J. E., Mizuuchi, K., and Gellert, M. (1988). Novel strand exchanges in V(D)J
recombination. *Cell* 55:1099–1107.

Li, M., Morzycka-Wroblewska, E., and Desiderio, S. V. (1989). NBP, a protein that specifically
binds an enhancer of immunoglobulin gene rearrangement: Purification and characterization. *Genes Develop.* 3:1801–1813.

Lieber, M. R., Hesse, J. E., Lewis, S., Bosma, G. C., Rosenberg, N., Mizuuchi, K., Bosma, M. J.,
and Gellert, M. (1988b). The defect in murine severe combined immune deficiency: Joining
of signal sequences but not coding segments in V(D)J recombination. *Cell* 55:7–16.

Lieber, M. R., Hesse, J. E., Mizuuchi, K., and Gellert, M. (1987). Developmental stage specificity
of the lymphoid V(D)J recombination activity. *Genes Develop.* 1:751–761.

Lieber, M. R., Hesse, J. E., Mizuuchi, K., and Gellert, M. (1988a). Lymphoid V(D)J recombination: Nucleotide insertion at signal joints as well as coding joints. *Proc. Natl. Acad. Sci. USA* 85:8588–8592.

Livant, D., Blatt, C., and Hood, L. (1986). One heavy chain variable region gene segment subfamily in the BALB/c mouse contains 500–1000 or more members. *Cell* 47:461–470.

Lutzker, S., Rothman, P., Pollock, R., Coffman, R., and Alt, F. W. (1988). Mitogen- and IL-4-regulated expression of germ-line Ig gamma-2b transcripts: Evidence for directed heavy chain class switching. *Cell* 53:177–184.

Manz, J., Denis, K., Witte, O., Brinster, R., and Storb, U. (1988). Feedback inhibition of immunoglobulin gene rearrangement by membrane mu, but not by secreted mu heavy chains. *J. Exp. Med.* 168:1363–1381.

Mather, E., and Perry, R. (1981). Transcriptional regulation of immunoglobulin V genes. *Nucleic Acids Res.* 9:6855–6867.

Matsunami, N., Kuze, K., Kangawa, K., Matsuo, H., Kawaichi, M., and Honjo, T. (1989). A protein binding to the J$_\kappa$ recombination sequence of immunoglobulin genes contains a sequence related to the integrase motif. *Nature* 342:934–937.

Max, E. E., Seidman, J., and Leder, P. (1979). Sequences of five potential recombination sites encoded close to an immunoglobulin kappa constant region gene. *Proc. Natl. Acad. Sci. USA* 76:3450–3454.

McCormack, W. T., Tjoelker, L. W., Carlson, L. M., Petryniak, B., Barth, C. F., Humphries, E. H., and Thompson, C. B. (1989). Chicken Ig$_L$ gene rearrangement involves deletion of a circular episome and addition of single nonrandom nucleotides to both coding segments. *Cell* 56:785–791.

Menetski, J. P., and Gellert, M. (1990). V(D)J recombination activity in lymphoid cell lines is increased by agents that elevate cAMP. *Proc. Natl. Acad. Sci. USA* 87:9324–9328.

Meyer, K. B., and Neuberger, M. S. (1989). The immunoglobulin kappa locus contains a second, stronger B-cell-specific enhancer which is located downstream of the constant region. *EMBO J.* 8:1959–1964.

Mombaerts, P., Iacomini, J., Johnson, R. S., Herrup, K., Tonegawa, S., and Papaioannou, V. E. (1992). RAG-1-deficient mice have no mature B and T lymphocytes. *Cell* 68:869–877.

Morzycka-Wroblewska, E., Lee, F.E.H., and Desiderio, S. V. (1988). Unusual immunoglobulin gene rearrangement leads to replacement of recombinational signal sequences. *Science* 242:261–263.

Mueller, M. M., Ruppert, S., Schaffner, W., and Matthias, P. (1988). A cloned octamer transcription factor stimulates transcription from lymphoid-specific promoters in non-B-cells. *Nature* 336:544–551.

Nolan, G. P., Ghosh, S., Liou, H.-C., Tempst, P., and Baltimore, D. (1991). DNA binding and IκB inhibition of the cloned p65 subunit of NF-κB, a *rel* related polypeptide. *Cell* 64:961–969.

Nussenzweig, M. C., Shaw, A. C., Sinn, E., Danner, D. G., Holmes, K. L., Morse, C., and Leder, P. (1987). Allelic exclusion in transgenic mice that express the membrane form of immunoglobulin mu. *Science* 236:816–819.

Oettinger, M. A., Schatz, D. G., Gorka, C., and Baltimore, D. (1990). RAG-1 and RAG-2, adjacent genes that synergistically activate V(D)J recombination. *Science* 248:1517–1523.

Ott, D., Alt, F. W., and Marcu, K. B. (1987). Immunoglobulin heavy chain switch region recombination within a retroviral vector in murine pre-B cells. *EMBO J.* 6:577–584.

Perlmutter, R. M., Crews, S. T., Douglas, R., Sorensen, G., Johnson, N., Nivera, N., Gearhart, P. J., and Hood, L. (1984). The generation of diversity in phosphorylcholine-binding antibodies. *Adv. Immunol.* 35:1–37.

Pernis, B., and Chiappino, G. (1964). Identification in human lymphoid tissues of cells that produce group 1 or group 2 gamma-globulins. *Immunology* 7:500–510.

Pernis, B., Chiappino, G., Kelus, A. S., and Gell, P.G.H. (1965). Cellular localization of immunoglobulins with different allotypic specificities in rabbit lymphoid tissues. *J. Exp. Med.* 122:853–875.

Peterson, C., Orthe, K., and Calame, K. L. (1986). Binding in vitro of multiple cellular proteins to immunoglobulin heavy-chain enhancer DNA. *Mol. Cell. Biol.* 6:4168–4178.

Pettersson, S., Cook, G. P., Bruggemann, M., Williams, G. T., and Neuberger, M. S. (1990). A second B cell-specific enhancer 3′ of the immunoglobulin heavy-chain locus. *Nature* 344:165–168.

Porter, R. R. (1967). The structure of the heavy chain of immunoglobulin and its relevance to the nature of the antibody-combining site. *Biochem. J.* 105:417–426.

Reth, M. G., and Alt, F. W. (1984). Novel immunoglobulin heavy chains are produced from DJ$_\mu$ gene segment rearrangements in lymphoid cells. *Nature* 312:418–423.

Reth, M., M. G., Ammirati, P. A., Jackson, S. J., and Alt, F. W. (1985). Regulated progression of a cultured pre-B-cell line to the B-cell stage. *Nature* 317:353–355.

Reth, M., Gehrmann, P., Petrac, E., and Wiese, P. (1986a). A novel V$_H$ to V$_H$DJ$_H$ joining mechanism in heavy-chain-negative (null) pre-B cells results in heavy chain production. *Nature* 322:840–842.

Reth, M., Jackson, S., and Alt, F. W. (1986b). V$_H$DJ$_H$ formation and DJ$_H$ replacement during pre-B differentiation: Non-random usage of gene segments. *EMBO J.* 5:2131–2138.

Reth, M., Petrac, E., Wiese, P., Lobel, L., and Alt, F. W. (1987). Activation of V$_\kappa$ gene rearrangement in pre-B cells follows the expression of membrane-bound immunoglobulin heavy chains. *EMBO J.* 6:3299–3305.

Ritchie, K. A., Brinster, R. L., and Storb, U. (1984). Allelic exclusion and control of endogenous immunoglobulin gene rearrangement in kappa transgenic mice. *Nature* 312:517–520.

Rogers, J., Early, P., Carter, C., Calame, K., Bond, M., Hood, L., and Wall, R. (1980). Two mRNAs with different 3′ ends encode membrane-bound and secreted forms of immunoglobulin mu chain. *Cell* 20:303–312.

Rogers, J., and Wall, R. (1984). Immunoglobulin RNA rearrangements in B lymphocyte differentiation. *Adv. Immunol.* 35:39–59.

Rosner, M. H., Vigano, M. A., Ozato, K., Timmons, P. M., Poirier, F., Rigby, P.W.J., and Staudt, L. M. (1990). A POU-domain transcription factor in early stem cells and germ cells of the mammalian embryo. *Nature* 345:686–692.

Sakano, H., Huppi, K., Heinrich, G., and Tonegawa, S. (1979). Sequences at the somatic recombination sites of immunoglobulin light-chain genes. *Nature* 280:288–294.

Sakano, H., Maki, R., Kurosawa, Y., Roeder, W., and Tonegawa, S. (1980). Two types of somatic recombination are necessary for the generation of a complete immunoglobulin heavy-chain gene. *Nature* 286:676–683.

Schatz, D. G., Oettinger, M. A., and Baltimore, D. (1989). The V(D)J recombination activating gene, RAG-1. *Cell* 59:1035–1048.

Scheidereit, C., Heguy, A., and Roeder, R. G. (1987). Identification and purification of a human lymphoid-specific octamer-binding protein (OTF-2) that activates transcription of an immunoglobulin promoter in vitro. *Cell* 51:783–793.

Schlissel, M. S., and Baltimore, D. (1989). Activation of immunoglobulin kappa gene rearrangement correlates with induction of germline kappa gene transcription. *Cell* 58:1001–1007.

Schroeder, H. W., Jr., Walter, M. A., Hufker, M. H., Ebers, A., VanDijk, K., Liao, L. C., Cox, D. W., Milner, C. B., and Perlmutter, R. M. (1988). Physical linkage of a human immunoglobulin heavy chain variable region gene segment to diversity and joining region elements. *Proc. Natl. Acad. Sci. USA* 85:8196–8200.

Seidman, J. G., and Leder, P. (1978). The arrangement and rearrangement of antibody genes. *Nature* 276:790–795.

Seidman, J., Max, E. E., and Leder, P. (1979). A kappa-immunoglobulin gene is formed by site-specific recombination without further somatic mutation. *Nature* 280:370–375.

Sen, R., and Baltimore, D. (1986a). Inducibility of kappa immunoglobulin enhancer-binding protein NF-κB by a posttranslational mechanism. *Cell* 47:921–928.

Sen, R., and Baltimore, D. (1986b). Multiple nuclear factors interact with the immunoglobulin enhancer sequences. *Cell* 46:705–716.

Sheriff, S., Silverton, E. W., Padlan, E. A., Cohen, G. H., Smith-Gill, J., Finzel, B. C., and Davies, D. R. (1987). Three dimensional structure of an antigen-antibody complex. *Proc. Natl. Acad. Sci. USA 84*:8075–8079.

Shimuzu, A., Takahashi, N., Yaoita, Y., and Honjo, T. (1982). Organization of the constant-region gene family of the mouse immunoglobulin heavy chain. *Cell 28*:499–506.

Shinkai, Y., Rathbun, G., Lam, K.-P., Oltz, E. M., Stewart, V., Mendelsohn, M., Charron, J., Datta, M., Young, F., Stall, A. M., and Alt, F. W. (1992). RAG-2-deficient mice lack mature lymphocytes owing to inability to initiate V(D)J rearrangement. *Cell 68*:855–867.

Singh, H., Sen, R., Baltimore, D., and Sharp, P. (1986). A nuclear factor that binds to a conserved sequence motif in transcriptional control elements of immunoglobulin genes. *Nature 319*:154–158.

Snapper, C. M., and Paul, W. E. (1987). Interferon-gamma and B cell stimulatory factor-1 reciprocally regulate Ig isotype production. *Science 236*:944–947.

Staudt, L., Singh, H., Sen, R., Wirth, T., Sharp, P. A., and Baltimore, D. (1986). A lymphoid-specific protein binding to the octamer motif of immunoglobulin genes. *Nature 323*:640–643.

Staudt, L. M., and Lenardo, M. J. (1991). Immunoglobulin gene transcription. *Annu. Rev. Immunol. 9*:373–398.

Stenzel-Poore, M. P., and Rittenberg, M. B. (1987). Immunoglobulin variable region heptamer-nonamer sequence joined to rearranged D-J segment: Implications for the immunoglobulin recombinase mechanism. *J. Immunol. 138*:3055–3059.

Storb, U., O'Brien, R. L., McMullen, M. D., Gollahon, K. A., and Brinster, R. L. (1984). High expression of cloned immunoglobulin kappa gene in transgenic mice is restricted to B lymphocytes. *Nature 310*:238–241.

Sturm, R., Baumrucker, B. R., Franza, J. R., and Herr, W. (1987). A 100-kD HeLa cell octamer binding protein (OBP 100) interacts differently with two separate octamer-related sequences within the SV40 enhancer. *Genes Develop. 1*:1147–1160.

Sutton, R. E., and Boothroyd, J. D. (1986). Evidence for trans splicing in trypanosomes. *Cell 47*:527–535.

Takeda, S., Masteller E. L., Thompson, C. B., and Buerstedde, J-M. (1992). RAG-2 expression is not essential for chicken immunoglobulin gene conversion. *Proc. Natl. Acad. Sci. U.S.A. 89*:4023–4027.

Tiselius, A., and Kabat, E. A. (1939). An electrophoretic study of immune sera and purified antibody preparations. *J. Exp. Med. 69*:119–132.

Tonegawa, S. (1983). Somatic generation of antibody diversity. *Nature 302*:575–581.

Turka, L. A., Schatz, D. G., Oettinger, M. A., Chun, J.J.M., Gorka, C., Lee, K., McCormack, W. T., and Thompson, C. B. (1991). Thymocyte expression of RAG-1 and RAG-2: Termination by T cell receptor cross-linking. *Science 253*:778–781.

Van Ness, B. G., Weigert, M., Coleclough, C., Mather, E. L., Kelley, D. E., and Perry, R. P. (1981). Transcription of the unrearranged mouse C_κ locus: sequence of the initiation region and comparison of activity with a rearranged V_κ-C_κ gene. *Cell 27*:593–602.

Wang, J. C., Caron, P. R., and Kim, R. A. (1990). The role of DNA topoisomerases in recombination and genome stability: A double-edged sword? *Cell 62*:403–406.

Wu, T. T., and Kabat, E. A. (1970). An analysis of the sequences of the variable regions of Bence-Jones proteins and myeloma light chains and their implications for antibody complexity. *J. Exp. Med. 132*:211–250.

Yancopoulos, G. D., and Alt, F. W. (1985). Developmentally controlled and tissue-specific expression of unrearranged V_H gene segments. *Cell 40*:271–281.

Yancopoulos, G. D., Blackwell, K., Suh, H., Hood, L., and Alt, F. W. (1986). Introduced T cell receptor variable region gene segments in pre-B cells: Evidence that B and T cells use a common recombinase. *Cell 44*:251–259.

7

Developmental Control of T Cell Antigen Receptor Gene Rearrangement and Expression

MITCHELL KRONENBERG,

ROBERT BRINES

Cell-mediated immunity is carried out by T (thymus-derived) lymphocytes, which have classically been divided into at least two subcategories based on function. Helper T cells act on B cells to stimulate immunoglobulin (Ig) production, and, depending on the lymphokines they secrete, helper T cells may also act on other cell types, including T lymphocytes and macrophages. Cytotoxic effector cells directly eliminate virus-infected or other antigenic cells. Both helper and cytotoxic T cells recognize antigens that are processed or degraded peptide fragments of larger protein antigens (Shimonkevitz et al., 1983). The peptide fragments are bound to major histocompatibility complex (MHC)-encoded cell surface molecules and are thereby presented to T cells (Bjorkman et al., 1987).

The T cell antigen receptor (TCR) responsible for recognition of antigen plus MHC was first defined in 1983 (Allison et al., 1982; Haskins et al., 1983; Meuer et al., 1983). It is a heterodimer, consisting of disulfide-bonded α (acidic isoelectric point) and β (basic isoelectric point) chains (Fig. 7–1). On SDS-polyacrylamide gels, the α and β polypeptides are rather similar in molecular size, between 38 and 49 kD, the differences in molecular size depending in part on the degree of glycosylation. The first cDNA clones encoding the TCR α and β chains were isolated at roughly the same time (Chien et al., 1984; Hedrick et al., 1984; Saito et al., 1984b; Yanagi et al., 1984). These were found using either subtractive hybridization or differential screening; in both cases, hybridization was used to define the small number of cDNA clones that distinguish T lymphocytes from B lymphocytes. The selected cDNA clones were then further screened for those clones that encode genes that undergo gene rearrangement in T cells. In this way, not only were clones encoding the TCR α and β chains isolated, but clones encoding a third polypeptide, γ, were also found (Saito et al., 1984a). In 1986 the first cells that express γ polypeptides were described in the human. The γ chain is coexpressed disulfide-bonded to a fourth TCR polypeptide called δ (Fig. 7–1) (Bank et al., 1986; Brenner et al., 1986). In 1987 the first genomic clones containing TCR δ gene sequences were isolated (Chien et al., 1987a).

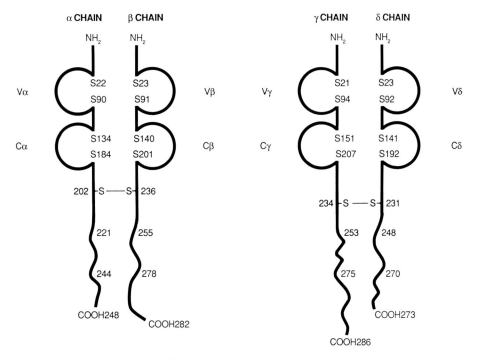

Figure 7–1.   Representative two-dimensional model of the $\alpha\beta$ and $\gamma\delta$ T cell receptor heterodimers. Each chain has two Ig homology units, a variable (V) homology unit at the N-terminal and a constant (C) homology unit toward the C-terminal. Each homology unit contains an internal disulfide (S–S) bond between cysteine residues spaced 50–73 amino acids apart. The two chains are linked by an S–S bond. The proteins are anchored into the cell membrane by hydrophobic transmembrane sequences that end in short, hydrophilic, carboxyl-terminal intracytoplasmic sequences. All four chains are glycosylated to varying degrees (not shown).

   The expression of TCR is subject to isotypic exclusion. This means that nearly all T cells express either α/β TCR or γ/δ TCR, but not both (Chen et al., 1988). In addition, it is probable that mixed heterodimers composed of α/γ or β/δ cannot be efficiently expressed at the cell surface (Saito et al., 1988) although at least one example of a β/δ TCR has been described (Hochstenbach and Brenner, 1989). T cells that express γ/δ TCR generally constitute less than 5% of the circulating, splenic, and lymph node T cells in mouse and human (Lanier et al., 1987; Lanier and Weiss, 1986; Pardoll et al., 1988). In the chicken, a much greater fraction of the T cells in these sites are γ/δ cells (Chen et al., 1988; Sowder et al., 1988). In all species studied, T cells that express γ/δ TCR arise earlier in ontogeny than do T cells that express α/β TCR (Pardoll et al., 1987a; Sowder et al., 1988). In mice and the chicken, they are also abundant in the intestinal epithelium (Bucy et al., 1988; Goodman and Lefrancois, 1988), and in mice they are predominant in the epidermis (Kuziel et al., 1987). It is not yet certain whether γ/δ TCR recognize antigens bound to MHC molecules in a fashion similar to α/β TCR. There are many unresolved questions about cells that express γ/δ TCR. However, their origin early in ontogeny, and their presence in tissues such as skin and the intestinal epithelium, which might be the first to contact pathogens, have given rise to new hypotheses about the development and function of the immune system.

In this chapter we will briefly review the structure of the T cell antigen receptor gene families. Three major issues will then be considered. *First, to what extent are the rearrangements of TCR genes ordered during ontogeny?* In the immunoglobulin (Ig) gene families, it has been found that gene rearrangements are highly ordered (reviewed in Yancopoulos and Alt, 1986). For example, in developing B cells, Ig heavy chain (H) genes rearrange before Ig light (L) chain genes. In T lymphocytes, four gene families can potentially rearrange. Is there order in this process too? In addition, within the repertoire of Ig H variable (V_H) gene segments, certain V gene segments rearrange preferentially early in the ontogeny of the immune system (Yancopoulos and Alt, 1986). The preference depends on chromosomal position: those V_H gene segments that are proximal to the joining (J) gene segments and constant region (C) genes rearrange first. This suggests a scanning model for rearrangement, whereby the recombination machinery in precursor B cells somehow scans the Ig V_H gene locus and rearranges the Ig V_H gene segments sequentially. Do the TCR V gene segments rearrange in a similar, ordered fashion? *Second, are T lymphocytes that express γ/δ TCR a separate lineage from those that express α/β TCR?* Because of the early appearance in ontogeny of cells that express γ/δ, original hypotheses included suggestions that γ/δ T cells might be precursors of α/β T cells (Reinherz, 1987). A variety of experiments argue against this model. It is likely, however, that the first TCR gene rearrangements occur in a common precursor to both γ/δ and α/β T cells. We will consider the evidence for and against this idea in the section on T lymphocytes and lineage that follows. *Third, is the γ/δ T cell population split into subpopulations that rearrange their TCR genes in the thymus at different times, and that go on to seed different sites in the organism?*

In discussing these issues, we will emphasize studies of mouse lymphocytes, although crucial information derived from experiments using human and chicken lymphocytes will also be considered.

T CELL ANTIGEN RECEPTOR GENE STRUCTURE

A diagram outlining the organization of the four T cell antigen receptor gene families in the mouse is shown in Figure 7–2 (for review, see Davis and Bjorkman, 1988; Kronenberg et al., 1986). Except for the γ gene locus, the human TCR gene families are very similar to those of the mouse in their overall gene organization. Although the number and organization of the elements vary between families, several features are common to all of them. In both the TCR and Ig gene families the V regions of the polypeptide, encompassing approximately 110 amino acids (Fig. 7–1), are encoded in the germline by separate gene segments. These include a V gene segment that encodes the 90–95 N-terminal amino acids and a J gene segment that encodes the 14–20 C-terminal amino acids of the V region. In the Ig H and the TCR β and δ gene families, a third diversity (D) gene segment that encodes a few amino acids between V and J is also present. In principle, any V gene segment within a gene family can recombine with any J (or D) gene segment within that family, thus substantially amplifying the diversity of possible V-region sequences. Gene rearrangements occur only in developing lymphocytes; for T cells, however, most, if not all, of the TCR gene rearrangements occur in the thymus.

A few features of TCR gene organization and rearrangement are relevant to the questions that we have defined in the introduction. First, V gene segments that are closely related to one another (>75%–80% nucleotide sequence similarity) can be conveniently

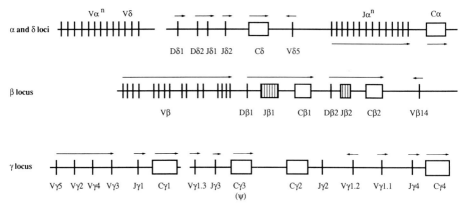

Figure 7–2. Genomic organization of the α, β, γ, and δ T cell receptor loci in the mouse (not to scale). Variable (V), joining (J), and diversity (D) gene segments are shown as vertical lines; flanking sequences are indicated by a horizontal line; and the C genes are shown by an open box. For simplicity, the separate exons and introns of the C genes are not indicated. The orientation of transcription (where known) is indicated with arrows. With the exception of the Vα and Vδ gene segments, coding sequences that have been physically linked are connected by a continuous horizontal line; a break in the line denotes groups of sequences that have not yet been physically linked. ψ indicates a pseudogene from which a functional coding sequence cannot be transcribed. For further details see the references in Davis and Bjorkman (1988) and Kronenberg et al. (1986).

grouped into subfamilies. In a widely used nomenclature, members of a subfamily are designated by the same digit(s), followed by a decimal point and a second number to designate the individual gene segments. Thus, according to this system, V1.1, V1.2, and V1.3 are all members of the V1 gene segment subfamily. The numbering is arbitrary. Second, the TCR α, β, and γ genes are encoded on separate chromosomes. However the α and δ gene loci are linked, the Dδ and Jδ gene segments and the Cδ gene are located between the Jα gene segments and the Cα gene segments on one side and the Vα gene segments on the other (Fig. 7–2). Rearrangement of Vα to Jα will therefore delete these δ gene sequences. Third, in the mouse, the number of known functional germline Vγ gene segments is quite low, 6–7. The number in humans is similarly low. For the δ locus, the number of germline V gene segments is probably also low, although an exact figure is difficult to determine. There are at least 80 mouse Vα gene segments grouped into 17 subfamilies. As the α and δ J gene segments are closely linked and in the same orientation, in principle, any of these Vα gene segments should be able to rearrange to the Dδ and Jδ gene segments as well as to Jα. In practice, however, there are a relatively small number of V gene segments in this locus that have been found to rearrange preferentially to Dδ and Jδ (Chien et al., 1987b; Elliot et al., 1988). Fourth, the rearrangement mechanism is imprecise. For example, when a germline V gene segment is joined to a germline J gene segment, nucleotides are often deleted from the ends of the gene segments (Fig. 7–3). In addition, non-germline elements or N regions are frequently found included in V-J or V-D-J junctions. These short sequences (1–15 base pairs) are nucleotides that are not encoded in the genome but are added between the joined gene segments (Fig. 7–3). The enzyme terminal deoxynucleotidyl transferase, which can add nucleotides to the 3' hydroxyl end of DNA in a template-independent manner, may be responsible for creating N regions (Desiderio et al., 1984). There are two consequences of this imprecision in gene rearrangement. The first, which is presumably beneficial to the organism, is increased diversity of V-region sequences. This mechanism may be very important for the generation of diversity of γ/δ TCR in particular because there is a very limited repertoire of germline Vγ and Vδ gene segments, and therefore most diversity is a result of gene rearrangement. However, because the number of added and/or deleted nucleotides is probably random, many of the rearranged TCR genes will be shifted in their reading frame so that there are in-frame stop codons. Genes that are not in the proper translational reading frame, and which therefore encode truncated and presumably nonfunctional proteins, are referred to as nonproductively rearranged.

The TCR genes are related to the Ig genes and therefore have many features in common with the Ig gene families. These include the presence of V, D, and J gene segments and C genes, and a rearrangement mechanism that leads to the generation of sequence diversity in the V-J junction. Nevertheless, a number of features specific to TCR gene organization and rearrangement probably have an important influence on T cell differentiation and function. Some of these include the location of the δ locus within the TCR α locus, the very large number of Jα gene segments, the limited number of germline Vγ gene segments, and the tendency of Vγ gene segments to rearrange to particular J gene segments.

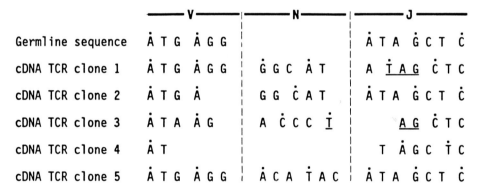

Figure 7–3.   Generation of junctional diversity during T cell receptor gene recombination. Nucleotide sequences of hypothetical T cell receptor cDNA clones in the region of the V-J junctions are shown. Each sequence results from the rearrangement of the identical germline V and J gene segments. Nucleotides encoded by the germline V segment are indicated on the left, and those encoded by the germline J segment are on the right. The central region, between the vertical broken lines, indicates additional N-region nucleotides not encoded by the germline V or J genes. A dot above a nucleotide indicates the first nucleotide in each triplet codon of the reading frame generated in each case. In clone 1 the addition of five N-region nucleotides between the V and J coding sequences causes a frame shift, leading to the generation of a stop codon (underlined). In clone 2 the same five N-region nucleotides are added, but a coincident deletion of two nucleotides from the V segment generates a functional reading frame. Conversely, in clone 3 the addition of N-region nucleotides combines with deletions in both V and J codons to generate a stop codon. Finally, the deletion (clone 4) or addition (clone 5) of nucleotides at the V-J junction generates functional reading frames in both cases.

ARE TCR GENE REARRANGEMENTS ORDERED?

Ordered Rearrangements of Different Gene Families

As noted above, previous work on Ig genes had shown that gene rearrangements in B lymphocytes are highly ordered with, for example, Ig H gene rearrangements occurring before light chain (Ig L) gene rearrangements. Because gene rearrangements usually delete the intervening DNA between the joined V, (D), and J gene segments, rearrangements are irreversible events. Therefore, Ig gene rearrangement is a valuable tool that can be used to assess the maturity of B cells. For instance, cells that have Ig V_H but not Ig κ gene rearrangements are judged to be less mature than those that have rearrangements in both gene families. For this reason, it was of interest to determine whether TCR genes also rearrange in order. Furthermore, an ordered rearrangement of TCR genes would imply that the α/β and γ/δ lineages are interconnected.

When the TCR γ gene family was discovered, it was found that γ gene transcripts appear in the fetal thymus before functional, full-length α and β gene transcripts (Haars et al., 1986; Raulet et al., 1985; Snodgrass et al., 1985). In addition, γ/δ TCR are expressed earlier than α/β TCR in ontogeny (Pardoll et al., 1987a; Sowder et al., 1988). These results, derived from analysis of the whole thymus, do not directly address the question of ordered rearrangement, however. It is possible, for example, that some cells in the fetal thymus rearrange γ and δ TCR genes and that others rearrange α and β only. Alternatively, some cells may rearrange γ/δ before α/β, but a significant minority could do the opposite. To attack this issue directly, cloned cells that represent different stages of differentiation are required. Most studies of ordered gene rearrangements in T cells therefore have employed two different types of material. First, hybridomas made by fusing mouse fetal thymocytes or immature adult thymocytes to the T lymphoma BW5147 have been characterized. Second, the surface phenotype and gene rearrangements present in a variety of mouse and human tumors have been determined.

Several conclusions can be drawn from the studies of hybridomas.

1. Fetal hybridomas generally have a stable phenotype with regard to gene rearrangement. In addition, hybridomas derived from early stages of ontogeny have more germline genes and have undergone fewer gene rearrangements than those derived from later stages. In this regard, the hybridomas are likely to be representative of the larger pool of normal thymocytes from which they are derived, as similar results have been observed when Southern blots of total thymus DNA from various stages were hybridized with TCR gene probes (Born et al., 1985; Haars et al., 1986).

2. γ and δ genes generally rearrange earliest. In some cells, δ genes rearrange before γ genes; in others, the γ gene family rearranges first (Brines and Kronenberg, unpublished data; Chien et al., 1987b).

3. The early Vγ and Vδ gene segment rearrangements are not random; most involve joining of Vγ3 to Jγ1 (Havran and Allison, 1988) and of Vδ1 to Dδ2 (Chien et al., 1987b). In the γ gene locus, the only possible rearrangements join V and J gene segments. The δ gene locus, however, contains V, D, and J gene segments, and the formation of a complete Vδ gene requires several steps that do not occur simultaneously. The earliest δ gene rearrangements can either join Dδ to Jδ, can fuse to two Dδ gene segments, or can join Vδ to Dδ gene segments (Chien et al., 1987b).

4. The β gene locus also has V, D, and J gene segments. As was described previously for the Ig H locus, the first rearrangements join a Dβ gene segment to one of the Jβ gene

segments. V gene segment rearrangement to the joined D-J follows as a discrete second step in both Ig V_H and $V\beta$ gene formation (Born et al., 1985). Rearrangements of V to D, without prior D-J gene segment joining, are not observed in this gene family. In the hybridomas, $D\beta$-$J\beta$ joining can occur as early as day 14, in some cases even before γ and/or δ gene rearrangement (Born et al., 1986; Brines and Kronenberg, unpublished data). $V\beta$ gene segment rearrangement nearly always occurs following $V\gamma$-$J\gamma$ joining (Born et al., 1986; Brines and Kronenberg, unpublished data). This suggests that TCR γ gene segment rearrangement may be an obligatory step in T cell differentiation before $V\beta$ gene segment rearrangement. However, there are noteworthy exceptions in which a hybridoma seems to have joined $V\beta$-$D\beta$-$J\beta$ without $V\gamma$ gene segment rearrangement. In these cases, there are several technical reasons why a $V\gamma$-$J\gamma$ gene rearrangement could have been missed in the hybridoma. First, many of the common $V\gamma$-$J\gamma$ gene segment rearrangements give rise to similarly sized restriction fragments. Because the BW5147 fusion partner also has several of these common γ gene rearrangement, a $V\gamma$-$J\gamma$ restriction fragment derived from the fetal thymocyte fusion partner in some cases cannot be distinguished from one derived from BW5147. Second, hybridomas with a tetraploid or near tetraploid chromosome content frequently lose chromosomes. A hybrid could therefore lose a chromosome that contains rearranged γ genes and retain only the one that does not. In summary, from these data it is not clear whether an early rearrangement of $V\gamma$ gene segments is an absolute requirement for subsequent $V\beta$ gene segment rearrangement, or whether it is only a tendency that a vast majority of differentiating T cells follow.

5. Alpha gene segment rearrangement is difficult to analyze on conventional Southern blots because there are many $J\alpha$ gene segments (Fig. 7–2) that are dispersed over 60 kb of DNA. Successful attempts to analyze $V\alpha$-$J\alpha$ gene segment rearrangement have been made using either pulsed-field gels or panels of probes that cover the $J\alpha$ region (Lindsten et al., 1987; Sangster et al., 1986). In the case of the hybridomas, the presence of α gene rearrangements was determined by hybridizing the genomic DNA with TCR δ probes (Brines and Kronenberg, unpublished data). For example, if $V\alpha$-$J\alpha$ gene segments had rearranged on both chromosomes, the $C\delta$ gene would be deleted. Similarly, the presence of two δ gene rearrangements indicates that TCR α gene rearrangement could not have occurred. By this analysis, it was judged that $V\alpha$ gene segment rearrangements occurred after $V\beta$-$D\beta$-$J\beta$ joining.

The results from studies on a variety of human lymphoid malignancies tend to be concordant with the fetal hybridoma results (reviewed in Greaves, 1986; Toyonaga and Mak, 1987). Where TCR α gene rearrangement or transcription was analyzed, the data were consistent with α gene rearrangement occurring as a late event after β gene rearrangement. Most of the T-lineage leukemias have rearrangements of both TCR β and γ genes. However, among the minority of malignancies that lack gene rearrangement in one of these gene families, cells that have TCR γ without β gene rearrangement are found more frequently than are cells with TCR β gene rearrangement only. In some of the cases where TCR β gene rearrangement seems to have occurred first, $D\beta$-$J\beta$ gene segment rearrangements were not distinguished from V-D-J joining. Therefore, the pattern is similar to that in the fetal hybridomas: TCR γ genes usually, if not always, rearrange first. This pattern was also observed when TCR gene rearrangements in B cell malignancies were studied. Although many B cell leukemias have germline TCR genes, TCR γ gene rearrangements are frequent in some types of B cell malignancies. In one study, 16% of B-lineage lymphocytic leukemias had TCR γ gene rearrangement and germline TCR β genes, whereas no cells had the opposite pattern of TCR β gene rearrangement with germline γ genes (Goorha et al., 1987). This suggests that some B lymphocytes may carry out the very early gene rearrangements that are part of the normal T cell differentiation pathway.

Ordered Rearrangement of V Gene Segments in a Gene Family

Are there any TCR V gene segments that rearrange preferentially early in ontogeny? If so, are these located proximal to the J gene segments and C gene(s) in that family? As noted above, it has been shown in the mouse Ig V_H family that the frequency of rearrangement of V_H gene segments at a particular time of development depends on chromosomal position (Yancopolous and Alt, 1986), suggesting that the rearrangement mechanism is capable of scanning the chromosome. This type of ordered mechanism may be needed to ensure that the entire V gene segment repertoire is eventually rearranged and expressed. Alternatively, those V gene segments expressed early might be needed for the response to certain critical pathogens. Finally, according to the network hypothesis of Jerne (1974), it is possible that generation and selection of the expressed V repertoire is influenced not by antigen, but by other V regions. If this were true, then early expression of certain V regions might be necessary for proper selection and development of the repertoire.

Several techniques have been employed to order the V gene segments in the TCR gene families. One powerful technique is Southern blot analysis of cloned cell lines that have TCR gene rearrangements. If a cell line has rearrangements on both chromosomes, and these rearrangements occur through deletion as is generally the case, then the intervening DNA will be deleted. The deleted DNA will contain V gene segments most proximal to the J gene segments and C gene segments, while more distal V gene segments will be retained in the genome. By analyzing a variety of cell lines with different rearrangements, a deletion map of the V gene locus can be constructed. This gives the gene order, but not the physical distance between V gene segments. To obtain the physical distance, analysis of genomic clones derived from libraries in cosmid vectors and/or pulsed-field gels can be used.

It is clear that chromosomal position can affect the timing and frequency of gene rearrangement in the TCR as well as in the Ig gene families. In the γ gene family, certain V gene segments preferentially rearrange to certain J gene segments. With few exceptions, the Vγ1.1 gene segment rearranges to Jγ4; Vγ1.2 rearranges to Jγ2; and Vγ3, Vγ4, Vγ2, and Vγ5 all rearrange to the Jγ1 gene segment (Fig. 7–2). The predominant rearrangements are between those V and J gene segments that are closest to one another (Woolf et al., 1988). Among the Vγ gene segments that rearrange to Jγ1, Vγ3 is the dominant early rearrangement (Havran and Allison, 1988), while Vγ2 and Vγ5 rearrange later (Garman et al., 1986). All of this is consistent with chromosomal position affecting rearrangement and is similar to the findings derived from the study of Ig V_H gene segments.

For the TCR β gene locus, the known Vβ gene segments have been mapped by the techniques described above (Fig. 7–4). A number of Vβ gene segments could be detected on Northern blots of day 16 fetal thymus RNA, including transcripts derived from the Vβ1, Vβ4, Vβ5, and Vβ8 gene segment subfamilies (Brines and Kronenberg, unpublished data). Day 16 is soon after the time when Vβ gene segment rearrangements first occur. Vβ14, which is Jβ-Cβ proximal but in the opposite transcriptional orientation to the other β gene segments (Fig. 7–2), is not detectably transcribed. Analysis of Vβ gene segment

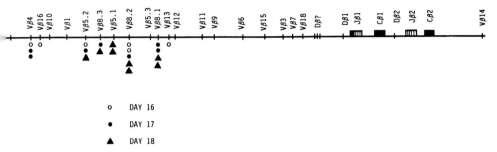

Figure 7–4. Organization of mouse TCR Vβ gene segments (not drawn to scale). Distribution over three days of V$_\beta$ gene usage in fetal thymocyte hybridomas with V$_\beta$-D$_\beta$-J$_\beta$ rearrangements. (Data for the Vβ gene segments are from Behlke et al., 1986; Lai et al., 1987; and Lee and Davis, 1988.) If a Vβ gene segment is found to be rearranged in one of the fetal hybridomas, it is indicated below the line (Brines and Kronenberg, unpublished data). The solid bars depict the exons of the C$_\beta$1 and C$_\beta$2 genes, and the striped bars denote the two clusters of tightly linked J$_\beta$ gene segments.

rearrangements in fetal hybridomas gave similar results (Fig. 7–4) (Brines and et al., Kronenberg, unpublished data). Although not all Vβ gene segments are rearranged and transcribed to the same extent at this stage, there is no preference for the expression of Jβ proximal V gene segments. By contrast, at day 16 in the fetal liver, nearly all the rearranged Ig V$_H$ gene segments in differentiating B cells are members of the Ig V$_H$ 7183 gene segment subfamily (Yancopolous and Alt, 1986). The fetal thymus has also been analyzed for Vβ gene segment expression by in situ hybridization (Pardoll et al., 1987b). On day 14, there was greater expression of the Vβ3 gene segment than the more distal Vβ5.1 gene segment, but at later stages of fetal development a greater number of thymocytes transcribed the Vβ5.1 gene segment (Fig. 7–4). This suggests that at the very earliest stages of T cell differentiation there might be some effect of chromosomal position on Vβ gene segment rearrangement and expression. It must be noted, however, that the transcription of Vβ gene segments from germline as opposed to rearranged gene segments could have been detected on day 14. Transcription from unrearranged Ig V$_H$ gene segments in immature β cells has been reported (Yancopolous and Alt, 1986).

TCR δ genes are transcribed several days earlier than are α genes (Chien et al., 1987a). For the TCR Vα/Vδ locus, most of those V gene segments that rearrange preferentially to Dδ and Jδ, as opposed to Jα, gene segments are J-C gene proximal. As noted above, the predominant early rearrangements involve Vδ1, which, with the exception of Vδ5, is closest to the J gene segments and C genes (Chien et al., 1987b; J. Klotz, K. Wang, L. Hood, and M. Kronenberg, unpublished manuscript). Vδ5 is similar to Vβ14 because it is in the opposite orientation compared to the J gene segments and C genes, and it must therefore rearrange by inversion (Fig. 7–2). It is commonly rearranged in adult CD4$^-$ CD8$^-$ thymocytes (Iwashima et al., 1988). Several mouse V gene segments can rearrange to either Jα or Jδ gene segments, and most of these V gene segments are also rather proximal to the J gene segments (J. Klotz, K. Wang, L. Hood, and M. Kronenberg, unpublished manuscript). On the other hand, some J gene segment-proximal Vα gene segments are dispersed among the Vδ gene segments that rearrange to Jα and apparently only rarely or never rearrange to Jδ (J. Klotz, K. Wang, L. Hood, and M. Kronenberg, unpublished manuscript). Similar results have been reported for the human α/δ locus (Satyanarayana et al., 1988). At least some Vδ gene segments are J-C proximal, but so are several Vα gene segments. Finally, the preference for proximal V gene segments to rearrange to Jδ gene segments is more pronounced for fetal than for adult thymocytes. In the adult thymus, more distal V gene segments have been found rearranged to Dδ-Jδ gene segments (Elliot et al., 1988; Korman et al., 1988; J. Klotz, K. Wang, L. Hood, M. Kronenberg, unpublished manuscript). The data from the α/δ locus is therefore broadly consistent with a scanning type model, although the position of a V gene segment is not the sole determinant of the frequency of rearrangement at a given stage of ontogeny.

ARE T LYMPHOCYTES THAT EXPRESS γ/δ TCR A SEPARATE LINEAGE FROM THOSE THAT EXPRESS α/β?

Several of the original experiments on TCR γ genes and expression of γ/δ antigen receptors suggested that these cells could be precursors of the lymphocytes that express α/β TCR. According to this view (Fig. 7–5a), the γ/δ TCR is expressed in thymus and perhaps plays some role in T cell differentiation there. Once the thymocytes mature, they would then switch to expression of α/β TCR. Several findings argue against this possibility. First, although γ/δ T cells are predominant in fetal thymus and are present in adult thymus, they are also found to a significant extent in the peripheral lymphoid organs and epithelial layers. The peripheral γ/δ cells are functional lymphocytes that secrete lymphokines and

PRECURSOR MODEL

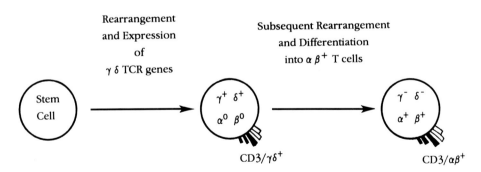

Figure 7–5a.   Precursor model for γδ T cell function. The number of differentiation steps between hemopoietic stem cells and cells entering the thymus and committed to the T cell lineage is unknown. Cells enter the thymus and rearrange and express γδ TCR genes. After subsequent development and selection in the thymus, these precursor cells rearrange and express their αβ TCR genes, allowing for the final stage of thymic education. A zero superscript denotes germline genes; a plus superscript represents productively rearranged genes; and a minus superscript represents nonproductively rearranged genes.

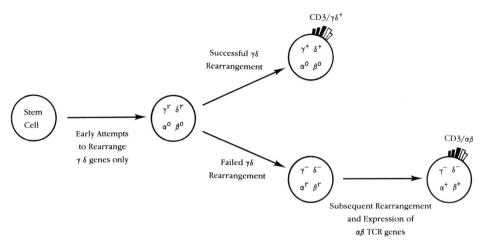

Figure 7–5b.   Model of progressive thymocyte ontogeny involving initial programmed rearrangement of γ and δ genes. Cells enter the thymus and attempt to rearrange γ and δ gene segments. Productive rearrangement at both loci commits the thymocyte to γδ expression, and Vα and Vβ gene segment rearrangements therefore do not occur. If nonproductive γ and δ gene rearrangements occur, α and β gene rearrangement and expression of αβ TCR can follow.

INDEPENDENT REARRANGEMENT MODEL

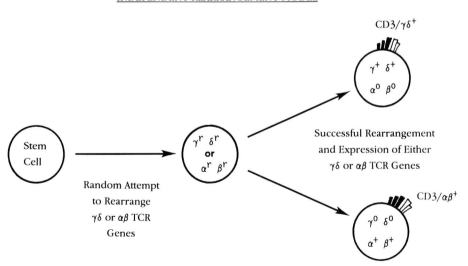

Figure 7–5c.   Model of random TCR gene rearrangement involving independent rearrangement of either γ and δ or α and β genes. This model proposes that cells attempt to rearrange γδ or αβ genes at random; the first set of genes to productively rearrange would go forward to form the functional receptor and exclude expression of the alternative dimer. The TCR genes that are not expressed could be germline (as indicated in the figure), or they could be nonproductively rearranged.

that have demonstrable cytotoxic activity (Borst et al., 1987; Matis et al., 1987). These data indicate that the γ/δ TCR are not solely expressed by immature precursors. Second, although most α/β T lymphocytes have TCR γ gene rearrangements (Heilig et al., 1985), in the majority of cases these rearrangements are nonproductive (Reilly et al., 1986). It is difficult to understand how a productive γ gene rearrangement postulated in the model, could become nonproductive later on. Because of these results, the precursor model is probably not correct.

Several alternative hypotheses for the origin of separate α/β and γ/δ T lymphocyte populations have been put forward. A progressive rearrangement model (Lanier and Allison, 1987; Pardoll et al., 1987a), based on the expression of TCR in fetal thymus and the finding of mostly nonproductive γ gene rearrangements in cells that express α/β TCR, is shown in Figure 7–5b. According to this hypothesis, lymphocytes arriving in the thymus attempt to rearrange the TCR γ and δ genes first. If a productive γ and δ gene rearrangement takes place, then the T lymphocyte expresses a γ/δ TCR and will remain a γ/δ T cell. On the other hand, if nonproductive γ or δ gene rearrangements take place on both chromosomes, then the cell will progress to rearrange its α and β genes.

Another alternative model proposes that random TCR gene rearrangements play a role in determining whether a lymphocyte will express α/β or γ/δ TCR (Fig. 7–5c). According to this hypothesis, although biases in the frequency of gene rearrangement in different TCR gene families may exist, the different gene families can rearrange in any order. Productive gene rearrangement(s) in one or several (i.e., α/β or γ/δ) gene families will then shut off further rearrangement. If this is the case, then an obvious point to exert control is the rearrangement of V gene segments in the TCR α/δ locus. Vα-Jα gene segment rearrangement deletes the δ locus, eliminating the possibility of TCR δ gene expression. A frequent rearrangement in human thymocytes has been described (de Villartay et al., 1988) that achieves this δ gene deletion, before Vα-Jα joining, by fusing a non-coding sequence upstream of Dδ1 to a 5' or Cα-distal Jα gene segment. It has been proposed that this rearrangement could be involved in regulating TCR expression in differentiating T lymphocytes.

At first glance, the independent or random model is not consistent with the data derived from fetal hybridomas and malignant lymphocytes discussed above. However, if we assume that the tendency to rearrange Vγ before Vβ gene segments is not absolute and not relevant to the determination of T cell fate, a critical issue that might distinguish between the progressive and random rearrangement models is whether cells that have rearranged Vα gene segments already had a Vδ gene segment rearrangement. If they always did, this would further support the progressive rearrangement model. If not, then the random model would be corroborated. In principle, this issue could be resolved by examining the circular DNA that is present in thymocytes (Okazaki et al., 1987). The circular DNA is the excision product that results from V-J or V-D-J joining, the chromosomal deletion of the intervening DNA, and the formation of a circle by the reciprocal joining of the two breakpoints (Fig. 7–6). Circular DNA present in thymocytes could be examined for the presence of germline TCR Dδ and Jδ gene segments; germline DNA should not be present if the progressive rearrangement model is correct. Two recent studies of the circular excision products found in thymocyte DNA reached very different conclusions. One group (Takeshita et al., 1989) found that the excision products contain mostly rearranged TCR δ genes, in support of the progressive model. A second group (Winoto and Baltimore, 1989b) found mostly germline δ genes in the circular excision

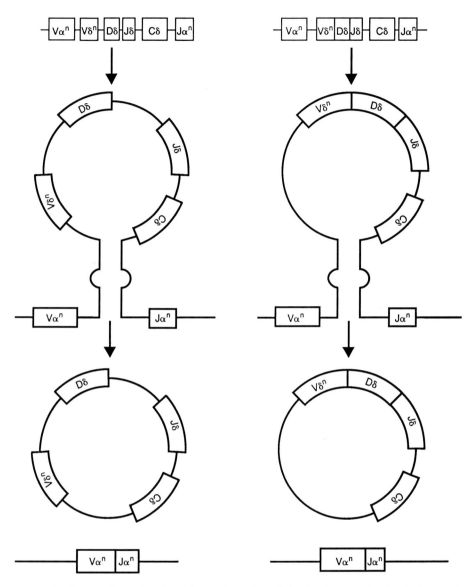

Figure 7–6. Schematic model of excision and creation of circular DNA molecules during TCR gene rearrangement. (a) Germline α genes initiate a $V\alpha$ to $J\alpha$ gene rearrangement, looping out the intervening germline δ genes. The final excision and recombination process generates a contiguous $V\alpha$-$J\alpha$ coding sequence and a circular product containing the δ genes in their germline configuration. Such a product should not be found if the progressive rearrangement model (Fig. 7–5b) is correct. (b) After an initial nonproductive attempt to rearrange a functional V-D-J at the δ locus, there is a subsequent successful $V\alpha$ to $J\alpha$ rearrangement. The circular excision product contains the δ genes in a rearranged configuration.

products. There is no simple way to account for this discrepancy, although it can be explained in part by technical factors related to the isolation and cloning of the circular DNA.

The progressive and random TCR gene rearrangement models both postulate that the gene product of a productive TCR gene rearrangement will effect a feedback inhibition on further TCR gene rearrangements. Finally, it is possible that there is no such feedback inhibition. In this case, there are a variety of hypothetical mechanisms that could account for isotypic exclusion. For example, TCR molecules are assembled with and coexpressed with the invariant CD3 molecule (reviewed in Clevens et al., 1988). Both assembly with CD3 and competition for CD3 polypeptides could affect isotypic exclusion.

Currently no data exist that permit a definitive resolution of the issues raised by these competing models. Several T cell clones and hybridomas have been described that express γ/δ TCR and that have Vβ gene segment rearrangements and full-length β gene transcripts (Asarnow et al., 1988; Brines and Kronenberg, unpublished manuscript). If we exclude the possibility that the observed TCR Vβ gene segment rearrangements are an abnormal event resulting from in vitro culture or fusion of the T cells, none of these cases are consistent with the progressive rearrangement model outlined in Figure 7–5b. As shown in the figure, a T cell with productive gene rearrangements, and therefore able to synthesize a γ/δ TCR, should not ordinarily rearrange its Vβ gene segments. Perhaps productive γ and δ TCR gene rearrangements and γ/δ TCR expression have an effect only on TCR Vα gene segment rearrangement.

On the other hand, if the progressive rearrangement model is not correct, it is difficult to understand why TCR γ gene rearrangements occur early and why they are mostly nonproductive in α/β T cells. Transgenic mice that have productively rearranged TCR genes in their germline DNA could provide a critical test of these models. Such mice could be tested for effects on the differentiation of the α/β and γ/δ T cell populations. So far however, there are two conflicting reports on TCR γ gene rearrangement in transgenic mice that contain a rearranged Vβ8 gene in their genome (Fenton et al., 1988; Uematsu et al., 1988). Although the animals generated in both laboratories express relatively high levels of TCR-containing Vβ8, in one case inhibition of γ gene rearrangement was reported (Fenton et al., 1988), whereas in the other case it was not (Uematsu et al., 1988). Recent experiments (Winoto and Baltimore, 1989a) suggest that transcriptional silencers could play an important role in the decision to rearrange and express a particular TCR gene locus. For example, silencers in the TCR α locus prevent the expression of TCR α genes in $\gamma\delta$ T cells. They do this by negatively regulating the action of the TCR α enhancer. Silencers in the α and perhaps also in other TCR loci could determine the ability of the chromatin to assume an open configuration, which may be required for transcription and gene rearrangement.

ARE THERE SUBPOPULATIONS OF γ/δ T LYMPHOCYTES?

Increasing evidence suggests that the set of T lymphocytes that express γ/δ TCR may be split into further subpopulations. These different γ/δ subpopulations may rearrange their TCR genes and attain immunocompetence at distinct stages of development; they may express different sets of TCR γ and δ V gene segments; they may migrate to different peripheral sites; and, finally, they may recognize entirely different types of antigens.

Subpopulations in the Thymus

In the mouse, the earliest cells that express TCRs are found at about day 14 of gestation. Nearly all of these cells express a TCR that is encoded by the Vγ3 gene segment (Havran and Allison, 1988; Ito et al., 1989), and many also have Vδ1 gene segment rearrangements (Chien et al., 1987b). The rearrangements at this stage exhibit very few deleted nucleotides and non-germline elements (Chien et al., 1987b). Therefore the rearranged gene in any one T cell is likely to be similar or identical in sequence to the rearrangement in any other. By days 16–17 gestation, the majority of fetal thymocytes that express TCR still have γ/δ TCR, but the proportion of these cells that express the Vγ3 gene segment has declined markedly (Havran and Allison, 1988). A greater frequency of rearrangement and expression of the Vγ2, Vγ4, Vγ1.2, and Vγ1.1 gene segments is observed (Garman et al., 1986; Houlden et al., 1988; Ito et al., 1989). Some of these fetal TCR also contain the Vδ5 gene segment (Ito et al., 1989), although productive rearrangement of other Vδ gene segments is observed at this time (Chien et al., 1987b). Still later in development, predominant rearrangement and expression of the Vγ4 gene segment together with the Vδ1 gene segment is frequent in neonatal thymocytes (Ito et al., 1989).

In the adult mouse, in the small population of thymocytes ($<3\%$) that express neither CD4 nor CD8, there is a subpopulation of lymphocytes that express γ/δ TCR (Lanier and Weiss, 1986). These cells have rearrangements predominantly of the Vγ1.1, Vγ2, and Vγ4 gene segments for the γ gene locus, and Vδ5 is the most common V gene segment rearranged in the δ gene locus (Elliot et al., 1988; Houlden et al., 1988; Iwashima et al., 1988, Korman et al., 1988). These adult γ/δ thymocytes therefore are somewhat similar to the fetal thymocytes that appear around days 16–17.

In conclusion, profound changes are exhibited in the types of γ/δ TCR expressed during thymus development. It is unclear exactly how many subpopulations exist, but the earliest cells at days 14–15 are quite distinct from the later γ/δ T cells. The neonatal γ/δ hybridomas (Ito et al., 1989) may also constitute a third, distinct category. The observed changes do not only involve the expression of V gene segments, but include J gene segments and C genes as well. The early fetal thymocyte Vδ1 gene segment rearrangements join Vδ1 to the Dδ2 and Jδ2 gene segments (Chien et al., 1987b; Ito et al., 1989). The resulting, complete Vδ gene is composed of the Vδ1-Dδ2-Jδ2 gene segments. The later rearrangements usually include two Dδ gene segments and the Jδ1 gene segment (Elliot et al., 1988). The resulting Vδ gene is therefore composed of four gene segments: Vδ-Dδ1-Dδ2-Jδ1. Because deleted nucleotides and N regions are found at each of the three junctions in these rearrangements, the potential for sequence diversity resulting from these later δ gene rearrangments is enormous (Chien et al., 1987b; Davis and Bjorkman, 1988). This is in diametrical contrast to the early Vδ1 gene segment rearrangements where N-region diversity is virtually absent. Similarly, because Vγ gene segments tend to rearrange to the nearest Jγ gene segment and C gene, the shift in the γ/δ T cell population from cells that mostly express the Vγ3 and Vγ4 gene segments to cells that express Vγ1.1 and Vγ1.2 also results in a shift from Cγ1 gene expression to expression of the Cγ2 and Cγ4 genes (Houlden et al., 1988).

Subpopulations in Peripheral Sites

As discussed earlier, in mammals γ/δ T cells usually constitute a minority ($<5\%$) of the T lymphocytes in the circulatory system, spleen, and lymph nodes (Lanier et al., 1987;

Lanier and Weiss, 1987b). In the chicken, they constitute a much greater fraction (20%–50%) of the total lymphocytes in these sites (Sowder et al., 1988). However, γ/δ T cells are the predominant lymphoid population in the mouse intestinal epithelium (Bonneville et al., 1988; Goodman and Lefrancois, 1988). In the mouse, they are also the major lymphocyte population in the skin (Kuziel et al., 1987). Extrapolating from these results, it is postulated that they are perhaps present in other epithelia, including the lung, and in the mammary gland (Janeway et al., 1988), but experiments to demonstrate this have not been published.

The γ/δ T cells found in the mouse skin are Thy-1$^+$, but CD4$^-$ and CD8$^-$. They have a dendritic morphology and are therefore referred to as dendritic epidermal cells (DECs). The DECs have rearrangements of Vγ3 to Jγ1 and Vδ1 to Jδ2 (Asarnow et al., 1988). These rearrangements generally lack both nucleotide deletions and non-germline elements. In all these features, the DECs are similar to the very early (days 14–15) fetal thymus γ/δ T cells. Furthermore, DECs with γ and δ TCR gene rearrangements are not present in athymic mice (Nixon-Fulton et al., 1988). This suggests that the very early fetal thymocyte population leaves the thymus and colonizes the skin.

By contrast, the γ/δ T cells in the mouse spleen express one of several TCR γ chains, composed of either Vγ1.2-Jγ2-Cγ2, Vγ2-Jγ1-Cγ1, or Vγ1.1-Jγ4-Cγ4 (Cron et al., 1988; Jones et al., 1986). There is less information on Vδ gene segment rearrangements in the mouse peripheral γ/δ T cells, but in the human most of the TCR δ chains in peripheral blood lymphocytes contain a particular Vδ gene segment (Loh et al., 1989). All these rearrangements in peripheral mouse and human T cells have significant diversity in the V-J or V-D-J junctions. In this regard, the mouse T cells are similar to the γ/δ thymocytes found late in fetal development and in the CD4$^-$/8$^-$ subpopulation of thymocytes in the adult. A thymic dependence of these cells is likely, as they are depleted in young, athymic mice (Pardoll et al., 1988). We may speculate therefore that the splenic γ/δ cells in the mouse derive from thymic precursors that arise relatively late in development (day 16 onward).

A third population of γ/δ T lymphocytes found in the intestinal epithelium has also been characterized for TCR γ gene expression. The intestinal epithelial lymphocytes (IELs) have transcripts and productive rearrangements of the Vγ5 and to a lesser extent, of the Vγ4 gene segments. (Bonneville et al., 1988). As was observed for the splenic γ/δ T cells, the TCR γ gene rearrangements in the IELs exhibit both deleted nucleotides and non-germline elements (Asarnow et al., 1989). The thymus dependence of these cells has not been determined. γ/δ T cells with this phenotype cannot be demonstrated in the intestinal epithelium until several weeks after birth (R. Levy, and M. Kronenberg, unpublished manuscript).

SUMMARY AND PROSPECTS

The discovery of TCR γ and δ genes, and the subsequent discovery of T cells that express γ/δ TCR, has led to new hypotheses about the ontogeny of the immune system and the role that gene rearrangement plays in the T cell differentiation process. In addition, the distribution of γ/δ T lymphocytes has led to new investigations of immune surveillance in epithelia. None of the data considered in this chapter directly address the question of the ligand recognized by γ/δ TCR. Is it processed antigen presented by MHC class I and class II molecules, as is true for cells that express an α/β TCR? The complete lack of sequence diversity in the DECs indicates that this is not likely for the epidermal subpopulation. Instead, these lymphocytes may recognize a monomorphic stress protein induced by infection or some other insult to the cells (Asarnow et al., 1988; Janeway et al., 1988). The molecular structure of this putative distress signal is unknown, although it has been suggested that it could be similar either to one of the heat shock–induced proteins or to MHC class I genes. Those γ/δ T cells that have significant TCR sequence diversity in the V-J and V-D-J junctions on account of rearrangement, however, may recognize a totally different, highly diverse set of antigens. These speculations are consistent with the hypothesis that there are subpopulations of γ/δ T cells that not only can be distinguished by their TCR gene expression, the time when they differentiate, and their location, but that also differ with regard to the diversity and types of antigens they recognize.

If γ/δ T cells are further divided into subpopulations, then a differentiating T cell not only has the choice of whether to become an α/β or γ/δ lymphocyte, but also it must be determined what type of γ/δ T cell it will become. If this is the case, then the questions raised earlier about the role of gene rearrangement in this decision process will have to be reformulated, in light of the greater complexity of γ/δ phenotypes seen among mature T cells. Although most of these developmental issues remain unresolved, we can anticipate some progress from the results of experiments in progress that involve TCR γ and δ gene transgenic mice. The recent development of antibodies that recognize native TCR is also of great importance. As sophisticated tools are being applied to these problems, we can anticipate a much deeper understanding of T cell differentiation in the near future.

Acknowledgments

We thank Phil Mixter for help with preparation of the figures, Mary Stepp for help with preparation of the manuscript, and our colleagues for stimulating discussions. Supported by NIH grant RO1 CA 45956. R. Brines was supported by the California Institute of Cancer Research and is currently a fellow of the California Cancer Research Coordinating Committee.

References

Allison, J. P., McIntyre, B. W., and Block, D. (1982). Tumour specific antigen of murine T lymphoma defined with monoclonal antibody. *J. Immunol. 129*:2293–2300.

Asarnow, D. M., Kuziel, W. A., Bonyhadi, M., Tigelaar, R. E., Tucker, P. W., and Allison, J. P. (1988). Limited diversity of $\gamma\delta$ antigen receptor genes of Thy-1[+] dendritic epidermal cells. *Cell 55*:837–847.

Asarnow, D. M., Goodman, T., Lefrancois, L., and Allison, J. P. (1989). Distinct antigen receptor repertoires of two classes of murine epithelium-associated T cells. *Nature 341*:60–62.

Bank, I., DePinho, R. A., Brenner, M. B., Cassimeris, J., Wal, F. W., and Chess, L. (1986). A functional T3 molecule associated with a novel heterodimer on the surface of immature human thymocytes. *Nature 322*:177–181.

Behlke, M., Chou, H., Huppi, K., and Loh, D. (1986). Murine T cell receptor mutants with deletions of β-chain variable region genes. *Proc. Natl. Acad. Sci. USA 83*:767–771.

Bjorkman, P. J., Saper, M. A., Samraoui, B., Bennett, W. S., Strominger, J. L., and Wiley, D. C. (1987). Structure of the human class I histocompatibility antigen, HLA-A2. *Nature 329*:506–512.

Bonneville, M., Janeway, Jr., C. A., Ito, K., Haas, W., Ishida, I., Nakanishi, N., and Tonegawa, S. (1988). Intestinal intraepithelial lymphocytes are a distinct set of $\gamma\delta$ T cell. *Nature 336*:479–481.

Born, W., Rathbun, G., Tucker, P., Marrack, P., and Kappler, J. (1986). Synchronized rearrangement of T cell gamma and beta genes in fetal thymocyte development. *Science 234*:479–482.

Born, W., Yague, J., Palmer, E., Kappler, J., and Marrack, P. (1985). Rearrangement of T cell receptor β chains during T cell development. *Proc. Natl. Acad. Sci. USA* 82:2925–2929.

Borst, J., van de Griend, R. J., van Oostveen, J. W., Ang, S. L., Melief, C. J., Seidman, J. G., and Bolhuis, R.L.H. (1987). A T cell receptor γ/CD3 complex found on cloned functional lymphocytes. *Nature* 325:683–688.

Brenner, M. B., McLean, J., Dialynas, D. P., Strominger, J. L., Smith, J. A., Owen, F. L., Seidman, J. G., Ip, S., Rosen, F., and Krangel, M. S. (1986). Identification of a putative second T cell receptor. *Nature* 322:145–149.

Bucy, R. P., Chen, C.-L. H., Cihak, J., Losch, U., and Cooper, M. D. (1988). Avian T cells expressing γδ receptors localize in the splenic sinusoids and the intestinal epithelium. *J. Immunol.* 7:2200–2205.

Chen, C. H., Cihak, J., Losch, U., and Cooper, M. D. (1988). Differential expression of two T cell receptors, TcR1 and TcR2, on chicken lymphocytes. *Eur. J. Immunol.* 18:539–543.

Chien, Y., Becker, D. M., Lindsten, T., Okamura, M., Cohen, D. I., and Davis, M. M. (1984). A third type of murine T-cell receptor gene. *Nature* 312:31–35.

Chien, Y., Iwashima, M., Kaplan, K., Elliot, J. F., and Davis, M. M. (1987a). A new T-cell receptor gene located within the alpha locus and expressed early in T-cell differentiation. *Nature* 327:677–682.

Chien, Y., Iwashima, M., Wettstein, D. A., Kaplan, K. B., Elliot, J. F., Born, W., and Davis, M. M. (1987b). T-cell receptor δ gene rearrangements in early thymocytes. *Nature* 330:722–727.

Clevens, H., Alarcon, B., Wileman, T., and Terhorst, C. (1988). The T cell receptor/CD3 complex: A dynamic protein ensemble. *Annu. Rev. Immunol.* 6:629–662.

Cron, R. Q., Konin, F., Maloy, W. L., Pardoll, D., Coligan, J. F., and Bluestone, J. A. (1988). Peripheral murine CD3⁺, CD4⁻ T lymphocytes express novel T cell receptor γδ structures. *J. Immunol.* 141:1074–1082.

Davis, M. M., and Bjorkman, P. J. (1988). T-cell antigen receptor genes and T-cell recognition. *Nature* 344:395–401.

Desiderio, S. V., Yancopoulos, G. D., Paskind, M., Thomas, E., Boss, M. A., Landau, N., Alt, F. W., and Baltimore, D. (1984). Insertion of N regions into heavy chains is correlated with expression of terminal deoxytransferase in B cells. *Nature* 311:752–755.

de Villarty, J.-P., Hockett, R. D., Coran, D., Korsmeyer, S. J., and Cohen, D. I. (1988). Deletion of the human T-cell receptor δ-gene by a site-specific recombination. *Nature* 335:170–174.

Elliot J. F., Rock, E. P., Patten, P. A., David, M. M., and Chien, Y.-H. (1988). The adult T-cell receptor δ-chain is diverse and distinct from that of fetal thymocytes. *Nature* 331:627–631.

Fenton, G. F., Marrack, P., Kappler, J. W., Kanagawa, O., and Seidman, J. G. (1988). Isotypic exclusion of γδ T cell receptors in transgenic mice bearing a rearranged β-chain gene. *Science* 241:1089–1092.

Garman, R. D., Doherty, P. J., and Raulet, D. H. (1986). Diversity, rearrangement and expression of murine T cell gamma genes. *Cell* 45:733–742.

Goodman, T. and Lefrancois, L. (1988). Expression of the γδ T cell receptor on intestinal CD8⁺ intraepithelial lymphocytes. *Nature* 333:855–858.

Goorha, R., Bunin, N., Mirro, Jr., J., Murphy, S. B., Cross, A. H., Behm, F. G., Quartermous, T., Seidman, J., and Kitchingman, G. R. (1987). Provocative pattern of rearrangements of the genes for the γ and β chains of the T-cell receptor in human leukemias. *Proc. Natl. Acad. Sci. USA* 84:4547–4551.

Greaves, M. F. (1986). Differentiation-linked leukemogenesis in lymphocytes. *Science* 234:697–704.

Haars, R., Kronenberg, M., Gallatin, M. G., Weissman, I. L., Owen, F. L., and Hood, L. (1986). Rearrangement and expression of T cell antigen receptor and γ genes during thymic development. *J. Exp. Med.* 164:161–184.

Haskins, K., Kubo, R., White, J., Pigeon, M., Kappler, J., and Marrack, P. (1983). The major histocompatibility complex-restricted antigen receptor on T cells: I. Isolation with a monoclonal antibody. *J. Exp. Med.* 157:1149–1169.

Havran, W. L., and Allison, J. P. (1988). Developmentally ordered appearance of thymocytes expressing different T-cell antigen receptors. *Nature 335*:443–445.

Hedrick, S. M., Cohen, D. I., Nielsen, E. A., and Davis, M. M. (1984). Isolation of cDNA clones encoding T cell-specific membrane-associated proteins. *Nature 308*:149–153.

Heilig, J. S., Glimcher, L. H., Kranz, D. M., Clayton, L. K., Greenstein, J. L., Saito, H., Maxam, A. M., Burakoff, S. J., Eisen, H. N., and Tonegawa, S. (1985). Expression of the T-cell-specific γ gene is unnecessary in T cells recognizing class II MHC determinants. *Nature 317*:68–70.

Hochstenbach, F., and Brenner, M. B. (1989). T-cell receptor delta-chain can substitute for alpha to form a beta-delta heterodimer. *Nature 340*:562–565.

Houlden, B. A., Cron, R. Q., Coligan, J. E., and Bluestone, J. A. (1988). Systematic development of distinct T cell receptor-γδ T cell subsets during fetal ontogeny. *J. Immunol. 141*:3753–3759.

Ito, K., Bonneville, M., Takagaki, Y., Nakanishi, N., Kanagawa, O., Krecko, E. G., and Tonegawa, S. (1989). Different γδ T-cell receptors are expressed on thymocytes at different stages of development. *Proc. Natl. Acad. Sci. USA 86*:631–635.

Iwashima, M., Green, A., Davis, M. M., and Chien, Y.-H. (1988). Variable region (V$_\delta$) gene segment most frequently utilized in adult thymocytes is 3' of the constant (C$_\delta$) region. *Proc. Natl. Acad. Sci. USA 85*:8161–8165.

Janeway, Jr., C. A., Jones, B., and Hayday, A. (1988). Specificity and function of T cells bearing γδ receptors. *Immunol. Today 9*:73–76.

Jerne, N. K. (1974). Towards a network theory of the immune system. *Ann. Immunol.* (Institut Pasteur) *125C*:373–384.

Jones, B., Mjolsness, S., Janeway, Jr., C. A., and Hayday, A. C. (1986). Transcripts of functionally rearranged gamma genes in primary T cells of adult immunocompetent mice. *Nature 323*:635–638.

Korman, A. J., Marusic-Galesic, S., Spencer, D., Kruisbeek, A. M., and Raulet, D. H. (1988). Predominant variable region gene usage by γ/δ T cell receptor-bearing cells in the adult thymus. *J. Exp. Med. 168*:1021–1040.

Kronenberg, M., Siu, G., Hood, L. E., and Shastri, N. (1986). The molecular genetics of the T cell antigen receptor and T cell antigen recognition. *Annu. Rev. Immunol. 4*:529–591.

Kuziel, W. A., Takashima, A., Bonyhadi, M., Bergstresser, P. R., Allison, J. P., Tigelaar, R. E., and Tucker, P. W. (1987). Regulation of T-cell receptor γ-chain RNA expression in murine Thy-1$^+$ dendritic epidermal cells. *Nature 328*:263–266.

Lai, E., Barth, R. K., and Hood, L. (1987). Genomic organization of the mouse T-cell receptor β-chain gene family. *Proc. Natl. Acad. Sci. USA 84*:3846–3850.

Lanier, L. L., and Allison, J. P. (1987). The T cell antigen receptor gamma gene: Rearrangement and cell lineages. *Immunol. Today 8*:293–296.

Lanier, L. L., Serafini, A. T., Ruitenberg, J. J., Cwirla, S., Federspiel, N. A., Phillips, J. H., Allison, J. P., and Weiss, A. (1987). The gamma T-cell antigen receptor. *J. Clin. Immunol. 7*:429–439.

Lanier, L. L., and Weiss, A. (1986). Presence of Ti (WT31) negative T lymphocytes in normal blood and thymus. *Nature 324*:268–270.

Lee, N. W., and Davis, M. M. (1988). T cell receptor β-chain genes in BW5147 and other AKR tumors. *J. Immunol. 140*:1665–1675.

Lindsten, T., Fowlkes, B. J., Samelsen, L. E., Davis, M. M., and Chien, Y.-H. (1987). Transient rearrangements of the T cell antigen receptor α locus in early thymocytes. *J. Exp. Med 166*:761–765.

Loh, E. Y., Elliott, J. F., Cwirla, S., Lanier, L. L., and Davis, M. M. (1989). Polymerase chain reaction with single-sided specificity: Analysis of T cell receptor δ chain. *Science 243*:217–220.

Matis, L. A., Cron, R., and Bluestone, J. A. (1987). MHC-linked specificity of γδ receptor bearing T lymphocytes. *Nature 330*:262–265.

Meuer, S. C., Cooper, D. A., Hodgon, J. C., Hussey, R. E., Fitzgerald, K. A., Schlossman, S., and

Reinherz, E. L. (1983). Identification of the receptor for antigen and major histocompatibility complex on human inducer T lymphocytes. *Science 222*:1239–1242.

Nixon-Fulton, J. L., Kuziel, W., Santerse, B., Bergstresser, P. R., Tucker, P. W., and Tigelaar, R. E. (1988). Thy-1$^+$ epidermal cells in nude mice are distinct from their counterparts in thymus-bearing mice. *J. Immunol. 141*:1897–1903.

Okazaki, K., Davis, D. D, and Sakano, H. (1987). T cell receptor β gene sequences in the circular DNA of thymocyte nuclei: Direct evidence for intramolecular DNA deletion in V-D-J joining. *Cell 49*:477–485.

Pardoll, D. M., Fowlkes, B. J., Bluestone, J. A., Kruisbeek, A., Maloy, W. L., Coligan, J. E., and Schwartz, R. H. (1987a). Differential expression of two distinct T-cell receptors during thymocyte development. *Nature 326*:79–81.

Pardoll, D. M., Fowlkes, B. J., Lechler, R. I., Germain, R. N., Schwartz, R. H. (1987b). Early events in T cell development analyzed by in situ hybridization. *J. Exp. Med 165*:1624–1638.

Pardoll, D. M., Fowlkes, B. J., Lew, A. M., Maloy, W. L., Weston, M. A., Bluestone, J. A., Schwartz, R. H., Coligan, J. E., and Kruisbeek, A. M. (1988). Thymus-dependent and thymus-independent developmental pathways for peripheral T cell receptor-$\gamma\delta$-bearing lymphocytes. *J. Immunol. 140*:4091–4096.

Raulet, D. H., Garman, R. D., Saito, H., and Tonegawa, S. (1985). Developmental regulation of T cell receptor gene expression. *Nature 314*:103–107.

Reilly, E. B., Kranz, D. M., Tonegawa, S., and Eisen, H. N. (1986). A functional γ gene formed from known γ-gene segments is not necessary for antigen-specific responses of murine cytotoxic T lymphocytes. *Nature 321*:878–880.

Reinherz, E. L. (1987). T Cell receptors: Who needs more? *Nature 325*:660–663.

Saito, T., Hochtenbach, F., Marosic-Galesi, S., Kruisbeek, A. M., Brenner, M., and Germain, R. M. (1988). Surface expression of only γ/δ and/or α/β T cell receptor heterodimers by cells with four (α, β, γ, δ) functional receptor chains. *J. Exp. Med. 168*:1003–1020.

Saito, H., Kranz, D. M., Takagaki, Y., Hayday, A. C., Eisen, H. N., and Tonegawa, S. (1984a). Complete primary structure of a heterodimeric T-cell receptor deduced from cDNA sequences. *Nature 309*:757–762.

Saito, H., Kranz, D. M., Takagaki, Y., Hayday, A. C., Eisen, H. N., and Tonegawa, S. (1984b). A third rearranged and expressed gene in a clone of cytotoxic T lymphocytes. *Nature 312*:36–40.

Sangster, R. N., Minowada, J., Suciu-Foca, N., Minden, M., and Mak, T. W. (1986). Rearrangement and expression of the α, β and γ chain T cell receptor genes in human thymic leukemia cells and functional T cells. *J. Exp. Med. 163*:1491–1508.

Satyanarayana, K., Hata, S., Devlin, P., Roncarolo, M. G., De Vries, J. E., Spits, H., Strominger, J. L., and Krangel, M. S. (1988). Genomic organization of the human T-cell antigen-receptor α/δ locus. *Proc. Natl. Acad. Sci. USA 85*:8166–8170.

Shimonkevitz, R., Kappler, J. W., Marrack, P., and Grey, H. M. (1983). Antigen recognition by H-2 restricted T cells: I. Cell free antigen processing. *J. Exp. Med 158*:303–316.

Snodgrass, H. R., Kisielow, P., Keifer, M., Steinmetz, M., and von Boehmer, H. (1985). Ontogeny of the T-cell antigen receptor within the thymus. *Nature 313*:592–595.

Sowder, J. T., Chen, C. H., Ager, L. L., Chan, M. M., and Cooper, M. D. (1988). A large subpopulation of avian T cells express a homologue of the mammalian T γ/δ receptor. *J. Exp. Med 167*:315–332.

Takeshita, S., Toda, M., and Yamagishi, H. (1989). Excision products of the T cell receptor gene support a progressive rearrangement model at the alpha/delta locus. *EMBO J 8*:3261–3270.

Toyonaga, B. and Mak, T. W. (1987). Genes of the T-cell antigen receptor in normal and malignant T cells. *Annu. Rev. Immunol. 5*:585–620.

Uematsu, Y., Ryser, S., Dembic, Z., Borgulya, P., Krimpenfort, P., Berns, A., von Boehmer, H., and Steinmetz, M. (1988). In transgenic mice the introduced functional T cell receptor β gene prevents expression of endogenous β genes. *Cell 52*:831–841.

Winoto, A., and Baltimore, D. (1989a). Alpha beta lineage-specific expression of the alpha T cell receptor gene by nearby silencers. *Cell 57*:649–655.

Winoto, A., and Baltimore, D. (1989b). Separate lineages of T cells expressing the $\alpha\beta$ and $\gamma\delta$ receptors. *Nature 338*:430–432.

Woolf, T., Lai, E., Kronenberg, M., and Hood, L. (1988). Mapping genomic organization by field inversion and two-dimensional gel electrophoresis: Application to the murine T-cell receptor γ gene family. *Nucleic Acids Res. 16*:3863–3875.

Yanagi, Y., Yoshikai, Y., Leggett, K., Clark, S. P., Aleksander, I., Mak, T. W. (1984). A human T cell-specific cDNA clone encodes a protein having extensive homology to immunologlobulin chains. *Nature 308*:145–149.

Yancopolous, G. D., and Alt, F. W. (1986). Regulation of the assembly and expression of variable region genes. *Annu. Rev. Immunol. 4*:338–368.

III

ONTOGENY OF
THE IMMUNE SYSTEM

8

Stem Cells

EDWIN L. COOPER,

CHRISTA E. MÜLLER-SIEBURG,

GERALD J. SPANGRUDE

Blut ist ein besonderer Saft.
Faust by J. W. Goethe

In ancient times blood was considered a substance of magic and vitality. More recently we have begun to identify some of these "magical" components. Early on, for instance, it became apparent that blood contains many morphologically and functionally distinct cells, which can be regenerated throughout an individual's life span. This feature sets blood cells apart from most other organs. If mature hematopoietic cells are removed from an animal, for example by extensive bleeding, these mature cells are subsequently replenished. Furthermore, many mature cells are short-lived and need to be replaced regularly. These early discoveries led to the concept of pluripotent hematopoietic stem cells that constantly regenerate the mature cell populations. Thus, the hematopoietic system is developmentally active throughout an animal's life span and constantly gives rise to a wide variety of functionally diverse cells. These features render the hematopoietic system particularly suitable as a model for the analysis of problems related to developmental biology, including lineage relationships and regulation of differentiation.

Considerable effort has been spent to identify pluripotent hematopoietic cells, and this area remains one of active investigation. Certain regions in the bone marrow and spleen show a high number of proliferating cells in response to physiological stress. Consequently, numerous experiments have sought to analyze the morphology of cells in these transient regions with the aim of characterizing stem cells. Based on morphological criteria and localization of the mature cells, it has been possible to construct developmental charts for the lineage-relationships of hematopoietic cells. In the 1950s and 1960s a number of assays for clonogenic hematopoietic progenitors and precursors were developed. These included the spleen-colony forming unit (CFU-S) assay introduced by Till and McCulloch (1961) and an in vitro assay for hematopoietic progenitors, the colony forming unit (CFU) culture assay (CFU-C) (Bradley and Metcalf, 1966; Ichikawa et al., 1966). The radioprotection assay measures the capacity of cells to rescue and reconstitute the hematopoietic lineages in a lethally irradiated animal and thus provides a measure of stem cell activity (Ford et al., 1956; Micklem et al., 1966; van Bekkum and de Vries, 1967). At about this time the concept developed that hematopoietic precursors and stem cells can be identified ONLY by their functions and, in fact, are defined by their function. Now it is firmly established that a single stem cell can give rise to all hematopoietic lineages, lymphocytes, myeloid and erythroid cells. Pluripotent stem cells, which are found predominantly in the bone marrow, commit sequentially to more restricted progenitors and precursors. Figure 8–1 depicts the current state of knowledge of lineage relationships within the hematopoietic system.

While great progress has been made in identifying precursors and progenitors functionally and in determining their lineage relationships, little is known of the processes that

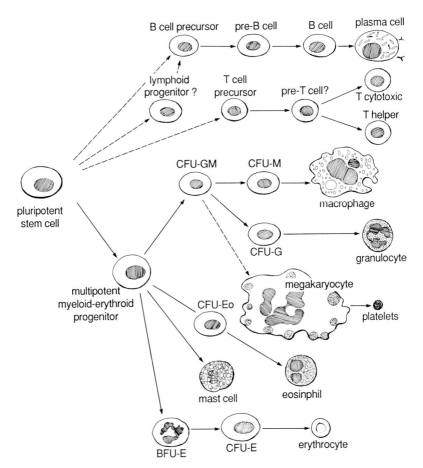

Figure 8–1.   Differentiation pathways of hematolymphoid cells. Pluripotent stem cells give rise to multipotent myeloid-erythroid progenitors that commit to restricted precursors of the myeloid and erythroid lineages. Restricted erythroid-myeloid progenitors and precursors are primarily defined through their capacity to give rise to a colony upon stimulation by a defined cytokine and are designated colony forming unit (CFU). The letter(s) after CFU refer to the lineage restriction: E, erythrocytes; M, macrophages; G, granulocytes; Eo, eosinophils. The BFU-E (burst forming unit erythroid) is an early precursor restricted to the erythroid lineage. Precursors give rise to a colony containing a single mature cell type. Colonies derived from progenitors can be composed of two or more cell types. Lymphoid precursors are believed to separate from myeloid-erythroid progenitors early in the differentiation cascade. Both T and B lymphocyte differentiation depend on regulatory stromal cells found in thymus and bone marrow.

regulate this differentiation cascade. Stem cells not only constantly give rise to mature cells but also are supposed to self renew to maintain a constant pool of themselves. The mechanisms that regulate this balance between induction of differentiation and self-maintenance are poorly understood. Furthermore, we have little information about what regulates commitment—or how a stem cell decides to differentiate along the various lineages.

Several factors complicate the analysis of early hematopoietic differentiation. For one thing, the beauty of the hematopoietic system for studying differentiation also poses a significant drawback. The system is highly complex and the frequency of relevant precursors and stem cells is low, often less than a fraction of one percent. Furthermore, mature cells can profoundly influence the differentiation of earlier stage cells. For example, T cells are known to secrete cytokines such as interleukin 3 (IL-3) that stimulate progenitors, including those for the myeloid-erythroid lineages (Ihle et al., 1983). In addition, bone marrow stromal cells can influence the differentiation and proliferation of stem cells and other precursors. Therefore, a major focus in studying hematopoiesis over the last 15 years has been to identify and isolate cells at distinct stages in this differentiation tree. Future progress will depend upon the availability of purified populations of progenitors and precursors, which will help to elucidate the complex interactions that regulate hematopoiesis.

DEFINING HEMATOPOIETIC STEM CELLS AND PROGENITORS: RELEVANT ASSAY SYSTEMS

Unfortunately, the terms stem cell, progenitor, and precursor are often used interchangeably. To prevent further confusion, we will use *stem cell* only for the pluripotent hematopoietic stem cells. Pluripotent stem cells are sometimes referred to as totipotent, a designation that should, in our opinion, be reserved for zygotes. A *progenitor* is a cell that usually has limited self-renewal capacity and can give rise to several, but not all, hematopoietic lineages. Finally, we define *precursor* as a cell that is restricted to a single lineage, e.g., a B cell precursor. B cell precursors differentiate into pre-B cells and subsequently into B cells.

The differentiation and proliferation potentials of cells at most early hematopoietic differentiation stages can be detected in vivo by analyzing their progeny. In addition, several in vitro assays have been developed to detect progenitors and precursors of the myeloid, erythroid, and B cell lineages. Because stem cells, progenitors, and precursors are very infrequent, they are often defined by specific assays. Thus, we will briefly describe the relevant assay systems.

Stem Cells

There are few assays that can accurately measure pluripotent stem cells. All are based on assessment of repopulation by donor cells in hematologically compromised host animals. Some groups have used genetically anemic mice as hosts and have demonstrated repopulation of all cell lineages after transfer of cells from congenic mice (Mintz et al., 1984; Nakano et al., 1989). This elegant model system obviates radiation damage to the microenvironment. However, repopulation of all lineages may also be achieved when a mixture of committed progenitors for all lineages is injected. In fact, Heimfeld et al. (1990) have shown that a certain bone marrow–derived subpopulation could reconstitute all cell lineages, but it failed to allow survival of lethally irradiated mice. This observation suggests that this repopulation assay may identify other hematopoietic progenitors in addition to stem cells. Similar concerns have also been raised for competitive repopulation assays in which lethally irradiated animals are reconstituted simultaneously with equal numbers of cells from two genetically distinct donor populations. The underlying concept is that stem cells from each population will compete with one another for repopulation (Harrison et al., 1989). However, if one population contained stem cells and the other contained only committed progenitors, both might appear to be sources of stem cells.

The radioprotection assay defines pluripotent stem cells based on their capacity to completely repopulate all hematopoietic lineages in lethally irradiated animals and thus procure their long-term survival (Fig. 8–2). Survival of these animals is strictly dependent on transferring a source of stem cells. Mice that receive bone marrow depleted of stem cells, but which contains lymphoid, myeloid, and erythroid precursors, will die (Müller-Sieburg, 1989). It is of advantage to use a genetic polymorphism between host and donor to allow assessment of the degree of chimerism in each of the different cell lineages. Stem cell activity can be titrated by injecting limiting numbers of cells, but it is difficult to measure the specific frequency of clonogenic cells in this assay.

Development of a reliable assay for stem cells should be a high priority in this field. The ideal assay should be linear, and it should measure a single limiting stem cell. An in vitro assay would be preferable, as reconstitution models are not available for a number of species.

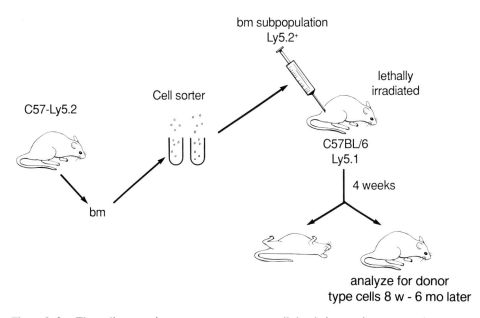

Figure 8–2.   The radioprotection assay measures stem cells by their capacity to repopulate a lethally irradiated animal. Lethal irradiation for C57BL/6 mice: 1,140 rads. (bm, bone marrow.)

Myeloid-Erythroid Progenitors and Precursors

The most frequently employed technique to measure restricted myeloid-erythroid cells at the clonal level is the CFU-C assay. This in vitro assay defines progenitors and precursors by their capacity to proliferate and differentiate upon stimulation by defined growth factors, thus giving rise to a colony (Bradley and Metcalf, 1966; Ichikawa et al., 1966). Progenitors that have the capacity to give rise to multiple lineages can be identified when cultured in the presence of the cytokine IL-3 or to a lesser extent, by the granulocyte-macrophage colony stimulating factor (GM-CSF). Myeloid or erythroid precursors restricted to a single lineage respond to macrophage-CSF, granulocyte-CSF, erythropoietin (EPO), and interleukin 5. In the murine system, multipotent progenitors and restricted myeloid precursors express different cell surface markers and thus can be separated (Müller-Sieburg et al., 1988).

Myeloid-erythroid–restricted progenitors can also be assessed by the colony forming unit spleen assay (CFU-S) developed by Till and McCulloch (1961). This assay measures the capacity of bone marrow– or spleen-derived progenitors (the CFU-S) to home to the spleen and there give rise to colonies consisting of erythroid and/or myeloid cells. CFU-S are frequently and erroneously equated to pluripotent hematopoietic stem cells. However, experiments by Paige et al. (1981) showed that at least day 9 spleen colonies do not contain lymphoid precursors and thus are restricted to the myeloid-erythroid lineages. Furthermore, it was shown that transplantation of a single stem cell with extensive proliferative potential, rather than transplantation of multiple CFU-S, was critical for reconstitution of hematopoiesis (Nakano et al., 1989).

T Cell Precursors

T cell precursors can be assayed in vivo as thymus-homing cells that differentiate into mature T cells and subsequently migrate to peripheral lymphoid organs, including the blood, the spleen, Peyer's patches, and lymph nodes (Spangrude et al., 1988b). If limiting numbers of bone marrow cells are injected, the thymic lobes become clonally repopulated and the frequency of T cell precursors can then be deduced (Ezine et al., 1984). Differentiation into mature T cells can be assessed by following the acquisition of cell surface markers such as CD4, CD8, and CD5. Differentiation into T cells, as assessed by acquisition of the CD8 and CD4 cell surface markers, can also be determined following intrathymic injection of precursor populations (for a review, see Adkins et al., 1987).

B Cell Precursors

The more mature stages of B-lineage differentiation, beginning with the pre-B cell, express characteristic markers such as the B220 antigen and immunoglobulin (Ig) that allow their reliable detection by immunofluorescence. The elucidation of the early stages of B-lineage differentiation has been facilitated by the development of culture systems that provide microenvironments for early B cell lymphopoiesis in vitro. Perhaps the earliest precursor restricted to the B cell lineage can be detected by its capacity to establish long-term Whitlock-Witte cultures (Whitlock and Witte, 1982; Müller-Sieburg et al.,

1986). In these primary bone marrow cultures, an adherent stromal layer allows the proliferation and differentiation of predominantly B-lineage cells. Lymphoid cells in a well-established culture have all rearranged their Ig heavy chain (Ig H) genes. About 50% of the cells express the B220 antigen and roughly 15% bear surface IgM. Recently, we have developed an in vitro limiting dilution culture system that directly measures the frequency of clonogenic B cell precursors. Uniform conditions throughout the cultures are ensured by the use of a monoclonal stromal cell line that, like the heterogeneous stromal layer, supports B-lineage differentiation (Whitlock et al., 1987).

Another assay system measures antibody-secreting cells in colonies derived from either a subset of B220-positive pre-B cells or a B220-negative precursor found in fetal liver. This agar culture system depends on a feeder layer (obtained from fetal liver cells) for the differentiation of B cells and has been useful for the study of the antibody repertoire at a clonal level (Paige et al., 1984). Studies of the immunoglobulin variable gene (Ig V) repertoire have been greatly facilitated by the availability of transformed clonal B cell lines corresponding to defined differentiation stages. Cells immortalized by somatic cell fusion or by transformation with Abelson murine leukemia virus have been used extensively in this work (reviewed in Paige and Wu, 1989).

EVOLUTIONARY ASPECTS OF HEMOPOIESIS

All coelomate invertebrates possess hemopoietic/lymphoid tissues in selected sites, which generate cells that participate in immune responses. Almost all terrestrial vertebrates possess hemopoietic bone marrow. Beginning with amphibians, this tissue becomes more and more specialized during evolution, through reptiles, birds, and mammals (Cooper et al., 1985). One may question why hemopoiesis occurs in bone marrow only in terrestrial vertebrates (Cooper et al., 1980). A facile explanation could be based on the presence of long bones only in tetrapods; however, this does not account for the distribution of hemopoietic sites in animals that possess neither marrow nor long bones. For example, bone marrow is absent in fishes and other ectothermic vertebrates in which there is osteologic organization. However, hemopoietic bone marrow–like microenvironments are well developed in Osteichtyes, Chondrichthyes, and Cyclostomata and are found in the esophageal wall, kidneys, supraneural organs, spleen, Leydig organ, and liver. This section reviews our knowledge of hemopoiesis in invertebrates and ectothermic vertebrates. The situation in ectotherms, birds, and, to some extent, monotremes and marsupials was recently surveyed in considerable detail (Zapata and Cooper, 1990). Cell nomenclature is complex and we should keep in mind how these cells may have given rise to stem cells in placental mammals. Further consideration should also be given to the evolutionary concepts of convergence and divergence, as well as to homology and analogy.

Blood Cell Origins in Selected Invertebrates

Ratcliffe et al. (1985) provided an excellent review of cells and tissues of invertebrate immune systems. In their view, comparative immunologists are only just beginning to recognize that numerous blood cell types participate in immunodefense reactions in invertebrates; this situation is much more complex than that which obtains in mammals.

Cells of the invertebrate immune system can be subdivided into two main groups—namely, the freely circulating blood cells, or coelomocytes, and a variety of fixed cells. Fixed cells may be scattered throughout tissues or may be localized to hemopoietic/phagocytic organs. As in vertebrates, however, highly variable terminology has been used by different groups to refer to particular cells and tissues, and there is an urgent need for a standardized nomenclature.

For example, progenitor cells in invertebrates have been variously referred to as hemoblasts, hemocytoblasts, proleukocytes, stem cells, and lymphocyte-like cells (see definitions for mammals, below). Morphologically, these are generally small cells, 4–10 μm in diameter, with a high nuclear:cytoplasmic ratio. The thin rim of cytoplasm is usually undifferentiated and contains numerous free ribosomes, a few profiles of rough endoplasmic reticulum, and an occasional Golgi complex. The pluripotentiality of such cells has not been confirmed for most invertebrates. Although these cells may be morphologically similar, there are clearly several functionally defined subpopulations that are already committed to particular lines of differentiation. Because of the morphological similarity between invertebrate progenitor cells and vertebrate lymphocytes, there is great controversy over the evolutionary relationships of lymphocytes. In tunicates, which are probably the immediate precursors of vertebrates, stem cells and lymphocyte-like cells are morphologically distinct (Wright, 1981; Wright and Ermak, 1982), in that stem cells but not lymphocyte-like cells possess a nucleolus.

The structure and location of hemopoietic (hemocytopoietic) tissues vary greatly within invertebrates, and in some the sites of blood cell formation are unknown. For example, in acoelomate invertebrates, such as sponges and coelenterates, the origin of amoebocytes and other functional cell types is not clear. Still, it would be highly unlikely to find well-developed hemopoietic tissues in animals with such a simple body organization. In coelomate invertebrates, the morphology and distribution of hemopoietic sites is often well understood, but the lineages and interrelationships of coelomocyte and blood cell types are still obscure. The processes of hemopoiesis as defined in the most commonly encountered invertebrates are summarized in Table 8–1.

Are There Equivalents of the Bone Marrow in Cyclostomes?

The Pronephros and Intestine of Hagfish

The main lympho-hemopoietic organs of adult hagfish are the pronephros and intestine, as a thymus equivalent has not been discovered. The pronephros, or head kidney, is a complex structure—a small, paired organ suspended in the wall of the pericardial cavity. Ultrastructural studies have shown a possible relationship to bone marrow of higher vertebrates because the central mass is composed of lympho-hemopoietic elements. These include blast cells, lymphocyte-like cells, plasma cells, spindle cells, and erythroid cells. A high mitotic activity in the central mass suggests that the cell accumulations are foci of active genesis of blood cells. Within the intestinal lympho-hemopoietic tissue, neutrophils are the most abundant cell population and are present in various stages of development. Small non-granulated cells, probably derived from reticular cells, have been proposed as the precursors of granulocytes (Tomonaga et al., 1973). High mitotic activity in this tissue suggests a high rate of hemopoietic activity, which includes cells of granulocytic, erythrocytic, and non-granular lineages.

Table 8–1.   Hemopoietic/Lymphoid Tissues in Selected Coelomate Invertebrates

Group	Cell Types	Hemopoietic Site(s)
Annelids	Amebocytes	Coelomic lining cells
	Chloragogen cells (eleocytes)	Chloragogen tissue around intestine
	Blood cells	"Blood glands" in hemocoelic system
		From coelomic lining (?)
Mollusks	Blood cells	Varies, some gastropods have localized hemopoietic sites, although none reported for bivalves
		Specialized hemopoietic tissue, the white bodies, exist behind the eyes in cephalopods
Insects	Blood cells	Varies, either in distinct hemopoietic areas or by division of freely circulating cells
Crustaceans	Blood cells	Foci in various sites, including base of rostrum, dorsal and lateral walls of foregut, and vicinity of ophthalmic artery
Echinoderms	Coelomocytes	Coelomic lining cells
		Axial organ
		Tiedemann's bodies
		Polian vesicles
Urochordates	Blood cells	"Lymph nodules," in pharyngeal wall, around digestive tract, and in the body wall
Cephalochordates	Coelomocytes	Coelomic lining cells

Modified from Ratcliffe et al. (1985).

The Typhlosole, Opisthonephros, and Supraneural Body of Lampreys

Lampreys have a long and complex life cycle. During their larval stages, the ammocoetes remain buried in river silt; after metamorphosis, the young adults migrate to the sea, where they parasitize teleost fish by attaching to their integument. Lympho-hemopoiesis and a number of other processes are profoundly altered during metamorphosis. According to Ardavin et al. (1984) and Ardavin and Zapata (1987), the hemopoietic sites of the anadromous sea lamprey, *Petromyzon marinus,* are the equivalents of vertebrate bone marrow.

Lympho-hemopoietic tissue in the typhlosole consists of foci of mature and developing blood cells apparently interspersed randomly between connective tissue elements that comprise the stroma. Differentiation of any or all of the erythrocytic, lymphocytic, monocytic, or granulocytic lineages can occur within a single focus. Although full development of the cell lineages generally occurs outside the vascular compartment, and is followed by migration of mature cells through the endothelial walls of the blood vessels, immature blood cells can occasionally be found in the blood sinuses. The presence of these cells, which belong chiefly to the granulocytic and erythrocytic lineages, suggests that immature cells can cross vascular endothelia, or, alternatively, that the latter stages of erythro-, granulo-, and monocytopoiesis in *Petromyzon marinus* can take place in the bloodstream. A similar situation has been described for cells of the plasmacytic lineage because plasmablasts, proplasmacytes, and mature plasma cells occur in the typhlosolar lympho-hemopoietic tissue and in the blood sinuses.

In the opisthonephros, the lympho-hemopoietic tissue of the nephric fold contributes to the production and differentiation of erythrocytes, lymphocytes, monocytes, and granulocytes. The sinusoids within the lympho-hemopoietic masses contain numerous plasmablasts, immature and mature plasma cells, and large macrophages, whose projections make contact with both the endothelial cells and circulating lymphoid elements. During the larval period, the opisthonephros retains its lympho-hemopoietic character, which is presumably determined by the maintenance of a suitable cell microenvironment. However, during metamorphosis, there is total degeneration of the larval opisthonephros, giving rise to the adult kidney that develops from the nephrogenic tissue of the posterior nephric fold.

The supraneural body, which has classically been proposed as the phylogenetic precursor of bone marrow, has essentially the same structure as the lympho-hemopoietic tissue in the typhlosole and nephric fold. In young adult lampreys (nearly 5 years of age), numerous lympho-hemopoietic aggregates appear in the supraneural body, which persists as the most important hemopoietic organ during adult stages of development. These aggregates contain mature and developing blood cells of several lineages, including lymphocytes and plasma cells, together with abundant blasts and other actively dividing cells.

What Are the Bone Marrow Equivalents in Modern Fish?

In order to gain an appreciation for the evolution of hemopoietic tissues in fish, it is necessary to briefly describe the different phylogenetic groupings. Within the vertebrates, modern fish constitute an abundant group, which can be divided into two main categories: (1) Chondrichthyes, including Holocephali and Elasmobranchi, and (2) Osteichthyes, which comprise Crossopterygii, Dipnoi, Chondrostei, Polypteriformes, and Teleostei. Despite this enormous diversity, the bulk of immunological investigations have been done on the teleosts, although a few studies have been reported in cartilaginous elasmobranchs. Taken together, this work has identified a wide array of lympho-hemopoietic tissues that, although they are located in diverse organs, share one important feature: a common histological organization, which resembles the bone marrow of higher vertebrates.

Hemopoietic Sites in Chondrichthyes, Dipnoi, Chondrosteans, and Holosteans

The organ of Leydig in cartilaginous fishes is a lympho-hemopoietic tissue that forms white masses in the orbit of the eye, in the preorbital canal of the cranial cartilage, and in a depression in the *basis cranii.* The stroma is made up of reticular cells, which form a network in which differentiation of granulocytes and some lymphocytes occurs. Granulocytic cells include neutrophils and eosinophils, which may show varying degrees of cytoplasmic granularity. Numerous mature and immature plasma cells and some macrophages are also present. The intestinal spiral fold contains a rich infiltration of lymphocytes, which suggests that it is a site of lymphocytopoiesis. Erythropoiesis has been shown to occur in other regions, including the cardiac epithelium. The epigonal organs usually occupy the parenchyma of the ovary and testis, and, in some species, the gonads are embedded in and partially covered by lympho-hemopoietic tissue. Study of the lympho-hemopoietic system of lungfishes *(Protopterus ethiopicus)* has shown that granulopoietic tissue is present caudal to the spleen, within the spiral intestine, kidney, and gonads. Lympho- and erythropoiesis occur mainly in the spleen, whereas granulopoiesis is localized to the kidney and gonads. In chondrosteans and holosteans, two characteristic lympho-hemopoietic tissues are present, which closely resemble the bone marrow of higher vertebrates. The main sites of granulopoiesis are in the meninges of the cranial cavity and occasionally in foci at the base of the heart. In all these regions, the hemopoietic tissue is organized into lobes that consist of closely packed cells, granulocytes, and lymphocytes separated by venous or lymphatic spaces.

The Pronephros of Teleosts

The pronephros consists of two distinct segments—one anterior and one comprising the middle and posterior portions. Although both regions contain lympho-hemopoietic activity, it is present at higher levels in the anterior cephalic kidney where the renal tubules have disappeared. The structure of the lympho-hemopoietic tissue is equivalent in both areas, however (Zapata et al., 1981). Phagocytic activity is present in the reticular cells, as well as in the endothelial cells that line blood sinusoids; phagocytosis is an important function for recognition and processing of antigens. All lineages of hemopoietic differentiation have been observed, including lympho- and plasmacytopoiesis; pluripotent hemopoietic stem cells are also present. The coexistence of these functional activities within a single tissue has prompted speculation as to whether the pronephros more closely resembles the lymph node or bone marrow. Yet it has important immune capacities that are essential characteristics of both tissues. The pronephros is considered to be a postembryonic source of hemopoietic stem cells. It also contains antigen binding and antibody-producing cells whose responses to various mitogens have been assessed. Evidence for the presence of T cells has come from studies using mitogens such as lipopolysaccharide (LPS), concanavalin A (Con A), and phytohemagglutinin (PHA) and various blocking antisera, chiefly directed against determinants expressed on brain tissue. Other characteristics of lymphoid cells from the pronephros include the ability to react in mixed lymphocyte reaction and to exhibit cytotoxic responses.

The Liver and Pronephros in Amphibians

Amphibians show active lympho-hemopoiesis in the liver and kidney, mainly during embryonic and larval life before the development of hemopoietic bone marrow. In these tissues, immature and mature granulocytes (chiefly neutrophils and eosinophils), as well as lymphocytes and plasma cells, are interspersed with fibroblasts and bundles of collagenous fibers. The pronephros and mesonephros later become centers for granulopoiesis and possibly lymphopoiesis in embryonic leopard frogs *(Rana pipiens)* (Carpenter and Turpen, 1979). Hemopoietic precursors derived from the lateral plate mesoderm migrate to these organs, where they differentiate into blood cells, possibly under the influence of local cell microenvironments (Turpen and Knudson, 1982). Moreover, cell generation seems to occur within the kidney of larval leopard frogs during premetamorphic stages (Zettergren et al., 1980). During this phase and later during stage II, surface and cytoplasmic—$sIgM^-$ $cIgM^+$—cells can be found in the pronephros and mesonephros. During premetamorphosis, these cells are more abundant in the kidney than in other tissues. These pre-B cells begin to express sIgM by the time the young frogs begin to feed. $sIgM^+$ lymphocytes seem to be precursors for $sIgM^+$ $sIgG^+$ B lymphocytes. Finally, $sIgM^-$ $cIgG^+$ B lymphocytes appear to be more prominent early in ontogeny than during later stages.

Hematopoiesis During Embryogenesis in Anuran Amphibians

In anuran amphibians, as exemplified by *Xenopus laevis,* the primitive hemopoietic stem cells, which differentiate during early embryogenesis, are derived from a defined region of ventral mesoderm known as the Ventral Blood Island (VBI). The VBI of the 20-hour-old *Xenopus* embryo consists of approximately 2,300 mesodermal and ectodermal cells and contains pluripotential hemopoietic stem cells that are capable of giving rise to erythroid, myeloid, and T and B lymphocyte lineages (Smith et al., 1989). In an extension of these studies, Turpen and Smith (1989) examined the influence of location within the VBI on the ability of single stem cells to give rise to particular lymphoid populations. Pieces of diploid VBI mesoderm, each containing an average of one hemopoietic stem cell, were transplanted to central or peripheral locations within the defined boundaries of the VBI of triploid stage-matched embryos. The proportion of animals with diploid donor-derived lymphoid cells was markedly increased when stem cells were grafted to the periphery of the VBI. In three cases, transplanted stem cells were shown to contribute to lymphoid but not to erythroid populations. These data are consistent with the existence either of lymphoid stem cells or of restricted B and T lymphocyte precursors. These data also suggest that during embryogenesis, stochastic differentiation of hemopoietic stem cells is influenced by regional differences in the VBI microenvironment.

Bone Marrow and Reconstitution of T and B Cell Functions in Anuran Amphibians

Lymphoid accumulations occur in several anatomical regions, including the bone marrow, in the leopard frog (Cooper et al., 1980). The cells that make up the bone marrow in anurans have been elegantly summarized in an atlas of hematology (Hadji-Azimi et al., 1987). Transfer of bone marrow can restore the capacity to reject allogeneic skin grafts, an effect mediated by putative T cells (Cooper and Schaeffer, 1970), and antibody-producing cells can be induced in marrow after stimulation with the antigen TSN-LPS (Eipert et al., 1970). In addition, bone marrow is more effective than spleen cells in maintaining individual viability. Marrow cells can also be stimulated to proliferate by Con A and to respond to sheep red blood cells with the generation of plaque-forming cells (Ramirez et al., 1983).

Reptilian Bone Marrow

Morphological studies have shown the presence of most types of blood cells in bone marrow from reptiles (reviewed in Cooper et al., 1985). Some of the more interesting aspects of these studies concern variations in the numbers of blood cells as a result of seasonal conditions, sex, age, mating, oviduct gravidity, and diseases. The functions of reptilian bone marrow have not been investigated fully, however, and only scanty data are available regarding its hemopoietic activities. In *Phrynosoma solare* the spleen is the principal blood-forming organ, although bone marrow predominates in most lizards; in turtles, on the other hand, the contributions are more evenly divided between the two tissues (Jordan, 1938). In this regard, reptiles seem to represent a transitional group between amphibians, in which the spleen is the main erythropoietic organ, and birds, in which the bone marrow is practically the only blood-forming tissue (Andrew, 1965; Jordan, 1938; LeDouarin, 1966).

Bone marrow structure in *Podarcis hispanica* (Zapata et al., 1981) is similar to that described for the Plethodontidae (Curtis et al., 1979) and anuran amphibians (Campbell, 1970; Cooper et al., 1980), except that erythropoiesis also occurs in the bone marrow. The marrow stroma, which is composed of reticular cells and sinusoidal lining cells, is a site of granulocytopoiesis and erythropoiesis. Moreover, despite previous reports that suggest a lack of lymphoid tissue in reptilian bone marrow (Borysenko and Cooper, 1972; LeFevre et al., 1973; Sidky and Auerbach, 1968), small lymphocytes, plasma cells, and hemopoietic stem-like cells have more recently been identified (Zapata et al., 1981). The immune functions of reptilian bone marrow are one aspect of comparative immunology about which little is known. Investigation of bone marrow immunobiology in relation to B cell origin and immune memory in reptiles is therefore of much interest insofar as it may allow us to decipher the evolution of stem cell function.

Seeding of Stem Cells and Lineage Interrelationships in Birds

Over the years, many now classical concepts about the development of the immune system have been elucidated by analyses of cell pathways in birds, including those for stem cells and their differentiated products. We recognize the dichotomy between humoral and cellular immunity that is related to the existence of T and B cells. Other properties of avian lymphoid development include the existence of the unique B cell differentiation organ (the bursa of Fabricius), the extrinsic origin of lymphoid progenitors, and the periodic colonization of the thymus by progenitors (Dieterlen-Lièvre, 1989). With respect to stem cells, avian interspecies yolk sac chimeras have been used to detect stem cells that are formed in the embryo itself; these cells are responsible for seeding both the primary lymphoid organs and the definitive hemopoietic organs (Dieterlen-Lièvre, 1984). These intraembryonic stem cells appear to develop in the region of the aortic arch.

Through use of the unique and now classic chick-quail embryo chimera system, the aortic region was shown to contain diffuse hemopoietic cell foci in E6 to E8 chick or quail embryos (Dieterlen-Lièvre and Martin, 1981). The E3 or E4 thoracic segment of the aorta gives rise to the differentiation of hemopoietic cells of various lineages when grafted in vivo or plated in vitro (Cormier et al., 1986; Cormier and Dieterlan-Lièvre, 1988; Dieterlen-Lièvre, 1984). The E4 quail aorta when associated with an E6 chick thymus (at its first receptive period) and co-transplanted into a chick host, provides colonizing progenitors for the thymus (Dieterlen-Lièvre, 1989). These experiments confirm the predictions of LeDouarin (1978), who suggested that the cyclical renewal of hemopoietic cells that colonize the thymus was governed by a chemotactic mechanism. Clearly the use of this very special avian system and monoclonal antibodies holds great promise for developing unified concepts of avian and mammalian hemopoiesis, through identification of stem cell progenitors that colonize peripheral centers in the developing immune system.

STEM CELLS IN MAMMALS

Blood cells develop in essentially the same manner in embryos of all mammals. The primary anatomical site where hemopoiesis is first observed in mammalian embryos is extraembryonic, in the numerous blood islands of the yolk sac. The mesenchymal covering of the yolk sac is essentially a portion of the primitive gastrointestinal tract (Maximow and Bloom, 1952). Progenitor cells that initiate formation of the blood islands probably migrate to the yolk sac from the caudal portion of the early primitive streak during gastrulation (Bloom, 1938). Later, the proliferating cells of the blood islands differentiate along two distinct pathways: one to form the endothelial cell boundaries of the first blood vessels at the periphery of the blood islands, and the other to give rise to primitive blood cells in the center of the islands. Thus, the common mesenchymal ancestry of endothelium and of blood cells can be traced to a relatively late period during their ontogeny.

The primordia of blood vessels in the embryo are not usually hemopoietic, although the occasional blood island can be observed. Thus, while the first intraembryonic vessels (and the heart) arise in situ rather than as extensions of developing blood vessels of the extraembryonic yolk sac, most early hemopoietic processes occur in the yolk sac. It is these cells that first enter circulation in the embryonic vascular system (Moore and Metcalf, 1970). Primitive blood cells of the yolk sac develop first into large, nucleated erythrocytes, which are eventually replaced by typical mammalian erythrocytes after hemopoiesis begins in the fetal liver.

Mammalian hemopoiesis can be divided into three distinct phases—mesoblastic (or vitelline), hepatic, and myeloid (Wintrobe, 1967). During the mesoblastic phase, most or all blood cells are formed outside the embryo in the blood islands of the yolk sac. This stage persists for about 10 weeks in human embryos (about 12 days in the mouse), after which the fetal liver begins to assume the major responsibility for blood formation. The hepatic period encompasses the phase of highest hemopoietic activity, and it extends from 6 or 8 weeks' gestation in human embryos (10–12 days in the mouse) until shortly before birth. In humans, a transient period of hemopoiesis in the spleen precedes the onset of myeloid hemopoiesis at 10–12 weeks of gestation. Up to this point, most embryonic hemopoiesis is erythroid. Hemopoietic activity in the spleen is transposed to the bone marrow beginning at around 11–12 weeks (15–16 days in the mouse), and by 20 weeks essentially all blood cell formation in human embryos occurs in the bone marrow. While splenic hemopoiesis is only a transient stage during human development, the spleen remains hemopoietically active throughout the adult life of the mouse. Hemopoiesis begins in mouse bone marrow at about 15 days of gestation, and the marrow subsequently remains the primary site of blood formation in adults.

MICROENVIRONMENTS OF HEMOPOIESIS

Bone Marrow

Bone is a calcified extracellular matrix of collagen and glycosaminoglycans that are synthesized by osteoblasts, or bone-forming cells. The medullary cavity of bone may be hemopoietically active and contain so-called red marrow, or it may be predominantly inactive and filled with fat cells (white marrow). The medullary cavity and haversian canals of bone, which house blood vessels, are lined by a membrane called the endosteum. Occasional osteoclasts, which destroy bone, are found in the walls of the medullary cavity and are associated with areas of bone resorption. Osteoclasts are multinucleated giant cells, and may result from fusion of osteoblasts. The endosteum with its associated osteoclasts is of particular interest because the most primitive stem cells for blood formation appear to be localized near the walls of the medullary cavity (Gong, 1978; Lord et al., 1975). The matrix of the medullary cavity includes structural elements of the blood vascular system, nerve fibers, and a system of reticular cells and fibers. This matrix is established during embryogenesis before initiation of hemopoiesis. Thus, it provides a specialized microenvironment that supports hemopoietic cells within the parenchyma (Fliedner and Calvo, 1978).

The bone marrow microenvironment probably has several distinct functions with regard to hemopoiesis. First, it must provide conditions to maintain pluripotent hemopoietic stem cells in a primitive state throughout an animal's lifetime, thus ensuring an adequate supply of the seeds of hemopoiesis. Second, it must provide appropriate inductive signals for primitive stem cells to direct hemopoiesis toward the erythroid, myelomonocytic, and B lymphoid lineages. The processes of maintenance and differentiation of stem cells must be balanced in order to sustain functionally mature blood cell populations without depleting the pluripotent hemopoietic stem cell pool. The concept of self-renewal, through which a pluripotent hemopoietic stem cell can undergo mitosis to give rise to two equally primitive stem cells or to one primitive stem cell and a cell destined to differentiate, is not universally accepted (Micklem et al., 1983). Some researchers believe that each cycle of mitosis is accompanied by differentiation, to the extent that the number of mitotic cycles that can occur before differentiation is limited. Whatever mechanisms may be involved, however, the role of regulating the maintenance and differentiation of hemopoietic stem cells is carried mostly by stromal elements.

The distinct functions of hemopoietic stem cells and the microenvironments that support them are clearly seen in two mutant mouse strains that have resulted from mutations at the "dominant spotting" (W) and "steel" (Sl) loci on chromosomes 5 and 10, respectively (Shultz and Sidman, 1987). These mutations cause deficiencies in pigment-forming cells (the basis for the original identification of the mutants and their corresponding strain designations) as well as in reproductive capacity and hemopoietic function. Interestingly, hemopoietic abnormalities in W-locus mutants are due to a defect in stem cells, while Sl-locus mutants provide a defective microenvironment for normal stem cells (McCulloch et al., 1965). Animals carrying either mutation have severe macrocytic anemia and show increased sensitivity to radiation. Doubly heterozygous animals carrying a single mutant allele at each locus are near normal phenotypically, and injection of bone marrow cells from Sl-locus mutants into lightly irradiated W-locus mutants cures the anemia. These observations can also be replicated in in vitro co-cultures of bone marrow from the two strains (Dexter and Moore, 1977). The molecular basis for both the W and Sl mutations has recently been defined experimentally (reviewed in Witte, 1990).

Spleen

In mice, where hemopoietic activity is maintained in the spleen throughout adult life, the microenvironments of the bone marrow and spleen seem to be markedly different. Although both organs support erythropoiesis and granulopoiesis, the spleen is predominantly involved in erythroid development, whereas granulopoiesis exceeds erythropoiesis in the bone marrow. Implantation of bone fragments into irradiated spleens allows the two types of microenvironments to exist simultaneously. In such cases, individual colonies of hemopoietic cells that arise at the border of the two microenvironments are frequently granulocytic in the vicinity of the bone fragment and erythroid in the splenic area (Wolf and Trentin, 1968). Results from these experiments argue that microenvironmental influences can dictate developmental pathways followed by individual multipotent cells.

Hemopoiesis in Culture

The establishment of long-term cultures of bone marrow stromal elements has provided further evidence that specialized stromal cells contribute to microenvironments that support hemopoiesis. In addition, culture systems have been developed that selectively support either stem cell maintenance with myeloid and erythroid development (Dexter and Lajtha, 1974) or B lymphocyte development (Whitlock and Witte, 1982). These culture systems involve the maintenance of a complex bone marrow–derived stromal cell feeder layer, which consists of multiple cell types including fibroblasts, epithelial cells, reticular-dendritic cells, adipocytes, and mononuclear cells. Early observations of the myeloid cultures developed by Dexter and his colleagues indicated that adipocytes were most closely associated with hemopoiesis in vitro. The subsequent development of stromal cell lines from Whitlock-Witte cultures (Hunt et al., 1987; Whitlock et al., 1987) has strengthened these findings; the stromal cell lines that support in vitro growth of early lymphoid and myeloid cells, and which maintain early spleen colony forming cells, are adipocyte-like and resemble adventitial reticular cells. These latter are fibroblastoid cells that form a cellular network, or reticulum, within the bone marrow cavity. They accumulate neutral lipid deposits under certain conditions and are secretory, producing collagen and other proteins (Hunt et al., 1987). The bone marrow culture systems hold much promise for providing important clues that may help clarify the roles of cell–cell interactions and of soluble growth factors in regulating hemopoiesis.

THE T LYMPHOCYTE LINEAGE

Thymus-derived (T) lymphocytes play a critical role in vertebrate immune responses, both in mediating cellular immunity and in regulating cellular and humoral immune responses. The structural framework of the thymus arises from epithelial cells, which form the third and fourth pharyngeal pouches. In mammals, this thymic anlage descends through the developing thoracic inlet and eventually comes to be located within the thoracic cavity. In the mouse, the epithelial anlage is colonized by bone marrow–derived stem cells at about days 10–11 of gestation (Owen and Ritter, 1969); in birds, three successive waves of lymphoid stem cells colonize the thymus during embryogenesis (Jotereau and LeDouarin, 1982). Similar wave-like colonization may also occur during development of the mouse thymus (Jotereau et al., 1987). In contrast, during adult life very few bone marrow–derived cells enter the thymus, and most of the T cells that develop in the thymus are derived from intrathymic progenitor cells. However, the thymus probably remains dependent on occasional influx of bone marrow–derived progenitor cells to allow it to continue producing T lymphocytes throughout adult life (Scollay et al., 1986).

The developmental stage of the bone marrow–derived cells that initiate lymphoid development in the thymic microenvironment is not clear. In the normal steady-state condition of a mature mammal, the bone marrow stem cells that migrate to the thymus and give rise to large populations of T lymphocytes have not been identified. It is therefore difficult to ascertain whether a developmental sequence that begins in the bone marrow causes a multipotent stem cell to restrict its development to the T cell lineage and thus to seek out the thymic microenvironment (Mulder et al., 1984), or whether the rare multipotent cell that seeds to the thymus is subsequently induced to differentiate along a T lymphocytic pathway (Moore and Owen, 1967). Although it is clear that pluripotent hemopoietic stem cells purified from bone marrow are able to develop into T lymphocytes if they are artificially introduced into the thymus (Spangrude et al., 1988a), it is possible that under normal conditions the commitment to T cell development begins in the bone marrow, before the migration of relevant cells into the thymus.

THE B LYMPHOCYTE LINEAGE

The existence of a lymphocyte-restricted progenitor that gives rise to both T and B lymphocytes but not to myeloid or erythroid cells is still controversial. The long-term bone marrow culture system described by Dexter and Lajtha (1974) appears to maintain a cell that fits the criteria for a lymphocyte progenitor (Fulop and Phillips, 1989). On the other hand, experiments using freshly explanted bone marrow have failed to demonstrate the existence of such cells (Snodgrass and Keller, 1987). Perhaps lymphocyte-restricted progenitors are extremely rare in bone marrow and can be demonstrated only under favorable culture conditions. It is possible that several differentiation pathways lead to production of mature lymphocytes, but that only one proceeds via a common progenitor. In fact, independent B cell lineages have been proposed (Hayakawa et al., 1985; Mosier et al., 1977).

The earliest restricted B lineage precursor may be contained in a bone marrow population that expresses low levels of Thy-1 antigen and lacks markers associated with mature lymphocytes and myeloid cells. This Thy-1^{lo} Lin$^-$ population is also enriched for stem cells, T cell precursors, and myeloid-erythroid progenitors (Müller-Sieburg et al., 1986, 1988; Spangrude et al., 1988a, b). B cell precursors with less extensive proliferative capacity are found in two bone marrow populations—one characterized by the lack of Thy-1 and mature cell markers (Thy-1$^-$ Lin$^-$ cells), while the other coexpresses low levels of Thy-1 and B220 (Thy-1^{lo} B220$^+$ cells). The latter are a subset of large pre-B cells and perhaps represent the earliest identifiable pre-B cells (Tidmarsh et al., 1989). Pre-B cells express cytoplasmic μ (cμ) chains and have already rearranged both of their Ig gene alleles. Small pre-B cells start to rearrange their Ig light chains (Ig L) and are the direct precursors of surface Ig-bearing B cells. Both pre-B and B cells express the B220 antigen, a B-lineage–specific form of the T200 glycoprotein (Coffman, 1983).

During ontogeny, pre-B cells can first be detected in the mouse fetal liver at day 11 of gestation. Pre-B cells appear in the spleen around day 15 and in bone marrow at day 19 of gestation (Verlardi and Cooper, 1984). Most pre-B cells in adult animals are found in the bone marrow. B cells that express surface Ig are first detected simultaneously in liver, spleen, and bone marrow of day 17 mouse embryos (Verlardi and Cooper, 1984). In adult animals, B cells leave the bone marrow and home to spleen, Peyer's patches, and lymph nodes. In these organs B cells can undergo Ig class switching and mature to plasma cells upon appropriate stimulation by antigen and T lymphocytes.

Development of B-lineage cells in birds differs from that in mammals as it depends on the bursa of Fabricius. The bursa is colonized between days 7 and 15 of incubation by immature hematopoietic cells. Surface Ig-bearing B cells appear first at day 13. Bursectomy in embryonic or newborn chicks results in pronounced long-term deficiencies in B-lineage cells (reviewed in Kincade, 1984). It has been reported that the main function of the bursa is to provide a microenvironment in which Ig V_L genes undergo somatic mutation, leading to expansion of the antibody repertoire (Reynaud et al., 1987).

While the differentiation cascade of B-lineage cells seems to progress similarly in fetal, newborn, and adult mice, there is a difference in the Ig V_H repertoire that is expressed. Comparison of the Ig V_H make-up of B cells in fetal and adult mice indicates that the repertoire develops in a step-wise manner. A complete Ig heavy chain is constructed by joining one member each from four different gene families: constant segment (C) genes, joining segment (J) genes, diversity segment (D) genes, and variable segment (V) genes. Early during ontogeny (day 13 fetal liver) a large proportion of B-lineage cells use the Ig V_H genes which lie closest to the joining region genes. As the animal ages, more distant Ig V_H genes are used, until, in an adult mouse, cells that express the Ig V_H genes most proximal to the joining region genes are absent (Yancopoulos and Alt, 1986). Interestingly, the fetal repertoire contains a high proportion of antibodies that recognize each other. Some of these anti-idiotypic antibodies could either enhance or suppress the corresponding idiotype and profoundly change the adult repertoire (Kearney and Vakil, 1986). It is possible that developmentally regulated changes in repertoire reflect the learning process that leads to immunological tolerance.

Acknowledgments

We thank Dr. C. Gritzmacher for critical reading of the manuscript, parts of which are reprinted with the permission of the publisher, Marcel Dekker. C. Müller-Sieburg is a Leukemia Society of America Scholar, supported by NIH grant DK41214. G. Spangrude is a Special Fellow of the Leukemia Society of America.

References

Adkins, B., Müller, C., Okada, C. Y., Reichert, R. A., Weissman, I. L., and Spangrude, G. J. (1987). Early events in T cell maturation. *Annu. Rev. Immunol.* 5:325–365.

Andrew, W. (1965). *Comparative Hematology.* New York: Grune and Stratton.

Ardavin, C. F., Gomariz, R. P., Barrutia, M. G., Fonfria, J., and Zapata, A. (1984). The lympho-hemopoietic organs of the anadromous sea lamprey *Petromyzon marinus:* A comparative study throughout its life span. *Acta Zool. Stockh.* 65:1–15.

Ardavin, C. F., and Zapata, A. (1987). Ultrastructure and changes during metamorphosis of the lympho-hemopoietic tissue of the larval anadromous sea lamprey *Petromyzon marinus. Dev. Comp. Immunol.* 11:79–93.

Bloom, W. (1938). The embryogenesis of mammalian blood. In: *Handbook of Hematology* (Downey, H., ed.). New York: Paul B. Hoebar, pp. 863–922.

Borysenko, M., and Cooper, E. L. (1972). Lymphoid tissue in the snapping turtle, *Chelydra serpentina. J. Morphol.* 138:487–498.

Bradley, T. R., and Metcalf, D. (1966). The growth of bone marrow cells in vitro. *Aust. J. Exp. Biol. Med. Sci.* 44:287–300.

Campbell, F. (1970). Ultrastructure of the bone marrow of the frog. *Am. J. Anat.* 129:329–356.

Carpenter, K. L., and Turner, J. B. (1979). Experimental studies on hemopoiesis in the pronephros of *Rana pipiens. Differentiation* 14:167–174.

Coffman, R. L. (1983). Surface antigen expression and immunoglobulin gene rearrangement during mouse pre-B cell development. *Immunol. Rev.* 69:5–23.

Cooper, E. L. (1976). Evolution of blood cells. *Ann. Immunol. (Inst. Pasteur)* 127C:817–825.

Cooper, E. L., Klempau, A. E., Ramirez, J. A., and Zapata, A. G. (1980). Source of stem cells in evolution. In: *Development and Differentiation of Vertebrate Lymphocytes* (Horton, J. D., ed.). pp. 3–14. Amsterdam: Elsevier/North Holland.

Cooper, E. L., Klempau, A. E., Zapata, A. G. (1985). Reptilian immunity. In *Biology of the Reptilia* (Gans, C., Billett, F., and Maderson, P.F.A., eds.). New York: Wiley, pp. 600–678.

Cooper, E. L., and Schaeffer, D. W. (1970). Bone marrow restoration of transplantation immunity in the leopard frog *Rana pipiens. Proc. Soc. Exp. Biol. Med. 135*:406–411.

Cormier, F., De Paz, P., and Dieterlen-Lièvre, F. (1986). In vitro detection of cells with monocytic potentiality in the wall of the chick embryo aorta. *Dev. Biol. 118*:167–175.

Cormier, F., and Dieterlen-Lièvre, F. (1988). The wall of the chick embryo aorta harbors M-CFC, G-CFC, GM-CFC and BFU-E. *Development 102*:279–285.

Curtis, S. K., Cowden, R. R., and Knagel, J. W. (1979). Ultrastructure of the bone marrow of the salamander, *Plethodon glutinosus* (Caudata: Plethodontidae). *J. Morphol. 159*:151–184.

Dexter, T. M., and Lajtha, L. G. (1974). Proliferation of haemopoietic stem cells in vitro. *Brit. J. Haematol. 28*:525–530.

Dexter, T. M., and Moore, M.A.S. (1977). In vitro duplication and 'cure' of haemopoietic defects in genetically anaemic mice. *Nature 269*:412–414.

Dieterlen-Lièvre, F. (1984). Blood in chimeras. In *Chimeras in Developmental Biology* (LeDouarin, N., and McLaren, A., eds.). London: Academic Press, p. 133.

Dieterlen-Lièvre, F., and Martin, C. (1981). Diffuse intraembryonic hemopoiesis in normal and chimeric avian development. *Dev. Biol. 88*:180–191.

Dieterlen-Lièvre, F., Pardanaud, L., Yassine, F., and Cormier, F. (1988). Early haemopoietic stem cells in the avian embryo. *J. Cell Sci.* (suppl.) *10*:29–44.

Eipert, E. G., Klempau, A. E., Lallone, R. L., and Cooper, E. L. (1970). Bone marrow as a major lymphoid organ in *Rana. Cell. Immunol. 46*:275–280.

Ezine, S., Weissman, I. L., and Rouse, R. V. (1984). Bone marrow cells give rise to distinct clones within the thymus. *Nature 309*:629–631.

Fliedner, T. M., and Calvo, W. (1978). Hematopoietic stem-cell seeding of a cellular matrix: A principle of initiation and regeneration of hematopoiesis. In *Differentiation of Normal and Neoplastic Hematopoietic Cells* (Clarkson, B., Marks, P. A., and Till, J. E., eds.). Cold Spring Harbor, N.Y.: Cold Spring Harbor Laboratory, pp 757–773.

Ford, C. E., Hamerton, J. L., Barnes, D.W.M., and Loutit, J. (1956). Cytological identification of radiation-chimeras. *Nature (Lond.) 177*:452–454.

Fulop, G. M., and Phillips, R. A. (1989). Use of *scid* mice to identify and quantitate lymphoid-restricted stem cells in long-term bone marrow cultures. *Blood 74*:1537–1544.

Gong, J. K. (1978). Endosteal marrow: A rich source of hematopoietic stem cells. *Science 199*:1443–1445.

Hadji-Azimi, I., Coosemans, V., and Canicatti, C. (1987). Atlas of adult *Xenopus laevis laevis* hematology. *Dev. Comp. Immunol. 11*:807–874.

Harrison, D. E., Astle, C. M., and Stone, M. (1989). Number and function of transplantable primitive immunohematopoietic stem cells. *J. Immunol. 142*:3833–3840.

Hayakawa, K., Hardy, R. R., Herzenberg, L. A., and Herzenberg, L. A. (1985). Progenitors to Ly-1 B cells are distinct from progenitors for other B cells. *J. Exp. Med 161*:1554–1568.

Heimfeld, S., Guidos, C. J., Holzmann, B., Siegelman, M. H., and Weissman, I. L. (1990). Developmental analysis of the mouse hematolymphoid system. *Cold Spring Harbor Symposia LIV*:75–85.

Hunt, P., Robertson, D., Weiss, D., Rennick, D., Lee, F., and Witte, O. N. (1987). A single bone marrow–derived stromal cell type supports the in vitro growth of early lymphoid and myeloid cells. *Cell 48*:997–1007.

Ichikawa, Y., Plutznick, D. H., and Sachs, L. (1966). In vitro control of the development of macrophage and granulocyte colonies. *Proc. Natl. Acad. Sci. USA 56*:488–495.

Ihle, J. N., Keller, J., Oroszlan, S., Henderson, L. E., Copeland, T. D., Fitch, F., Prystowsky, M. B., Goldwasser, E., Schrader, J. W., Palaszynsky, E., Dy, K. M., and Lebel, B. (1983). Biological properties of homogeneous interleukin 3: I. Demonstration of WEHI-3 growth factor activity, mast cell growth factor activity, P cell stimulating factor activity, colony-stimulating factor activity, and histamine-producing cell-stimulating factor activity. *J. Immunol. 131*:282–287.

Jordan, H. E. (1938). Comparative hematology. In *Handbook of Hematology,* vol. 2 (Downey, H., ed.). New York: Paul B. Hoeba, pp. 700–862.

Jordan, H. E., and Flippin, J. C. (1913). Hematopoiesis in Chelonia. *Folia Haematol. 15*:1–24.

Jotereau, F., Heuze, F., Solomon-Vie, V., and Gascan, H. (1987). Cell kinetics in the fetal mouse thymus: Precursor cell input, proliferation, and emigration. *J. Immunol. 138*:1026–1030.

Jotereau, F. V., and LeDouarin, N. M. (1982). Demonstration of a cyclic renewal of the lymphocyte precursor cells in the quail thymus during embryonic and perinatal life. *J. Immunol. 129*:1869–1879.

Kearney, J. F., and Vakil, M. (1986). Idiotype-directed interactions during ontogeny play a major role in the establishment of the adult B cell repertoire. *Immunol. Rev. 94*:39–50.

Kincade, P. W. (1984). Formation of B lymphocytes in fetal and adult life. *Adv. Immunol. 31*:177–245.

LeDouarin, N. (1966). L'hematopoiese dans les formes embryonnaires et jeunes des vertébrés. *L'Anneé Biol. 5*:105–171.

LeDouarin, N. (1978). Ontogeny of hematopoietic organs studied in avian embryo interspecific chimeras. In *Differentiation of Normal and Neoplastic Hematopoietic Cells* (Clarkson, B., Marks, P. A., and Till, J. E., eds.). Cold Spring Harbor, N.Y.: Cold Spring Harbor Laboratory, p. 5.

LeFevre, M. E., Reincke, U., Arbas, R., and Gennaro, J. F. (1973). Lymphoid cells in the turtle bladder. *Anat. Rec. 176*:111–120.

Lord, B. I., Testa, N. G., and Hendry, J. H. (1975). The relative spatial distributions of CFUs and CFUc in the normal mouse femur. *Blood 46*:65–72.

Maximow, A. A., and Bloom, W. (1952). Blood cell formation and destruction. In *Textbook of Histology.* Philadelphia: Saunders, pp 75–104.

McCulloch, E. A., Siminovitch, L., Till, J. E., Russell, E. S., and Bernstein, S. E. (1965). The cellular basis of the genetically determined hemopoietic defect in anemic mice of genotype Sl/Sl^d. *Blood 26*:399–410.

Micklem, H. S., Ansell, J. D., Wayman, J. E., and Forrester, L. (1983). The clonal organization of hematopoiesis in the mouse. In *Progress in Immunology,* vol. 5 (Yamamura, Y., and Tada, T., eds.). Japan: Academic Press, pp. 633–644.

Micklem, H. S., Ford, C. E., Evans, E. P., and Gray, J. (1966). Interrelationship of myeloid and lymphoid cells: Studies with chromosome marked-cells transfused into lethally irradiated mice. *Proc. R. Soc. Lond. B., (Biol. Sci.) 165*:78–102.

Mintz, B., Anthony, K., and Litwins, S. (1984). Monoclonal derivation of mouse myeloid and lymphoid lineages from totipotent hematopoietic stem cells experimentally engrafted in fetal hosts. *Proc. Natl. Acad. Sci. USA 81*:7835–7839.

Moore, M.A.S., and Metcalf, D. (1970). Ontogeny of the haemopoietic system: Yolk sac origin of in vivo and in vitro colony forming cells in the developing mouse embryo. *Brit. J. Haematol. 18*:279–296.

Moore, M.A.S., and Owen, J.J.T. (1967). Stem-cell migration in developing myeloid and lymphoid systems. *Lancet ii*:658–659.

Mosier, D. E., Zitron, I. M., Mond, J. J., Ahmed, A., Scher, I., and Paul, W. E. (1977). Surface immunoglobulin D as a functional receptor for a subclass of B lymphocytes. *Immunol. Rev. 37*:89–103.

Mulder, A. H., Bauman, J.G.J., Visser, J.W.M., Boersma, W.J.A., and van den Engh, G. J. (1984). Separation of spleen colony-forming units and prothymocytes by use of a monoclonal antibody detecting an H-2K determinant. *Cell Immunol. 88*:401–410.

Müller-Sieburg, C. E. (1989). Isolation and characterization of hematopoietic stem cells and progenitors. In *Progress in Immunology,* vol. 7 (Melchers, F. et al. eds.). Japan: Academic Press, pp. 331–338.

Müller-Sieburg, C. E., Townsend, K., Weissman, I. L., and Rennick, D. (1988). Proliferation and differentiation of highly enriched mouse hematopoietic stem cells and progenitor cells in response to defined growth factors. *J. Exp. Med.* *167*:1825–1840.

Müller-Sieburg, C. E., Whitlock, C. A., and Weissman, I. L. (1986). Isolation of two early B lymphocyte progenitors from mouse bone marrow: A committed pre-pre-B cell and a clonogenic Thy-1^{lo} hematopoietic stem cell. *Cell* *44*:653–662.

Nakano, T., Waki, N., Asai, H., Hidekazu, A., and Kitamura, Y. (1989). Lymphoid differentiation of the hematopoietic stem cell that reconstitutes total erythropoiesis of a genetically anemic W/W^v mouse. *Blood* *73*:1175–1179.

Owen, J.J.T., and Ritter, M. A. (1969). Tissue interaction in the development of thymus lymphocytes. *J. Exp. Med* *129*:431–442.

Paige, C. J., Gisler, R. H., McKearn, J. P., and Iscove, N. N. (1984). Differentiation of murine B cell precursors in agar culture. Frequency, surface marker analysis and requirements for growth of clonable pre-B cells. *Eur. J. Immunol.* *14*:979–987.

Paige, C. J., Kincade, P. W., Shinfeld, L. A., and Sato, V. L. (1981). Precursors of murine B lymphocytes: Physical and functional characterization and distinction from myeloid cells. *J. Exp. Med.* *153*:154–165.

Paige, C. J., and Wu, G. E. (1989). The B cell repertoire. *FASEB J.* *3*:1818–1824.

Ramirez, J. A., Wright, R. K., and Cooper, E. L. (1983). Bone marrow reconstitution of immune responses following irradiation in the leopard frog *Rana pipiens*. *Dev. Comp. Immunol.* *7*:303–312.

Ratcliffe, N. A., Rowley, A. F., Fitzgerald, S. W., and Rhodes, C. P. (1985). Invertebrate immunity: Basic concepts and recent advances. *Int. Rev. Cytol.* *97*:183–350.

Reynaud, C.-A., Anquez, V., Grimal, H., and Weill, J.-C. (1987). A hyperconversion mechanism generates the chicken light chain preimmune repertoire. *Cell* *48*:379–388.

Scollay, R., Smith, J., and Stauffer, V. (1986). Dynamics of early T cells: Prothymocyte migration and proliferation in the adult mouse thymus. *Immunol. Rev.* *91*:129–157.

Shultz, L. D., and Sidman, C. L. (1987). Genetically determined murine models of immunodeficiency. *Annu. Rev. Immunol.* *5*:367–404.

Sidky, Y. A., and Auerbach, R. (1968). Tissue culture analysis of immunological capacity of snapping turtles. *J. Exp. Zool.* *167*:187–196.

Smith, P. B., Flajnik, M. F., and Turpen, J. B. (1989). Experimental analysis of ventral blood island hematopoiesis in *Xenopus* embryonic chimeras. *Dev. Biol.* *131*:302–312.

Snodgrass, R., and Keller, G. (1987). Clonal fluctuation within the haematopoietic system of mice reconstituted with retrovirus-infected stem cells. *EMBO J.* *6*:3955–3960.

Spangrude, G. J., Heimfeld, S., and Weissman, I. L. (1988a). Purification and characterization of mouse hematopoietic stem cells. *Science 241*:58–62.

Spangrude, G. J., Müller-Sieburg, C. E., Heimfeld, S., and Weissman, I. L. (1988b). Two rare populations of mouse Thy-1^{lo} bone marrow bone cells repopulate the thymus. *J. Exp. Med.* *167*:1671–1683.

Tidmarsh, G. F., Heimfeld, S., Whitlock, C. A., Weissman, I. L., and Müller-Sieburg, C. E. (1989). Identification of a novel bone marrow-derived B-cell progenitor population that coexpresses B220 and Thy-1 and is highly enriched for Abelson leukemia virus targets. *Mol. Cell. Biol.* *9*:2665–2671.

Till, J. E., and McCulloch, E. A. (1961). A direct measurement of the radiation sensitivity of normal bone marrow cells. *Radiation Res.* *14*:213–222.

Tomonaga, S., Hirokane, T., Shinohara, H., and Awaya, K. (1973). The primitive spleen of the hagfish. *Zool. Mag.* *82*:215–217.

Turpen, J. B., and Knudson, C. M. (1982). Ontogeny of hemopoietic cells in *Rana pipiens:* Precursor cell migration during embryogenesis. *Dev. Biol.* *89*:138–151.

Turpen, J. B., and Smith, P. B. (1989). Location of hemopoietic stem cells influences frequency of lymphoid engraftment in *Xenopus* embryos. *J. Immunol.* *143*:3455–3460.

van Bekkum, D. W., and de Vries, M. J. (1967). *Radiation Chimeras.* London: Logos/Academic Press.

Verlardi, A., and Cooper, M. D. (1984). An immunofluorescence analysis of the ontogeny of myeloid, T, and B lineage cells in mouse hematopoietic tissues. *J. Immunol. 133*:672–677.

Whitlock, C. A., Tidmarsh, G. F., Müller-Sieburg, C., and Weissman, I. L. (1987). Bone marrow stromal cell lines with lymphopoietic activity express high levels of a pre-B neoplasia-associated molecule. *Cell 48*:1009–1021.

Whitlock, C. A., and Witte, O. N. (1982). Long-term culture of B lymphocytes and their precursors from murine bone marrow. *Proc. Natl. Acad. Sci. USA 79*:3608–3612.

Wintrobe, M. M. (1967). The origin and development of the cells of the blood in the embryo, infant, and adult. In *Clinical Hematology.* Philadelphia: Lea and Febiger, pp. 1–62.

Witte, O. N. (1990). Steel locus defines new multipotent growth factor. *Cell 63*:5–6.

Wolf, N. S., and Trentin, J. J. (1968). Hemopoietic colony studies. V. Effect of hemopoietic organ stroma on differentiation of pluripotent stem cells. *J. Exp. Med. 127*:205–214.

Wright, R. K. (1981). Urochordates. In *Invertebrate Blood Cells,* vol. 2 (Ratcliffe, N. A., and Rowley, A. F., eds.). New York: Academic Press, pp. 565–626.

Wright, R. K., and Ermak, T. H. (1982). Cellular defense systems of the protochordata. In *The Reticuloendothelial System: A Comprehensive Treatise: Vol. 3. Ontogeny and Phylogeny* (Cohen, N., and Sigel, M. M., eds.). New York: Plenum Press, pp. 283–320.

Yancopoulos, G. D., and Alt, F. W. (1986). Regulation of the assembly and expression of variable-region genes. *Annu. Rev. Immunol. 4*:339–368.

Zapata, A. G., and Cooper, E. L. (1990). *The Immune System: Comparative Histo-physiology.* Chichester, England: Wiley and Sons.

Zapata, A., Leceta, J., Villena, A. (1981). Ultrastructure of reptilian bone marrow: A study in the Spanish lizard, *Lacerta hispanica. J. Morphol. 168*:137–149.

Zettergren, L. D., Kubagawa, H., and Cooper, M. D. (1980). Development of B cells in *Rana pipiens.* In *Phylogeny of Immunological Memory* (Manning, M. J., ed.). Amsterdam: Elsevier/North-Holland, pp. 117–186.

9

Monocytes, Macrophages, and Accessory Cells

SHIGERU MURAMATSU

About a century ago, Metchnikoff discovered phagocytes when he observed that tangerine thorns inserted into the blastcoel of transparent starfish gastrulae and spores of fungi (Monospora) growing in water fleas *(Daphnia)* were encapsulated by numerous, free amoeboid cells. Later investigations revealed that phagocytes, which take up or encapsulate foreign materials, including inorganic and organic particulates, microbes, and effete cells or cell debris of hosts, are ubiquitous in the animal kingdom and are always found in inflammatory reactions (Metchnikoff, 1892). The term "macrophage" was coined by Metchnikoff in 1901 for the free and fixed, large, mononuclear phagocytes in tissues to distinguish them from the smaller, polymorphonuclear phagocytes.

Such a definition of macrophages may be somewhat ambiguous because it is not based on the developmental origin of macrophages but only on their function and morphology, even though we know that macrophages are distributed widely and vary greatly in morphological and cytochemical characteristics. It thus appears appropriate to define macrophages as the progeny of blood monocytes according to the concept of the mononuclear phagocyte system, which will be described later, in homothermic vertebrates. Even this is not precise, however, because it is impossible to derive some mononuclear phagocytes from monocytes although they share many characteristics with monocyte-derived macrophages; they therefore cannot be called macrophages. In addition, whether the same principle can be applied to poikilothermic vertebrates, to invertebrates, and also to the embryo of homothermic vertebrates is uncertain. Considering these problems, it still seems convenient and flexible to use the term macrophage to represent mononuclear phagocytes irrespective of their developmental origins.

MACROPHAGES IN INVERTEBRATES

Macrophages, or mononuclear phagocytes, are the oldest cells in the animal kingdom. The origin of macrophages can be traced to the protozoans, from which all existing animals have evolved over billions of years. Protozoans were, and those that exist today still are, unicellular organisms that ingest particulate food from the environment. Metazoans (multicellular animals) probably originated from a colony of ancient protozoans, which was composed of food-getting outer zooids and non-food-getting but diversely specialized inner zooids. In existing metazoans, the outer zooids are analogous to the gut-associated, food-ingesting epithelial cells of the poriferans, coelenterates, and platyhelminths, and the inner zooids are analogous to differentiated cells in different tissues. Among these inner cells, there is a conserved cell population that retains its original protozoan characteristics. Such cells may be considered as the progenitors of macrophages because they do not participate directly in food getting but reside in tissues to phagocytose not only exogenous "foreign" materials but also endogenous "denatured" materials.

In existing metazoans, poriferans (sponges) are classified in the lowest phylum, Porifera. Their bodies are composed of an outer layer (pinacoderm), an inner layer (gastral layer), and a mesoglea (mesohyl or mesenchymal gel layer) that fills up the gap between the other two layers. The gastral layer is lined by flagellated cells (choanocytes) that create water currents by flagellar movements. In higher sponges, the gastral layer invaginates in an intricate manner into the mesoglea to form flagellated chambers (choanocyte chambers). Choanocytes take up food particles from the water current and transfer them to the mesoglea. Archeocytes in the mesoglea are strikingly large cells when compared to several other cell types (Fig. 9–1). They not only receive and digest foods from choanocytes but also clean up the matrix by decomposing dead cells. Thus, from the phylogenetic viewpoint, archeocytes can be considered the oldest example of tissue macrophages in existing metazoans. Furthermore, archeocytes are known to be totipotent stem cells; isolated archeocytes can differentiate into any cell type to reconstitute sponge tissue structure (Cohen and Sigel, 1982; de Sutter and van de Vyver, 1979). Thus archeocytes can be compared to primeval, evolved protozoans that acquired the ability to form colonies and to differentiate into various phenotypes, but which, at the same time, conserved their original properties as phagocytes.

Coelenterates also lack mesodermal development, but they are ranked higher than poriferans due to the possession of the diffuse nervous system (nerve net). The phylum

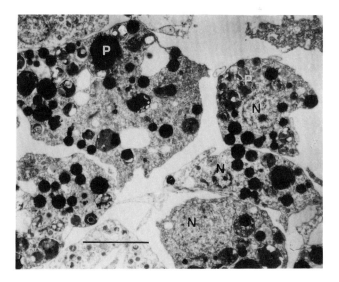

Figure 9–1.   Archeocytes of a freshwater sponge, *Ephydatia fluviatilis.* Each cell possesses numerous phagosomes (P). (N, nucleus.) Bar represents 5 μm. (Courtesy of Dr. Yoko Watanabe, Department of Biology, Ochanomizu University, Tokyo, Japan.)

Coelenterata consists of three classes: Hydrozoa (hydroids), Scyphozoa (true jellyfish), and Anthozoa (sea anemones, sea fans, and corals). Amoebocytes in syphozoans and anthozoans are phagocytic and pluripotent, if not totipotent, progenitors like archeocytes in sponges. Interstitial cells in hydroids are also pluripotent progenitors, but they lack phagocytic capacity (Cohen and Sigel, 1982). The neoblast in platyhelminths (flat worms) is a similar case. These animals are ranked higher than coelenterates, but they are the lowest in the animals having mesodermal development and a central nervous system (Cohen and Sigel, 1982). Perhaps, originally the interstitial cells and neoblasts were phagocytic but lost this ability during evolution.

Macrophages, or mononuclear phagocytes, in invertebrates are morphologically diversified as a result of either convergent or divergent evolution, or both, and they have been given various names according to differences in animal species and the investigators who classify them (Table 9-1). For instance, the granular cell in insects is a typical mononuclear phagocyte (Fig. 9-2), but the term granular cell may be confused with granulocytes (polymorphs) in vertebrates. Ratcliffe and Rowley (1979) took a general view of blood cells in eucoelomate invertebrates—e.g., insects, crustaceans, mollusks, annelids, tunicates, and echinoderms—and argued that phagocytic amoebocytes are ubiquitous and probably represent the descendants of primitive scavenger cells of lower invertebrates. They also stated that they should be regarded as invertebrate macrophages.

MACROPHAGES IN VERTEBRATES

Reticuloendothelial System

Although Metchnikoff was a pioneer in macrophage research, he only suggested vague pluralism in their origin. Kiyono (1914) discovered free macrophages in connective tissues (termed "histiocytes") by vital staining of experimental animals with lithium carmine. He assumed that histiocytes that stained strongly with carmine might originate from cells in the reticuloendothelium of sinusoids in the bone marrow, liver, spleen, and lymph nodes. Later, Aschoff (1924) proposed the concept of the reticuloendothelial sys-

Table 9-1. Macrophages or Mononuclear Phagocytes in Invertebrates

Phylum	Examples of the Nomenclature
Porifera	Archeocyte
Coelenterata (except Hydrozoa)	Amoebocyte
Platyhelminthes	?
Nemertea	Macrophage
Sipunculida	Granular cell; large hyalocyte
Annelida	Amoebocyte; granulocyte
Mollusca	Amoebocyte; hemacyte
Arthropoda	
Crustacea	Hyaline cell; granulocyte
Insecta	Plasmatocyte; granular cell
Echinodermata	Phagocytic amoebocyte
Protochordata	Granular leukocyte (macrophage); hyaline leukocyte

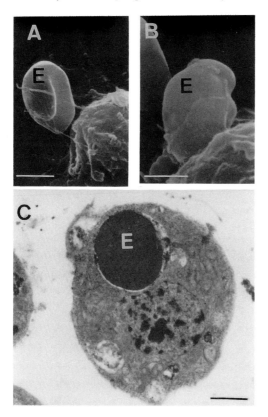

Figure 9–2.   Granular cells of a silkworm, *Bombyx mori,* phagocytosing sheep erythrocyte (E). (A) Elongation of many filopodia toward one E; (B) Extension of veil-like membrane processes to internalize E; and (C) Formation of phagosome including ingested E. Each bar represents 2 μm. (Courtesy of Dr. Haruhisa Wago, Department of Bacteriology, Saitama Medical School, Saitama, Japan.)

tem (RES) based on results from joint research work with Kiyono aimed at systematizing those phagocytes other than granulocytes. The RES theory argues that cells that strongly ingest vital dyes, such as histiocytes and monocytes (endothelial leukocytes or blood histiocytes) may be derived from cells that weakly take up vital dyes, such as the reticulum cells of lymphoid tissues or organs, the reticuloendothelial cells of lymph node sinuses, and those of sinusoids in the liver, spleen, bone marrow, adrenal cortex, and pituitary. Briefly, macrophages were assumed to be derived from mesenchymal cells independent of the hemopoietic system.

Mononuclear Phagocyte System

In the 1960s, more advanced methods in cell biology, such as cytochemistry, autoradiography, and use of chromosomal markers, brought salient progress in studying cell lineages, and monocytes became highlighted as the precursors of macrophages. Indeed, monocytes are found in hagfish blood, the most primitive existing vertebrate belonging

Table 9-2. Hemopoietic Organs or Tissues in Vertebrates

Class	Organs or Tissues
Agnatha	
Hagfish	Hemopoietic foci (in lamina propria of gut); pronephros (or head kidney)
Lamprey	Primitive spleen (as an invagination of anterior gut tissue); primitive bone marrow-like tissue (in fibrocartilaginous protovertebral arch)
Chondrichthyes	Spleen; cranial lymphomyeloid tissue; organ of Leydig (of esophagus and epigonal organ)
Osteichthyes	Spleen; pronephros; mesonephros ?
Amphibia	Spleen; kidney; liver; bone marrow (anuran)
Reptilia	Bone marrow; spleen
Aves	Bone marrow; spleen
Mammalia	Yolk sac (fetus); liver (fetus); bone marrow; spleen

to the class Agnatha (Linthicum, 1975), along with that of more advanced fish and, of course, the higher vertebrates (Cohen and Sigel, 1982; Rijkers, 1982). Monocytes seem to be formed in hemopoietic tissues or organs and, after release, they circulate in the blood (Table 9–2).

To reexamine the validity of the RES theory and to establish a new unified interpretation of the cellular lineage of mononuclear phagocytes, the first conference on mononuclear phagocytes was held in Leiden, The Netherlands, in 1969 (van Furth, 1970). There, van Furth and other investigators proposed that all highly phagocytic mononuclear cells and their precursors can be placed in one system, termed the mononuclear phagocyte system (MPS). According to the MPS concept, mononuclear phagocytes, or macrophages, in adult animals are derived solely from blood monocytes, which are the progeny of promonocytes and monoblasts that originated from hemopoietic stem cells in

(Bone marrow) Stem cell → Committed stem cell →

Monoblasts → Promonocytes → Monocytes →

(Peripheral blood) Monocytes →

(Tissues) Macrophages

Normal state	Inflammation
Connective tissue (histiocyte)	Exudate macrophage
Liver (Kupffer cell)	Exudate-resident macrophage
Lung (alveolar macrophages)	Epithelioid cell
Lymph node, Spleen (free and fixed macrophages)	Multinucleated giant cell (Langerhans type and foreign-body type)
Bone marrow (fixed macrophage)	
Serous cavities (pleural and peritoneal macrophages)	
Bone (osteoclast)	
Nervous tissue (microglial cell)	
Skin (histiocyte)	
Synovia (type A cell ?)	
Other organs (tissue macrophage)	

Figure 9–3. Mononuclear phagocyte system proposed by van Furth (1970, 1980).

bone marrow. The schematic representation of MPS by van Furth (1980) is shown in Figure 9–3.

Monoblasts and promonocytes are identifiable in bone marrow. They share various characteristics with blood monocytes. Thus, they express receptors for the Fc region of immunoglobulin G (IgG) (FcR) and for complement, mainly for the activated third component (C3bR). They also possess phagocytic and pinocytic capacities (van Furth, 1980). One stem cell generates one monoblast; the division of one monoblast produces two promonocytes (Fig. 9–4A); and division of one promonocyte produces two monocytes (Fig. 9–4B). Thus, from stem cells to monocytes, there is a four-fold amplification. As a rule, monocytes do not divide further but infiltrate into various tissues to differentiate into exudate macrophages (Fig. 9–5A). Exudate macrophages may shift to exudate-resident macrophages and finally become resident macrophages (Fig. 9–5B–5D).

The intracellular localization of peroxidase activity changes with the differentiation of MPS cells (Figs. 9–4, 9–5; Table 9–3). The differentiation order, indicated by arrows in Table 9–3, is based on changes in intracellular peroxidase localization in MPS cells differentiating from mouse bone marrow cells in vitro and rat macrophages under sterile inflammatory conditions in vivo (van der Meer et al., 1979; van Furth, 1980). Thus, peroxidase localization seems to present a convenient index for defining differentiation stages of MPS cells.

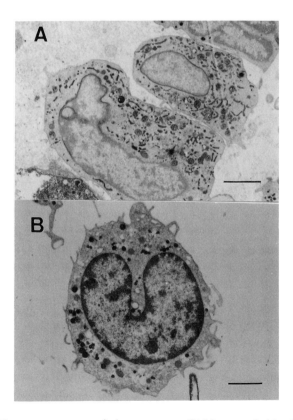

Figure 9–4.   (**A**) Human promonocyte in bone marrow; (**B**) Monocyte in blood tested for peroxidase. Each bar represents 2 μm. (Courtesy of Dr. Makoto Naito, Department of Pathology, Kumamoto University School of Medicine, Kumamoto, Japan.)

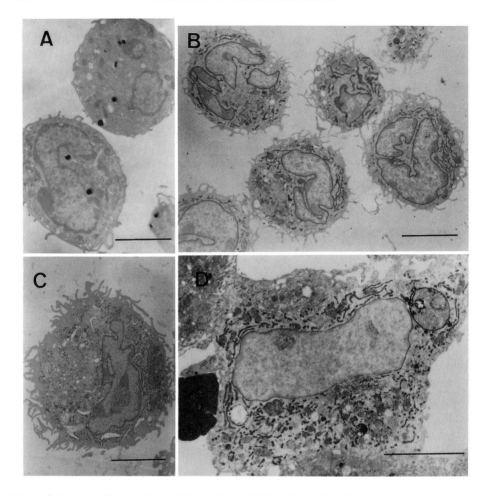

Figure 9–5.   Localization of peroxidase activity of (**A**) peritoneal exudate macrophage (**B**), resident peritoneal, and (**C**) alveolar macrophages of guinea pig, and (**D**) resident liver macrophage (Kupffer cell) of rat. Each bar represents 5 μm. (Courtesy of Dr. Makoto Naito.)

According to estimations by van Furth and colleagues (1982a,b), in the normal steady state in adult rodents, about 56% of blood monocytes migrate to the liver, 16% to lungs, 8% to peritoneal cavity, 5% to intestine, and the remaining 15% to other tissues. A low percentage of promonocytes seems to reach various tissues directly and divide once there. Thus less than 5% of tissue mononuclear phagocytes divide under in vitro conditions (van Furth et al., 1982a,b). This may result in contributions by local proliferation for the maintenance of pulmonary macrophage number, of which about 25% are assumed to develop locally. In contrast, about 25% of murine blood monocytes are estimated to enter the spleen at the rate of 15×10^3 cells/hour whereas splenic macrophages leave at 27×10^3 cells/hour. Thus the balance of 12×10^3 cells/hour may be compensated for by local production of macrophages (van Furth and Diesselhoff-den Dulk, 1984). It is still uncertain whether this is ensured by one or more divisions of blood-borne promonocytes and/or monocytes, or by the autonomous formation of macrophages in spleen independently of the influx of blood cells.

Table 9-3. Localization of Peroxidase Activity in Mononuclear Phagocytes[a]

	Nuclear Envelope	Rough Endoplasmic Reticulum	Lysosomal Granules	Golgi System
Monoblast ↓	±	±	+	
Promonocyte ↓	+	+	+	±
Monocyte ↓	−	−	+	−
Exudate (early) macrophage ↓	−	−	+	−
Exudate-resident (transitional) macrophage ↓	+	+	+	−
Resident (mature) macrophage ↓	+	+	−	−
Peroxidase activity—negative macrophage	−	−	−	−

[a] ± means very low; blank = unknown.

MACROPHAGES IN ONTOGENY

It seems beyond controversy that virtually all adult mammalian tissue macrophages are derived from blood monocytes or their precursors. However, it is still debatable whether exceptional macrophages exist that are derived from tissue cells other than those included in the MPS. In fact, macrophages appear in various fetal tissues before initiation of hemopoiesis in the bone marrow (Fig. 9-6) (Naito and Wisse, 1977; Takahashi et al., 1980, 1983). This may possibly support the binary theory for the origin of macrophages and lead us to assume that the RES theory might be preferable to the MPS theory during fetal stages. Such a binary theory, however, seems to lack evidence that fetal macrophages are not derived from hemopoietic stem cells, or from progenitors of monocytes, because fetal hemopoiesis occurs first in yolk sac, then in liver, and finally in bone marrow at a late fetal stage. Though the MPS theory modestly holds that all macrophages, or highly phagocytic mononuclear cells, of adult animals originate from stem cells in bone marrow, extrapolation of the MPS theory to fetal stages thus proposes that macrophages formed earlier than by bone marrow hemopoiesis are derived from stem cells in the yolk sac and liver. Therefore, if one intends to maintain the existence of hemopoietic stem cell–independent macrophages in the fetus, macrophages must be observed before hemopoiesis begins in the yolk sac or, at least, before vascularization of tissues.

So far, the origin of fetal macrophages has not been investigated as extensively as that of adult macrophages, but circumstantial evidence strongly suggests that even fetal macrophages originate from hemopoietic stem cells—in other words, there is no concrete evidence to support the RES origin of fetal macrophages. Thus, an ontogenetic study of human fetuses revealed that macrophages in tissues are found only after vascularization (Andersen and Matthiessen, 1966). In rodents, yolk sac hemopoiesis begins at about the 9th day of gestation and immature macrophages can be found there, first at the 10th or 11th day of gestation (Fig. 9-6A), although macrophages in other tissues and in the circulation appear at the 11th day or later (Cline, 1975; Cline and Moore, 1972; Takahashi et al., 1980) (Figs. 9-6B–6E). Moreover, in vitro cultures of subdermal mesenchymal cells

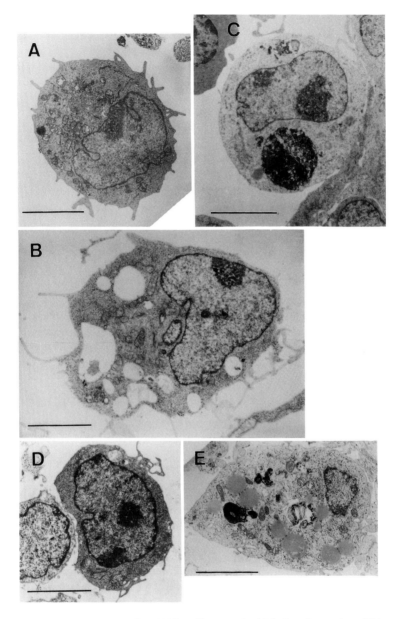

Figure 9-6.   Fetal macrophages of rat (**A**) in yolk sac at the 11th day of gestation, (**B**) in subepidermal mesenchyme at the 12th day, (**C**) in liver at the 13th day, (**D**) in placenta at the 14th day, and (**E**) in brain at the 16th day. Each bar represents 5 μm. (Courtesy of Dr. Makoto Naito.)

of rat fetuses resulted in no favorable evidence for macrophage formation from undifferentiated mesenchymal cells (Bard et al., 1975; van Furth, 1980). Taken together, these obesrvations suggest that fetal macrophages are generated first from hemopoietic stem cells in yolk sac, not from primitive mesenchymal cells, and are distributed to tissues through the circulation after vascularization. Liver hemopoiesis begins at the 13th day of gestation in mouse, after which macrophages are generated in the liver and disseminated to various tissues until birth.

One of the distinguishing traits of fetal macrophages is their high proliferative capacity (Naito et al., 1982; Takahashi et al., 1980). In contrast, typical promonocytes and monocytes cannot be found until the late phases of fetal development. Therefore, in a narrow sense, the MPS theory, which argues that macrophages are derived from blood monocytes, appears inapplicable to the developmental process of fetal macrophages. The MPS theory, however, seems also still valid for the fetus, if the theory is interpreted in a broad sense and assuming that macrophages are derived from hemopoietic stem cells.

Macrophages play an important role during embryonic-fetal development. As was pointed out by Saunders (1966), the genetically programmed death of certain cells in tissues and phylogenetically vestigial organs is a prerequisite for adequate histogenesis and organogenesis. The resulting necrotic cells must therefore be scavenged and cleaned up by macrophages. In certain cases, macrophages appear to cut and trim tissues to ensure morphogenesis. For example, Figure 9–7 shows the contribution of macrophages to the formation of digits from the limb bud in a rat fetus. At any rate, this enormous activity of fetal macrophages—i.e., clearing away denatured, or sometimes transformed, cells— is laborious. It is not too unreasonable to assume therefore that fetal macrophages may be in a somewhat activated state in comparison with those of adult animals. That the phagocytic activity and antitumor activity of newborn mouse macrophages is remarkably higher than those in adult mice may be a developmental relic derived from the activated state of fetal macrophages (Ido et al., 1984, 1985; Inaba et al., 1982; Nakano, et al., 1978).

ACCESSORY CELLS IN THE INITIATION OF IMMUNE RESPONSES

Accessory Cells for Helper (CD4$^+$8$^-$) T Cells Bearing Class II MHC Antigens

The activation of helper T cells is a prerequisite for initiating most immune responses, the so-called T cell–dependent immune responses. Helper T cells cannot respond directly to antigens and therefore require the mediation of other cells, called, almost interchangeably, accessory cells (A cells) or antigen presenting cells (APC). This was first discovered by Mosier (1967) in the in vitro antibody response of murine spleen cells. Since then, investigations by many immunologists have elucidated that A cell activity usually resides in a cell population that readily adheres to glass, plastic, and other solid substrates. Depletion of cells bearing class II MHC antigens (Ia antigens) from this adherent cell population abrogates A cell activity. On the other hand, several immunologists have indicated that the ontogenetic development of A cell activity parallels the increase in Ia-bearing (Ia$^+$) cell content of adherent cell populations. These findings indicate that the A cells are Ia$^+$-adherent cells.

Until the early 1980s, most immunologists believed that A cells are Ia$^+$ macrophages (Shevach, 1984; Unanue, 1984) because macrophages are major cells in the adherent cell population and they can express Ia antigens in response to T cell–derived γ interferon (Steeg et al., 1982). In more recent years, however, the identity of A cells has become a subject of controversy as extensive studies of Ia$^+$ non-macrophage cells with the morphology of dendritic-type cells (D cells) reveal that these cells serve as potent A cells even in the absence of Ia$^+$ macrophages (Möller, 1980). In addition, some researchers reported that B lymphoma cells, B lymphoblasts, and normal B cells, which are all Ia$^+$ cells, behave

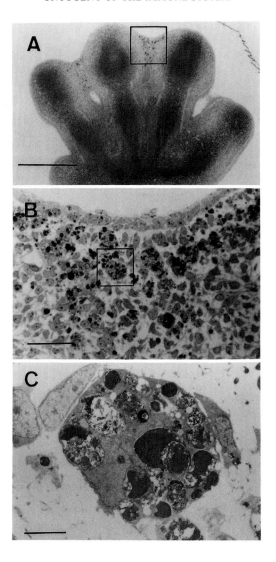

Figure 9–7.   Fetal macrophages at the site of digit formation of rat at the 16th day of gestation. Figures **B** and **C** are high power views of the squares in **A** and **B**, respectively. Bars represent 500 μm (A), 50 μm (B), and 5 μm (C). (Courtesy of Dr. Makoto Naito.)

as APC in the antigen-specific proliferative response of immune T cells (Ashwell et al., 1984; Shevach, 1984).

Characteristics of D Cells

The term D cell was given by the American Reticuloendothelial Society Committee on Nomenclature to those cells that are irregularly shaped but distinct from macrophages, reticulum cells, and fibroblasts in various tissues, including lymphoid tissues, and probably participate in immune responses (Tew et al., 1982). In the committee report, the following cells are classified as D cells: follicular dendritic cell(s) (FDC), lymphoid dendritic cell(s) (LDC), interdigitating cell(s) (IDC), and Langerhans cell(s) (LC). Figure 9–8 represents the morphology of LDC and LC in electron microscopy, and Table 9–4 indicates the various characteristics of D cells in comparison with those of macrophages; it is thus evident that D cells are apparently distinguishable from macrophages. In addition to these four cell types, the committee report mentions that the following three other

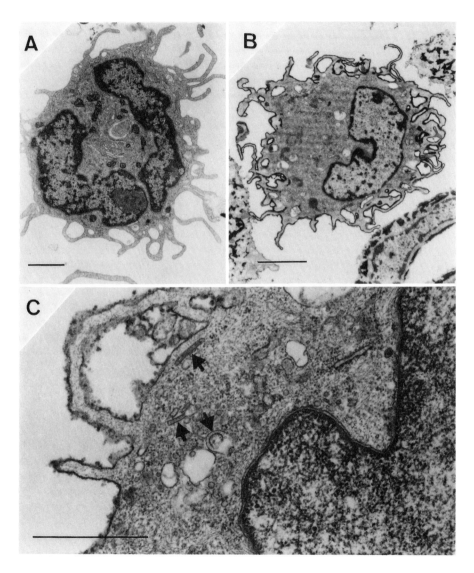

Figure 9–8. (**A**) Lymphoid dendritic cell; (**B, C**) Langerhans cell. Ia antigens on Langerhans cells are visualized with an immunoelectron micrography reagent as electron dense products on the cell surface. Arrows (in C) indicate Langerhans (Birbeck) granules. Bars represent 2 μm (A, B), and 1 μm (C). (Reproduced from *The Journal of Experimental Medicine 165:* 526–546, 1985, by copyright permission of The Rockefeller University Press.)

Table 9–4.   Characteristics of D cells and Macrophages

	Follicular Dendritic Cell	Lymphoid Dendritic Cell	Interdigitating Cell	Langerhans Cell	Macrophage
Endocytosis					
Phagocytosis	−	−	−	+	+ + +
Pinocytosis	±	±	+	+	+ + +
Receptors for					
Fc (IgG)	+	−	−	+	+
C3b	+	+ or −[a]	−	+	+
Surface markers					
Class II MHC (Ia)	+	+	+	+	+ or −[b]
Thy-1	−	−	−	−	−
Ig	−	−	−	−	−
Adherence	±	+ ~ −	?	+	+ ~ + + +
Enzymes					
ATPase	+ ~ −	−	+	+	+
Nonspecific esterase	±	−	+ ~ −	+	+
Peroxidase	−	−	+ ~ −	−	+
Acid phosphatase	−	±	+ ~ −	±	+
Bone marrow origin	−	+	?	+	+
Langerhans granules	−	−	−	+	−

[a] +, human; −, mouse and rat.
[b] +, IFN-stimulated; −, unstimulated.

sources of D cells are likely to be related to some of the D cells described above: (1) thymic D cells: bone marrow–derived cells expressing Ia antigens, and perhaps identical to "nurse cells" identified by Wekerle et al. (1980); (2) D cells in solid or nonlymphoid tissues (interstitial dendritic cells): bone marrow–derived Ia+ cells identified by Hart and Fabre (1981) in several rat organs other than brain; (3) D cells in lymph, especially afferent lymph: identified by Kelly et al. (1978) as "veiled cells" in afferent lymph, and probably identical to LDC.

Expression of Ia antigens on FDC is still under debate. One report documents that Ia antigens are detectable on isolated FDC (Schnizlein et al., 1985), while another describes that B lymphocytes, which also are Ia+ cells, are frequently wrapped in the dendritic processes of FDC (Humphrey et al., 1984). FDC have also been termed dendritic reticular cells and Nossal's "follicular dendritic reticulum cells." They localize at lymphoid follicles and germinal centers in the spleen and the lymph nodes, and they trap on their surfaces, but do not ingest, the immune complexes transported by B lymphocytes (Humphrey et al., 1984).

LDC, which are usually simply called dendritic cells, were described by Steinman and Cohn (1973) as a novel cell type in lymphoid organs. LDC from mouse spleen and lymph nodes are adherent to the vessel bottom in in vitro cultures for an initial few hours, but most of them lose the ability to adhere in overnight cultures. This observation has been used for separating and purifying LDC from macrophages. Rat LDC and mouse Peyer's patch LDC are poorly adherent even shortly after isolation. Human LDC can be prepared from peripheral blood as short-term adherent cells and are useful for in vitro experiments. They account for 0.5% or more of blood mononuclear cells (van Voorhis et al., 1982).

IDC, first described by Veldman in 1970, are found in the T cell region of lymphoid organs (Veldman and Kaiserling, 1980). They extend long dendritic processes, which make contact with adjacent lymphocytes. Thymocyte proliferation is most abundant in those cells that attach to processes. Close relations of IDC with LC is suggested by the presence of Langerhans granules in IDC-like cells.

LC were first discovered by Langerhans more than a century ago in the epidermis of skin sections that had been stained with gold chloride. LC appear very similar morphologically to other D cells, but they are distinguished by the presence of Langerhans granules (Birbeck granules) in the cytoplasm that are visible by electron microscopy. Originally, LC were defined as D cells as constant components in the epidermis, accounting for about 500–1,000 LC per mm^2 in humans. However, D cells that possess Langerhans granules are also found in various tissues, such as the lymphatics, lymph nodes, thymic medulla, mucous membranes of oral and pharyngeal tissues, gastric mucosa, or uterine cervix. LC may be an immature form of LDC. Thus, the in vitro culture of LC, isolated from murine epidermis, for a few days, results in the loss of Langerhans granules, Fc receptors, and cytochemical reactivity for nonspecific esterase and membrane ATPase, as well as in the ultrastructural and phenotypical resemblance with LDC (Schuler and Steinman, 1985). In accordance with this, A cell activity tends to increase toward the level of A cell activity of freshly prepared splenic LDC.

Ontogeny of D Cells

Most D cells appear to be present at birth. However, typical FDC are not found in the neonatal spleen, and mouse LDC do not reach adult levels until several weeks of age (Tew et al., 1982). The ontogeny of LDC seems to parallel that of differentiation and maturation of immunocompetent lymphocytes (Inaba et al., 1982). The earliest appearance of D cells takes place in the thymus just after formation of the thymic rudiment; this is at gestation day 14 in mice, week 6–7 in humans, and 12-day embryo in chickens (Kyewski et al., 1986; Oliver and LeDouarin, 1984). Thereafter, D cells are found in several tissues, including the intestine and epidermis (Mayrhofer et al., 1983). Thus LDC appear as D cells throughout the body in different lymphoid tissues.

Identity of Accessory Cells

The identity of A cells in immune responses to foreign antigens has been investigated by the in vitro experiment in which responding lymphocytes depleted of adherent cells are supplemented with a syngeneic cell population comprising presumptive A cells. This methodology is applicable for primary and secondary antibody responses, for the antigen-specific, proliferative responses of immune T cells collected from in vivo primed animals, and for the induction of cytotoxic T lymphocytes (CTL) for allogeneic cells or hapten-modified syngeneic cells. In the proliferative response of T cells to concanavalin A (Con A) or phytohemagglutinin (PHA), and to the so-called oxidative mitogens, such as sodium periodate or neuraminidase plus galactose oxidase, mitogens are added to the culture in place of foreign antigens.

A cells not only mediate the immune response to foreign antigens or mitogens but also trigger the response of allogeneic helper T precursor cells recognizing class II MHC (Ia) antigens on A cells. The proliferative response has been called the mixed lymphocyte (or leukocyte) reaction (MLR). When the responding T cell population contains those cells which recognize that allogeneic class I MHC antigens (CD4$^-$8$^+$ T cells) are contained in the responding T cell population, they will proliferate and differentiate into allo-reactive CTL, not only autonomously but also with the aid of helper T cells (Inaba et al., 1987). Therefore, the identity of A cells can also be determined as the stimulator of CD4$^+$8$^-$ T cell proliferation for MLR and CTL induction.

As mentioned before, Ia$^+$ macrophages have been assumed to be representative of A cells. This seems to be questionable as abundant evidence has been accumulated which shows that LDC are far more adequate as A cells than macrophages in virtually all kinds of T cell responses (Austyn et al., 1983; Inaba et al., 1983; Klinkert et al., 1980; Komat-subara et al., 1985; Miyazaki and Osawa, 1983; Naito et al., 1984; Nussenzweig et al., 1980; Röllinghof et al., 1982; Steinman et al., 1983; Sunshine et al., 1980). Among other types of D cells, LC can be isolated from epidermis for experimental use without being contaminated with macrophages, and their MLR-stimulatory capacity (Stingl et al., 1978), allogeneic CTL-inducibility (Steiner et al., 1985; Tsuchida et al., 1984), A cell activity for induction of CTL to haptenated syngeneic cells (Steiner et al., 1985), and APC activity for immune T cells (Stingl et al., 1981) have been elucidated. Two other sources of D cells, FDC and IDC, are difficult to prepare as sufficiently purified cell populations for experiments on in vitro immune responses, so that the information on their A cell activity is still scanty.

Care should be exerted in assigning to macrophages the role of potent A cells because this conclusion was derived from experiments in which LDC were not intentionally removed from macrophage populations or responding lymphocyte populations. Thus even if cultures appear to be comprised almost exclusively of lymphocytes and macro-phages, the possible existence of a small number of LDCs may cause high-level immune responses in the presence of macrophages. In other words, macrophages, especially Ia$^+$ macrophages, markedly amplify the otherwise undetectably low level of immune responses triggered by low-dose DC (Inaba et al., 1981, 1982; Kawai et al., 1987; Komat-subara et al., 1985; Naito et al., 1984).

Almost all of the reliable data indicating that macrophages and B cells, or their trans-formed cell lines, could serve as A cells have been derived from experiments in which T cell clones or hybridomas established from antigen-specifically triggered T cells of primed animals were employed as responding cells (Ashwell et al., 1984; Shevach, 1984). This seems, however, only to prove that macrophages and B cells can replace LDC after T cells have been triggered. As a matter of fact, though resting T cells are triggered only by LDC in MLR, antigen-primed T cell blasts can be restimulated by B blasts, resting B cells, and macrophages, as well as by LDC (Inaba and Steinman, 1984). On the other hand, resting CTL precursors (CD4$^-$8$^+$ T cells) can be directly activated in recognizing class I MHC antigens not only by LDC (Inaba et al., 1987) but also artificially by neuraminidase-treated macrophages plus exogenous interleukin 2 to stimulate proliferation and differ-entiation into CTL (Hirayama et al., 1988).

Taken together, it seems reasonable to assume that autonomous, potent, and stable A cells, which trigger the resting CD4$^+$8$^-$ T cell are D cells, at least LDC and LC. Mac-rophages and B cells can serve, in most cases, as successors of D cells for activated T cells triggered initially by D cells.

FINAL COMMENT

Macrophages, or mononuclear phagocytes, are the oldest cells and are ubiquitous in the animal kingdom. Not only self-defense mechanisms but also development and evolution cannot be considered adequately without understanding macrophages. In addition, we must remember another category of cells, generically termed D cells. Among several types of D cells, lymphoid dendritic cells and Langerhans cells are known to be potent accessory cells in the initiation of immune responses in mammals. Perhaps they are relatives of macrophages and exist also in other classes of vertebrates. More extensive studies on their ontogeny and phylogeny are anticipated.

Acknowledgments

For their kindness in providing several photographs, I wish to thank Dr. Yoko Watanabe of Ochanomizu University, Dr. Haruhisa Wago of Saitama Medical School, Dr. Makoto Naito of Kumamoto University, and Dr. Ralph Steinman and Dr. Gerald Schuler of The Rockefeller University. Thanks also to Dr. Naito for his valuable suggestions and advice in writing the section on macrophages in ontogeny.

References

Andersen, H., and Matthiessen, M. E. (1966). The histiocyte in human fetal tissues. *Z. Zellforsch.* 72:193–211.

Aschoff, L. (1924). Das Reticuloendotheliale System. *Ergebn. inn. Med. Kinderheilk.* 26:1–118.

Ashwell, J. D., DeFranco, A. L., Paul, W. E., and Schwartz, R. H. (1984). Antigen presentation by resting B cells: Radiosensitivity of the antigen-presentation function and two distinct pathways of T cell activation. *J. Exp. Med.* 159:881–905.

Austyn, J. M., Steinman, R. M., Weinstein, D. E., Granelli-Piperno, A., and Palladino, M. A. (1983). Dendritic cells initiate a two-stage mechanism for T lymphocyte proliferation. *J. Exp. Med.* 157:1101–1115.

Bard, J.B.L., Hay, E. D., and Meller, S. M. (1975). Formation of the endothelium of the avian cornea: A study of cell movement in vivo. *Dev. Biol.* 42:334–361.

Cline, M. J. (1975). Ontogeny of mononuclear phagocytes. In *Mononuclear Phagocytes in Immunity, Infection and Pathology* (van Furth, R., ed.). Oxford: Blackwell, pp. 263–267.

Cline M. J., and Moore, M.A.S. (1972). Embryonic origin of the mouse macrophage. *Blood* 39:842–849.

Cohen, N., and Sigel, M. M. (ed). (1982). *The Reticuloendothelial System: Vol. 3. Phylogeny and Ontogeny.* New York: Plenum Press.

de Sutter, D., and van de Vyver G. (1979). Isolation and recognition of some defomote sponge cell types. *Dev. Comp. Immunol.* 3:389–397.

Hart, D.N.J., and Fabre, J. W. (1981). Demonstration and characterization of Ia-positive dendritic cells in the interstitial connective tissues of rat heart and other tissues, but not brain. *J. Exp. Med.* 153:347–361.

Hirayama, Y., Inaba, K., Inaba, M., Kato, T., Kitaura, M., Hosokawa, T., Ikehara, S., and Muramatsu, S. (1988). Neuraminidase-treated macrophages stimulate allogeneic CD8[+] T cells in the presence of exogenous interleukin 2. *J. Exp. Med.* 168:1443–1456.

Humphrey, J. H., Grennan, D., and Sundaram, V. (1984). The origin of follicular dendritic cells in the mouse and the mechanism of trapping immune complexes on them. *Eur. J. Immunol.* 14:859–864.

Ido, M., Uno, K., Inaba, K., Aotsuka, Y., and Muramatsu, S. (1984). Ontogeny of "macrophage" function: IV. Newborn macrophages strongly suppress tumor cell growth and readily acquire cytolytic activity in comparison with adult macrophages. *Immunology* 52:307–317.

Ido, M., Uno, K., Inaba, K., Komatsubara, S., Hosono, M., and Muramatsu, S. (1985). Ontogeny of "macrophage" function: V. Differential effect of prostaglandin E$_2$ on the activation of newborn and adult mouse macrophages. *Dev. Comp. Immunol.* 9:719–725.

Inaba, K., Granelli-Piperno, A., and Steinman, R. M. (1983). Dendritic cells induce T lymphocytes to release B cell-stimulating factors by an interleukin-2–dependent mechanism. *J. Exp. Med.* 158:2040–2057.

Inaba, K., Masuda, T., Miyama-Inaba, M., Aotsuka, Y., Kura, F., Komatsubara, S., Ido, M., and Muramatsu, S. (1982). Ontogeny of "macrophage" function: III. Manifestation of high accessory cell activity for primary antibody response by Ia$^+$ functional cells in newborn mouse spleen in collaboration with Ia$^-$ macrophages. *Immunology* 47:449–457.

Inaba, K., Nakano, K., and Muramatsu, S. (1981). Cellular synergy in the manifestation of accessory cell activity for in vitro antibody response. *J. Immunol.* 127:452–461.

Inaba, K., and Steinman, R. M. (1984). Resting and sensitized T lymphocytes exhibit distinct stimulatory (antigen-presenting cell) requirement for growth and lymphokine release. *J. Exp. Med.* 160:1717–1735.

Inaba, K., Steinman, R. M., van Voorhis, W. C., and Muramatsu, S. (1983). Dendritic cells are critical accessory cells for thymus-dependent antibody responses in mouse and in man. *Proc. Natl. Acad. Sci. USA* 80:6041–6045.

Inaba, K., Young, J. W., and Steinman, R. M. (1987). Direct activation of CD8$^+$ cytotoxic T lymphocytes by dendritic cells. *J. Exp. Med.* 166:182–194.

Kawai, J., Inaba, K., Komatsubara, S., Hirayama, Y., Naito, K., and Muramatsu, S. (1987). Role of macrophages as modulators but not as autonomous accessory cells in the proliferative response of immune T cells to soluble antigen. *Cell. Immunol.* 109:1–11.

Kelly, R. H., Balfoar, B. M., Armstrong, J. A., and Griffith, S. (1978). Functional anatomy of lymph nodes: II. Peripheral lymph-borne mononuclear cells. *Anat. Rec.* 190:5–21.

Kiyono, K. (1914). *Die vitale Karminspeicherung.* Jena: Fischer Verlag.

Klinkert, W.E.F., LaBadie, J. H., O'Brien, J. P., Beyer, C. F., and Bowers, W. E. (1980). Rat dendritic cells function as accessory cells and control the production of soluble factor required for mitogenic response of T lymphocytes. *Proc. Natl. Acad. Sci. USA* 77:5414–5418.

Komatsubara, S., Hirayama, Y., Inaba, K., Naito, K., Yoshida, K., Kawai, J., and Muramatsu, S. (1985). Role of macrophages as modulators but not as autonomous accessory cells in primary antibody response. *Cell. Immunol.* 95:288–296.

Kyewski, B. A., Fathman, C. G., and Rouse, R. V. (1986). Intrathymic presentation of circulating non-MHC antigens by medullary dendritic cells: An antigen-dependent microenvironment for T cell differentiation. *J. Exp. Med.* 163:231–246.

Linthicum, D. S. (1975). Ultrastructure of hagfish blood leukocytes. In *Immunologic Phylogeny* (Hildemann, W. H., and Benedict, A. A., ed.). New York: Plenum Press, pp. 241–250.

Mayrhofer, G., Pugh, C. W., and Barclay, A. N. (1983). The distribution, ontogeny and origin in the rat of Ia-positive cells with dendritic morphology and of Ia antigen in epithelia, with special reference to the intestine. *Eur. J. Immunol.* 13:112–122.

Metchnikoff, E. (1892/1968). *Leçon sur la Pathologie Comparée de L'Inflammation* (Lectures on the Comparative Pathology of Inflammation). Paris: Inst. Pasteur; New York: Dover.

Metchnikoff, E. (1901). *Immunity in Infective Diseases.* Cambridge: Cambridge University Press.

Miyazaki, H., and Osawa, T. (1983). Accessory functions and mutual cooperation of murine macrophages and dendritic cells. *Eur. J. Immunol.* 13:984–989.

Möller, G., ed. (1980). Accessory cells in the immune response. *Immunol. Rev.* 53.

Mosier, D. E. (1967). A requirement for two cell types for antibody formation in vitro. *Science* 158:1573–1575.

Naito, K., Komatsubara, S., Kawai, J., Mori, K., and Muramatsu, S. (1984). Role of macrophages as modulators but not as stimulators in primary mixed leukocyte reaction. *Cell. Immunol.* 88:361–373.

Naito, M., Takahashi, K., Takahashi, H., and Kojima, M. (1982). Ontogenetic development of Kupffer cells. In *Sinusoidal Liver Cells* (Knook, D. L., and Wisse, E., eds.). Amsterdam: Elsevier, pp. 155–164.

Naito, M., and Wisse, E. (1977). Observation on the fine structure and cytochemistry of sinusoidal cells in fetal and neonatal rat liver. In *Kupffer Cells and Other Liver Sinusoidal Cells* (Wisse, E., and Knook, D. L., eds.). Amsterdam: Elsevier, pp. 497–505.

Nakano, K., Aotsuka, Y., and Muramatsu, S. (1978). Ontogeny of macrophage function: II. Increase of A-cell activity and decrease of phagocytic activity of peritoneal macrophages during ontogenetic development of immune responsiveness in mice. *Dev. Comp. Immunol.* 2:679–688.

Nussenzweig, M. C., Steinman, R. M., Gutchnov, B., and Cohn, Z. A. (1980). Dendritic cells are accessory cells for the development of anti-trinitrophenyl cytotoxic T lymphocytes. *J. Exp. Med.* 152:1070–1084.

Oliver, P. D., and LeDouarin, N. M. (1984). Avian thymic accessory cells. *J. Immunol. 132*:1748–1755.

Ratcliffe, N. A., and Rowley, A. F. (1979). A comparative synopsis of the structure and function of the blood cells of insects and other invertebrates. *Dev. Comp. Immunol. 3*:189–221.

Rijkers, G. T. (1982). Non-lymphoid defense mechanisms in fish. *Dev. Comp. Immunol. 6*:1–13.

Röllinghof, M., Pfizenmaier, K., and Wagner, H. (1982). T–T cell interaction during cytotoxic T cell responses: IV. Murine lymphoid dendritic cells are powerful stimulators for helper T lymphocytes. *Eur. J. Immunol. 12*:337–342.

Saunders, J. W. (1966). Death in embryonic systems. *Science 154*:604–612.

Schnizlein, C. T., Kosco, M. H., Szakal, A. K., and Tew, J. G. (1985). Follicular dendritic cells in suspension: Identification, enrichment, and initial characterization indicating immune complex trapping and lack of adherence and phagocytic activity. *J. Immunol. 134*:1360–1368.

Schuler, G., and Steinman, R. M. (1985). Murine epidermal Langerhans cells mature into potent immunostimulatory dendritic cells in vitro. *J. Exp. Med. 161*:526–546.

Shevach, E. M. (1984). Macrophage and other accessory cells. In *Fundamental Immunology* (Paul, W. E., ed.). New York: Raven Press, pp. 71–107.

Steeg, P. S., Moore, R. N., Johnson, H. M., and Oppenheim, J. J. (1982). Regulation of murine macrophage Ia antigen expression by a lymphokine with interferon activity. *J. Exp. Med.* 156:1780–1793.

Steiner, G., Wolff, K., Pehamberger, H., and Stingl, G. (1985). Epidermal cells as accessory cells in the generation of allo-reactive and hapten-specific cytotoxic T lymphocyte (CTL) responses. *J. Immunol. 134*:736–741.

Steinman, R. M., and Cohn, Z. A. (1973). Identification of a novel cell type in peripheral lymphoid organs of mice: I. Morphology, quantitation, tissue distribution. *J. Exp. Med. 137*:1142–1162.

Steinman, R. M., Gutchinov, B., Witmer, M. D., and Nussenzweig, M. C. (1983). Dendritic cells are the principal stimulators of the primary mixed leukocyte reaction in mice. *J. Exp. Med. 157*:613–627.

Stingl, G., Gazze-Stingl, L. A., Aberre, W., and Wolff, K. (1981). Antigen presentation by murine epidermal Langerhans cells and its alteration by ultraviolet B light. *J. Immunol. 127*:1707–1713.

Stingl, G., Katz, S. I., Clement, L., and Shevach, E. M. (1978). Immunological functions of Ia-bearing epidermal Langerhans cells. *J. Immunol. 121*:2005–2013.

Sunshine, G. H., Katz, D. R., and Feldmann, M. (1980). Dendritic cells induce T cell proliferation to synthetic antigens under Ir gene control. *J. Exp. Med. 152*:1817–1822.

Takahashi, K., Sakuma, H., Naito, M., Yaginuma, Y., Takahashi, H., Asano, S., Hojo, H., and

Kojima, M. (1980). Cytological characters, transformation, and ontogenesis of dermal histiocytes and fibroblasts of rats. *Acta Pathol. Japan 30*:743–766.

Takahashi, K., Takahashi, H., Naito, M., Sato, T., and Kojima, M. (1983). Ultrastructural and functional development of macrophages in the dermal tissue of rat fetuses. *Cell. Tissue Res. 232*:539–552.

Tew, J. G., Thorbecke, J., and Steinman, R. M. (1982). Dendritic cells in the immune response: Characteristics and recommended nomenclature (A report from the Reticuloendothelial Society Committee on Nomenclature). *J. Reticulendothel. Soc. 31*:371–380.

Tsuchida, T., Iijima M., Fujiwara, H., Pehamberger, H., Sheare, G. M., and Katz, S. I. (1984). Epidermal Langerhans cells can function as stimulator cells but not as accessory cells in CTL induction. *J. Immunol. 132*:1163–1168.

Unanue, E. R. (1984). Antigen-presenting function of the macrophage. *Annu. Rev. Immunol. 2*:395–423.

van der Meer, J.W.M., Beelen, R.M.J., Fluitsma, D. M., and van Furth, R. (1979). Ultrastructure of mononuclear phagocytes developing in liquid bone marrow cultures. A study on peroxidase activity. *J. Exp. Med. 149*:17–26.

van Furth, R., ed. (1970). *Mononuclear Phagocytes.* Oxford: Blackwell.

van Furth, R., ed. (1980). *Mononuclear Phagocytes: Functional Aspects.* The Hague: Martinus Nijhoff.

van Furth, R., and Diesselhoff-den Dulk, M. M. C. (1984). Dual origin of mouse spleen macrophages. *J. Exp. Med. 160*:1273–1283.

van Furth, R., Goud, Th. J.L.M., van der Meer, J.W.M., Blussé van Oud Albas, A., Diesselhoff-den Dulk, M.M.C., and Schadewijk-Nieuwstad, M. (1982a). Comparison of the in vivo and in vitro proliferation of monoblasts, promonocytes, and the macrophage cell line J774. In *Macrophages and Natural Killer Cells: Regulation and Function* (Norman, S. J., and Sorkin, E., eds.). New York: Plenum Press, pp. 175–186.

van Furth, R., van der Meer, J.W.M., Blussé van Oud Albas, A., and Sluiter, W. (1982b). Development of mononuclear phagocytes. In *Self-Defense Mechanisms: Role of Macrophages* (Mizuno, D., Cohn, Z. A., Takeya, K., and Ishida, N., eds.). Amsterdam: Elsevier, pp. 25–41.

van Voorhis, W. C., Hair, L. S., Steinman, R. M., and Kaplan, G. (1982). Human dendritic cells. Enrichment and characterization from peripheral blood. *J. Exp. Med. 155*:1172–1187.

Veldman, J. E., and Kaiserling, E. (1980). Interdigitating cells. *The Reticuloendothelial System: Vol. 1. Morphology.* In: Carr, I., and Daems, W. T. (eds.). New York: Plenum Press, pp. 381–416.

Wekerle, H., Ketelsen, U.-P., and Ernst, M. (1980). Thymic nurse cells. Lymphoepithelial cell complexes in murine thymuses: Morphological and serological characterization. *J.Exp. Med. 151*:925–944.

10

T Cell Development and Repertoire Selection

HUNG-SIA TEH

B lymphocytes produce specific antibodies that are essential for counteracting infectious organisms and their toxins in the extracellular milieu. However, antibodies are ineffective against organisms that exist as intracellular pathogens. To deal with intracellular pathogens the immune system has evolved a different set of immunocompetent lymphocytes called T lymphocytes (or T cells). T lymphocytes are morphologically similar to B lymphocytes and clonally express specific antigen receptors. During the development of B and T lymphocytes, DNA gene segments that are required for the synthesis of receptor molecules for both lymphocyte classes undergo rearrangement in somatic cells. However, T lymphocytes differ from B lymphocytes in many respects. For instance, T lymphocytes do not bind directly to soluble antigens; rather, they are guided to recognize foreign antigen on surfaces of living cells by molecules encoded by the major histocompatibility complex (MHC). This form of antigen recognition, referred to as MHC-restriction, enables T lymphocytes to perform two vital immune functions—namely, the destruction of pathogen-infected host cells by effector T lymphocytes and the delivery of helper signals to B lymphocytes by helper T lymphocytes (T_H). In this manner T_H cells can control the quantity and class of specific antibodies made by B cells.

Despite the requirement for self MHC molecules in antigen recognition by T cells, they do not respond to self MHC antigens per se or to self MHC plus other self antigens. This is referred to as self tolerance. However, T cells do show a marked propensity to respond to allogeneic MHC molecules, a phenomenon referred to as alloreactivity. The mechanisms by which T cells learn to recognize foreign antigens in the context of self-MHC while remaining tolerant to self-MHC and self antigens are the primary focus of this chapter. The next two sections introduce the developmental biology of T cells. The references cited in these two sections are incomplete, and the interested reader is referred to the many review articles cited for a more comprehensive list.

SPECIFICITY AND FUNCTION OF T LYMPHOCYTES

Surface Markers Defining T Lymphocyte Subsets

All murine T lymphocytes express the Thy-1 antigen (Raff, 1971), a 24-kD glycoprotein identified by a number of monoclonal antibodies (mAb). Recently, the CD3 complex was described as a more definitive marker for T lymphocytes (reviewed in Clevers et al., 1988). It consists of at least six different proteins that are associated with the T cell receptor (TCR) and which can be detected by a mAb (Leo et al., 1987). Two other surface markers, referred to as CD4 (formerly L3T4 in mice) and CD8 (formerly Lyt-2 in mice), serve to divide mature T lymphocytes into two major subpopulations with the CD4$^+$8$^-$ or CD4$^-$8$^+$ cell surface phenotype (reviewed in Littman, 1987). These cells are referred to as single-positive T cells.

The murine Lyt-2 molecule is the homologue of the CD8 molecule in humans. In mice the Lyt-2 molecule is normally coexpressed with the Lyt-3 molecule on T lymphocytes. The Lyt-2/3 molecule is a 70-kD heterodimer, comprising two covalently linked chains, an α chain (35–38 kD) expressing epitopes detected by anti-Lyt-2 mAb and a β chain (30–34 kD) expressing Lyt-3 epitopes. The Lyt-2 and Lyt-3 chains are both transmembrane glycoproteins and are encoded by two closely linked genes on chromosome 6. There are two types of Lyt-2 chains, α and α', of slightly different size, which result from alternative modes of mRNA splicing. The biological function of the Lyt-2/3 heterodimer is largely determined by the Lyt-2 chain, a conclusion that is substantiated by transfection of the Lyt-2 gene into CD8$^-$ T cells (Dembic et al., 1987). The murine L3T4 (CD4) molecule exists as a monomeric cell surface glycoprotein with a molecular size of about 55 kD.

Function and Specificity of T Lymphocyte Subsets

CD4$^+$8$^-$ and CD4$^-$8$^+$ T lymphocytes recognize foreign peptides associated with self class II or class I MHC molecules, respectively. In the mouse the MHC complex is encoded by a region of between 2,000 to 4,000 kb in length on chromosome 17 and is referred to as the H-2 complex (reviewed in Hood et al., 1983). Class I MHC molecules consist of a highly polymorphic 45-kD heavy chain encoded by the H-2 complex and a nonpolymorphic 12-kD light chain, referred to as β_2-microglobulin, which is encoded by chromosome 2 in the mouse. The polymorphic 45-kD chains are encoded by the K and D regions of the H-2 complex. The 45-kD chain has three external domains of about 90 amino acid residues each; the transmembrane region comprises about 40 residues, and the cytoplasmic domain is about 30 residues in length. Extensive polymorphism is apparent in the first and second external domains. The third domain shows relatively little polymorphism. Class I MHC molecules are expressed on most nucleated cells.

In mice there are two sets of class II MHC molecules, termed I-A and I-E. Each class II molecule consists of an α and a β chain in noncovalent association. The α chain has a molecular size of about 35 kD, whereas the β chain has a molecular size of about 29 kD. Both chains are transmembrane proteins and are encoded by the I region of the H-2 complex. There are at least eight class II MHC genes in mice: three in the Aβ region, three in the Eβ region, and one Aα and one Eα gene. The order of class I and class II MHC genes in the H-2 complex is K, Aβ, Aα, Eβ, Eα, D. Aα and Aβ chains associate to form I-A molecules, whereas Eα and Eβ chains associate to form I-E molecules. I-A and I-E molecules are also referred to as Ia antigens. Both α and β chains are composed of two external domains of about 90 amino acid residues in length, a transmembrane region of about 30 residues, and a short cytoplasmic region of about 10–15 residues. In contrast to class I MHC molecules, class II molecules have a more restricted tissue distribution and are expressed primarily on the surfaces of B lymphocytes, macrophages, and dendritic cells and on the epithelial cells in the thymus.

The T cell receptor (TCR) on CD4$^+$8$^-$ T lymphocytes binds to foreign peptides associated with class II MHC molecules. These peptides are derived from foreign proteins, which are endocytosed by macrophages, dendritic cells, or B lymphocytes and are cleaved by proteases in lysosomes of these cells. It has been hypothesized that class II MHC molecules contain a groove created by the interaction of the first domain of the α and β chains and that peptides of the right size and specificity will bind in this groove before class II molecules are expressed on the cell surface (Brown et al., 1988). This hypothesis was based on X-ray crystallographic studies of a human class I molecule in which it was shown that the first two domains of the heavy chain of class I molecules can interact to form a peptide-binding groove (Bjorkman et al., 1987). Association of class I molecules with self or foreign peptides can take place in the cytosol. Alternatively, binding of foreign peptides to class I molecules can occur extracellularly. Recognition of foreign peptides in association with class I MHC molecules is mediated by T lymphocytes of the CD4$^-$8$^+$ phenotype. These are usually cytotoxic T lymphocytes (CTL) whose primary function is to seek out and destroy host cells that harbor intracellular pathogens, in particular cells that are infected with viruses. In contrast, T lymphocytes of the CD4$^+$8$^-$ phenotype, which recognize foreign peptides in the context of class II MHC molecules, are usually T$_H$. The primary functions of T$_H$ cells are to help antigen-stimulated B cells to proliferate and differentiate into antibody-producing cells and also to help antigen-stimulated CTL precursors to proliferate and to differentiate into killer cells. These helper functions are largely mediated by lymphokines that are produced by the T$_H$ cells.

In addition to the CD4$^+$8$^-$ and CD4$^-$8$^+$ T lymphocytes, a very small proportion of peripheral T lymphocytes express the Thy-1$^+$ CD3$^+$ CD4$^-$8$^-$ phenotype (reviewed in Brenner et al., 1988). Unlike the single positives, the antigen specificity of these double-negative CD3$^+$ T lymphocytes is unknown. However, like the single-positive cells, they can, upon activation, secrete lymphokines and acquire cytolytic function.

The T Cell Receptor

Unlike B lymphocytes, T lymphocytes do not use immunoglobulin (Ig) genes for encoding their antigen receptors, although T cell receptors are structurally related to immunoglobulins. The first direct information on the structure of TCR molecules came from studies in the early 1980s with mAb that react with clonotypic structures on T cells (Haskins et al., 1983; Meuer et al., 1983). These studies revealed that the TCR on single-positive T cells consists of a disulfide-linked, glycosylated heterodimer consisting of an α and a β chain. Each chain is a transmembrane protein and, in mice, has a molecular size of 40 to 43 kD when glycosylated. The first TCR genes were cloned in mice and humans by subtractive or differential cDNA-mRNA hybridization (Hedrick et al., 1984; Yanagi et al., 1984) and were shown to correspond to TCR β genes. The TCR β gene complex in mice is located on chromosome 6 and contains a cluster of about 21 V$_\beta$ (variable) gene segments and two consecutive clusters of D$_\beta$ (diversity), J$_\beta$ (joining), and C$_\beta$ (constant) segments. Each cluster has one D region and six J regions; the two C$_\beta$ gene segments are almost identical. The subtractive cDNA hybridization method was also used to isolate cDNA clones for mouse TCR α genes (Chien et al., 1984; Saito et al., 1984). The TCR α complex is located on chromosome 14 in mice. There are about 50 V$_\alpha$ segments and about 50 J$_\alpha$ segments spread over more than 50 kb, which lie upstream of a single C$_\alpha$ segment. No D$_\alpha$ segment has been found. Rearrangements of V, D, and J elements are mediated by heptamer and nonamer site-specific recombination sequences similar to those used by B cells in rearranging Ig genes. Additional diversity of TCR $\alpha\beta$ is generated at the joining sites by variable deletions and/or additions of bases. The TCR $\alpha\beta$ heterodimer is expressed on T cells in association with the CD3 complex.

One may expect a priori that because $CD4^+8^-$ and $CD4^-8^+$ T lymphocytes perform different functions and are restricted to class II or class I MHC molecules, respectively, they might use different genes to encode their TCR. However, this is not the case, and gene transfection studies have indicated that the $\alpha\beta$ heterodimer is used by both $CD4^+8^-$ and $CD4^-8^+$ T cells to recognize antigen in the context of class II or class I MHC molecules (Dembic et al., 1986; Saito et al., 1987). The fairly strict correlations observed between class I MHC restriction and expression of CD8 and between class II MHC restriction and expression of CD4 on T lymphocytes led to the proposal that the invariant CD4 and CD8 molecules bind to the nonpolymorphic portion of class II and class I MHC molecules, respectively (Dialynas et al., 1983; Ledbetter and Herzenberg, 1979; Swain, 1983). This hypothesis has been largely substantiated by gene transfection studies in which direct interactions between MHC and CD4/8 accessory molecules were demonstrated and these interactions were shown to augment the binding of T cells to their respective targets (Dembic et al., 1987; Gabert et al., 1987; Gay et al., 1987; Norment et al., 1988).

In addition to enhancing the binding of T cells to their target cells, CD4/8 accessory molecules may also be involved in transducing activation signals to T cells. The discovery that the cytoplasmic domains of CD4 and CD8 molecules are associated with the protein tyrosine kinase, p56^{lck} is consistent with this hypothesis (Rudd et al., 1988; Veillette et al., 1988). For instance, submitogenic concentrations of anti-TCR mAb can synergize with anti-CD4 or anti-CD8 mAb in activating T cells (Eichmann et al., 1987). Furthermore, heteroconjugates of anti-CD3 and anti-CD4 mAb's are extremely effective in activating $CD4^+8^-$ T cells (Ledbetter et al., 1988). Thus cross-linking of TCR to CD4 or CD8, a function normally performed by MHC molecules, may constitute a physiological signal for T cell activation.

In contrast to the $CD4^+8^-$ and $CD4^-8^+$ T lymphocytes, $CD3^+CD4^-8^-$ T cells use different TCR genes, referred to as γ and δ, to encode their antigen receptors (reviewed in Brenner et al., 1988). Like TCR $\alpha\beta$, TCR $\gamma\delta$ is coexpressed with CD3. In the mouse, TCR $\gamma\delta$ exists as a disulfide-linked heterodimer, while in humans, TCR $\gamma\delta$ can exist as a disulfide-linked or nonlinked heterodimer, depending on which TCR γ constant gene segment is used. There are only six to eight functional V_γ segments and fewer J_γ segments. Germline TCR δ diversity is similar to γ, if not more limited. Most of the TCR $\gamma\delta$ diversity is generated in the functional region. It is unclear whether the TCR $\gamma\delta$ recognizes antigen in the context of certain restriction elements. The lack of expression of CD4 or CD8 on $\gamma\delta$-expressing T lymphocytes would argue against the use of MHC molecules as restricting elements for antigen recognition.

T LYMPHOCYTE DEVELOPMENT ON THE THYMUS

Surface Markers Defining Thymocyte Subsets

Two additional markers are useful for identifying thymocyte subsets. One is identified by the J11d mAb and is a 50-kD protein; it is also referred to as the heat-stable antigen (HSA), and it reacts with two other mAb's, namely M1/69 and B2A2 (Scollay et al., 1984). The other marker is referred to as Pgp-1 and is expressed on pre-T cells that home to the thymus (Trowbridge et al., 1982). In the thymus, four major thymocyte subsets can be identified on the basis of CD4 and CD8 expression. The major subset expresses both CD4 and CD8 (CD4$^+$8$^+$) and comprises about 80% of all thymocytes. The single-positive cells (CD4$^+$8$^-$ and CD4$^-$8$^+$) make up 15% of all thymocytes, and there are about twice as many CD4$^+$8$^-$ thymocytes as CD4$^-$8$^+$ thymocytes. About 5% of thymocytes express neither marker (CD4$^-$8$^-$). The CD4$^+$8$^+$ thymocytes are J11d$^+$ and are found in the thymic cortex. About half of the CD4$^+$8$^+$ thymocytes express a low density of CD3. The single-positive cells are located primarily in the medulla and express high levels of CD3 and are J11d$^-$. The CD4$^-$8$^-$ thymocytes can be subdivided into three subpopulations, namely CD3$^-$J11d$^+$ (70%), CD3$^+$ TCR$\gamma\delta^+$ J11d$^+$ (10%), and CD3$^+$ TCR $\alpha\beta^+$ J11d$^-$ (20%).

Functional Potential of Thymocyte Subsets

The single-positive thymocytes are functionally similar to peripheral T lymphocytes, and they can be induced by antigen or T cell mitogens to become T$_H$ or CTL. However, the frequency of T cells that can be activated is somewhat less than that of peripheral T cells. This suggests that the functional competence of single positives is probably acquired after the thymocytes have switched to this phenotype from their immediate progenitor cells, which are likely to be immunoincompetent. The double-positive thymocytes are functionally incompetent despite the fact that 50% of these cells express a low level of the TCR as detected by the CD3 mAb.

The double-negative thymocytes are functionally diverse. The CD3$^+$ TCR$\gamma\delta^+$ CD4$^-$8$^-$ J11d$^+$ subset can be induced by lectins and lymphokines to become cytotoxic cells and can also be induced to produce lymphokines. Thus, this double-negative population behaves like mature single-positive T cells that express TCR $\alpha\beta$. The major CD3$^-$ CD4$^-$8$^-$ J11d$^+$ subset is an actively cycling population, some of which express the interleukin 2 (IL-2) receptor. However, studies on the role of IL-2 in controlling the proliferation of these cells have yielded conflicting results, and further studies are required to define the growth requirements for these cells. The function of the CD3$^+$ CD4$^-$8$^-$ TCR $\alpha\beta^+$ J11d$^-$ subset is unknown. These cells appear late in ontogeny (birth to day 5 in mice) and show a large bias to usage of the V$_\beta$8 gene (Crispe et al., 1987; Fowlkes et al., 1987).

Thymic Migrants

The thymus is continuously colonized by bone marrow–derived hemopoietic stem cells. In the mouse, colonization by stem cells starts at about the 11th day of gestation. These thymic immigrants also constitute a very minor subset in the adult thymus and have the Thy-1$^-$ CD3$^-$ CD4$^-$8$^-$ Pgp-1$^+$ phenotype. The export of cells from the thymus has been studied by intrathymic labeling of thymocytes with fluorescein and subsequent analysis of labeled cells in the periphery. These studies indicate that recent thymic emigrants are like medullary thymocytes and have either the CD4$^+$8$^-$ or CD4$^-$8$^+$ phenotype. The migrants are also MEL-14$^+$, a marker that is associated with a receptor required for homing of thymic migrants to peripheral lymphoid organs (Gallatin et al., 1983). However, only half of the medullary single-positive cells stain with the MEL-14 mAb. This suggests tha the MEL-14 marker may be acquired only after differentiation of the single positives from their progenitors has occurred and that only single positives that are also MEL-14$^+$ can exit the thymus. About 2×10^6 emigrants leave the thymus per day in the mouse.

Turnover of Cortical and Medullary Thymocytes

It has been known since the 1960s that the thymus contains many actively proliferating cells and has a high rate of cell turnover. ^{3}H-thymidine labeling studies were used to determine the differential turnover rates of thymocyte subsets (Shortman and Scollay, 1984). Cortical thymocytes of the CD4$^+$8$^+$ phenotype were found to be entirely labeled over a 3-day labeling period. Because the murine thymus contains approximately 1.5×10^8 CD4$^+$8$^+$ thymocytes, it can be concluded that about 5×10^7 CD4$^+$8$^+$ thymocytes are produced per day. Using the same method, it was found that the single-positive medullary thymocytes labeled at a much slower rate, indicating that the majority of these cells are not in cycle. It is estimated that about 2×10^6 single-positive thymocytes are produced per day. Because the thymus exports mainly single-positive cells it is clear that the vast majority, if not all, of the CD4$^+$8$^+$ thymocytes have to die intrathymically. Biochemical "visualization" of intrathymic cell death was achieved by measuring the rate of reutilization of ^{3}H-thymidine and ^{125}I-UdR in the thymus. If all thymocytes leave the thymus, both labels should exit from the thymus at the same rate. If most thymocytes die intrathymically, and their degraded DNA nucleosides are reutilized, then reutilized ^{3}H-thymidine should be retained much more efficiently than the poorly reutilized ^{125}I-UdR, and this was observed (Scollay and Shortman, 1984). In summary, the CD4$^+$8$^+$ cortical thymocytes are produced at a rapid rate, and the vast majority of these cells die intrathymically. Single-positive medullary thymocytes are produced at a low rate, and these cells serve as a major source of thymic emigrants.

Precursors of Cortical and Medullary Thymocytes

In the early embryonic murine thymus ($\leq$ 14th day of gestation), most of the lymphoid cells are lymphoblasts of the CD3$^-$ CD4$^-$8$^-$ J11d$^+$ Thy-1$^+$ phenotype. In organ cultures of 14th-day, fetal thymus, it can be shown that the production of the CD4$^+$8$^+$ thymocytes is preceded by the appearance of CD4$^-$8$^+$ J11d$^+$ thymocytes (Kisielow et al., 1984). This CD4$^-$8$^+$ J11d$^+$ population is also found in the adult thymus. These cells lack MEL-14 and CD3 expression and are functionally immature (Crispe and Bevan, 1987; Wilson et al., 1987). Functional single-positive cells always appeared after the nonfunctional double-positive cells in fetal thymus organ cultures.

    Fetal thymus organ cultures can be completely depleted of lymphoid cells by incubation with deoxyguanosine (Jenkinson et al., 1982). Such depleted thymuses can be recolonized by a single lymphoblast isolated from 14-day fetal thymus (Kingston et al., 1985). The recolonized thymus contains all thymocyte subsets. Thus, one can conclude that the CD4$^-$8$^-$ J11d$^+$ lymphoblasts contain all the precursors for all the other subpopulations and that a single intrathymic precursor is sufficient to give rise to both cortical double-positive and medullary single-positive thymocytes. Lymphoblasts of the same phenotype with similar precursor potential as fetal thymus lymphoblasts have also been isolated from the adult thymus. Thus, one may assume that the production of cortical and medullary thymocytes is a continuing process and progresses well into adulthood.

The developmental sequence of thymocytes during ontogeny is from $CD4^-8^-$ $J11d^+ \rightarrow CD4^-8^+ J11d^+ \rightarrow CD4^+8^+ J11d^+ \rightarrow CD4^+8^- J11d^-$ or $CD4^-8^+ J11d^-$. Theoretically, the immediate precursors of single-positive cells could be any of the three preceding populations. Several recent studies have yielded data consistent with the conclusion that the single-positive thymocytes are derived directly from the cortical double-positive thymocytes. The rationale behind these experiments is based on the expectation that if single-positive thymocytes are derived from double-positive precursor cells, then the development of CD4 single-positive cells will be affected by both anti-CD4 and anti-CD8 mAb's. Similarly, the development of CD8 single positive cells will also be affected by both mAb's. In one series of experiments, irradiated recipients were repopulated with two different sets of stem cells, which gave rise to T lymphocytes expressing different CD8 and Thy-1 alleles. The reconstituted recipients were then continuously infused with an allele-specific anti-CD8 mAb. By double staining with anti-CD4 and an allele-specific Thy-1 mAb, it was possible to determine the stem cells from which the single-positive CD4 cells were derived. The results of this experiment showed that reduction in the number of CD4 single positives occurred only in the lineage of cells expressing the CD8 allele against which the injected anti-CD8 mAb was directed. This suggests that the CD4 single-positive cells were derived from $CD8^+$ precursor cells, presumably the double-positive thymocytes (Smith, 1987).

Further evidence favoring the above model was obtained by following the expression of certain V_β gene products on $CD4^+8^+$ thymocytes and on single-positive cells that are specific for either I-E antigens (Fowlkes et al., 1988) or antigens encoded by the *a* allele of the minor lymphocyte stimulatory (*Mls^a*) locus (MacDonald et al., 1988a). In mouse strains that express I-E class II MHC molecules, the $V_\beta17a$ determinant is expressed on a proportion of functionally incompetent $CD4^+8^+$ TCR^{lo} thymocytes, but it is almost totally absent in mature $CD4^+8^-$ and $CD4^-8^+$ T cells. On the other hand, in I-E$^-$ mouse strains, the $V_\beta17a$ determinant is expressed at high levels on a proportion of CD4 and CD8 single-positive thymocytes as well as on double-positive thymocytes (Kappler et al., 1987). In vivo blocking of the CD4 molecule with anti-CD4 mAb in irradiated I-E$^+$ mouse strains that were reconstituted with syngeneic bone marrow cells led to failure to develop CD4 single-positive cells. More interestingly, CD4 blocking also resulted in the appearance of $CD4^-8^+$ $V_\beta17a^+$ T lymphocytes in these animals. The simplest explanation for the data is that both CD4 and CD8 single positives that are $V_\beta17a^+$ must have come from a common progenitor, namely the $CD4^+8^+$ $V_\beta17a^{lo}$ cells and that this population is targeted for deletion in I-E$^+$ mice. Furthermore, the deletion process appears to require the participation of the CD4 molecule.

In the Mls system, T cell reactivity to antigens encoded by the *Mls^a* locus is associated with the expression of either $V_\beta6$ or $V_\beta8.1$ gene products on T lymphocytes. $V_\beta6^+$ or $V_\beta8.1^+$ T lymphocytes are deleted in mouse strains that express *Mls^a* (Kappler et al., 1988; MacDonald et al., 1988c). Responses by T lymphocytes to *Mls^a* were restricted by polymorphic class II MHC molecules and were mediated primarily by $CD4^+8^-$ $V_\beta6^+$ T cells and not by $CD4^-8^+$ $V_\beta6^+$ T cells. Thus, it was somewhat surprising to find that both $CD4^+8^-$ $V_\beta6^+$ and $CD4^-8^+$ $V_\beta6^+$ T lymphocytes were deleted in *Mls^a* mouse strains, even though only the CD4$^+$ subset was reactive to *Mls^a*. The most likely explanation for these data is that the $V_\beta6$-expressing single-positive cells were derived from $CD4^+8^+$ $V_\beta6^+$ precursor cells and that these double-positive precursor cells are the target of deletion in *Mls^a* mice. Deletion of the double positives will also lead to the lack of development of $CD4^-8^+$ $V_\beta6^+$ T cells even though they are not *Mls^a*-reactive. The observation that treatment of Mls^a mice from birth with anti-CD4 mAb resulted in the appearance of $V_\beta 6^+$ cells in the $CD4^-8^+$ T lymphocyte subset is consistent with this hypothesis. It also emphasizes the importance of the CD4 molecule in the deletion process.

Studies involving the use of TCR transgenic mice expressing a receptor of known specificity have also provided compelling evidence favoring the conclusion that the immediate precursors of single-positive T lymphocytes are $CD4^+8^+$ thymocytes that express a low level of the transgenic TCR (see the discussion of positive and negative selection of T cells that follows below).

Ontogeny of the T Cell Receptor in the Thymus

TCR gene rearrangement and expression in the murine fetal thymus follow a highly ordered sequence (reviewed in Brenner et al., 1988). Transcription of TCR γ genes can be detected as early as the 14th day of gestation and attains substantial levels by day 15–16. TCR δ genes are rearranged and expressed at the same time as TCR γ genes. In contrast, functional TCR β and α gene rearrangements and transcription lag behind and then largely replace those of TCR γ and δ in the latter part of thymic development in the mouse. Although β rearrangements can be detected as early as day 14 of gestation, full-length β transcripts are not easily detected until the 16th day. Transcription of full-length TCR α mRNA transcripts begins after the 16th day of gestation. Using mAb or antiserum against the different TCRs, it is possible to correlate the rearrangement and transcription data with the appearance of cell surface TCR proteins during development. TCR$\gamma\delta$-CD3 protein complexes can be detected on day-15 fetal thymocytes; they peak at about day 16 of gestation and decline to a very low level by day 20 (time of birth). TCR $\alpha\beta$ complexes were detectable on cell surfaces by the 17th day of gestation, and they became more abundant through day 20.

It is pertinent to note that TCR $\gamma\delta$-bearing lymphocytes are also found in athymic mice. Furthermore, Thy-1$^+$ dendritic epidermal cells in the skin of normal mice also express TCR $\gamma\delta$. These observations suggest that TCR$\gamma\delta^+$ lymphocytes can also develop extrathymically.

POSITIVE SELECTION OF THE T CELL REPERTOIRE BY THYMIC MHC

Influence of Thymic MHC on T Cell Restriction Specificity

Because T lymphocytes acquire their antigen receptors and differentiate into functional, mature cells in the thymus, it is reasonable to expect that this is also the organ in which T cells learn to discriminate between self and foreign MHC. Although most of the data gathered during the 1970s favored this view, there were also observations that some felt to be inconsistent with imprinting of MHC restriction specificity on T cells by the thymus (reviewed in Sprent and Webb, 1987).

The MHC restriction phenomenon was discovered in the early 1970s. In antibody responses to T-dependent antigens, it was found that T_H cells only collaborated effectively with MHC-compatible B cells. The genetic loci that control this interaction were mapped to class II MHC loci. Class II restriction was also observed for the presentation of antigen to T_H cells by macrophages. Shortly thereafter it was found that virus-infected mice produced cytotoxic T lymphocytes (CTL) that could lyse virus-infected target cells, provided the target cells expressed the same class I MHC molecules as the CTL. Class I MHC restriction was also observed for CTLs that were specific for minor histocompatibility antigens or for chemical haptens. The next major development was the discovery that MHC restriction was independent of the MHC molecules expressed by the T cells but, rather, was dependent on the MHC molecules to which the T cells were exposed during their differentiation from hemopoietic stem cells to mature T cells. For instance, in a $\rightarrow$ (a $\times$ b) F_1 bone marrow chimeras (a and b denote animals with different MHC haplotypes, the chimeras were constructed by injecting T cell–depleted bone marrow cells of 'a' genotype into lethally irradiated (a $\times$ b) F_1 hybrid animals), mature T cells derived from the chimeras exhibited both a and b MHC-restriction specificities even though they expressed only 'a' MHC molecules. Such coaxing of T cells to interact with B cells, macrophages, or target cells expressing otherwise foreign MHC molecules has been termed adaptive differentiation. A more dramatic illustration of adaptive differentiation is provided by the construction of (a $\times$ b) F_1 $\rightarrow$ a or b (parent) bone marrow chimeras (Bevan, 1977). In this situation, mature T cells from chimeric animals were restricted to the MHC type of parent in which the (a $\times$ b) F_1 stem cells differentiated despite the fact that these T cells expressed both a and b MHC molecules. Direct evidence favoring the hypothesis that thymic MHC determines the MHC restriction specificity of T cells was obtained through use of the following type of radiation chimeras: (a $\times$ b) F_1 animals were thymectomized, lethally irradiated, and reconstituted with (a $\times$ b) F_1 bone marrow cells plus irradiated thymus grafts of either the a or b MHC type. It was found that T cells derived from such chimeras were primarily restricted to the MHC type of the thymus graft (Fink and Bevan, 1978; Zinkernagel et al., 1978).

The absoluteness of this thymic selection process was brought into question, however, when it was found that strong restriction by chimeric T cells to thymic MHC was not generally observed in totally MHC incompatible a $\rightarrow$ b bone marrow chimeras (Matzinger and Mirkwood, 1978). In the case of virus-specific CTL, the MHC restriction pattern of T cells from these chimeras ranged from complete restriction to host MHC to complete restriction to donor MHC, or total unresponsiveness, depending on the virus and the strain combination used (Zinkernagel et al., 1984). The data from studies with nude (congenitally athymic) mice grafted with MHC-different thymuses were equally confusing, in that T cells from such animals often showed complete restriction to host (non-thymic) MHC (Zinkernagel et al., 1980). Furthermore, T cells from strain a animals acutely depleted of b alloreactivity by the thoracic duct cannulation method or rendered neonatally tolerant to b MHC antigens were found to contain significant numbers of b MHC-restricted T cells (Forman and Streilein, 1979; Teh et al., 1982; Wagner et al., 1981). These findings apply particularly to class I restricted T cells and are not generally observed for class II restricted cells. However, certain in vitro studies are consistent with the interpretation that there is a lack of thymic selection for $CD4^+8^-$ T cells. Thus strain a T cells, depleted of b alloreactivity by the bromodeoxyuridine and light method, can collaborate with antigen presenting cells of the b MHC haplotype (Ishii et al., 1982). However, these findings are not easily reproducible, and the efficacy of the bromodeoxyuridine method in depleting alloreactive cells is questionable. On the other hand, the existence in normal or neonatally tolerized mice of $CD4^-8^+$ T cells that are restricted to foreign class I MHC molecules is less easily refuted. If there is no overlap of antigenic specificities between a and b MHC-restricted T cells, then this observation may be used to argue for a lack or incompleteness of thymic selection for self MHC–restricted T cells. In this regard, cross-reactivity of the [a MHC + X] = [b MHC + Y] (Hunig and Bevan, 1981) and of the [a MHC + X] = [b MHC + X] (Teh et al., 1982) types had been observed for CTL clones, where X and Y refer to non–cross-reacting extrinsic antigens. Thus, the presence of a small proportion of T cells in normal mice that are apparently restricted to foreign MHC cannot be used as an argument against the completeness of the thymic selection process.

Nature of the Selecting Cell

If one accepts the view that the thymus plays a central role in instructing T cell MHC-restriction specificity, then how is this restriction imposed and which component of the thymus is responsible for this selection process? On the basis of data from $F_1 \rightarrow$ parent chimeras and reconstitution experiments with irradiated thymuses, it was tacitly assumed that a radiation-resistant component of the thymus, presumably thymic epithelial cells, was responsible for the selection process. Direct evidence in support of this view was obtained by reconstituting thymectomized (a × b) F_1 animals with parental 14-day fetal thymuses that had been cultured for 5 days in deoxyguanosine. This treatment completely depleted the fetal thymus of lymphoid cells and of cells of the macrophage/dendritic cell lineage, which are also derived from hemopoietic stem cells; the treated thymus consisted largely of epithelial cells. Two weeks after thymus engraftment the recipients were heavily irradiated with 1,100 rads and then reconstituted with F_1 bone marrow cells. Two months after reconstitution, T cells derived from these animals showed strong restriction to the MHC type of the fetal thymus grafts (Lo and Sprent, 1986). This was taken as irrefutable evidence that radiation-resistant thymic epithelial cells are responsible for the selection process. However, experiments by Longo and Schwartz (1980) yielded data suggesting that bone marrow–derived macrophages and/or dendritic cells were responsible for the selection process. They observed that in $F_1 \rightarrow$ parent chimeras the rate of reconstitution of the parental thymuses by macrophages and/or dendritic cells of donor origin depended on the radiation dose used. If the recipients received only 900 rads, donor-derived macrophages/dendritic cells were not evident until 2 months after reconstitution. In contrast, for recipients that received 1,200 rads, donor-derived macrophages/dentritic cells were easily detected in the thymus by 3 weeks post-reconstitution. If 900 rad–irradiated recipients were left for several months and then treated with cortisone and anti-thymocyte serum to remove mature T cells, the new wave of T cells that was produced in these animals showed restriction to both parental MHC types. On the other hand, 1,200 rad–irradiated recipients did not show preferential restriction to thymic MHC, regardless of whether these animals were treated with cortisone and anti-thymocyte serum. Furthermore, they observed that mature T cells derived from (a × b) F_1 nude mice that were engrafted with thymuses from long-term b $\rightarrow$ a chimeras, which should contain strain a epithelial cells and strain b macrophages/dendritic cells, were restricted to b, rather than a, MHC antigens (Longo et al., 1985). Until more definitive experiments are performed, the issue of whether epithelial or bone marrow–derived cells is responsible for thymic selection must be considered unsettled.

Positive Selection of T Cells in Thymus by Restricting MHC Molecules

Until recently, little was known about the mechanisms by which T cells become restricted to self MHC molecules. However, because it is well known that class I and class II MHC-restricted T cells are of the CD4$^-$8$^+$ and CD4$^+$8$^-$ phenotypes, respectively, and because a strong case can be made for CD4$^+$8$^+$ thymocytes being the immediate precursors of single-positive thymocytes, the following scheme for positive selection of T lymphocytes was proposed (von Boehmer, 1986, 1988). CD4$^+$8$^+$ precursors express a random TCR $\alpha\beta$ repertoire. Those double-positive cells that do not express a TCR $\alpha\beta$ as a result of nonfunctional β and/or α rearrangements, as well as those that express a TCR $\alpha\beta$ with no specificity for either self class I or class II MHC molecules, will die from programmed cell death because they cannot interact with self MHC. Those precursors that express TCR $\alpha\beta$ that can interact with thymic MHC molecules will be rescued from programmed cell death. In the absence of nominal antigen, the interaction of the TCR with class II or class I thymic MHC molecules will direct the differentiation of CD4$^+$8$^+$ precursors into CD4$^+$8$^-$ and CD4$^-$8$^+$ mature T cells, respectively. It was further proposed by Von Boehmer (1986, 1988) that class I or class II MHC molecules serve to cross-link the TCR $\alpha\beta$ on CD4$^+$8$^+$ thymocytes to either CD8 or CD4, respectively. Cross-linking of TCR $\alpha\beta$ to CD4 leads to downregulation of CD8, and the resulting mature T cells will have a CD4$^+$8$^-$ phenotype and be restricted to class II MHC antigens. Similarly, cross-linking the TCR $\alpha\beta$ to CD8 by class I MHC will produce class I–restricted T cells of the CD4$^-$8$^+$ phenotype.

To test the above hypothesis, transgenic mice expressing a TCR of known MHC and antigen specificity were constructed (Bluthmann et al., 1988). The functionally rearranged α and β TCR genes were isolated from a CTL clone of the CD4$^-$8$^+$ phenotype that expressed TCR $\alpha\beta$ specific for the male (H-Y) antigen in association with H-2D^b. Previous studies have shown that expression of the functionally rearranged β transgene requires an enhancer sequence located 4 kb downstream of C$_\beta$2 (Krimpenfort et al., 1988). Although it has not been formally shown that the transcription of the α transgene is similarly regulated, genomic fragments of the same size as the β transgene were included upstream and downstream of the functionally rearranged α gene in the construction of the α transgene. Expression of the β transgene (V$_\beta$8.2) on T cell surfaces could be monitored with a previously described mAb, F23.1 (Staerz et al., 1985). Detection of the α transgene required the production of a new mAb, referred to as T3.70 (Teh et al., 1988, 1989). The effect of thymic MHC on selection of T cells from immature precursors was assessed by monitoring transgenic TCR expression in female H-2^b transgenic mice (Teh et al., 1988). In these mice it was observed that all the CD4$^+$8$^+$ thymocytes were $\alpha_T^{lo}\beta_T^{lo}$ (α_T = transgenic α; β_T = transgenic β). In contrast, mature CD4$^+$8$^-$ T cells were $\alpha_T^-\beta_T^{high}$ and mature CD4$^-$8$^+$ T cells were $\alpha_T^{high}\beta_T^{high}$, and one in three of these cells responded to H-2^b male antigens (Kisielow et al., 1988a). It was previously shown that the β transgene could completely inhibit functional rearrangement of the endogenous β genes (Uematsu et al., 1988). Consequently, in the $\alpha\beta$ transgenic mice, all T cells express the β transgene only.

The above observation indicates that although the α-transgene was expressed in CD4$^+$8$^+$ and CD4$^-$8$^+$, it was not expressed in CD4$^+$8$^-$ T cells. Because the β transgene needs to be coexpressed with an α chain in order to be detected by F23.1, it was concluded that the TCR on CD4$^+$8$^-$ T cells were $\alpha_E^{high}\beta_T^{high}$, where α_E is encoded by a functionally rearranged endogenous α gene. Thus, unlike the β transgene, expression of the α transgene does not appear to prevent the functional rearrangement and expression of endogenous α genes. It was also observed that H-2^b transgenic female mice contain a disproportionately large number of CD4$^-$8$^+$ T cells: there appear to be twice as many CD4$^-$8$^+$ T cells as CD4$^+$8$^-$ T cells, a situation which is the reverse of that in non-transgenic H-2^b

male or female mice. These results were interpreted as follows. All CD4$^+$8$^+$ precursor cells expressed a low level of the transgenic TCR. This population corresponded to the CD4$^+$8$^+$ TCR^{lo} thymocytes observed in normal mice. In H-2^b female mice, the transgenic receptor can be positively selected by H-2D^b, resulting in the production of a disproportionately large number of CD4$^-$8$^+$ T cells. This process probably requires cross-linking of the transgenic TCR to CD8 by H-2D^b. On the other hand, production of CD4$^+$8$^-$ T cells requires functional rearrangements of endogenous α genes. Only those CD4$^+$8$^+$ thymocytes that expressed $\alpha_E\beta_T$ TCR, and which can bind to H-2^b class II MHC molecules, can be rescued from programmed cell death and differentiate into CD4$^+$8$^-$ T cells. This presumably occurs through cross-linking of the $\alpha_E\beta_T$ TCR to CD4 by H-2^b class II MHC molecules.

By backcrossing the founder H-2$^{b/d}$ transgenic mouse to normal H-2^d mice it was possible to produce transgenic mice that were homozygous for H-2^d. When thymocytes from female H-2^d transgenic mice were analyzed, it was found that CD4$^+$8$^+$ thymocytes expressed a low level of $\alpha_T\beta_T$ TCR, as observed for female H-2^b trangenic mice. However, in contrast to female H-2^b transgenic mice, both CD4$^+$8$^-$ and CD4$^-$8$^+$ T cells were of the $\alpha_E^{high}\beta_T^{high}$ phenotype. Furthermore, there were more CD4$^+$8$^-$ T cells than CD4$^-$8$^+$ T cells in H-2^d transgenic mice (Teh et al., 1988). Thus, in H-2^d transgenic mice, CD4$^-$8$^+$ T cells have no selective advantage over CD4$^+$8$^-$ T cells because the transgenic TCR does not bind to H-2^d class I MHC molecules. In this case the development of both CD4$^-$8$^+$ and CD4$^+$8$^-$ T cells would depend on functional rearrangements and expression of endogenous α genes. In a second series of experiments, it was found that both the increase in the number of CD4$^-$8$^+$ T cells and the exclusive expression of the transgenic α chain on CD4$^-$8$^+$ T cells could be mapped to the H-2D^b locus (Kisielow et al., 1988b). This was done by constructing radiation bone marrow chimeras where bone marrow cells from female H-2^b transgenic mice were injected into lethally irradiated female mice of various recombinant MHC haplotypes. The data indicate that expression of the D^b locus was essential and sufficient for the increase in the number of CD4$^-$8$^+$ T cells in the thymus and for expression of the transgenic α chain on these CD4$^-$8$^+$ T cells. These experiments provide definitive evidence for the positive selection of T cells by the restricting MHC molecule in the absence of nominal antigen.

To test the hypothesis that the production of mature T cells that did not express the transgenic α chain was due to endogenous α rearrangements, the transgenic mice were backcrossed to mice with the *scid* (severe combined immunodeficiency) mutation (Scott et al., 1989). *Scid* mice are unable to functionally rearrange their endogenous Ig and TCR genes (Schuler et al., 1986). Thus, *scid* mice that carry the α and the β transgenes can only express these two TCR genes. Transgenic *scids* of the H-2$^{d/b}$ and H-2$^{d/d}$ MHC types were constructed. It was expected that female H-2$^{d/b}$ transgenic *scids* could produce only CD4$^-$8$^+$ single positives and no CD4$^+$8$^-$ T cells because the formation of this latter population requires endogenous α rearrangements. Similarly, in H-2$^{d/d}$ *scids* no single positives could be produced. These predictions were fulfilled. Thus, female H-2$^{d/b}$ transgenic *scids* contained three populations of thymocytes—CD4$^+$8$^+$$\alpha_T^{lo}\beta_T^{lo}$; CD4$^-8^+$$\alpha_T^{high}\beta_T^{high}$; and CD4$^-8^-$$\alpha_T^{high}\beta_T^{high}$. No CD4$^+8^-$ T cells were found in these mice. In female H-2$^{d/d}$ transgenic *scids*, only two thymocyte populations were observed—CD4$^+$8$^+$$\alpha_T^{lo}\beta_T^{lo}$ and CD4$^-$8$^-$$\alpha_T^{high}\beta_T^{high}$. No single-positive cells were found. These results provided definitive evidence that the production of CD4$^+$8$^-$ and CD4$^-$8$^+$ T cells in female non-*scid* H-2^d transgenic mice depended on endogenous α gene rearrangements. Furthermore, this study showed that positive selection of single-positive T cells occurred in the absence of suppressor T cells or an antibody-idiotype network because both functional T cells and B cells are absent in *scid* mice. The data also indicate that the functional α and β transgenes enable pre-T cells to develop to the $CD4^+8^+$ precursor stage, but that development beyond the double-positive stage requires interaction with MHC molecules. However, the mechanism by which the large number of $CD4^-8^-\alpha_T^{high}\beta_T^{high}$ thymocytes is produced is not known, although it is clear that their development does not depend on selection by thymic MHC because they are also found in large numbers in $H-2^d$ transgenic *scids*.

The data favoring positive selection of T cells by thymic MHC in transgenic mice that express the male-specific receptor are in agreement with those observed in transgenic mice that express a TCR specific for an allogeneic class I MHC product (Sha et al., 1988b). In this case, the α and β transgenes were isolated from a $CD4^-8^+$ CTL clone derived from a Balb.B ($H-2^b$) mouse; this CTL clone was specific for the allogeneic $H-2L^d$ class I MHC molecule. It was found that, in $H-2^b$ transgenic mice expressing this pair of transgenes, $CD4^-8^+$ T cells were produced in large numbers and that they expressed high levels of the α and β chains and were highly specific for L^d. $CD4^+8^-$ T cells were produced in lower numbers, and they expressed only high levels of the β transgene but not the α transgene. Furthermore in $H-2^s$ transgenic mice there were few $CD4^-8^+$ T cells, and they expressed only high levels of the β transgene (Sha et al., 1988a). Recent observations in normal mice are also consistent with the model for positive selection of T cells by thymic MHC. It was observed that $CD4^+8^-$ T cells failed to develop in mice that were neonatally treated with anti-class II MHC mAb (Kruisbeek et al., 1985). Similarly, mice that were treated with large doses of anti-class I MHC mAb from birth failed to develop $CD4^-8^+$ T cells (Marusic-Galesic et al., 1988). It was mentioned earlier that T cell reactivity to Mls^a-encoded antigens was associated with the expression of the $V_\beta 6$ gene product on $CD4^+8^-$ T cells. In Mls^a mice, both $CD4^+8^-V_\beta 6^+$ and $CD4^-8^+V_\beta 6^+$ T cells were absent. The intrathymic deletion of $V_\beta 6^+$ T cells was shown to be controlled by polymorphic class II MHC determinants. In mice that lacked expression of Mls^a, the same class II MHC loci were found to control the frequency of $V_\beta 6^+$ cells among mature $CD4^+8^-$ T lymphocytes. In mouse strains that expressed these class II MHC molecules, $CD4^+8^-V_\beta 6^+$ T cells were two- to three-fold more frequent than in mice that did not express these molecules (MacDonald et al., 1988b). These observations were interpreted as favoring positive selection of the $CD4^+8^-V_\beta 6^+$ T cells by restricting class II MHC molecules. In summary, these observations suggest that positive selection of T cells by thymic MHC occurs via similar mechanisms in both TCR transgenic mice and in normal mice.

Studies with TCR transgenic mice that express the male-specific receptor yielded the rather unexpected observation that, unlike the β transgene, the functional α transgene appears to have little or no effect on functional rearrangement and expression of endogenous α transgenes. However, the vast majority of T cell clones analyzed so far appear to express only one functional α and one functional β TCR gene. One would have expected that a similar situation could also exist in normal mice, due to the fact that the α transgene does not prevent the expression of endogenous α genes. Thus, one would expect to find at least some T cells to have one functional β and two functional α chains. In this regard it is of interest to note that a T cell clone with one functional β and two functionally rearranged α chains has been derived from a normal mouse; only one of these $\alpha\beta$ heterodimers was used for self MHC-restricted antigen recognition (Malissen et al., 1988). It is therefore conceivable that a significant proportion of $CD4^+8^+$ thymocytes in normal mice may have two productively rearranged α genes. If one of these two α chains pairs with β and yields an MHC-selectable TCR, then the resulting single-positive cell should express both the MHC-restricted $\alpha\beta$ heterodimer as well as the alternative heterodimer that need not be restricted to self MHC. However, preferential association between the β chain and one of the two α chains may lead to the expression of only one type of $\alpha\beta$ heterodimer on the cell surface.

In summary, the experiments described in this section provide definitive evidence that all mature $CD4^+8^-$ and $CD4^-8^+$ T cells are positively selected by thymic MHC in the absence of foreign or nominal antigen. Furthermore, the CD4/CD8 phenotype of T cells is predetermined by the MHC specificity of the $\alpha\beta$ heterodimer on immature $CD4^+8^+$ precursor cells. This selective mechanism ensures that all $CD4^+8^-$ and $CD4^-8^+$ mature T cells are restricted to self MHC and are potentially able to recognize foreign antigen in the context of self MHC. It also provides a reasonable explanation for the high rate of death of $CD4^+8^+$ thymocytes in normal mice because this mechanism predicts that only the very few calls that express low affinity for self MHC can be protected from programmed cell death and allowed to differentiate into mature single-positive T cells. This preoccupation with self MHC also provides an explanation for the high frequency of T cells that are reactive against allogeneic MHC. It is reasonable to expect a high degree of cross-reactivity between the self MHC–selected T cell repertoire and allogeneic MHC antigens because of the high degree of structural homology between self and allogeneic MHC molecules. These experiments also suggest a central role for CD4 and CD8 accessory molecules in the positive selection process: cross-linking of the $\alpha\beta$ TCR to CD4 or CD8 by class II or class I MHC molecules, respectively, appears to be a critical signal for differentiation of $CD4^+8^+$ thymocytes into single-positive T cells.

NEGATIVE SELECTION OF AUTOSPECIFIC T CELLS IN THE THYMUS

Deletion of Autospecific T Cells in Normal Mice

The mechanisms underlying the development and maintenance of T cell tolerance to self antigens are not clear, but they are generally believed to involve either deletion, anergy, or suppression of self-reactive clones (reviewed in Nossal, 1983). Until recently there was no direct evidence for any of these mechanisms. One of the difficulties in designing experiments to test these hypotheses has been our inability to identify lymphocytes at various stages of T cell ontogeny that express receptors for a particular anti-self specificity. This difficulty was partly overcome by the judicious choice of certain antigenic systems. As mentioned in the preceding discussion of precursors of cortical and medullary thymocytes, responses to I-E encoded molecules were associated with expression of the $V_\beta 17a$ determinant on T cells. In I-E$^-$ strains, a small percentage (usually $< 10\%$) of mature T cells was found to be $V_\beta 17a^+$. However, $V_\beta 17a^+$ cells were absent from the pool of peripheral T cells and medullary thymocytes but were present among cortical $CD4^+8^+$ thymocytes (Kappler, et al., 1987). These results can best be explained by the deletion of autospecific cells, and that deletion takes place somewhere during the transition stage from $CD4^+8^+$ thymocytes to mature single-positive cells. This finding was subsequently extended to the Mls system (Kappler et al., 1988; MacDonald et al., 1988c). As previously discussed, Mls^a-reactivity is associated with the expression of the $V_\beta 8.1$ or $V_\beta 6$ determinants on $CD4^+8^-$ T cells. T cell responses to Mls^a are restricted to polymorphic class II MHC molecules. It was found that, in mouse strains which are Mls^a and which express the same class II MHC–restricting genes, T cells that are $V_\beta 8.1^+$ or $V_\beta 6^+$ are deleted from the peripheral and medullary T cell pools but not from the cortical thymocytes. In contrast, mouse strains that are Mls^b contained a high frequency of $V_\beta 8.1^+$ or $V_\beta 6^+$ T cells in their peripheral and medullary T cell pools. Thus, single-positive cells that express $V_\beta 8.1$ or $V_\beta 6$ are deleted in Mls^a mice.

Deletion of Autospecific T Cells and Clonal Anergy in TCR Transgenic Mice

The development of TCR transgenic mice offers an alternative approach to analyze the mechanisms of self tolerance. In female transgenic mice that expressed a TCR specific for the H-Y antigen, which was restricted to H-2D^b, CD4$^+$8$^+$ thymocytes were positively selected by thymic H-2D^b and differentiated into CD4$^-$8$^+$ T cells (Kisielow et al., 1988b; Teh et al., 1988). By analyzing TCR expression in male transgenic H-2^b mice, one can determine the mechanisms by which tolerance to self antigens is induced at the T cell level (Kisielow et al., 1988a). It was found that the thymuses of male transgenic mice that expressed the H-Y antigen were about 1/10th to 1/20th the size of thymuses obtained from female H-2^b transgenic mice. This drastic reduction in the size of the thymus was due to deletion of the vast majority of CD4$^+$8$^+$ thymocytes, all of which had been shown to express a low level of the transgenic TCR in female transgenic mice. The majority of the thymocytes that escaped this deletion process were of the CD4$^-$8$^-\alpha_T^{high}\beta_T^{high}$ phenotype. The phenotypes of T cells in the spleen and lymph nodes of male transgenic mice were also very different from those observed in female transgenic mice. More than 50% of peripheral T cells were of the CD4$^-$8$^-\alpha_T^{high}\beta_T^{high}$ phenotype. The CD8-expressing cells were of the CD4$^-$8$^{lo}\alpha_T^{high}\beta_T^{high}$ phenotype, and they were about five times more frequent than the CD4$^+$8$^-$ T cells, which generally did not express the α and β transgenes (Kisielow et al., 1988a; Teh et al., 1989). These results were interpreted as follows. Immature CD4$^+$8$^+$ thymocytes that expressed a low level of the transgenic TCR and a high level of CD8 were deleted in male transgenic mice. This accounts for the absence of CD4$^+$8$^+$ cells and the low total number of thymocytes in these mice. CD4$^+$8$^+$ thymocytes that expressed a low level of CD8 were spared from deletion because they interacted less effectively with the cells that induced the deletion. Nevertheless, these thymocytes could interact with sufficient affinity with the thymic selecting cell to allow their differentiation into CD4$^-$8$^{lo}\alpha_T^{high}\beta_T^{high}$ T cells. A small number of these cells could leave the thymus and accumulate in the periphery. As mentioned previously, the production of CD4$^-$8$^-\alpha_T^{high}\beta_T^{high}$ cells did not depend on positive selection by thymic MHC because these cells were also found in large numbers in H-2^d transgenic *scids*. These cells could also leave the thymus and accumulate in the periphery.

The very few CD4$^+$8$^-$ T cells that are found in the periphery of male transgenic mice provide independent support that these cells are derived from CD4$^+$8$^+$ precursor cells. Because the CD4$^+$8$^+$ precursor cells that express the trangenic TCR are deleted very efficiently, there is insufficient time for these precursor cells to rearrange and express endogenous α genes that can be selected by class II MHC molecules. Hence very few CD4$^+$8$^-$ T cells are produced. The observation that the majority of these cells do not express either the transgenic β or the transgenic α chain suggests that most of these cells are the progeny of CD4$^+$8$^+$ thymocytes that must have deleted both transgenes. Deletion of both transgenes in these mice has been observed (Bluthmann et al., 1988). It would be predicted that CD4$^+$8$^+$ thymocytes that have deleted both transgenes should survive sufficiently long to produce functional endogenous α and β gene rearrangements; some of these $\alpha_E\beta_E$ heterodimers could then be positively selected by class II MHC molecules. Deletion of $CD4^+8^+$ thymocytes that expressed the transgenic TCR was also observed in male H-$2^{d/b}$ *scid* mice. This indicates that the deletion of self-reactive precursors does not require suppressor T cells or anti-idiotypic mechanisms because these cells and mechanisms are absent from *scid* mice.

The peripheral T cells that express the transgenic TCR could not be activated by H-2^b male stimulator cells even in the presence of excess T cell lymphokines (Kisielow et al., 1988a; Teh et al., 1989). This observation is in line with previous transfection experiments that indicate a major role of CD8 molecules in the interaction of T cells with their targets. Thus, in male transgenic mice, tolerance was mediated by two major mechanisms: clonal deletion and clonal anergy. Clonal deletion was achieved by the deletion of $CD4^+8^+$ precursor cells that expressed autospecific TCR with the assistance of CD8 molecules. Clonal anergy was associated with a low level or lack of CD8 expression on T cells such that they were not antigen-reactive, despite the fact that they expressed a high level of the autospecific receptor.

Similar mechanisms for tolerance induction also appeared to operate in transgenic mice that expressed the L^d-specific TCR (Sha et al., 1988a). Transgenic mice bearing the autoantigen were produced by mating the founder H-$2^{b/b}$ mouse with normal H-$2^{b/d}$ mice to produce F_1 mice of the H-$2^{b/b}$ and H-$2^{b/d}$ MHC types. Type A H-$2^{b/d}$ transgenic mice, which bear four copies of the α transgene and two copies of the β transgene (similar to transgenic mice for the male-specific TCR), had small thymuses and peripheral T cells that expressed both α and β transgenes, and they had a low level of accessory molecules. However, in type B H-$2^{b/d}$ transgenic mice, which carried one copy of the α transgene and eight copies of the β transgene, the size of the thymus was not significantly different from that of their H-$2^{b/b}$ littermates, although the percentage of $CD4^+8^+$ thymocytes was less than that in their H-$2^{b/b}$ littermates. In both type A and B transgenic mice all peripheral $CD4^+8^-$ and $CD4^-8^+$ T cells were $\beta_T^+\alpha_T^-$, and expression of the α transgene was limited to $CD4^-8^-$ T cells. It remains to be established whether the failure to see a reduction in the number of $CD4^+8^+$ thymocytes in type B mice was related to the fact that these mice carried only one copy of the α transgene. Alternatively, the $CD4^+8^+$ thymocyte population may not express a sufficiently high level of the transgenic TCR to mediate deletion.

The incomplete deletion of $CD4^+8^+$ thymocytes expressing autospecific TCRs was also observed for TCRs that are specific for I-E or Mls[a] in nontransgenic mice (Kappler et al., 1988; MacDonald et al., 1988c). In this case one can argue that the apparent failure to delete $CD4^+8^+$ thymocytes expressing these class II MHC–restricted TCR may be due to the lack of expression of class II MHC molecules by cells in the outer cortex of the thymus. In this regard, it is of interest to note that recent experiments using these antigenic systems indicated that the target of deletion is the immature $CD4^+8^+$ thymocyte and that the CD4 accessory molecule plays a fundamental role in the deletion process (Fowlkes et al., 1988; MacDonald et al., 1988a). Thus there is good agreement on the mechanisms by which self tolerance is achieved in both TCR transgenic mice and in normal mice.

Nature of the Tolerance-Inducing Cell

Most of the experimental data favor the conclusion that tolerance to thymic MHC is induced by a bone marrow–derived cell and that thymic epithelial cells play a limited role in tolerance induction. Thus, nude mice grafted with MHC-disparate neonatal thymus grafts showed full tolerance to thymic MHC. In contrast, nude mice grafted with irradiated MHC-disparate adult thymuses were not tolerant to thymic MHC antigens (Zinkernagel et al., 1980). Deoxyguanosine-treated fetal thymus grafts that have been depleted of bone marrow–derived lymphoid cells can be transplanted across histocompatibility barriers into normal mice without being rejected, in spite of the continued expression of foreign class I and class II MHC antigens by the thymic grafts (Ready et al., 1984). However, both the intragraft and the mature peripheral T cell populations proliferated in vitro in response to the foreign MHC antigens of the graft, indicating that tolerance had failed to develop. Deoxyguanosine-treated fetal thymus grafts also failed to induce tolerance to thymic MHC when used to reconstitute T cell function in nude mice (von Boehmer and Schubiger, 1984). These results thus favor the conclusion that induction of tolerance to foreign MHC antigens depends primarily on a nonepithelial component of the thymus, probably the bone marrow–derived stromal cells such as the macrophages and/or dendritic cells. More recently, murine thymocytes, i.e., T cells, have been implicated in the induction of tolerance to MHC class I but not class II alloantigens (Shimonkevitz and Bevan, 1988).

FUTURE DIRECTIONS

The past two years have seen dramatic advances in our understanding of mechanisms leading to T cell tolerance to self antigens and T cell repertoire selection. The use of TCR transgenic mice and antigenic systems in which T cells specific for the relevant antigen are found in large numbers and can be visualized with specific TCR mAb's has been instrumental for this progress. However, much remains to be done with regard to the characterization of the cells that mediate positive or negative selection and the intracellular signals for these processes. The use of TCR transgenic mice also enables one to obtain large numbers of homogeneous antigen-specific T cells, and these cells should be useful for analyzing the mechanisms by which virgin T cells are activated. From the clinical standpoint it is important to understand the mechanisms by which mature single positive T cells can be rendered tolerant to their respective antigens. The availability in large numbers of homogeneous populations of antigen-specific T cells should also be useful for studying the various mechanisms that lead to autoimmunity.

Acknowledgments

I am indebted to Harald von Boehmer and his colleagues at the Basel Institute for Immunology (BII) for the privilege of working with the male-specific TCR transgenic mice. This work was done while I was on sabbatical at the BII in 1988. The BII was founded and is supported by F. Hoffmann-La Roche & Co. Ltd., Basel, Switzerland. I was a recipient of a University of British Columbia Killam Faculty Research Fellowship during my sabbatical. I thank Rosario Bauzon for typing the manuscript.

References

Bevan, M. J. (1977). In a radiation chimera, host H-2 antigens determine immune responsiveness of donor cytotoxic cells. *Nature* 269:417–418.
Bjorkman, P. J., Saper, M. A., Samraoui, B., Bennett, W. S., Strominger, J. L., and Wiley, D. C.

(1987). Structure of the human class I histocompatibility antigen, HLA-A2. *Nature 329*:506–512.

Bluthmann, H., Kisielow, P., Uematsu, Y., Malissen, M., Krimpenfort, P., Berns, A., von Boehmer, H., and Steinmetz, M. (1988). T cell specific deletion of T cell receptor transgenes allow functional rearrangement of endogenous α- and β-genes. *Nature 334*:156–159.

Brenner, M. B., Strominger, J. L., and Krangel, M. S. (1988). The γδ T cell receptor. *Adv. Immunol. 43*:133–192.

Brown, J. H., Jardetzky, T., Saper, M. A., Samraoui, B., Bjorkman, P. J., and Wiley, D. C. (1988). A hypothetical model of the foreign antigen binding site of class II histocompatibility molecules. *Nature 332*:845–850.

Chien, Y., Becker, D. M., Lindsten, T., Okamura, M., Cohen, D. I., and Davis, M. M. (1984). A third type of murine T-cell receptor cell. *Nature 312*:31–35.

Clevers, H., Alarcon, B., Wileman, T., and Terhorst, C. (1988). The T cell receptor/CD3 complex: A dynamic protein ensemble. *Annu. Rev. Immunol. 6*:629–662.

Crispe, I. N., and Bevan, M. J. (1987). Expression and functional significance of the J11d marker on mouse thymocytes. *J. Immunol. 138*:2013–2018.

Crispe, I. N., Moore, M. W., Husmann, L. A., Smith, L., Bevan, M. J., and Shimonkevitz, R. P. (1987). Differentiation potential of subsets of CD4⁻8⁻ thymocytes. *Nature 329*:336–339.

Dembic, Z., Haas, W., Weiss, S., McCubrey, J., Kiefer, H., von Boehmer, H., and Steinmetz, M. (1986). Transfer of specificity by murine α and β T-cell receptor gene. *Nature 320*:232–238.

Dembic, Z., Haas, W., Zamoyska, R., Parnes, J., Steinmetz, M., and von Boehmer, H. (1987). Transfection of the CD8 genes enhances T-cell recognition. *Nature 326*:510–511.

Dialynas, D. P., Quan, Z. S., Wall, K. A., Pierres, A., Quintans, J., Loken, M. R., Pierres, M., and Fitch, F. W. (1983). Characterization of the murine T cell surface molecule, designated L3T4, identified by monoclonal antibody GK1.5: similarity of L3T4 to the human Leu 3/T4 molecule. *J. Immunol. 131*:2445–2451.

Eichmann, K., Jonsson, J.-I., Falk, I., and Emmrich, F. (1987). Effective activation of resting mouse T lymphocytes by cross-linking submitogenic concentrations of the T cell antigen receptor with either Lyt-2 or L3T4. *Eur. J. Immunol. 17*:643–650.

Fink, P. J., and Bevan, M. J. (1978). H-2 antigen of the thymus determines lymphocyte specificity. *J. Exp. Med. 148*:766–775.

Forman, J., and Streilein, J. W. (1979). T cells recognize minor histocompatibility antigens on H-2 allogeneic cells. *J. Exp. Med. 150*:1001–1007.

Fowlkes, B. J., Kruisbeek, A. M., Ton-That, H., Weston, M. A., Coligan, J. E., Schwartz, R., and Pardoll, D. M. (1987). A novel population of T-cell receptor αβ-bearing thymocytes which predominantly expresses a single V$_\beta$ gene family. *Nature 329*:251–254.

Fowlkes, B. J., Schwartz, R. H., and Pardoll, D. M. (1988). Deletion of self-reactive thymocytes occurs at a CD4⁺8⁺ precursor stage. *Nature 334*:620–623.

Gabert, J., Langlet, C., Zamoyska, R., Parnes, J. R., Schmitt-Verhulst, A.-M., and Malissen, B. (1987). Reconstitution of MHC class I specificity by transfer of the T cell receptor and Lyt-2 genes. *Cell 50*:545–554.

Gallatin, W. M., Weissman, I. L., and Butcher, E. C. (1983). A cell-surface molecule involved in organ-specific homing of lymphocytes. *Nature 304*:30–34.

Gay, D., Maddon, P., Sekaly, R., Talle, M. A., Godfrey, M., Long, E., Goldstein, G., Chess, L., Axel, R., Kappler, J., and Marrack, P. (1987). Functional interaction between human T-cell protein CD4 and the major histocompatibility complex HLA-DR antigen. *Nature 328*:626–629.

Haskins, K., Kubo, R., White, J., Pigeon, M., Kappler, J., and Marrack, P. (1983). The major histocompatibility complex–restricted antigen receptor on T cells: I. Isolation with a monoclonal antibody. *J. Exp. Med. 157*:1149–1169.

Hedrick, S. M., Cohen, D. I., Nielsen, E. A., and Davis, M. M. (1984). Isolation of cDNA clones encoding T cell–specific membrane-associated proteins. *Nature 308*:149–153.

Hood, L., Steinmetz, M., and Malissen, B. (1983). Genes of the major histocompatibility complex of the mouse. *Annu. Rev. Immunol. 1*:529–568.

Hunig, T., and Bevan, M. J. (1981). Specificity of T cell clones illustrates altered self hypothesis. *Nature 294*:460–462.

Ishii, N., Nagy, Z. A., and Klein, J. (1982). Absence of Ir gene control of T cells recognizing foreign antigen in the context of allogeneic MHC molecules. *Nature 295*:531–533.

Jenkinson, E. J., Franchi, L. L., Kingston, R., and Owen, J.J.T. (1982). Effect of deoxyguanosine on lymphopoiesis in the developing thymus rudiment in vitro: Application in the production of chimeric thymic rudiments. *Eur. J. Immunol. 12*:583–587.

Kappler, J. W., Roehm, N., and Marrack, P. (1987). T cell tolerance by clonal elimination in the thymus. *Cell 49*:273–280.

Kappler, J. W., Staerz, U., White, J., and Marrack, P. C. (1988). Self-tolerance eliminates T cells specific for *Mls*-modified products of the major histocompatibility complex. *Nature 332*:35–40.

Kingston, R., Jenkinson, E. J., and Owen, J.J.T. (1985). A single stem cell can recolonize an embryonic thymus producing phenotypically distinct T cell populations. *Nature 317*:811–813.

Kisielow, P., Bluthmann, H., Staerz, U. D., Steinmetz, M., and von Boehmer, H. (1988a). Tolerance in T cell receptor transgenic mice involves deletion of nonmature CD4$^+$8$^+$ thymocytes. *Nature 333*:742–746.

Kisielow, P., Leiserson, W., and von Boehmer, H. (1984). Differentiation of thymocytes in fetal organ culture: Analysis of phenotypic changes accompanying the appearance of cytolytic and interleukin 2–producing cells. *J. Immunol. 133*:1117–1123.

Kisielow, P., Teh, H. S., Bluthmann, H., and von Boehmer, H. (1988b). Positive selection of antigen-specific T cells in thymus by restricting MHC molecules. *Nature 335*:730–733.

Krimpenfort, P., de Jong, R., Uematsu, Y., Dembic, Z., Ryser, S., von Boehmer, H., Steinmetz, M., and Berns, A. (1988). Transcription of T cell receptor β-chain genes is controlled by a downstream regulatory element. *EMBO J. 7*:745–750.

Kruisbeek, A. M., Mond, J. J., Fowlkes, B. J., Carmen, J. A., Bridges, S., and Longo, D. L. (1985). Absence of the Lyt-2$^-$, L3T4$^+$ lineage of T cells in mice treated neonatally with anti-I-A correlates with absence of intrathymic I-A-bearing antigen-presenting cell function. J. Exp. Med. *161*:1029–1047.

Ledbetter, J. A., and Herzenberg, L. A. (1979). Xenogeneic monoclonal antibodies to mouse lymphoid differentiation antigens. *Immunol. Rev. 47*:63–90.

Ledbetter, J.A., June, C. H., Rabinovitch, P. S., Grossmann, A., Tsu, T. T., and Imboden, J. B. (1988). Signal transduction through CD4 receptors: Stimulatory vs. inhibitory activity is regulated by CD4 proximity to the CD3/T cell receptor. *Eur. J. Immunol. 18*:525–532.

Leo, O., Foo, M., Sachs, D. H., Samelson, L. E., and Bluestone, J. A. (1987). Identification of a monoclonal antibody specific for a murine T3 polypeptide. *Proc. Natl. Acad. Sci. USA 84*:1374–1378.

Littman, D. R. (1987). The structure of the CD4 and CD8 genes. *Annu. Rev. Immunol. 5*:561–584.

Lo, D., and Sprent, J. (1986). Identity of cells that imprint H-2 restricted T cell specificity in the thymus. *Nature 319*:672–675.

Longo, D. L., Kruisbeek, A. M., Davis, M. L., and Matis, L. A. (1985). Bone marrow–derived thymic antigen-presenting cells determine self-recognition of Ia-restricted T lymphocytes. *Proc. Natl. Acad. Sci. USA 82*:5900–5904.

Longo, D. L., and Schwartz, R. H. (1980). T-cell specificity for H-2 and Ir gene phenotype correlates with the phenotype of thymic antigen presenting cells. *Nature 287*:44–46.

MacDonald, H. R., Hengartner, H., and Pedrazzini, T. (1988a). Intrathymic deletion of self-reactive cells prevented by neonatal anti-CD4 antibody treatment. *Nature 335*:174–176.

MacDonald, H. R., Lees, R. K., Schneider, R., Zinkernagel, R. M., and Hengartner, H. (1988b). Positive selection of CD4$^+$ thymocytes controlled by MHC class II gene products. *Nature 361*:471–473.

MacDonald, H. R., Schneider, R., Lees, R. K., Howe, R. C., Acha-Orbea, H., Festenstein, H., Zinkernagel, R. M., and Hengartner, H. (1988c). T-cell receptor V$_β$ use predicts reactivity and tolerance to *Mls^a*-encoded antigens. *Nature 332*:40–45.

Malissen, M., Tracy, J., Letourneur, F., Rebai, N., Dunn, D. E., Fitch, F. W., Hood, L., and Malissen, B. (1988). A T-cell clone expresses two T cell receptor α genes but uses one $\alpha\beta$ heterodimer for allorecognition and self MHC–restricted antigen recognition. *Cell* 55:49–59.

Marusic-Galesic, S., Stephany, D. A., Longo, D. L., and Kruisbeek, A. M. (1988). Development of CD4⁻CD8⁺ cytotoxic T cells requires interaction with class I MHC determinants. *Nature* 333:180–183.

Matzinger, P., and Mirkwood, G. (1978). In a fully H-2 incompatible chimera, T cells of donor origin can respond to minor histocompatibility antigens in association with either donor or host H-2 type. *J. Exp. Med.* 148:84–92.

Meuer, S. C., Hodgdon, J. C., Hussey, R. E., Protentis, J. P., Schlossman, S. F., and Reinherz, E. L. (1983). Antigen-like effects of monoclonal antibodies directed at receptors on human T cell clones. *J. Exp. Med.* 158:988–993.

Norment, A. M., Salter, R. D., Parham, P., Engelhard, V. H., and Littman, D. R. (1988). Cell–cell adhesion mediated by CD8 and MHC class I molecules. *Nature* 336:79–81.

Nossal, G.J.V. (1983). Cellular mechanisms of immunological tolerance. *Annu. Rev. Immunol.* 1:33–62.

Raff, M. C. (1971). Surface antigenic markers for distinguishing T and B lymphocytes in mice. *Transplant. Rev.* 6:52–80.

Ready, A. R., Jenkinson, E. J., Kingston, R., and Owen, J.J.T. (1984). Successful transplantation across major histocompatibility barrier of deoxyguanosine-treated embryonic thymus expressing class II antigens. *Nature* 310:231–233.

Rudd, C. E., Trevillyan, J. M., Dasgupta, J. D., Wong, L. L., and Schlossman, S. F. (1988). The CD4 receptor is complexed in detergent lysates to a protein-tyrosine kinase (pp58) from human T lymphocytes. *Proc. Natl. Acad. Sci. USA* 85:5190–5194.

Saito, H., Kranz, D. M., Takagaki, Y., Hayday, A. C., Eisen, H. N., and Tonegawa, S. (1984). A third rearranged and expressed gene in a clone of cytotoxic T lymphocytes. *Nature* 312:36–40.

Saito, T., Weiss, A., Miller, J., Norcross, M. A., and Germain, R. N. (1987). Specific antigen-Ia activation of transfected human T cells expressing murine Ti $\alpha\beta$-human T3 receptor complexes. *Nature* 325:125–130.

Schuler, W., Weiler, I. J., Schuler, A., Phillips, R. A., Rosenberg, N., Mak, T. W., Kearney, J. F., Perry, R. P., and Bosman, M. J. (1986). Rearrangement of antigen receptor genes is defective in mice with severe combined immune deficiency. *Cell* 46:963–972.

Scollay, R., and Shortman, K. (1984). Cell traffic in the adult thymus. In *Recognition and Regulation in Cell-Mediated Immunity* (Watson, James D. and Marbrook, John, eds.). New York and Basel: Marcel Dekker, pp. 3–30.

Scott, B., Bluthmann, H., Teh, H. S., and von Boehmer, H. (1989). The generation of mature T cells requires interaction of the $\alpha\beta$ T-cell receptor with major histocompatibility antigens. *Nature* 338:591–593.

Sha, W. C., Nelson, C. A., Newberry, R. D., Kranz, D. M., Russell, J. H., and Loh, D. Y. (1988a). Positive and negative selection of an antigen receptor on T cells in transgenic mice. *Nature* 336:73–76.

Sha, W. C., Nelson, C. A., Newberry, R. D., Kranz, D. M., Russell, J. H., and Loh, D. Y. (1988b). Selective expression of an antigen receptor on CD8-bearing T lymphocytes in transgenic mice. *Nature* 335:271–274.

Shimonkevitz, R. P., and Bevan, M. J. (1988). Split tolerance induced by the intrathymic adoptive transfer of thymocyte stem cells. *J. Exp. Med.* 168:143–156.

Shortman, K., and Scollay, R. (1984). Cortical and medullary thymocytes. In *Recognition and Regulation in Cell-Mediated Immunity* (Watson, James D. and Marbrook, John, eds.). New York and Basel: Marcel Dekker, pp. 31–60.

Smith, L. (1987). CD4⁺ murine T cells develop from CD8⁺ precursors in vivo. *Nature* 326:798–800.

Sprent, J., and Webb, S. R. (1987). Function and specificity of T cell subsets in the mouse. *Adv. Immunol. 41*:39–133.

Staerz, U., Rammensee, H., Benedetto, J., and Bevan, M. J. (1985). Characterization of a murine monoclonal antibody specific for an allotypic determinant on T cell antigen receptor. *J. Immunol. 134*:3994–4000.

Swain, S. L. (1983). T cell subsets and the recognition of MHC class. *Immunol. Rev. 74*:129–142.

Teh, H. S., Bennink, J., and von Boehmer, H. (1982). Selection of the T cell repertoire during ontogeny: Limiting dilution analysis. *Eur. J. Immunol. 12*:887–892.

Teh, H. S., Kishi, H., Scott, B., and von Boehmer, H. (1989). Deletion of autospecific T cells in T cell receptor (TCR) transgenic mice spares cells with normal TCR levels and low levels of CD8 molecules. *J. Exp. Med. 169*:795–806.

Teh, H. S., Kisielow, P., Scott, B., Kishi, H., Uematsu, Y., Bluthmann, H., and von Boehmer, H. (1988). Thymic major histocompatibility complex antigen and the $\alpha\beta$ T-cell receptor determine the CD4/CD8 phenotype of T cells. *Nature 335*:229–233.

Trowbridge, I. S., Lesley, J., Schulte, R., Hyman, R., and Trotter, J. (1982). Biochemical characterization and cellular distribution of a polymorphic murine cell surface glycoprotein expressed on lymphoid tissues. *Immunogenetics 15*:299–312.

Uematsu, Y., Ryser, S., Dembic, Z., Borgulya, P., Krimpenfort, P., Berns, A., von Boehmer, H., and Steinmetz, M. (1988). In transgenic mice the introduced functional T cell receptor β gene prevents expression of endogenous β genes. *Cell 52*:831–841.

Veillette, A., Bookman, M. A., Horak, E. M., and Bolen, J. B. (1988). The CD4 and the CD8 T cell surface antigens are associated with the internal membrane tyrosine protein-kinase p56[lck]. *Cell 55*:301–308.

von Boehmer, H. (1986). The selection of the α,β heterodimeric T-cell receptor for antigen. *Immunol. Today 7*:333–336.

von Boehmer, H. (1988). The developmental biology of T lymphocytes. *Annu. Rev. Immunol. 6*:309–326.

von Boehmer, H., and Schubiger, K. (1984). Thymocytes appear to ignore class I major histocompatibility complex antigens expressed on thymic epithelial cells. *Eur. J. Immunol. 14*:1048–1052.

Wagner, H., Hardt, C., Stockinger, H., Pfizenmaier, K., Bartlett, R., and Rollinghoff, M. (1981). Impact of thymus on the generation of immunocompetence and diversity of antigen-specific MHC-restricted cytotoxic T lymphocyte precursors. *Immunol. Rev. 58*:95–129.

Wilson, A., Scollay, R., Reichert, R. A., Butcher, E. C., Weissman, I. L., and Shortman, K. (1987). The correlation of lectin-stimulated proliferation and cytotoxicity in murine thymocytes with expression of the MEL-14 homing receptor. *J. Immunol. 138*:352–357.

Yanagi, Y., Yoshikai, Y., Leggett, K., Clark, S. P., Aleksander, I., and Mak, T. W. (1984). A human T cell–specific cDNA clone encodes a protein having extensive homology to immunoglobulin chains. *Nature 308*:145–149.

Zinkernagel, R. M., Althage, A., Waterfield, E., Kindred, B., Welsh, R. M., Callahan, G., and Pincetl, P. (1980). Restriction specificities, alloreactivity, and allotolerance expressed by T cells from nude mice reconstituted with H-2-compatible or -incompatible thymus grafts. *J. Exp. Med. 151*:376–399.

Zinkernagel, R. M., Callahan, G. N., Klein, J., and Dennert, G. (1978). Cytotoxic T cells learn specificity for self H-2 during differentiation in the thymus. *Nature 271*:251–253.

Zinkernagel, R. M., Sado, T., Althage, A., and Kamisaku, H. (1984). Anti-viral immune response of allogeneic irradiation bone marrow chimeras: Cytotoxic T cell responsiveness depends upon H-2 combination and infectious agent. *Eur. J. Immunol. 14*:14–23.

11

B Lymphocyte Generation as a Developmental Process

KENNETH S. LANDRETH

The hallmark of B lymphocyte lineage cells is their ability to synthesize and secrete immunoglobulin (Ig) molecules. Although additional distinguishing characteristics are continually being described, the synthesis of Ig remains one of the most useful indicators of commitment to this lineage in all species thus far studied. This will no doubt change as the characteristics of committed progenitors for B-lineage cells become better defined and studied. As described in detail later in this chapter, a number of monoclonal antibodies have already opened the way to experimental manipulation of developmentally early cells, and exciting advances should be close at hand.

It is historically important to remember that the functions of lymphocytes were largely unknown until the middle of the present century. Until 1960 lymphocytes were considered "end cells" of little importance. However, carefully documented studies of lymphopoiesis appeared as early as 1774 when William Hewson published his observations on generation of the "central particles" of the lymph in a comprehensive treatise dedicated to Benjamin Franklin (Gulliver, 1846). Hewson described the formation of a majority of the "central particles" or lymphocytes by the thymus, their circulation in the lymph, and subsequent entry into the blood. Thus experimental observations on the production and migration of T lymphocytes predate similar observations on B lymphocytes by over 150 years.

Paul Ehrlich proposed in 1900 that cells destined for antibody production were present before antigenic experience and that these cells displayed cell surface antibodies representative of the antibodies they were capable of making. Lymphocytes were not identified as the cells responsible for antibody production, however, for another 40 years. In 1939, it was suggested that lymphocytes, and not macrophages, were the antibody-forming cells (Rich et al., 1939). This was followed by evidence that efferent lymph contained antibody-producing cells (Harris et al., 1945; Harris and Harris, 1960), but the role of lymphocytes in antibody production remained arguable into the 1950s. In 1956 Bruce Glick reported that surgical removal of the bursa of Fabricius of hatchling chickens severely compromised antibody production to injected *Salmonella*-type O antigen (Glick et al., 1956). The conclusion that this lymphoid organ was central to antibody production revolutionized the study of lymphopoiesis and served as the cornerstone for all subsequent studies of B lymphocyte development.

The demonstration that "bursa-derived cells" (B cells) in birds (Kincade and Cooper, 1971) and their equivalents in mammals (Pernis et al., 1970; Raff, et al., 1970) could be identified by the presence of cell surface immunoglobulin (sIg) fulfilled the prediction of Ehrlich some 70 years earlier. With the ability to experimentally identify cells specialized to produce antibody molecules came increasing interest in the embryonic and hemopoietic origins of these cells. Studies of lymphopoiesis, however, have been complicated by the fact that lymphocytes can be stimulated to divide after a period of mitotic quiescence (Gowans and McGregor, 1963). It is therefore difficult to distinguish between sites responsible for the primary generation of B lymphocytes (primary or central lymphoid tissues) and those in which lymphocytes respond to mitogenic stimuli (secondary lymphoid tissues).

This chapter will summarize what is known about the phylogenetic, embryologic, and hemopoietic stages in the primary generation of B lymphocyte lineage cells. Most studies to date have focused on events in avian and mammalian B lymphocyte generation, and available phylogenetic information about poikilothermic vertebrates will be compared to that for the aves and mammalia. This chapter will also focus on unresolved questions that bear on the cell biology of B lymphocyte development and consider hypothetical, but testable, schemes of cell progression through this developmental process.

B LYMPHOCYTE DEVELOPMENT
IN POIKILOTHERMIC VERTEBRATES

Although cells with lymphoid morphology can be demonstrated in invertebrates (Cooper et al., 1966), expression of immune effector functions and specialization into subsets that mediate humoral and cell-mediated immunity have been confirmed only in vertebrates. Lymphocytes that express sIg have been described in many vertebrate species, but we know very little about primary sites of "B lymphocyte" generation in most of these organisms. The following review of lymphopoiesis in poikilothermic vertebrates is based on the premise that lymphopoiesis is an integral part of hemopoiesis and that the identification of sites of blood cell formation may help define primary lymphopoietic organs. This approach has been, for the most part, substantiated by available studies. This review should also help point out the evolutionary origins of the avian and mammalian prototypes and indicate areas in need of additional study.

Class Agnatha

Jawless fishes of the Class Agnatha are the earliest vertebrates found in fossil deposits, appearing some 450 million years ago. Present-day representatives of these early vertebrates include hagfish and lampreys. Lymphocytes that secrete antibody have been demonstrated in hagfish (Raison and Hildemann, 1984). We know little about the origins of these antibody-producing cells, and most studies have been complicated by the fact that virtually all lymphocytes in these species appear to express sIg (Fiebig and Ambrosius, 1976). This may suggest that functional specialization of lymphocytes has not occurred in these agnathan descendants. However, this conclusion must be considered tentative until passive binding of circulating Ig to a subset of non-Ig synthesizing cells has been ruled out. Hagfish hemopoiesis is localized to a splenic-equivalent tissue and to the anterior pronephros (Tavassoli and Yoffey, 1983). The splenic-equivalent consists of unorganized hemopoietic foci in the anterior wall of the foregut. Lymphocyte-like cells are found in both the splenic-equivalent tissue and the anterior pronephros, as well as in the blood. In the absence of definitive data, it has been assumed that generation of antibody-secreting lymphocytes in this species is multifocal.

Lampreys also have hemopoietic tissues in the anterior gut wall, which have been likened to a "primitive spleen" (Tavassoli and Yoffey, 1983). However, the origin of Ig-secreting cells in the lamprey has not been determined.

Class Chondrichthyes

Jawed fishes arose some 400 million years ago and are represented today by two groups of descendants, the cartilaginous fishes of Class Chondrichthyes and the bony fishes in the Class Osteichthyes. Present-day chondrichthyes include sharks and rays. Although sharks have circulating IgM of both 7 S and 19 S forms, the origin of antibody-secreting cells in these species has not been assessed. Sharks differ from jawless fishes in having organized splenic tissues with red and white pulp (Tavassoli and Yoffey, 1983), which indicate blood-rich and lymphocyte-rich areas, respectively. Hemopoietic foci have been described in the shark spleen, intestinal wall, and anterior pronephros.

Class Osteichthyes

The bony fishes are represented today by two main groups, the ray-finned and the lobe-finned fishes. Although the lobe-finned fishes probably have a closer evolutionary rela-tionship to terrestrial vertebrates, most studies of lymphopoiesis have centered on teleost species of the ray-finned group. Hemopoiesis in the paddlefish, a primitive representative of the actinopterygian fishes, occurs largely in an organized tissue that overlies the base of the heart. The spleen is present as a separate organ, with distinct red and white pulp (Tavassoli and Yoffey, 1983). Some ganoid fishes also have hemopoietic foci associated with the bones of the cranium (Rosse et al., 1978). These unusual hemopoietic sites orig-inate from the meninges of the fourth ventricle and develop in bony pockets. This is the earliest instance in which hemopoietic tissues are associated with bony sites. Whether this early form of "bone marrow" supports the generation of lymphocytes is totally unknown. Bone-associated hemopoiesis is subsequently seen in anuran amphibia, and the marrow becomes the dominant site for blood and lymphoid cell formation in reptiles, birds, and mammals.

In the Teleost fishes, hemopoiesis is largely confined to the pronephros or anterior kidney, although morphologically identifiable immature lymphoid forms are found in the spleen and intestinal submucosa (Fiebig and Ambrosius, 1976; Tavassoli and Yoffey, 1983). Cell surface IgM can be detected on the lymphocytes of bony fish (Cuchens et al., 1976; Lobb and Clem, 1982), although no definitive site of B lymphopoiesis has been identified in these species. It is of interest, however, that the pronephros contains large accumulations of lymphoid cells, along with other hemopoietic elements, but, unlike the spleen and thymus, the pronephros does not appear to produce significant amounts of specific antibody. This suggests that the anterior kidney is the site of primary B lympho-poiesis in the bony fishes, although additional work is clearly needed to resolve this point.

Class Amphibia

The amphibia were the first species capable of terrestrial existence, having evolved from freshwater fishes. The present-day representatives of this class are the urodela (mudpuppies, newts, and salamanders), the apoda (a group of legless species), and the anura (frogs and toads). The urodeles and apoda are considered more primitive species than anurans and, unfortunately, have not been studied as extensively as the latter. In urodele amphibia, the liver retains hemopoietic functions throughout life, as do the spleen and intestinal submucosa (Tavassoli and Yoffey, 1983). The perihepatic layer has been proposed as the source of B cells in Pleurodeles (Cohen, 1977), but it is not known if B lymphopoiesis is anatomically restricted in this or any other urodele species.

In anuran amphibia, lymphocytes can be characterized as B and T cells on the basis of sIg expression (Fiebig and Ambrosius, 1976; Klempau and Cooper, 1983). B cells appear to be generated in both the spleen and bone marrow during adult life (Turpen et al., 1982). It has not been conclusively established whether the marrow is the primary lymphoid organ and the spleen a site of secondary B cell proliferation in these species, as in mammals. However, the following observations support such a conclusion. Accumulations of lymphoid cells are obvious in both the parenchyma and vascular spaces of adult frog marrow. Evidence that lymphoid cells found in the marrow contribute to primary B lymphopoiesis has come from studies in which the femurs of leopard frogs were lead-shielded during exposure to high doses of X-irradiation (Ramirez et al., 1983). Under these conditions, which presumably destroy hemopoietic cells in unshielded anatomic sites, immune responsiveness and generation of plaque-forming cells were restored without adoptive cell transfer. This is the first instance in which bone marrow has been shown to have hemopoietic functions (except for the ganoid fishes; see above), and it is the first phylogenetic group in which the marrow is the site of primary B lymphopoiesis.

The pronephros and mesonephros are sites of embryonic hemopoiesis in both *Xenopus* and *Rana* (Turpen et al., 1982). Hemopoietic cells originating in the ventral blood islands and the dorsal mesoderm of the developing *Rana* tadpole appear to enter these hemopoietic site tissues during development. Colonization of the mesonephros by extrinsic hemopoietic stem cells has also been demonstrated in *Xenopus.* In *Xenopus,* hemopoietic cells appear to arise primarily in the dorsal mesoderm of the developing embryo (Kau and Turpen, 1983). Certain principles, including colonization of hemopoietic sites by extrinsic cells and a multifocal origin for these cells, appear to be conserved in hemopoiesis and are discussed further in connection with the aves (see below).

In *Rana,* cells expressing IgM are first detected in embryos of between 8 and 9 days of age. These cells have detectable amounts of cytoplasmic μ heavy chain of IgM ($c\mu$) but lack sIg and have only been found in the developing pronephros (Zettergren, 1982). Cells with $c\mu$ have subsequently been described in the mesonephros and liver up to the metamorphic period, although the numbers of these $c\mu^+$ cells were significantly higher in the embryonic mesonephros than in the liver at all times reported. Surface IgM-bearing cells (B cells) were first observed approximately 2 days after the appearance of the $c\mu^+$ cells in both the pronephros and liver. During the metamorphic period, IgM$^+$ cells gradually disappeared from the larval kidneys and appeared in the bone marrow. In the adult frog, cells with $c\mu$, but without sIgM, were found exclusively in the bone marrow. Cells expressing this phenotype ($c\mu^+$ sIg$^-$) have been studied extensively in several mammalian species and have been identified as the immediate precursors of B cells (pre-B cells) in that phylogenetic class. The above observations on B cell precursors in developing urogenital tissues appear compatible with phylogenetic aspects of the appearance of both organized hemopoietic tissues and lymphocytes.

Class Reptilia

Studies that address the development of lymphocytes in reptiles are sparse. This is particularly unfortunate because both modern birds and mammals arose directly from reptilian ancestors. As discussed below, comparative studies of avian and mammalian species have identified a number of major differences in the anatomy and progression of B lymphopoiesis between these phylogenetic classes. Knowledge of lymphoid cell development in reptilian species would therefore likely shed considerable light on the evolution of these distinct developmental pathways.

Immunoglobulin molecules have been demonstrated on a subpopulation of lymphocytes in both lizards and snakes (Fiebig and Ambrosius, 1976). However, the sites of lymphocyte formation in reptiles remain largely unknown. As in Anuran amphibia, the bone marrow is largely responsible for hemopoiesis, with the spleen playing a variable role in the development of some cell lineages, particularly in turtles and snakes (Tavassoli and Yoffey, 1983). Virtually nothing is known about B lymphocyte development in reptiles, and further studies, particularly of the crocodilia, which are the most closely related to birds, seem essential.

B LYMPHOCYTE DEVELOPMENT
IN HOMIOTHERMIC VERTEBRATES

The role of B lymphocytes in antibody production and the anatomic sites in which the primary development of these cells takes place has been studied extensively in birds and mammals. The role of lymphocytes in antibody production was first described in chickens, owing to the existence of an apparently unique site for development of these cells. These cells were therefore termed "bursa-derived" or B cells (Glick, 1977), reflecting the critical observations of Glick that launched the study of B lymphopoiesis. Based on studies conducted in chickens and subsequently on rodents and humans, much is now known about the sequence of embryonic events and identifiable cell types in B lymphocyte development, as well as the relationship of this process to hemopoietic stem cells and to the production of other blood cells.

Class Aves

Based on the important initial observations of Glick and his colleagues (Glick 1967; Glick et al., 1956), the bursa of Fabricius has long been considered the most important site of B lymphopoiesis in birds. Removal of the bursa during embryonic development or early after hatching severely restricts the development of specific antibody responses and of plasma cells and lymph node germinal centers (Cooper et al., 1966; Fitzsimmons et al., 1973; Glick, 1967; Glick et al., 1956; Lerman and Weidanz, 1970; Warner et al., 1962). Lymphocytes that develop in the bursal follicles acquire sIg expression (Grossi et al.,

1977; Kincade and Cooper, 1971; Lydyard et al., 1976) and subsequently migrate out of the bursa to peripheral lymphoid organs. This temporal staging of events in the embryonic emergence of chicken B lymphocytes is depicted in Figure 11-1.

The functions of the bursa in many ways parallel those of the thymus. Both organs are specialized for the development of functional subsets of lymphocytes, are primary sites of lymphopoiesis early in life, and have significantly declined in function by sexual maturity. It is only in birds, however, that a distinct anatomical site, virtually restricted to B lymphopoiesis, has been demonstrated. It should be noted that the bursa does support a wave of embryonic granulopoiesis, which has concluded by the time of hatching (LeDouarin et al., 1984). Whether these granulocyte precursors derive from the same emigrating hemopoietic cells as B lymphocytes is unknown.

Bursa of Fabricius

Colonization by extrinsic cells
The bursa of Fabricius arises as a dorsal diverticulum of the proctodeal region of the cloaca between days 4 and 5 of incubation (Glick, 1977) (Fig. 11-1). Blood-borne hemopoietic cells of mesenchymal origin enter the bursa at between days 8 to 14 of incubation in the chicken (Jotereau and Houssaint, 1977). These invading cells give rise to follicles of lymphocytes, which then express surface IgM by day 12 (Kincade and Cooper, 1971). The yolk sac was originally proposed to be the source of lymphocyte stem cells on the basis of transplantation experiments (Moore and Owen, 1967a,b). However, these early studies did not conclusively establish whether the stem cells were truly of yolk sac origin or had migrated there from other sites (LeDouarin et al., 1975).

More recent studies using interspecific chicken-quail chimeras, have demonstrated that the hemopoietic stem cells are of intraembryonic origin (Dieterlen-Lièvre and Martin, 1981). These latter experiments involved grafting of stage-matched quail blastoderms onto the extraembryonic tissue (yolk sac) of developing chicken embryos (Martin, 1972). The presence of emigrant cells in the developing bursa could be determined on the basis of species-specific nuclear morphology. Although some cells of yolk sac origin were transiently present in the embryos, all of the cells that permanently colonized the bursa were of the same species as the embryo. These data strongly suggest that although yolk sac cells have the potential to colonize the bursa, intraembryonic cells normally make up the major population which migrates to this organ. The intraembryonic sources of migrating pre-bursal cells have not been firmly established, but they may include the para-aortic foci of hemopoietic cells described by LeDouarin et al. (1984). Certainly, the existence of multiple embryonic sources of hemopoietic stem cells capable of giving rise to lymphocytes and other blood cells is characteristic of early events in hemopoiesis. The failure of one source would likely be compensated by another. This degree of developmental redundancy also suggests the considerable evolutionary importance of the hemopoietic system to the survival of vertebrate species.

It has not been established whether the cells that colonize the developing bursa are multipotential stem cells or are already committed to B lymphocyte development. Although this point remains controversial, some evidence suggests that a proportion of cells in a colonized bursa have at least the potential to enter into T lymphopoiesis. When 6.5-day embryonic thymus was cocultured with 11-day embryonic quail bursa, some bursal cells entered the "receptive" thymus where they differentiated to express thymus-specific cell surface antigens (Jotereau et al., 1980; LeDouarin et al., 1984). However, it is

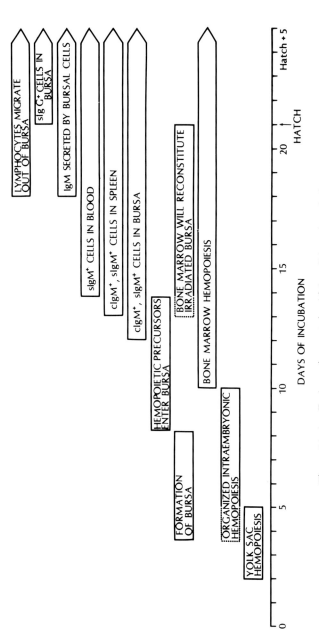

Figure 11–1.   Embryonic events in chicken B lymphopoiesis.

244

not known whether a subset of the cells that enter the bursa are truly pluripotential stem cells for the hemopoietic system.

Development of bursal follicles

Cells that contain IgM are found in the bursa beginning at day 12 of embryonation (Grossi et al., 1977; Kincade and Cooper, 1971; Ritter and Lebacq, 1977), some 4 days after the onset of bursal colonization by immigrating hemopoeitic stem cells. These cells express cytoplasmic IgM and small amounts of cell surface IgM. By day 12, bursal follicles each contain approximately 100 lymphoid cells (P. W. Kincade, personal communication). If each follicle is initially colonized by two to four hemopoietic precursors, as has been suggested (Pink et al., 1985), fewer than six cell divisions separate the putative hemopoietic stem cells from $sIgM^+$ B lymphocytes in this species.

Ig-bearing cells are not found in the bursal follicles on day 11 (Grossi et al., 1977), and it seems probable that sIg^+ cells arise directly from precursors that lack any Ig expression. This observation contrasts with those made in mammalian systems. In both mice and humans, the earliest Ig^+ cells in the fetal liver contain cytoplasmic μ heavy chains and can be detected several days before the expression of light chains and sIg (Gathings et al., 1977; Raff et al., 1970).

Growth and regression

B lymphocyte generation is a developmental rather than a regressive process; however, the stage at which B lymphopoiesis declines in primary lymphoid tissues may be important in relation to evolutionary selective pressures and the aging of the immune system. The bursa of Fabricius in the chicken continues to grow by active cell proliferation for at least 3 weeks after hatching (Glick, 1977; Landreth and Glick, 1973), but then it subsequently declines as a site of lymphopoiesis. By 4–6 months of age, the bursa regresses and loses its characteristic microarchitecture and expression of Ig (Grossi et al., 1977), and its involution is complete by the time of sexual maturity. In this respect, the regression of the bursa, and with it the production of newly generated B lymphocytes, closely resembles the thymus of both birds and mammals.

Molecular events in avian B cell maturation

Recent studies have contributed greatly to our understanding of events in B cell development in the avian bursa (reviewed in McCormack et al., 1991). It has been shown that the avian bursa is the site for expansion of B cell precursors during normal ontogeny and that the primary B cell repertoire is generated during embryonic life in the bursa. This contrasts with B cell development in mammals, in which precursors are generated continuously in the bone marrow throughout life (Kincade, 1981; Rosse, 1976).

Extensive characterization of the chicken Ig heavy (H) and light (L) chain genes has been reported. The Ig L locus has been shown to consist of a single functional light chain variable region (V_L) gene segment that is separated by 1.8 kb from a single functional light chain joining region (J_L) gene. The J_L gene, in turn, lies 2 kb upstream from a single functional light chain constant region (C_L) gene. Twenty-five V_L pseudogenes have been identified and lie upstream of the functional V_L gene (Reynaud et al., 1985, 1987). The Ig H locus similarly has limited potential for combinatorial diversity, consisting of a single heavy chain variable region (V_H) gene segment located 15 kb 5' of a single heavy chain joining region (J_H) segment. Approximately 15 heavy chain diversity region (D_H) segments are interposed between V_H and J_H, although there is little sequence variation between the D_H germline sequences. Approximately 100 V_H pseudogenes have been identified upstream of the functional V_H gene (Parvari et al., 1988; Reynaud et al., 1989).

V_L-J_L joining occurs through deletion of the intervening DNA, leading to the transient formation of a signal joint episome (McCormack et al., 1989c). The signal joint episome thus serves as a marker of recent V-J rearrangement. V-J rearrangement and the episome can be detected in bursal and extrabursal tissues between days 10–18 of embryogenesis, although the episome disappears by the time of hatching (McCormack et al., 1989b). The majority of V_L-J_L diversity appears to be generated by nonrandom nucleotide addition to the coding ends of V_L and J_L during rearrangement. Two-thirds of these Ig gene joining events that can be detected at days 10–12 of embryonic development are out of frame, but nearly all bursal B lymphocytes by day 18 have rearranged only a single Ig L allele, suggesting that the vast majority of rearrangements are productive. This contrasts with the situation in mammalian cells, in which rearrangements at both of the Ig H and/or Ig L alleles are common (Yancopoulos and Alt, 1986). Thus it appears that the majority of B cells in the chicken rearrange only a single Ig H and Ig L allele during the brief period between days 12–15 of embryogenesis. Lymphocytes that have made productive heavy and light chain rearrangements are subsequently selectively expanded in the bursa, probably through a positive selection mechanism, although the details of this process are unclear (McCormack et al., 1989b). According to at least one view, this process may be driven by autoreactivity of B cells against bursal antigens, cell proliferation occurring until the cell surface receptor no longer recognizes bursal antigens (McCormack et al., 1991). This is analogous to "positive selection" models of thymic T cell development. Taken together, these results suggest that a major function of the avian bursa is to expand clonal populations of B lymphocytes that have undergone functional Ig H and Ig L rearrangements. However, it does not appear that the bursa is required for Ig gene rearrangement, in that bursectomy at day 5 of embryological development did not prevent the development of sIg$^+$ B cells and Ig secretion, although it is asociated with limited clonal diversity (Granfors et al., 1982; Jalkanen et al., 1983a,b).

In addition, the bursa plays a major role in the somatic diversification of rearranged Ig genes through providing an appropriate microenvironment for diversification. The majority of the nucleotide substitutions in Ig L genes arise by multiple substitution events between V_L and the upstream ψV_L sequences (Reynaud et al., 1985, 1987). Similar substitutions have been shown to occur in chicken Ig H genes (Reynaud et al., 1989). The majority of these substitutions appear to represent intrachromosomal gene conversion events rather than double homologous recombination or other mechanisms (McCormack and Thompson, 1990a,b). A minor proportion of nucleotide substitutions represent single nucleotide point mutations and may be involved, for example, in affinity maturation of the antibody response.

Thus the Ig gene loci in the chicken have developed a distinct pattern of organization and associated novel mechanisms for generating somatic diversity. Similar organization of Ig L gene segments has been identified in a number of other avian species, including mallard duck, pigeon, quail, turkey, cormorant, and hawk (McCormack et al.,1989a). However, the Muscovy duck Ig L locus was found to contain up to seven V_L segments capable of rearranging to the (single) J_L segment. Thus, at least in this species, some degree of combinatorial diversity has developed, perhaps through duplication of a portion of the Ig L locus (McCormack et al., 1991).

Bone Marrow

The bone marrow is the most important postnatal site of B lymphocyte production in mammals (see the dicussion on postnatal generation of B cells below). However, a series of observations suggest that this is likely not the case in birds. Embryonically, 14-day chicken marrow can act as a rich source of cells that can colonize an early bursal rudiment (Fig. 11–1), but this capacity is lost after hatching (Honjo, et al., 1981; Weber and Maus-ner, 1977). This is particularly striking when compared to the ability of the posthatch marrow to colonize early thymic rudiments (Eskola and Toivanen, 1977). It would appear that there is either a qualitative or quantitative change in bone marrow hemo-poietic cells that can contribute to bursal B lymphopoiesis after hatching. There are also few morphologically immature lymphoid cells in the marrow of hatchling birds (Glick and Rosse, 1976). However, marrow from hatchling chicks has been shown to contain a small number of large B cells, which look like those found in the bursa from embryonic day 16 to several weeks after hatching (Grossi et al., 1977). Cells in hatchling chick mar-row that could respond to a "bursapoietin" and acquire a defined B-lineage cell surface antigen were demonstrated (Brand et al., 1983), but these cells were already B cells by our definition (i.e., sIg$^+$). These observations suggest that, normally, B lymphocyte precursors may exist in the marrow at hatching, and for a short time thereafter, but soon disappear.

In the adult chicken, bone marrow lymphocyte numbers increase with advancing age. These are small, postmitotic cells, and few or no immature forms are present (Glick and Rosse, 1976). These data contrast with observations made on mammalian marrow and argue that B lymphopoiesis in the aves is extramyeloid after hatching.

Other Anatomic Sites of B Lymphopoiesis

Glick was one of the first to propose that the bursa was not the only site of B lymphocyte differentiation in birds (Lerner et al., 1971). This was based on a number of studies in which precipitating antibodies were detectable in bursectomized chickens (Glick, 1983). Other investigators have demonstrated both Ig production and B lymphocytes expressing cell surface IgM, IgG, and IgA (Jalkanen et al., 1983b) in chickens after bursectomy at 60 hours of incubation, well before the appearance of bursal follicles. However specific immune responses to a panel of antigens were undetectable in birds bursectomized early in development (Jalkanen et al., 1983a). The significance of this observation has been clarified by the studies of the role of the bursa of Fabricius in antibody repertoire diver-sification discussed above. It is an inescapable conclusion that chickens deprived of their bursae early in embryonic development are still capable of producing B cells and immu-noglobulins, and that Ig isotype switching at the B cell level still occurs. Whether extra-bursal sites of B lymphocyte generation and differentiation occur normally or whether they are only active in the absence of bursa development is unknown.

Cells with membrane IgM have been reported in the yolk sac as early as 3 days of embryonation (LeBacq and Ritter, 1979); however, these observations have not been confirmed by another laboratory (Grossi et al., 1977). In addition, cells with the Cμ^+ sIgM$^-$ phenotype described in mammalian development were reported in the embryonic pronephros of the chicken between days 11 and 18 of incubation (Zettergren et al., 1980) similar to developmental observations on leopard frogs (Zettergren, 1982). The contri-bution of these apparently early B-lineage cells to bursa colonization or as accessory sites of B cell generation has not been determined.

Evolutionary Perspective

The avian bursa of Fabricius is in many ways unique among the vertebrates. It is only in avian species that B lymphopoiesis occurs in a unique anatomic location, in marked contrast to the situation in amphibians and mammals. Bursal structures have not been identified in reptiles, although the presence of organized patches of lymphoid cells associated with the urinary bladder and cloaca has been reported in alligators, turtles, and lizards (Cohen, 1977). Crocodilians are the nearest living reptilian relatives of birds, and the role of cloacal lymphopoiesis should be documented in these species.

If the bursa is indeed the principal anatomic site of B-lineage commitment in birds, this appears to have arisen as an experiment of nature in the aves as they evolved from their diapsid reptilian ancestors. According to this view, the bursa need not necessarily be present in mammals, which evolved early from primitive synapsid reptiles, well before and distinct from the evolution of avian species.

Class Mammalia

Advances in our understanding of mammalian B lymphocyte development have come primarily from two distinctly different lineages of investigators and subsequent convergence of their studies. One major school of thought, influenced by Glick's observations on the central role of the bursa of Fabricius in avaian B cell development, attempted to define equivalent anatomic sites in mammals. For nearly 10 years, the search for a mammalian "bursal equivalent tissue" was pursued by many independent laboratories. Despite a vast amount of work, however, no such unique anatomic site was identified (Owen et al., 1974). Nevertheless, this approach did lead to the finding that the earliest cells committed to the B lineage could be identified in the fetal liver of mice (Raff et al., 1976) and humans (Gathings et al., 1977) based on their expression of cytoplasmic μ chains of IgM.

The second major group of investigators was influenced by Yoffey and his detailed morphological studies of lymphocytes and lymphatic tissues (Yoffey 1931, 1933, 1977, 1981). Yoffey's studies of large and small lymphoid cells in the marrow led him to hypothesize that small lymphocytes which originated in extramyeloid sites subsequently emigrated to the marrow, where they enlarged and served as stem cells for other blood cell lineages. These enlarging lymphoid cells in the marrow compartment were termed transitional cells. Osmond and Everett who were, respectively, a student and a colleague of Yoffey's, were the first to demonstrate experimentally that small lymphocytes were continually produced in the mammalian bone marrow from larger lymphoid cells (Osmond and Everett, 1964)—quite the opposite of Yoffey's original hypothesis. These authors also pioneered the use of in vivo tritiated thymidine administration followed by autoradiography and microscopic observation of individual cells, an approach that proved to be particularly valuable for disecting the sequence of events in mammalian lymphopoiesis. This technique was also used in experiments which firmly established that the bone marrow was the primary site of B lymphocyte production in postnatal mammals (Osmond and Nossal, 1974; Rosse et al., 1978; Ryser and Vassalli, 1974).

Assays for B Lymphocyte Precursors

Before considering the developmental sequences that have been proposed for B-lineage cells in mammalian embryos (Kincade, 1987), it is appropriate to review the principles that underlie some of the more commonly used assays for these cells. Assays for B lymphocyte precursors can be divided into two main groups: those that rely on morphological criteria or phenotypic identification of cells based on binding of antibodies to lineage-related antigens, and those that assay the ability of particular cell populations to generate functional B lymphocytes. Observations from both types of assays have yielded essential information about the development of B-lineage cells, and particular insights have been gained when the two approaches are combined (Landreth et al., 1983). Unfortunately, in many cases, the same or similar terminologies have been used for cell populations that are defined using different criteria and which are functionally quite distinct. This is seen most clearly with the terms "pre-B cell" and "progenitor cell" as they relate to B lymphocyte differentiation. In the following sections, definitions will follow conventional hemopoietic terminology as closely as possible. Particular care will also be taken to distinguish between functionally and phenotypically defined cell populations, except where there is good evidence that they are equivalent.

Pre-B cells
The term pre-B cell was originally applied to the populations of large cells in mouse bone marrow that require extended residence in an irradiated host after adoptive transfer before giving rise to specific immune responses (LaFleur et al., 1971, 1972). These cells have since been shown to be sIg$^+$ and are in many respects equivalent to the cell population described by Shortman and Howard (1979) as "pre-progenitor cells." Neither of these cell populations fits our current definition of B cell precursors, however, because they should, by definition, lack sIg expression.

In 1976, Raff, Megson, Owen, and Cooper (Raff et al., 1976) described the presence of cells in mouse fetal liver which contained cytoplasmic μ chains of IgM in the absence of detectable light chain synthesis or expression of cell surface Ig (Fig. 11–2). These cells could be detected several days before the appearance of sIgM$^+$ cells in the same tissues (Owen, 1979). They were therefore termed "pre-B" cells, based on their appearance early in embryonic development and on the results of a number of cell repopulation and kinetic studies (Landreth et al., 1981; Opstelten and Osmond, 1983; Osmond and Owen, 1984). Similar cell populations were subsequently found in postnatal marrow (Cooper, 1981; Owen et al., 1977).

In this chapter, the term "pre-B" cell is reserved for cells that are the immediate precursors of B cells. The weight of evidence from a number of laboratories suggests that the $c\mu^+$ sIg$^-$ cells, which can be detected only in established sites of primary B lymphopoiesis, correspond to pre-B cells, and we will follow this definition. However, until these cells have been enriched to homogeneity and directly shown to give rise to B cells, some degree of scepticism is both healthy and appropriate.

Cell surface antigens
Pre-B cells can be detected by fluorescent antibody techniques only after centrifugal disruption and alcohol fixation, which is required for staining of cytoplasmic μ chains. As an alternative approach, which would preserve cell viability and thus allow functional assay of pre-B cell activity, several laboratories have produced monoclonal antibodies that recognize cell surface antigens expressed on pre-B cells (Greaves, 1986; Kincade, 1987). Among the most useful of these monoclonal antibodies are those that bind to the 220 kD component of the Ly-5 antigen system of the mouse (B220) (Kincade, 1987; Kincade et al., 1981; Park and Osmond, 1987; Scheid et al., 1982). These antibodies bind

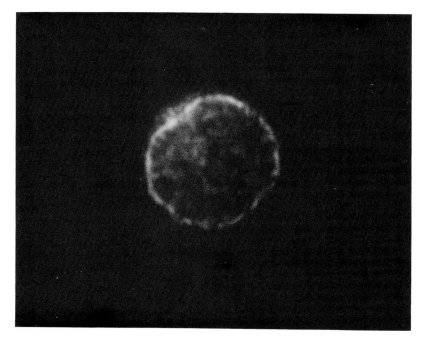

Figure 11–2.   Human pre-B cell. Cell surface Ig⁻ human bone marrow cells were cytocentrifuged, fixed, and stained with TRITC-anti-human IgM to detect cytoplasmic Ig.

preferentially to B-lineage cells and have had tremendous importance in functional studies. One of these monoclonal antibodies, designated 14.8 (Kincade et al., 1981b), has been used extensively in studies of B cell development in the fetal mouse.

It should be noted that expression of the 14.8 antigen (or determinant) has been reported on a subpopulation of peripheral T cells (Scheid et al., 1982). Conversely, pre-B cells in early embryos and some B-lineage lymphocytes grown in culture for prolonged periods lack the 14.8 epitope (Kincade et al., 1981b; Medlock et al., 1984; Velardi and Cooper, 1984; Witte et al., 1986). This does not diminish the usefulness of antibodies such as 14.8, but it does emphasize the importance of establishing a composite phenotype in assigning cell lineage. This is illustrated particularly well in autoimmune mice of the MRL lpr/lpr strain, in which a population of B220–positive T cells mediates the impressive lymphadenopathy observed in these animals (Scheid et al., 1982).

Functional assays
Functionally, B lymphocyte precursors can be identified by their ability to generate B cells from sIg⁻ cells in vivo (Burrows et al., 1978; Landreth et al., 1983; Paige et al., 1979, 1981) or in vitro (Kincade et al., 1981a; Landreth and Kincade, 1984). Pre-B cells, by definition, should quickly give rise to B cells in vitro under permissive culture conditions. B cells can be generated from fetal liver in organ fragment cultures (Owen et al., 1977), in dispersed cell liquid cultures (Hammerling et al., 1976; Kincade et al., 1981a; Lau et al., 1979; Melchers, 1977; Osmond and Nossal, 1974; Stocker, 1977; Whitlock et al., 1985), and in limiting dilution agar assays (Paige et al., 1984, 1985). In vitro assays are inherently more susceptible to experimental manipulation than are cell transfer studies. For this reason, the initial advances in our understanding of regulatory events in B cell differentiation

have come mainly from studies of cultured cells (Giri et al., 1984; Jyonouchi et al., 1983; Landreth et al., 1984a, 1985; Paige et al., 1985).

Adoptive transfer of B cell precursors in vivo into lethally irradiated or immunodeficient hosts apparently allows developmentally more primitive cells to differentiate into mature B cells than do in vitro assays (Landreth et al., 1984b). However, unless the transferred cell populations are homogeneous, it is difficult to interpret precursor-progeny relationships. Studies of regulatory events are even more difficult to interpret when in vivo adoptive transfer systems are used. This is particularly true when cells are transferred into irradiated recipients; under these conditions abscopal effects of irradiation on the host environment cannot be predicted.

In the following section, observations from phenotypic observations and functional studies are drawn together to produce a plausible model for the development of B-lineage cells during ontogeny. Figure 11–3 summarizes the temporal sequence of B cell ontogeny in the mouse and should help clarify the progression of events in this process.

Embryonic Emergence of B-Lineage Cells

As in birds, the earliest detectable hemopoietic tissue in mammals is the yolk sac. Yolk sac hemopoiesis can initially be detected in discrete blood islands by day 7 of gestation in the mouse and has virtually disappeared by day 13 (Moore and Metcalf, 1970; Moore and Owen, 1967b; Tavassoli and Yoffey, 1983). Although the yolk sac contains stem cells that can give rise to spleen colony forming units (CFU-S), granulocyte-macrophage progenitors (CFU-GM), or mixed colonies of granulocytes, erythrocytes, megakaryocytes,

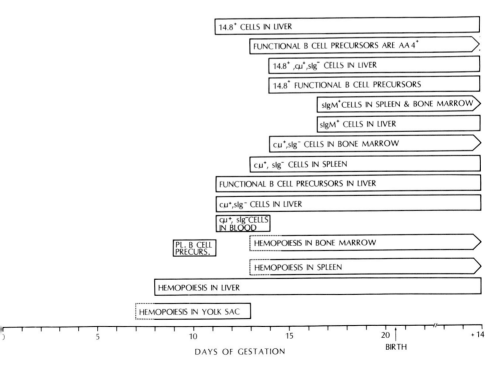

Figure 11–3.   Embryonic events in mouse B lymphopoiesis. (Pl: placenta.)

and macrophages in vitro, yolk sac cultures cannot be shown to contain pre-B cells as defined phenotypically or on the basis of vitro assays (Paige et al., 1979, 1981). However, yolk sac cells from 8–9-day gestation embryos can give rise to mature B lymphocytes when they are transferred to immunodeficient mice in vivo. Development of B cells required up to 8 weeks after adoptive transfer, suggesting that considerable proliferation and many cells interactions were involved. These results, however, are consistent with the hypothesis that hemopoietic stem cells present in the yolk sac of mammals may be truly pluripotent and may give rise to B lymphocytes, as well as to cells of myeloid lineages.

The mouse fetal liver contains accumulations of hemopoietic cells by day 8 of gestation, and this tissue remains a major site of hemopoiesis until about the time of birth, after which the blood-forming cells rapidly disappear. The earliest cells that synthesize detectable Ig (thus confirming their commitment to B-lineage development) can be found in low numbers in the fetal liver and blood of mice at 11 days of gestation (Andrew and Owen, 1978) and then increase to easily detectable numbers by day 13 (Velardi and Cooper, 1984). Cells that contain cytoplasmic μ chains of IgM in the absence of detectable light chain synthesis or expression of cell surface Ig are considered pre-B cells, as discussed earlier in this chapter.

Functionally, B-lineage cells appear to be present in the 9–11-day placenta and blood in that these tissues can give rise to B cells in culture (Antoine et al., 1979; Melchers, 1979), although they do not contain identifiable pre-B cells. This observation suggests that the fetal liver may not be the first embryonic site in which commitment to the B lineage occurs, and thus that it is not a "bursal equivalent tissue." However, it is also possible that cells that can mature along any of several hemopoietic pathways are able to develop in culture under the conditions provided.

In the liver, Ig⁻ cells, which can give rise to B cells, are found in organ explant cultures beginning at day 12 of gestation (Owen et al., 1977) and in cultures of dispersed fetal liver cells after day 14 (Kincade et al., 1981a). Maturation of 14-day fetal liver cells to express sIg and respond to lipopolysaccharide (LPS) in culture was delayed by 2 days, compared to 16-day fetal liver cells, suggesting that sIg expression is in some way intrinsically timed. B-lineage precursors—which could be induced to differentiate in agar cultures, to express sIg, to proliferate in response to the mitogen LPS, and to secrete antibody—were present as early as day 12 in fetal liver (Paige et al., 1984, 1985). The agar culture system used in these studies depends on the presence of fetal liver–adherent cells and detects progeny only after a secondary round of proliferative activity. Although indirect, this assay does allow study of the clonal progeny of developmentally early cells and is a promising approach to dissecting the regulatory events essential to B lymphocyte development. Fetal liver cells from 14-day embryos have also been shown to give rise to B lymphocytes following cell transfer in vivo, but B cells can be detected only after 3–4 weeks' residence in unirradiated, immunodeficient hosts (Paige et al., 1979). The difference between the time required to generate B cells in culture or in vivo after cell transfer suggests that different types of cells may give rise to B cells under these conditions. Alternatively, microenvironmental or other factors may be of critical importance in B cell development. It does seem certain, however, that the cells that give rise to B lymphocytes after cell transfer in vivo are distinct from myeloid precursors, as defined either by the splenic colony forming assay (CFU-S) of Till and McCullough (Curry and Menton, 1967) or on the basis of cell size and surface phenotype (Paige et al., 1981).

In mouse fetal liver, the 14.8 monoclonal antibody can be shown to bind to a small population of lymphoid cells by day 11 of gestation (Medlock et al., 1984). No 14.8$^+$ cells could be identified in any embryonic tissue before this time. Cells from 16-day fetal liver that could give rise to B cells in liquid cultures were 14.8$^+$ (Kincade et al., 1981a; Landreth et al., 1983), as were those cells from 14-day fetal liver that could be detected in the agar culture assay system described previously (Paige et al., 1984, 1985). Pre-B cells from 11–13-day fetal liver were 14.8$^-$, but this antigen was progressively expressed over the next 3 days so that by day 16, virtually all pre-B cells in fetal liver were 14.8$^+$ (Velardi and Cooper, 1984). Cells from 12–13-day fetal liver that can give rise to B cells in the agar culture assay were also 14.8$^-$, while pre-B cells from 19-day fetal liver were virtually all 14.8$^+$.

It is of interest that the determinant detected by 14.8 (and by other monoclonal antibodies to B220) is expressed on a considerably higher proportion of cells than detectable sIg. As described below, the majority of 14.8$^+$, sIg$^-$ cells in 16-day fetal liver have undergone Ig gene rearrangements, which suggests that they are capable of further development. However, the earliest detectable pre-B cells in embryonic development do not bind the 14.8 antibody. This latter observation suggests that B220 is expressed later than cμ in ontogeny. Considerable information has now accumulated on the genomic and protein structure of the Ly-5 (CD45) leukocyte common antigen family (Saga et al., 1986; Thomas et al., 1985). Multiple mRNA species are transcribed from a single large gene, and expression of particular Ly-5 molecules is regulated in a tissue-specific fashion. The functions of this family of transmembrane glycoproteins are not known, but their importance is suggested by substantial evolutionary conservation. It is possible that the observed variations in 14.8 binding to B-lineage cells during embryonic development reflect the expression of mulitple species of Ly-5 molecules or the existence of different pre-B cell subsets. Alternatively, post-transcriptional or post-translational modifications of the Ly-5 molecule may be important in the differentiation of pre-B cells. It is clear from all these data that the significance of Ly-5 expression on B-lineage cells has not yet been entirely clarified.

Other developmentally related cell surface antigens are expressed on B-lineage precursors. Lyb-2 is expressed on B-lineage precursors that respond in the agar assay, beginning at day 15 of gestation—one day later than 14.8 binding (Paige et al., 1985). Lyb-2 appears to be restricted to B-lineage cells (Kincade, 1981), although its functional importance remains unresolved. Monoclonal antibody AA4 (McKearn et al., 1984) binds to a cell surface molecule expressed on a number of hemopoietic cell types of fetal liver, and it can be detected on cells that generate B cells in the agar precursor assay, beginning at day 13 of gestation (Paige et al., 1985). Although this latter antibody does not recognize only lymphoid cells in the developing liver, the fact that it recognizes functional B lymphocyte precursors earlier in development than 14.8 gives it considerable utility (Cumano and Paige, 1992).

The developing spleen and bone marrow are active sites of hemopoiesis by day 13 of fetal life, and hemopoiesis persists in these two tissues throughout life (Tavassoli and Yoffey, 1983). Phenotypically defined pre-B cells can be found in the developing spleen by 13 days of gestation and in the marrow at day 14 (Velardi and Cooper, 1984). These cells are large (10–18 μm in cytocentrifuge preparations) and often have a lobulated nuclear outline (Landreth et al., 1983). Cells expressing membrane IgM appear simultaneously in the liver, spleen, and bone marrow between days 16 and 17 of gestation (Dorshkind, 1986; Raff et al., 1970), but they remain low in number and frequency until approximately the time of birth.

In humans, pre-B cells first appear in the fetal liver during the 8th week of gestation and in the marrow during week 12 (Gathings et al., 1977). Cells expressing sIg are found in the liver during the 9th embryonic week.

Postnatal Generation of B Cells in the Bone Marrow

The role of bone marrow as a site of blood cell formation was not appreciated until the pioneering studies of Bizzozero in the latter half of the 19th century (Bizzozero, 1868). Yoffey (1931, 1933), documented the presence of lymphocytes in the marrow and proposed that these cells were produced in the lymph nodes and gut-associated tissues and subsequently emigrated to the marrow from the blood. In this view, small lymphocytes underwent a process of maturation in the marrow, passing through a series of intermediate larger forms before being transformed into proerythroblasts or myeloblasts. Lymphocytes were therefore circulating stem cells for the hemopoietic system, and the intermediate forms in the marrow were termed transitional cells (Rosse, 1976; Yoffey, 1981).

In 1964, Osmond and Everett published radioautographic data, which showed that just the opposite was true. The majority of lymphocytes found in the bone marrow of the rat were recently generated postmitotic cells that had arisen from dividing "transitional cells." These results have subsequently been confirmed by a number of studies in rats, guinea pigs, and mice (Rosse, 1976). Transitional cells also include progenitors for erythroid cells and probably also progenitor cells for other hemopoetic lineages.

Lymphoid cells make up approximately 25% of nucleated cells in the marrow of adult mice, half of these being small lymphocytes. The majority of these cells are B lymphocytes that arise over a period of 1–2 days after the last cell division from a pool of postmitotic sIg$^-$ small lymphocytes (Osmond and Nossal, 1974; Ryser and Vassalli, 1974). B cells leave the marrow without respect to postmitotic age, entering venous sinusoids from which they are transported to the peripheral lymphoid tissues (Brahim and Osmond, 1973).

The marrow is also the sole source of pre-B cells during postnatal life in mice, rabbits, and humans (Cooper, 1981). In mice, bone marrow pre-B cells can be divided into large (10–18 μm) and small (6–9 μm) populations in cytocentrifuge preparations (Landreth et al., 1981; Osmond and Owen, 1984). Large pre-B cells have been shown to be a rapidly dividing population on the basis of in vitro (Owen et al., 1977), normal steady-state (Landreth et al., 1981), and stathmokinetic (Opstelten and Osmond, 1983) studies. As predicted on the basis of earlier studies, small pre-B cells are mitotically quiescent small lymphocytes.

Kinetic studies of pre-B cells generally support the hypothesis that large pre-B cells give rise to small postmitotic pre-B cells in the marrow by cell division, and that small pre-B cells gradually come to express sIg over a period of 1–2 days (Landreth et al., 1981; Osmond and Nossal, 1974). A summary of our current concepts of B lymphopoiesis in postnatal mouse marrow is presented in Figure 11–4. The relative placement of large and small pre-B cells in this scheme is based on studies from several laboratories. However, the possibility that cells which do not express cμ chains also contribute to B cell generation in the marrow has not been formally excluded. This point must be considered in light of functional studies which indicate that all pre-B cells in the marrow are 14.8$^+$ (Kincade et al., 1981b; Landreth et al., 1981). Marrow 14.8$^+$ sIg$^-$ cells are exclusively lymphoid and include all cμ^+ cells as well as cells without detectable Ig expression (Landreth et al., 1981). It therefore remains possible that both cμ^+ 14.8$^+$ and cμ^- 14.8$^+$ cells correspond

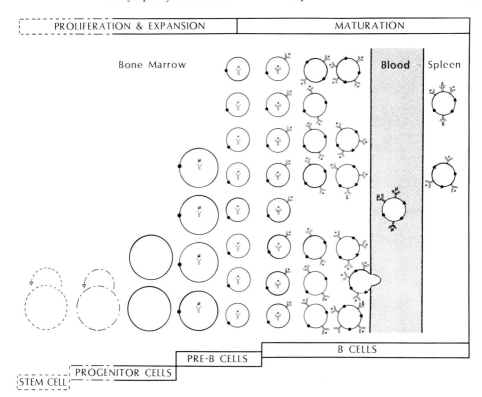

Figure 11-4. Murine bone marrow B lymphopoiesis. After an unknown number of cell divisions, committed B-lineage progenitor cells give rise to large pre-B cells that express B220 (circles) on their cell filled surfaces and cytoplasmic μ heavy chains of IgM. Large pre-B cells appear to be the final cell division in the lineage (see text). Small pre-B cells mature to express cell surface IgM, La (squares), Fc receptors (triangles), and IgD without cell division. During the final stages of this maturation process, B cells are randomly launched into the venous sinusoids of the marrow, and then migrate to the spleen.

to functional pre-B cells and may be developmentally related. Studies of the cell kinetics of 14.8^+ $c\mu^-$ cells in bone marrow suggest that this is, in fact, the case (Park and Osmond, 1987).

The scheme presented here also summarizes our limited knowledge of the role of proliferation and clonal expansion in development of the B-cell lineage. It appears from the available data that, although large pre-B cells divide rapidly, they also have a rapid rate of turnover. As indicated in Figure 11-4, pre-B cells are thought to divide only once on average before maturing to sIg expression (Landreth et al, 1981). This conclusion is also supported by studies of the kinetics of recovery of large and small pre-B cells in marrow after their ablation with cyclophosphamide or irradiation (Osmond and Owen, 1984). Following either of these myeloablative treatments, small pre-B cells reappear within 24 hours after the appearance of large pre-B cells in the marrow. Reaccumulation of B cells is delayed for several days after marrow ablation, although this probably relates more to the rate at which cells exit from the marrow than to the time required for post-mitotic maturation. Premature exit of B cells from the marrow under conditions of hemopoietic stress would also account for the appearance of pre-B cells in the spleens of cyclophosphamide-treated adult mice. The delayed appearance of B cells in the fetal liver, 5 days after pre-B cells are detected, may reflect patterns of cell migration in the embryo. Alternatively, embryonic events in B-lineage differentiation may be substantially different from those in postnatal marrow (Kincade, 1981).

Large pre-B cells do not normally self-renew and appear to be generated from proliferating B-lineage progenitor cells (Landreth and Kincade, 1984; Landreth et al., 1981). These same studies suggest that any substantial clonal expansion of cells with rearranged heavy chain genes must take place before the translation of detectable Ig protein. Although evidence for clonal expansion of B-lineage cells with restricted antigen binding capability has been presented (Klinman et al., 1966; Klinman and Stone, 1983), it is important to know the size of individual repertoire restricted clones of cells as a basis for understanding events in the progression of an immune response. Expansion of clones of cells that have rearranged Ig heavy chain genes must therefore occur at the level of B progenitor cells. The existence of these cells has been inferred in a number of studies (Landreth et al., 1981), and recent observations from several laboratories suggest that these progenitors are B220 negative and have not completed Ig gene rearrangement (Cumano and Paige, 1992; Hardy et al., 1991; Rolink et al., 1991). These progenitor cells do not appear to bear the determinant detected by monoclonal antibody 14.8, and critical studies of their proliferative characteristics in vivo must therefore await methods of recognizing or enriching for them.

Human B-lineage cells progress through a similar series of differentiative steps as murine cells (Pearl et al., 1978). The expression of cell surface antigens differs somewhat on human cells and has been extensively reviewed (Melink and Lebein, 1983). In particular, Ia molecules are expressed on the cell surface of human pre-B cells, appearing well before sIg (Cooper, 1981). Human small pre-B cells also pass through an additional differentiation stage in the marrow, when they express both heavy and light chains of $c\mu$ for a period before the appearance of sIg. A proportion of small pre-B cells in rabbits also contains detectable Ig light chains before sIg expression (Gathings et al., 1982).

Relationship of B Lymphoid Development to Hemopoietic Stem Cells

The relationship of B lymphopoiesis to the development of other cell lineages in the marrow, particularly of pluripotent hemopoietic stem cells, is both poorly understood and controversial. The existence of cells that are capable of giving rise to cells of both myeloid and lymphoid lineages after transfer to irradiated recipients has been demonstrated (Wu et al., 1968), but the contribution of these cells to the normal production of B cells is unknown. A number of different techniques, many relying on physical differences in bone marrow cells, including buoyant density and size as well as cell surface marker expression, have been used to enrich for multipotential stem cells (Müller-Sieburg et al., 1986, 1988; Visser et al., 1984). More recently, multiparameter cell sorting has been used to obtain virtually pure pluripotential stem cells (Spangrude et al., 1988). This should allow better definition of early differentiation stages of B lymphocytes, particularly in conjunction with defined stromal cell populations and defined cytokines. In addition, use of these techniques should allow elucidation of branched and/or parallel lineages of B cell differentiation (Hardy and Hayakawa, 1986; Herzenberg et al., 1986). In addition, W-anemic mice, which have deficits both of hemopoietic stem cells and myeloid-committed progenitors (McCulloch et al., 1964, Russell, 1979), have apparently normal numbers of pre-B and B cells (Landreth et al., 1984b; Mekori and Phillips, 1969).

There are even fewer data to support the existence of lymphoid stem cells or of stem cells committed to the B lymphoid lineage. However, it appears likely that such cells do exist as an intermediate stage between pluripotential hemopoietic stem cells and pre-B cells, at least during embryonic development. These technical barriers to studies of B lymphocyte development are no doubt temporary, and the development of techniques for long-term culture of lymphoid cells from the marrow (Whitlock et al., 1974; Whitlock and Witte, 1982) may help resolve questions about existence of lymphoid stem cells and their relationship to stem cells of other developing hemopoietic lineages.

Components of the Lymphopoietic Microenvironment

One of the goals of studies of B lymphopoiesis must be to define the cell compartments in the development of this lineage, to determine their proliferative kinetics, and to understand the regulatory mechanisms that affect differentiation. In vivo studies have been informative in these respects (Paige et al., 1979, 1981), but the development of clonal in vitro assays for hemopoietic cells has been critical to studies of commitment and regulation of defined hemopoietic cell populations. Assays in which cells are dispersed in agar do not replicate the anatomical relationships between hemopoietic cells and stromal elements that are found in vivo, and they preclude cell–cell contact that may be of critical importance in normal cellular development.

One approach to dissecting anatomical relationships and microenvironmental cell types has been through the establishment of long-term bone marrow cultures, as pioneered by Dexter (Dexter et al., 1977; Dexter and Testa, 1980). Bone marrow cells were allowed to establish an adherent "stromal layer" in culture, which then supported the generation of hemopoietic cells for extended periods of time. These hemopoietic cells were shown to require stromal cell contacts to allow differentiation (Dexter et al., 1977). Several groups showed that B cells and pre-B cells rapidly disappeared from these cultures, most likely as the culture conditions had been optimized to support the growth of myeloid cells (Aspinall and Owen, 1983; Dorshkind and Phillips, 1983). However, cells contained in the adherent layers were found to give rise to B cells after in vivo transfer into irradiated recipients, albeit with long latency (Dorshkind, 1986; Dorshkind and Phillips, 1983).

Whitlock and Witte developed modifications of this culture technique that allowed growth of lymphoid cells in long-term marrow cultures (Whitlock et al., 1984; Whitlock and Witte, 1982), and a number of technical advances have since been described (Dorshkind, 1986; Dorshkind et al., 1986; Whitlock et al., 1985). The cultures are in many ways similar to the original Dexter culture system. A suspension of bone marrow cells is allowed to establish an adherent stromal layer (Dorshkind et al., 1985). Lymphoid cells subsequently reappear, presumably from a limited number of surviving cells with considerable expansion potential. These lymphoid cells are heterogeneous in size and contain both pre-B and B cells. It is not yet clear whether these cultures contain stem cells or progenitors for the B lineage, which can repopulate the lymphoid system in vivo, but both nonadherent and adherent cells from these cultures can be shown to give rise to B cells in vivo (Kurland et al., 1984; Phillips et al., 1984).

The stromal cell layer in primary long-term bone marrow cultures is complex and includes macrophages, endothelial cells, adipocytes, and fibroblastic cells, as well as associated lymphohemopoietic cells (Fig. 11–5) (Allen and Dexter, 1984; Dorshkind et al., 1985; Kincade, 1987; Witte et al., 1987). It is therefore similar in composition to the nor-

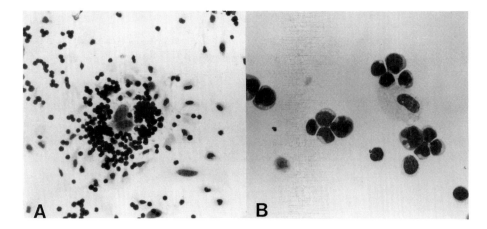

Figure 11–5.   Long-term cultures of bond marrow lymphoid cells: (**A**) Association of developing lymphoid cells with an adherent bone marrow stromal cell; (B) Higher magnification photomicrograph of the cells generated in these cultures. (Courtesy of Dr. P. Witte.)

mal adult bone marrow stroma, which is heterogeneous and of mesenchymal origin (Perkins and Fleischman, 1988; Tavassoli and Crosby, 1970).

In an attempt to simplify these culture systems, a number of investigators have derived clonal cell lines from adherent stromal populations (reviewed in Dorshkind, 1990). These cell lines can be subcloned repeatedly, and the resulting homogeneous populations can be used to define specific functions of the microenvironment. Stromal cell lines that support lymphopoiesis have a relatively consistent morphological appearance, with a large oval nucleus and highly spread cytoplasm of up to 150 μm in diameter.

Two laboratories showed independently that when Dexter-type long-term marrow cultures were changed to Witlock-Witte conditions, myeloid precursors rapidly disappeared and were replaced by lymphoid cells (Denis and Witte, 1986; Dorshkind, 1986). The newly generated lymphoid cells include pre-B and B cells, suggesting that the same stromal cell layer can support differentiation of both myeloid and lymphoid cells. Substantiation of this view has come from studies of stromal cell clones, which can sustain granulocytic, macrophage, and lymphocytic proliferation (Collins and Dorshkind, 1987; Hunt et al., 1987; Whitlock et al., 1987).

Regulation of B Lymphopoiesis

The last few years have brought about a considerable increase in our understanding of the hemopoietic regulatory interactions that affect the generation of B lymphocytes. Previously, the paucity of information about regulatory mechanisms for lymphopoiesis was in marked contrast to the abundance of studies that addressed regulation of myeloid and erythroid cell growth and differentiation. Many of the advances in this area have come about with the availability of cloned stromal cell lines as discussed above, and with the preparation of recombinant hemopoietic and lymphopoietic growth factors. Together, these approaches have facilitated elucidation of some of the regulatory factors that act on B lymphoid progenitors.

Sephadex G-10 adherent bone marrow cells are required to allow maturation of marrow pre-B cells to express sIg in vitro and to respond to mitogenic stimulation (Kincade et al., 1981a). These cells can be substituted by peritoneal macrophages, and it was therefore suggested that macrophages are essential regulatory cells. However, sustained in vitro B cell generation from marrow can be observed only in the presence of established marrow-adherent cells. Although initial experiments concluded that IL-1 was directly involved in sIg expression (Giri et al. 1984), more recent studies have shown that IL-1 acts through a cytokine cascade to stimulate IL-4 production (King et al. 1988). IL-4 appears to participate directly in sIg expression in the marrow (King et al., 1988; Simons and Zharhary, 1989).

A critical role for adherent cells in pre-B cell development has also been suggested by in vivo studies. Osmond's laboratory has reported that intraperitoneal injection of sheep red blood cells or other exogenous agents results in a transient increase in marrow pre-B cell numbers, but no alteration of their production rate (Fulop and Osmond, 1983a,b). This effect can be abrogated by in vivo treatment with silica, a known macrophage poison, or by splenectomy (Pietrangeli and Osmond, 1984). The apparent involvement of splenic macrophages in this process is intriguing, and further studies using in vitro techniques should help clarify the mechanisms involved. Certainly the observed effect of exogenous agents on pre-B cell generation may alter our view of primary B cell production as an antigen-independent process (Lawton and Cooper, 1977).

Adherent cells also appear to participate in embryonic B lymphopoiesis. Kincade and his colleagues (Kincade et al., 1981a) showed that maturation of fetal liver pre-B cells in culture could be enhanced by at least two kinds of regulatory cells that are present in adult mouse marrow. One of these regulatory cell types was shown to bind to Sephadex G-10. Gisler and colleagues (1984) also found that fetal liver B cell generation depended on the presence of an adherent stromal cell layer that was provided by bone marrow cells. B-lineage cells from 11–15-day fetal liver will mature in agar cultures to synthesize Ig in the presence of fetal liver–adherent layers (Paige et al., 1984, 1985). However, direct contact between these early embryonic precursors and the adherent cell layer is not required, suggesting that diffusible regulatory factors play a role in B-lineage development. Purification of these factors from fetal liver adherent cells has not been reported. Interestingly, several sources of myeloid colony stimulating factors (CSF) were found to potentiate the development of B cells in this agar culture system (Paige et al., 1985). Paige and colleagues showed that although these CSF-rich supernatants acted early in culture, their effects on pre-B cells were not direct and required multiple cell interactions. This conclusion was supported by the observation that B220 cells from mouse fetal liver could give rise to B cells in the presence of fetal liver adherent cells, but not when adherent cells were replaced by soluble CSF, including WEHI-3 supernatants that are known to contain IL-3.

A number of regulatory lymphokines are produced by cloned stromal cell populations under appropriate conditions. Virtually all stromal cells produce macrophage CSF (M-CSF), which can be detected using a variety of functional assays and mRNA analysis. Similarly, production of granulocyte-macrophage CSF (GM-CSF) and granulocyte CSF (G-CSF) by the majority of clones can be detected by bioassay and mRNA analysis (reviewed by Kincade et al., 1989). Interleukin 3 (IL-3) is a T cell–derived factor that can stimulate multipotential stem cells and more mature progenitors of several lineages, particularly in combination with IL-6 (Ikebuchi et al., 1987; Schrader, 1986). Although it might therefore be expected to play a role in hemopoiesis, clonal stromal cell lines have not been shown to secrete this factor, nor do long-term marrow culture supernatants contain this activity (Kincade et al., 1989).

Interleukin 7, a recently described cytokine, was isolated from a transformed marrow stromal cell line and produced in recombinant form (Namen et al., 1988a,b). It is a potent stimulator of growth of large pre-B cells, and IL-7 mRNA can be detected in some stromal cell clones. It may therefore play a significant role along with other lymphokines in inducing cell proliferation in long-term marrow cultures. The proliferative effect of IL-7 on pre-B cells is potentiated by the ligand for the *ckit* protooncogene (KL) (Billips et al., 1992; McNeice et al. 1991) or insulin-like growth factor 1 (IGF-1) (Landreth et al. 1992). Both of these molecules are produced by bone marrow stromal cell lines and are presumably essential regulatory components for in vivo B lymphopoiesis. It is also now clear that IL-7, KL, and IGF-1 act in vitro to expand a population of pro-B cells from fetal liver and bone marrow (Rolink et al. 1991; Petitte and Landreth, unpublished data). These data suggest an essential role for IL-7 in expansion of cells that are actively undergoing Ig gene rearrangement and repertoire diversification. Although both IGF-1 and KL act to potentiate the proliferative function of IL-7, only IGF-1 has been shown to participate in differentiation of these cells to $c\mu$ expression (Billips et al. 1992).

Other mediators produced by stromal cell clones include IL-6, which may enhance responses to IL-3 and other factors; transforming growth factor β (TGF-β) (Massague, 1987), neuroleukin (Gurney et al., 1986), and a unique factor that potentiated pre-B cell generation (Landreth and Dorshkind, 1988). TGF-β has recently been shown to be a potent inhibitor of lymphopoiesis and hemopoiesis (Lee et al., 1987). In addition to direct effects of cytokines on hemopoietic and lymphopoietic cells, other regulatory effects may be mediated through stromal cells themselves (Billips et al., 1990). Hemopoiesis is directly affected by infection and/or inflammation (Metcalf, 1971). This is seen most dramatically in the conversion of yellow to red bone marrow under conditions of hemopoietic stress (Kincade et al., 1989). The existence of paracrine regulatory networks involving stromal cells has therefore been proposed to account for at least some hemopoietic regulatory mechanisms (Billips et al. 1991).

Endotoxin can induce release of IL-1 and tumor necrosis factor (TNF) by macrophages (Thorens et al., 1987), and, in turn, TNF can stimulate the production of IL-1 by endothelial cells (Nawroth et al., 1986). Macrophages and endothelial cells can themselves synthesize various hemopoietic regulatory factors, including G-CSF, M-CSF, and GM-CSF (Broudy et al., 1986; Segal et al., 1987; Thorens et al., 1987). Bone marrow–derived stromal cell clones have been shown to synthesize a wide range of mediators either constitutively or after appropriate stimulation. These factors include M-CSF, G-CSF, GM-CSF, Il-6, IL-7, and a pre-B cell growth factor. In addition, these cells appear to express receptors for a wide variety of cytokines, including IL-6, TGF-β, Il-4, Il-7, TNF, and IL-1. Activated T lymphocytes and macrophages are known to synthesize a range of cytokines, which could potentially act on bone marrow stromal cells, including Il-4, Il-6, IFN-γ, myeloid CSFs, TNF, Il-1, and TGF-β (Miyajima et al., 1988). These data thus suggest at least a theoretical basis for paracrine regulation of stromal cell function.

Data from a number of studies have suggested that myelopoiesis and lymphopoiesis may be coordinately regulated. This is seen most clearly in the dysregulation of B lymphocyte production, which may be associated with cyclic neutropenia. In one study, factors derived from the urine of a patient with cyclic neutropenia were found to potentiate the generation of pre-B and B cells in cultures of human marrow (Landreth et al., 1985). These studies were based on observations made on the patient's marrow, which showed cyclical phases of exaggerated pre-B cell production that correlated with periods of absolute neutropenia (Engelhard et al., 1983). The active components of these urinary preparations also stimulated pre-B cell generation in cultures of normal mouse marrow (Landreth et al., 1984a). Two distinct activities were identified in the patient's urine on the basis of gel filtration analysis. These putative hemopoietic factors had apparent molecular mass of 15–17 and 45–47 kD (K. S. Landreth, M. Beare, and P. W. Kincade, unpublished data). Comparison of these data with studies of murine B cell differentiation suggest that this biological activity is due to IGF-1; however, this has not been formally demonstrated.

The identification of these regulatory factors under conditions of abnormal hemopoiesis and lymphopoiesis adds directly to our understanding of the differential regulation of hemopoiesis and may contribute to our management of a number of bone marrow disorders, as well as to the recovery of bone marrow function after transplantation. Based on the results discussed above, it seems reasonable to propose that the factors and regulatory cells described are elements of a regulatory cascade analogous to those described for other hemopoietic lineages. Development of a complete model of B lymphopoiesis that includes regulatory networks must await additional information. However, the recent technical advances discussed here should provide a framework for further study.

Cellular Aspects of Ig Gene Rearrangements

One of the most important advances in our understanding of B-lineage differentiation has been the demonstration that Ig genes are organized in discrete segments that must be rearranged from their germline configuration to be functionally transcribed (Honjo et al., 1981). The details of this process are beyond the scope of this review and are detailed elsewhere in this volume. However, some studies of normal cells have been described, and these results should be reviewed here.

Genetic rearrangements involving the variable (V), diversity (D), and joining (J) heavy chain gene clusters and V and J light chain gene clusters must precede transcription, translation, and expressin of complete Ig molecules (Alt et al., 1986; Hunkapiller and Hood, 1989). The stages through which B-lineage cells pass as they differentiate from uncommitted cells to cells expressing Ig therefore constitute a cellular "theater" in which these gene rearrangements progress. Of particular importance to our discussion is the definition of the cell populations in which specific Ig gene rearrangements occur and the relationship between these cell stages and B-lineage commitment, stem cells, clonal development, and expression of antibody repertoire. No less importantly, these observations may help us address the question of whether gene rearrangements occur only during embryonic development or are also a continuing process in postnatal marrow (Coffman and Weissman, 1983; Honjo et al., 1981).

Cells in 16-day fetal liver that express the B220 antigen contain $c\mu^+$ cells only at low frequency, although approximately 70% of Ig genes have undergone DJ_H rearrangements in these cells (P. W. Kincade, K. S. Landreth, G. Lee, and T. Kishimoto, unpublished data). Large $B220^+$ sIg^- cells from mouse marrow, detected using an antibody with almost the same reactivity as 14.8, have been shown to have rearranged virtually all of their J_H genes (Coffman and Weissman, 1983). These data suggest that DJ_H rearrangements precede the expression of B200, and that most fetal liver and marrow cells with the 14.8^+ $c\mu^-$ sIg^- phenotype have rearranged J_H gene segments. As discussed above, however, the contribution of these cells to B lymphocyte generation has not been entirely confirmed.

The sequence of heavy chain gene rearrangements has been proposed to begin with DJ_H followed by V_HDJ_H rearrangements (Sugiyama et al., 1983), and some fetal liver pre-B cell lines have been found to express a truncated C-region protein that lacks a V-region gene product (Reth and Alt, 1984). Abelson virus–derived pre-B cells lines that express a truncated μ protein on the cell surface also contain $c\mu$ as detected by immunofluorescence (C. J. Paige and K. S. Landreth, unpublished data). It has not been possible to define the proportion of large B220$^+$ cells in either normal marrow or fetal liver that lacks V-region gene rearrangements. Small B220$^+$ sIg$^-$ cells generally contain completely rearranged J_H gene segments, whereas their light chain J_κ genes are only partially rearranged. Conversely, large cells of the same phenotype lack detectable light chain gene rearrangements (Coffman and Weissman, 1983), suggesting that light chain gene rearrangement is an ongoing process in small pre-B cells.

These observations are consistent with previous cellular studies, which have shown that Ig heavy chain expression precedes light chain gene expression in pre-B cells (Burrows et al., 1979), and with molecular studies of Abelson virus–derived pre-B cell lines that contain rearranged heavy, but not light, chain genes (Maki et al., 1980; Reth and Alt, 1984; Sugiyama et al., 1982). Future studies should resolve whether rearrangement of Ig genes is integrally related to the regulatory mechanisms discussed above or if they are regulated by other intrinsic control processes. If gene rearrangements are in fact linked to regulatory signals, the elucidation of cellular mechanisms of signal reception and intracellular transmission will be critical to our understanding of the regulation of B-lineage development.

Newly Generated B Cells

Newly generated B cells in marrow express membrane IgM in increasing amounts with greater postmitotic age (Osmond and Nossal, 1974). In addition to IgM, a number of other cell surface antigens are sequentially expressed on B cells. In the mouse, Ia is expressed on newly formed marrow cells slightly later than IgM (Owen, 1979), and it also appears later in B cell ontogeny than IgM (Kearney et al., 1977). Ia, however, is expressed somewhat earlier in the differentiation of human pre-B cells (Cooper, 1981). Virtually all mouse bone marrow cells that form clones in soft agar after LPS stimulation are Ia$^+$ (Kincade et al., 1978), and this suggests that mitogen responsiveness may develop only after a period of quiescence as a sIg$^+$ B cell. The appearance of sIg on newly formed B cells presumably heralds the acquisition of specific antigen responsiveness, although it is unclear whether the capacity to bind antigen and to respond functionally to antigenic stimulation appear simultaneously. Certainly these newly formed cells in the marrow of mice are exquisitely sensitive to the induction of B cell tolerance by exposure to even low doses of antigen (Teale and Mandel, 1980). By contrast, tolerance induction in even slightly older cells requires extremely high antigen dosages and exposure times. This ease of tolerance induction in early B cells presumably reflects the critical importance of tolerizing or aborting clones of autoreactive cells that may arise during normal B lymphopoiesis. This hypothesis is based on the assumptions that (1) cells that come into contact with and bind antigen immediately after acquiring sIg are likely to be autoreactive, and (2) B cell specificities are generated randomly during ontogeny, and autoreactive clones thus occur normally during B cell diversification. This model is therefore similar to that proposed for thymic cells by Jerne (1971), and the possibility that Ig genes originated from autoreactive specificities, some of which may still exist in the germline, is intriguing.

Mouse B cells subsequently come to express receptors for the Fc portion of Ig and for complement components (Yang et al., 1978) before leaving the marrow. A proportion of newly formed B cells also appear to express detectable membrane IgD before they emigrate from the marrow. It is unknown if the expression of this second Ig isotype is related to antigen exposure. Development of B cells along multiple independent pathways has been proposed, based on functional criteria (Hardy et al., 1984; Kincade, 1977), but it is not clear how the above observations, including expression of sIgD, would relate to these models.

B cells leave the marrow randomly after sIg expression, independent of their post-mitotic age and therefore with variable expression of the different membrane antigens described. These lymphocytes migrate through the endothelium of venous sinusoids and accumulate there in substantially higher numbers than in the systemic circulation (Yoffey, 1977). This phenomenon of "lymphocyte loading" of marrow sinusoids has been considered in terms of partitioning of the local microcirculation. Once in the circulation, newly generated B lymphocytes are transported to the spleen and appear somewhat later in the lymph nodes (Brahim and Osmond, 1973) where they complete their maturation. It is not known how long a mature B cell can remain in the G_0 phase of the cell cycle and still be able to respond appropriately to antigenic stimulation.

CONCLUDING REMARKS

Throughout phylogeny, B-lineage precursors appear sequentially in the developing urogenital tissues, the gut-related tissues, and the bone marrow. At least in homiothermic vertebrates, B lymphopoiesis occurs not only during embryonic development, but also as a continuing postnatal developmental process associated either with organized gut-derived sites or with the bone marrow. The striking difference in the anatomic localization of B lymphopoiesis in birds and mammals presents a fascinating evolutionary puzzle, the resolution of which promises to teach us much about the selective pressures that affected the evolution of lymphopoiesis in early vertebrates. It is equally interesting that, although the marrow appears to be the sole site of B cell generation in both mouse and humans, this may not be true for all mammals. Certainly there is compelling evidence from both morphologic and kinetic studies that the primary sites of B lymphopoiesis in sheep are the Peyer's patches of the gut wall (Reynolds and Morris 1983, 1984a,b). Indeed, some investigators have proposed alternative sites of B lymphopoiesis in mice (Hayakawa et al., 1985), although none of the cells found in the extramedullary sites meet the current morphological or phenotypic criteria for pre-B cells.

With the availability of new monoclonal antibodies to identify cell types and differentiation stages, new cell culture systems to facilitate the study of B cell differentiation, and molecular probes with which to dissect stages in B cell development, this promises to be an exciting period in studies of B lymphopoiesis. These approaches, together with the in vivo and in vitro techniques already available, should give us a clearer understanding of the process of commitment of hemopoietic cells to B-lineage development. These advances should lead, as well, to a better understanding of the details of proliferative expansion of clones of related B-lineage cells, and the establishment of the functional repertoire of antibody binding specificities on which clonal selection (Burnet, 1959) operates.

Acknowledgments

Drs. Bruce Glick, Paul Kincade, and Kenneth Dorshkind provided stimulating discussion and critical comments during the preparation of this manuscript. I also thank Drs. Cornelius Rosse, Malcolm A. S. Moore, and Robert A. Good for continuing support and guidance.

References

Allen, T. D., and Dexter, T. M. (1984). The essential cells of the hemopoietic microenvironment. *Exp. Hematol. 12*:517–421.

Alt, F. W., Blackwell, T. K., DePinho, R. A., Reth, M. G., and Yancopoulos, G. D. (1986). Regulation of genome rearrangement events during lymphocyte differentiation. *Immunol. Rev. 89*:5–30.

Andrew, T. A., and Owen, J.J.T. (1978). Studies on the earliest sites of B cell differentiation in the mouse embryo. *Dev. Comp. Immunol. 2*:339–346.

Antoine, J.-C., Bleux, C., Avrameas, S., and Liacopoulos, P. (1979). Murine embryonic B lymphocyte development in the placenta. *Nature 277*:219–221.

Aspinall, R., and Owen, J.J.T. (1983). An investigation into the B lymphopoietic capacity of long-term bone marrow cultures. *J. Immunol. 48*:9–15.

Billips, L. G., Petitte, D., and Landreth, K. S. (1990). Bone marrow stromal cell regulation of B lymphopoiesis: Interleukin-1 (IL-1) and IL-4 regulate stromal cell support of pre-B cell production in vitro. *Blood 75*:611–619.

Billips, L. G., Petitte, D., Hostutler, R., Tsai, P., and Landreth, K. S. (1991). Suppression of bone marrow stromal cell function. *Annls NY Acad. Sci. 628*:313–322.

Billips, L. G., Petitte, D., Dorshkind, K. Narayanan, R., Chiu, C.-P., and Landreth, K. S. (1992). Differential roles of stromal cells, Interleukin 7, and *Kit*-ligand in the regulation of B-lymphopoiesis. *Blood 79*:1185–1192.

Bizzozero, G. (1868). Sulla funzione ematopoetica del midollo delle ossa. *Zentralbl. Med. Wissensch. 6*:885.

Brahim, F., and Osmond, D. G. (1973). The migration of lymphocytes from bone marrow to popliteal lymph nodes demonstrated by selective bone marrow labeling with ^{3}H-thymidine in vivo. *Anat. Rec. 175*:737.

Brand, A., Galton, J., and Gilmour, D. (1983). Committed precursors of B and T lymphocytes in chick embryo bursa of Fabricius, thymus, and bone marrow. *Eur. J. Immunol. 13*:449–455.

Broudy, V. C., Kaushansky, K., Segal, G. M., Harlan, J. M., and Adamson, J. W. (1986). Tumor necrosis factor type α stimulates human endothelial cells to produce granulocyte/macrophage colony-stimulating factor. *Proc. Natl. Acad. Sci. USA 83*:7467–7471.

Burnet, F. M. (1959). *The Clonal Selection Theory of Acquired Immunity.* Nashville: Vanderbilt University Press.

Burrows, P. D., Kearney, J. F., Lawton, A. R., and Cooper, M. D. (1978). Pre-B cells: Bone marrow persistence in anti-suppressed mice, conversion to B lymphocytes, and recovery following destruction by cyclophosphamide. *J. Immunol. 120*:1526–1531.

Burrows, P. D., LeJeune, D. M., and Kearney, J. F. (1979). Asynchrony of immunoglobulin chain synthesis during pre-B cell ontogeny: Evidence from cell hybridization that murine pre-B cells synthesize heavy chains but not light chains. *Nature 280*:838–841.

Coffman, R. L., and Weissman, I. L. (1981). A monoclonal antibody that recognizes B cells and B cell precursors in mice. *J. Exp. Med. 153*:269–279.

Coffman, R. L., and Weissman, I. L. (1983). Immunoglobulin gene rearrangement during pre-B cell differentiation. *J. Mol. Cell. Immunol. 1*:31–38.

Cohen, N. (1977). Phylogenetic emergence of lymphoid tissues and cells. In *The Lymphocyte–Structure and Function* (Marchalonis, J. J., ed.). New York: Marcel Dekker, pp. 149–202.

Collins, L. S., and Dorshkind, K. (1987). A stromal cell line from myeloid long term bone marrow cultures can support myelopoiesis and B lymphopoiesis. *J. Immunol. 138*:1082–1087.

Cooper, M. D. (1981). Pre-B cells: Normal and abnormal development. *J. Clin. Immunol. 1*:81–89.

Cooper, M., Peterson, R., South, M. A., and Good, R. A. (1966). The functions of the thymus system and the bursa system in the chicken. *J. Exp. Med. 123*:75–102.

Cuchens, M., McClean, E., and Clem, L. W. (1976). Lymphocyte heterogeneity in fish and reptiles. In *Phylogeny of Thymus and Bone Marrow Bursa Cells* (Wright, R. K., and Cooper, E. L., eds.). Amsterdam: Elsevier/North Holland, p. 205.

Cumano, A., and Paige, C. J. (1992). Enrichment and characterization of uncommited B-cell precursors from fetal liver at day 12 of gestation. *EMBO J. 11*:593–601.

Curry, J. L., and Trenton, J. J. (1967). Hemopoietic spleen colony studies. I. Growth and differentiation. *Dev. Biol. 15*:395–413.

Denis, K. A., and Witte, O. N. (1986). In vitro development of B lymphocytes from long-term cultured precursor cells. *Proc. Natl. Acad. Sci. USA 83*:441–445.

DePinho, M., Reth, G., and Yancopoulos, G. D. (1986). Regulation of genome rearrangement events during lymphocyte differentiation. *Immunol. Rev. 89*:5–30.

Dexter, T. M., Allen, T. D., and Lajtha, L. G. (1977). Conditions controlling the proliferation of haemopoietic stem cells in vitro. *J. Cell. Physiol. 91*:335–344.

Dexter, T. M., and Testa, N. G. (1980). In vitro methods in haemopoiesis and lymphopoiesis. *J. Immunol. Methods 38*:177–190.

Dieterlen-Lièvre, F., and Martin, C. (1981). Diffuse intraembryonic hemopoiesis in normal and chimeric avian development. *Dev. Biol. 88*:180–191.

Dorshkind, K. D. (1986). In vitro differentiation of B lymphocytes from primitive hemopoietic precursors present in long term bone marrow cultures. *J. Immunol. 136*:422–429.

Dorshkind, K. (1990). Regulation of hemopoiesis by bone marrow stromal cells and their products. *Annu. Rev. Immunol. 8*:111–137.

Dorshkind, K., Johnson, A., Collins, L., Keller, G. M., and Phillips, R. A. (1986). Generation of purified stromal cell cultures that support lymphoid and myeloid precursors. *J. Immunol. Methods 89*:37–47.

Dorshkind, K., and Phillips, R. A. (1981). Maturational state of lymphoid cells in long term bone marrow cultures. *J. Immunol. 129*:2444–2450.

Dorshkind, K., Schouest, L., and Fletcher, W. H. (1985). Morphologic analysis of long-term bone marrow cultures that support B-lymphopoiesis or myelopoiesis. *Cell Tissue Res. 239*:375–382.

Ehrlich, P. (1900). On immunity with special reference to cell life. *Proc. R. Soc. Lond. (Biol.) 66*:424–448.

Engelhard, D., Landreth, K. S., Kapoor, N., Kincade, P. W., DeBault, L. E., Theodore, A., and Good, R. A. (1983). Cycling of peripheral blood and marrow lymphocytes in cyclic neutropenia. *Proc. Natl. Acad. Sci. USA 80*:5734–5738.

Eskola, J., and Toivanen, P. (1977). Cell transplantation into immunodeficient chicken embryos: Reconstituting capacity of different embryonic cells. In *Avian Immunology* (Benedict, A. A., ed.). Amsterdam: Elsevier/North Holland, pp. 29–37.

Fiebig, H., and Ambrosius, H. (1976). Cell surface immunoglobulin of lymphocytes of lower vertebrates. In *Phylogeny of Thymus and Bone Marrow Bursa Cells* (Cooper, E. L., and Wright, R. K., eds.). Amsterdam: Elsevier/North Holland, pp. 195–203.

Fitzsimmons, R. C., Garrod, E., and Garnett, I. (1973). Immunological responses following early embryonic surgical bursectomy. *Cell. Immunol. 9*:377–383.

Fulop, G. M., and Osmond, D. G. (1983a). Regulation of bone marrow lymphocyte production: III. Increased production of B and non-B lymphocytes after administering systemic antigens. *Cell. Immunol. 75*:80–90.

Fulop, G., and Osmond, D. G. (1983b). Regulation of bone marrow lymphocyte production: IV. Cells mediating the stimulation of marrow lymphocyte production by sheep red blood cells: Studies in anti-IgM-suppressed mice, athymic mice and silica treated mice. *Cell. Immunol. 75*:91–102.

Gathings, W. E., Lawton, A. R., and Cooper, M. D. (1977). Immunofluorescent studies of pre-B cells, B lymphocytes and immunoglobulin isotype diversity in humans. *Eur. J. Immunol. 7*:804–810.

Gathings, W. E., Mage, R. G., Cooper, M. D., and Young-Cooper, G. O. (1982). A subpopulation of small pre-B cells in rabbit bone marrow expresses κ light chains and exhibits allelic exclusion of *b* locus allotypes. *Eur. J. Immunol. 12*:76–81.

Giri, J. G., Kincade, P. W., and Mizel, S. B. (1984). Interleukin 1–mediated induction of κ–light chain synthesis and surface immunoglobulin expression on pre-B cells. *J. Immunol.* *132*:223–228.

Gisler, R. H., Paige, C. J., and Hollander, G. (1984). Role of accessory cells in B lymphocyte progenitor differentiation. *Immunobiology 168*:441–452.

Glick, B. (1967). The bursa of Fabricius and the development of immunologic competence. In *The Thymus in Immunobiology* (Gabrielson, A. E., and Good, R. A., eds.). New York: Hoeber, pp. 345–358.

Glick, B. (1977). The bursa of Fabricius and immunoglobulin synthesis. *Int. Rev. Cytol. 48*:345–402.

Glick, B. (1983). Bursa of Fabricius. *Avian Biol. 7*:443–500.

Glick, B., Chang, T. S., and Jaap, R. G. (1956). The bursa of Fabricius and antibody production. *Poult. Sci. 35*:224–225.

Glick, B., and Rosse, C. (1976). Cellular composition of the bone marrow in the chicken: I. Identification of cells. *Anat. Rec. 185*:235–246.

Gowans, J. L., and McGregor, D. D. (1963). The origin of antibody-forming cells. In *Immunopathology: Proceedings of the Third International Symposium, La Jolla.* Basel: Schwabe, pp. 89–98.

Granfors, K., Martin, C., Lassila, O., Suvitaival, R., Toivanen, A., and Toivanen, P. (1982). Immune capacity of the chicken bursectomized at 60 hours of incubation: Production of the immunoglobulins and specific antibodies. *Clin. Immunol. Immunopathol. 23*:459–469.

Greaves, M. F. (1986). Differentiation-linked leukemogenesis in lymphocytes. *Science 234*:697–704.

Grossi, C. E., Lydyard, P. M., and Cooper, M. D. (1977). Ontogeny of B cells in the chicken: II. Changing patterns of cytoplasmic IgM expression and of modulation requirements for surface IgM by anti-μ antibodies. *J. Immunol. 119*:749–756.

Gulliver, G., ed. (1846). *The Works of William Hewson, F.R.S.* London: Sydenham Society.

Gurney, M. E., Apatoff, B. R., Spear, G. T., Baumel, M. J., Antel, J. P., Bania, M. B., and Reder, A. T. (1986). Neuroleukin: A lymphokine product of lectin-stimulated T cells. *Science 234*:574–581.

Hammerling, U., Chin, A. F., and Abbot, J. (1976). Ontogeny of murine B lymphocytes: Sequence of B-cell differentiation from surface-immunoglobulin-negative precursors to plasma cells. *Proc. Natl. Acad. Sci. USA 7*:2008–2012.

Hardy, R. R., Carmack, C. E., Shinton, S. A., Kemp, J. D., and Hayakawa, K. (1991). Resolution and characterization of pro-B and pre-pro-B cell stages in normal mouse bone marrow. *J. Exp. Med. 173*:1213–1225.

Hardy, R. R., and Hayakawa, K. (1986). Development and physiology of Ly-1 B and its human homolog, Leu-1 B. *Immunol. Rev. 93*:53–79.

Hardy, R. R., Hayakawa, K., Parks, D. R., Herzenberg, L. A., and Herzenberg, L. A. (1984). Murine B cell differentiation lineages. *J. Exp. Med. 159*:1169–1188.

Harris, T. N., Grimm, E., Mertens, E., and Ehrich, W. E. (1945). The role of the lymphocyte in antibody formation. *J. Exp. Med. 81*:73–83.

Harris, T. N., and Harris, S. (1960). Lymph node cell transfer in relation to antibody production. In *Cellular Aspects of Immunology* (Ciba Foundation Symposium). London: Churchill, p. 172.

Hayakawa, K., Hardy, R., Herzenberg, L., and Herzenberg, L. (1985). Progenitors for Ly-1 B cells are distinct from progenitors for other B cells. *J. Exp. Med. 161*:1554–1568.

Herzenberg, L. A., Stall, A. M., Lalor, P. A., Sidman, C., Moore, W. A., and Parks, D. R. (1986). The Ly-1 B cell lineage. *Immunol. Rev. 93*:81–102.

Honjo, T., Nakai, S., Nishida, Y., Kataoka, T., Yamawaki-Kataoka, Y., Takahashi, N., Obata, M., Shimizu, A., Yaoita, Y., Nikaido, T., and Ishida, N. (1981). Rearrangements of immunoglobulin genes during differentiation and evolution. *Immunol. Rev. 59*:33–67.

Houssaint, E., Torano, A., and Ivanyi, J. (1983). Ontogenic restriction of colonization of the bursa of Fabricius. *Eur. J. Immunol. 13*:590–595.

Hunkapiller, T., and Hood, L. (1989). Diversity of the immunoglobulin gene superfamily. *Adv. Immunol. 44*:1–63.

Hunt, P., Robertson, D., Weiss, D., Rennick, D., Lee, F., and Witte, O. N. (1987). A single bone marrow–derived stromal cell type supports the in vitro growth of early lymphoid and myeloid cells. *Cell 48*:997–1007.

Ikebuchi, K., Wong, G. G., Clark, S. C., Ihle, J. N., Hirai, Y., and Ogawa, M. (1987). Interleukin 6 enhancement of interleukin 3–dependent proliferation of multipotential hemopoietic progenitors. *Proc. Natl. Acad. Sci. USA 84*:9035–9039.

Jalkanen, S., Granfors, K., Jalkanen, M., and Toivanen, P. (1983a). Immune capacity of the chicken bursectomized at 60 hours of incubation: Failure to produce immune, natural, and autoantibodies in spite of immunoglobulin production. *Cell. Immunol. 80*:363–373.

Jalkanen, S., Granfors, K., Jalkanen, M., and Toivanen, P. (1983b). Immune capacity of the chicken bursectomized at 60 hours of incubation: Surface immunoglobulin and B-L (Ia-like) antigen bearing cells. *J. Immunol. 130*:2038–2041.

Jerne, N. K. (1971). The somatic generation of immune recognition. *Eur. J. Immunol. 1*:1–9.

Jotereau, F. V., and Houssaint, E. (1977). Experimental studies on the migration and differentiation of primary lymphoid stem cells in the avian embryo. In *Developmental Immunobiology* (Solomon, J. B., and Horton, J. D., eds.). Amsterdam: Elsevier/North Holland, pp. 123–130.

Jotereau, F. V., Houssaint, E., and Le Douarin, N. M. (1980). Lymphoid stem cell homing to the early thymic primordium of the avian embryo. *Eur. J. Immunol. 10*:620–627.

Jyonouchi, H., Kincade, P. W., Good, R. A., and Gershwin, M. E. (1983). B lymphocyte lineage cells in newborn and very young NZB mice: Evidence for regulatory disorders affecting B cell formation. *J. Immunol. 131*:2219–2225.

Kau, C.-L., and Turpen, J. B. (1983). Dual contribution of embryonic ventral blood island and dorsal lateral plate mesoderm during ontogeny of hemopoietic cells in *Xenopus laevis. J. Immunol. 131*:2262–2266.

Kearney, J. F., Cooper, M. D., Klein, J., Abney, E. R., Parkhouse, R.M.E., and Lawton, A. R. (1977). Ontogeny of Ia and IgD on IgM bearing B lymphocytes in mice. *J. Exp. Med. 146*:297–301.

Kincade, P. W. (1977). Defective colony formation by B lymphocytes from CBA/N and C3H/HeJ mice. *J. Exp. Med. 145*:249–263.

Kincade, P. W. (1981). Formation of B lymphocytes in fetal and adult life. *Adv. Immunol. 31*:177–245.

Kincade, P. W. (1987). Experimental models for understanding B lymphocyte formation. *Adv. Immunol. 41*:181–267.

Kincade, P. W., and Cooper, M. D. (1971). Development and distribution of immunoglobulin-containing cells in the chicken: An immunofluorescent analysis using purified antibodies to heavy and light chains. *J. Immunol. 106*:371–382.

Kincade, P. W., Lee, G., Paige, C. J., and Scheid, M. P. (1981a). Cellular interactions affecting the maturation of murine B lymphocyte precursors in vitro. *J. Immunol. 127*:255–260.

Kincade, P. W., Lee, G., Pietrangeli, C. E., Hayashi, S.-I., and Gimble, J. M. (1989). Cells and molecules that regulate B lymphopoiesis in bone marrow. *Annu. Rev. Immunol. 7*:111–143.

Kincade, P. W., Lee, G., Watanabe, T., Sun, L., and Scheid, M. P. (1981b). Antigens displayed on murine B-lymphocyte precursors. *J. Immunol. 127*:2262–2268.

Kincade, P. W., Paige, C. J., Parkhouse, R.M.E., and Lee, G. (1978). Characterization of murine colony forming B cells: I. Distribution, resistance to anti-immunoglobulin antibodies, and expression of Ia antigens. *J. Immunol. 120*:1289–1296.

King, A. G., Wierda, D., and Landreth, K. S. (1988). Bone marrow stromal cell regulation of B-lymphopoiesis. I. The role of macrophages, IL-1, and IL-4 in pre-B cell maturation. *J. Immunol. 141*:2016–2026.

Klempau, A. E., and Cooper, E. L. (1983). T-lymphocyte and B-lymphocyte dichotomy in anuran amphibians: I. T-lymphocyte proportions, distribution and ontogeny as measured by E-rosetting, nylon wool adherence, postmetamorphic thymectomy and nonspecific esterase staining. *Dev. Comp. Immunol. 7*:99–110.

Klinman, N. R., Pickard, A., Sigal, N. H., Gearhart, P. J., Metcalf, E. S., and Pierce, S. K. (1976). Assessing B cell diversification by antigen receptor and precursor cell analysis. *Ann. Immunol. (Inst. Pasteur) 127c*:487–502.

Klinman, N. R., and Stone, M. R. (1983). Role of variable region gene expression and environmental selection in determining the antiphosphorylcholine B cell repertoire. *J. Exp. Med. 158*:1948–1961.

Kurland, J. I., Zeigler, S. F., and Witte, O. N. (1984). Long term cultured B lymphocytes and their precursors reconstitute the B-lymphocyte lineage in vivo. *Proc. Natl. Acad. Sci. USA 81*:7554–7558.

LaFleur, L., Miller, R. G., and Phillips, R. A. (1971). A quantitative assay for the progenitors of bone marrow associated lymphocytes. *J. Exp. Med. 135*:1363–1374.

LaFleur, L., Underdown, B. J., Miller, R. G., and Phillips, R. A. (1972). Differentiation of lymphocytes: Characterization of early precursors of B lymphocytes. *Ser. Haemat. 5*:50.

Landreth, K. S., and Dorshkind, K. (1988). Pre-B cell generation potentiated by soluble factors from a bone marrow stromal cell line. *J. Immunol. 140*:845–852.

Landreth, K. S., Engelhard, D., Beare, M. H., Kincade, P. W., and Good, R. A. (1984a). Identification of a human pre-B cell differentiation activity associated with cyclic neutropenia. *Fed. Proc. 43*:1485.

Landreth, K. S., Engelhard, D., Beare, M., Kincade, P. W., Kapoor, N., and Good, R. A. (1985). Regulation of human B lymphopoiesis: Effect of a urinary activity associated with cyclic neutropenia. *J. Immunol. 134*:2305–2309.

Landreth, K. S., and Glick, B. (1973). Differential effect of bursectomy on antibody production in a large and small bursa line of New Hampshire chickens. *Proc. Soc. Exp. Biol. Med. 144*:501–505.

Landreth, K. S., and Kincade, P. W. (1984). Mammalian B lymphocyte precursors. *Dev. Comp. Immunol. 8*:773–790.

Landreth, K. S., Kincade, P. W., Lee, G., and Harrison, D. (1984b). B lymphocyte precursors in embryonic and adult W anemic mice. *J. Immunol. 132*:2724–2729.

Landreth, K. S., Lee, G., Kincade, P. W., and Medlock, E. S. (1983). Phenotypic and functional characterization of murine B lymphocyte precursors isolated from fetal and adult tissues. *J. Immunol. 132*:572–580.

Landreth, K. S., Rosse, C., and Clagett, J. (1981). Myelogenous production and maturation of B lymphocytes in the mouse. *J. Immunol. 127*:2027–2034.

Landreth, K. S., Narayanan, R., and Dorshkind, K. (1992). Insulin-like growth factor-I (IGF-I) regulates pro-B cell differentiation. *Blood*: in press.

Lau, C. Y., Melchers, F., Miller, R. G., and Phillips, R. A. (1979). In vitro differentiation of B lymphocytes from pre-B cells. *J. Immunol. 122*:1273–1277.

Lawton, A., and Cooper, M. D. (1977). Two new stages of antigen-independent B cell development in mice and humans. In *Development of Host Defenses* (Cooper, M. D., and Dayton, O. H., eds.). New York: Raven Press, p. 43.

LeBacq, A.-M., and Ritter, M. (1979). B-cell precursors in early chicken embryos. *Immunology 37*:123–134.

LeDouarin, N. M., Dieterlen-Lièvre, F., and Oliver, P. D. (1984). Ontogeny of primary lymphoid organs and lymphoid stem cells. *Am. J. Anat. 170*:261–299.

LeDouarin, N., Houssaint, E., Jotereau, F., and Belo, M. (1975). Origin of haemopoietic stem cells in the embryonic bursa of Fabricius and bone-marrow studied through interspecific chimeras. *Proc. Natl. Acad. Sci. USA 72*:2701–2705.

Lee, G., Ellingsworth, L. R., Gillis, S., Wall, R., and Kincade, P. W. (1987). β transforming growth factors are potential regulators of B lymphopoiesis. *J. Exp. Med. 166*:1290–1299.

Lerman, S. P., and Weidanz, W. P. (1970). The effect of cyclophosphamide on the ontogeny of the humoral immune response in chickens. *J. Immunol. 105*:614–619.

Lerner, K. G., Glick, B., and McDuffie, F. C. (1971). Role of the bursa of Fabricius in IgG and IgM production in the chicken: Evidence for the role of a non-bursal site in the development of humoral immunity. *J. Immunol. 107*:493–503.

Lobb, C. J., and Clem, L. W. (1982). Fish lymphocytes differ in the expression of surface immunoglobulin. *Dev. Comp. Immunol. 6*:473–479.

Lydyard, P. M., Grossi, C. E., and Cooper, M. D. (1976). Ontogeny of B cells in the chicken: I. Sequential development of clonal diversity in the bursa. *J. Exp. Med. 144*:79–97.

Maki, R., Kearney, J., Paige, C., and Tonegawa, S. (1980). Immunoglobulin gene rearrangement in immature B cells. *Science 209*:1366–1369.

Martin, C. (1972). Technique d'explantation in ovo de blastodermes d'embryons d'oiseau. *C. R. Soc. Biol. 166*:283–285.

Messague, J. (1987). The TGF-β family of growth and differentiation factors. *Cell 49*:437–438.

McCormack, W. T., Carlson, L. M., Tjoelker, L. W., and Thompson, C. B. (1989a). Evolutionary comparison of the avian IgL locus: Combinatorial diversity plays a role in the generation of the antibody repertoire in some avian species. *Int. Immunol. 1*:332–341.

McCormack, W. T., and Thompson, C. B. (1990a). Chicken IgL variable region gene conversions display pseudogene donor preference and 5' to 3' polarity. *Genes Dev. 4*:548–558.

McCormack, W. T., and Thompson, C. B. (1990b). Somatic diversification of the chicken immunoglobulin light-chain gene. *Adv. Immunol. 48*:41–67.

McCormack, W. T., Tjoelker, L. W., Barth, C. F., Carlson, L. M., Petryniak, B., Humphries, E. H., and Thompson, C. B. (1989b). Selection for B cells with productive IgL gene rearrangements occurs in the bursa of Fabricius during chicken embryonic development. *Genes Dev. 3*:838–847.

McCormack, W. T., Tjoelker, L. W., Carlson, L. M., Petryniak, B., Barth, C. F., Humphries, E. H., and Thompson, C. B. (1989c). Chicken IgL gene rearrangement involves deletion of a circular episome and addition of single nonrandom nucleotides to both coding segments. *Cell 56*:785–791.

McCormack, W. T., Tjoelker, L. W., Thompson, C. B. (1991). Avian B cell development: Generation of an immunoglobulin repertoire by gene conversion. *Annu. Rev. Immunol. 9*:219–241.

McCulloch, E. A., Siminovitch, L., and Till, J. E. (1964). Spleen colony formation in anemic mice of genotype *W/Wv*. *Science 144*:844–846.

McKearn, J. P., Baum, C., and Davie, J. M. (1984). Cell surface antigens expressed by subsets of pre-B cells and B cells. *J. Immunol. 132*:332–339.

Medlock, E. S., Landreth, K. S., and Kincade, P. W. (1984). Putative B lymphocyte lineage precursor cells in early murine embryos. *Dev. Comp. Immunol. 8*:887–894.

Mekori, T., and Phillips, R. A. (1969). The immune response in mice of genotypes *W/Wv* and *Sl/Sld*. *Proc. Soc. Exp. Biol. Med. 132*:115–119.

Melchers, F. (1977). B lymphocyte development in fetal liver: I. Development of reactivities to B cell mitogens in vivo and in vitro. *Eur. J. Immunol. 7*:476–481.

Melchers, F. (1979). Murine embryonic B lymphocyte development in the placenta. *Nature 277*:219–221.

Melink, G. B., and Lebein, T. W. (1983). Construction of an antigenic map for human B cell precursors. *J. Clin. Immunol. 3*:260–267.

Metcalf, D. (1971). Acute antigen-induced elevation of serum colony stimulating factor (CSF) levels. *Immunology 21*:427–436.

Miyajima, A., Miyatake, S., Schreurs, J., De Vries, J., Arai, N., Yokota, T., and Arai, K.-I. (1988).

Coordinate regulation of immune and inflammatory responses by T cell–derived lymphokines. *FASEB J.* 2:2462–2473.

Moore, M.A.S., and Metcalf, D. (1970). Ontogeny of the haemopoietic system: Yolk sac origin of in vivo and in vitro colony forming cells in the developing mouse embryo. *Brit. J. Immunol.* 18:279–296.

Moore, M.A.S., and Owen, J.J.T. (1967a). Chromosome marker studies in the irradiated chick embryo. *Nature 215*:1081–1082.

Moore, M.A.S., and Owen, J.J.T. (1967b). Stem cell migration in developing myeloid and lymphoid systems. *Lancet 2*:658–659.

Müller-Sieburg, C. E., Townsend, K., Weissman, I. L., and Rennick, D. (1988). Proliferation and differentiation of highly enriched mouse hematopoietic stem cells and progenitor cells in response to defined growth factors. *J. Exp. Med. 167*:1825–1840.

Müller-Sieburg, C. E., Whitlock, C. A., and Weissman, I. L. (1986). Isolation of two early B lymphocyte progenitors from mouse marrow: A committed pre-pre-B cell and a clonogenic Thy-1^{lo} hematopoietic stem cell. *Cell 44*:653–662.

Namen, A. E., Lupton, S., Hjerrild, K., Wagnall, J., Mochizuki, D. Y., Schmierer, A., Mosley, B., March, C. J., Urdal, D., Gillis, S., Cosman, D., and Goodwin, R. G. (1988a). Stimulation of B-cell progenitors by cloned murine interleukin-7. *Nature 333*:571–573.

Namen, A. E., Schmierer, A. E., March, C. J., Overell, R. W., Park, L. S., Urdal, D., and Mochizuki, D. Y. (1988b). B cell precursor growth-promoting activity. Purification and characterization of a growth factor active on lymphocyte precursors. *J. Exp. Med. 167*:988–1002.

Nawroth, P. P., Bank, I., Handley, D., Cassimeris, J., Chess, L., and Stern, D. (1986). Tumor necrosis factor/cachectin interacts with endothelial cell receptors to induce release of interleukin 1. *J. Exp. Med. 163*:1363–1375.

Opstelten, D., and Osmond, D. G. (1983). Pre-B cells in the bone marrow: Immunofluorescence stathmokinetic studies of the proliferation of cytoplasmic µ-chain–bearing cells in normal mice. *J. Immunol. 131*:2635–2640.

Osmond, D. G., and Everett, N. B. (1964). Radioautographic studies of bone marrow lymphocytes in vivo and in diffusion chamber cultures. *Blood 23*:1–17.

Osmond, D. G., and Nossal, G.J.V. (1974). Differentiation of lymphocytes in mouse bone marrow: II. Kinetics of maturation and renewal of antiglobulin-binding cells studied by double labeling. *Cell. Immunol. 13*:132–145.

Osmond, D. G., and Owen, J.J.T. (1984). Pre-B cells in the bone marrow: Size distribution profile, proliferative capacity and peanut agglutinin binding of cytoplasmic µ chain–bearing cell populations in normal and regenerating bone marrow. *Immunology 51*:333–342.

Owen, J.J.T. (1979). Ontogeny of B lymphocytes. In *B Lymphocytes and the Immune Response* (Cooper, M. D., Mosier, D. E., Scher, I., and Vitetta, E. S., eds.). New York: Elsevier/North Holland, pp. 71–76.

Owen, J.J.T., Cooper, M. D., and Raff, M. C. (1974). In vitro generation of B lymphocytes in mouse fetal liver, a mammalian bursal equivalent. *Nature 249*:361–363.

Owen, J.J.T., Wright, D. E., Habu, S., Raff, M. C., and Cooper, M. D. (1977). Studies on the generation of B lymphocytes in fetal liver and bone marrow. *J. Immunol. 118*:2067–2072.

Paige, C. J., Gisler, R., McKearn, J., and Iscove, N. (1984). Differentiation of murine B cell precursors in agar culture: Frequency, surface marker analysis and requirements of growth of clonable pre-B cells. *Eur. J. Immunol. 14*:979–987.

Paige, C. J., Kincade, P. W., Moore, M.A.S., and Lee, G. (1979). The fate of fetal and adult B-cell progenitors grafted into immunodeficient CBA/N mice. *J. Exp. Med. 150*:548–563.

Paige, C. J., Kincade, P. W., Shinefield, L. A., and Sato, V. L. (1981). Precursors of murine B lymphocytes: Physical and functional characterization and distinctions from myeloid stem cells. *J. Exp. Med. 153*:154–165.

Paige, C. J., Skarvall, H., and Sauter, H. (1985). Differentiation of murine cell precursors in agar culture: II. Response of precursor-enriched populations to growth stimuli and demonstration that the clonable pre-B assay is limiting for the B cell precursor. *J. Immunol. 134*:3699–3704.

Park, Y. H., and Osmond, D. G. (1987). Phenotype and proliferation of early B lymphocyte precursor cells in mouse bone marrow. *J. Exp. Med. 165*:444–458.

Parvari, R., Avivi, A., Lentner, F., Ziv, E., Tel-Or, S., Burstein, Y., and Schechter, I. (1988). Chicken immunoglobulin γ-heavy chains: Limited V_H gene repertoire combinatorial diversification by D gene segments and evolution of the heavy chain locus. *EMBO J. 7*:739–744.

Pearl, E. R., Vogler, L. B., Okos, A., Crist, W. M., Lawton, III, A. R., and Cooper, M. D. (1978). B lymphocyte precursors in human bone marrow: An analysis of normal individuals and patients with antibody-deficiency states. *J. Immunol. 120*:1169–1175.

Perkins, S., and Fleischman, R. A. (1988). Hematopoietic microenvironment: Origin, lineage, and transplantability of the stromal cells in long-term bone marrow cultures from chimeric mice. *J. Clin. Invest. 81*:1072–1080.

Pernis, B., Forni, L., and Amante, L. L. (1970). Immunoglobulin spots on the surface of rabbit lymphocytes. *J. Exp. Med. 132*:1001–1018.

Phillips, R. A., Bosma, M., and Dorshkind, K. (1984). Reconstitution of immune deficient mice with cells from long term bone marrow cultures. In *Long Term Bone Marrow Culture* (Greenberger, J. S., Nemo, G. J., and Wright, D. G., eds.). New York: Alan R. Liss, pp. 309–321.

Pietrangeli, C. E., and Osmond, D. G. (1984). Regulation of bone marrow lymphocyte production: Macrophage involvement and role of the spleen in the response to exogenous stimuli. *Fed. Proc. 43*:1485.

Pink, J. R., Vainio, O., and Rijnbeek, A.-M. (1985). Clones of B lymphocytes in individual follicles of the bursa of Fabricius. *Eur. J. Immunol. 15*:83–87.

Raff, M. C., Megson, M., Owen, J. J., and Cooper, M. D. (1976). Early production of intracellular IgM by B-lymphocyte precursors in mouse. *Nature 259*:224–226.

Raff, M. C., Sternberg, M., and Taylor, R. B. (1970). Immunoglobulin determinants on the surface of mouse lymphoid cells. *Nature 225*:553–554.

Raison, R. L., and Hildemann, W. H. (1984). Immunoglobulin bearing blood leucocytes in the Pacific hagfish. *Dev. Comp. Immunol. 8*:99–108.

Ramirez, J. A., Wright, R. K., and Cooper, E. L. (1983). Bone marrow reconstitution of immune responses following irradiation in the leopard frog, *Rana pipiens. Dev. Comp. Immunol. 7*:303–312.

Reth, M. G., and Alt, F. (1984). Novel immunoglobulin heavy chains are produced from DJ_H gene segment rearrangements in lymphoid cells. *Nature 312*:419–423.

Reynaud, C.-A., Anquez, V., Dahan, A., and Weill, J.-C. (1985). A single rearrangement event generates most of the chicken immunoglobulin light chain diversity. *Cell 40*:283–291.

Reynaud, C.-A., Anquez, V., Grimal, H., and Weill, J.-C. (1987). A hyperconversion mechanism generates the chicken light chain preimmune repertoire. *Cell 48*:379–388.

Reynaud, C.-A., Dahan, A., Anquez, V., and Weill, J.-C. (1989). Somatic hyperconversion diversifies the single V_H gene of the chicken with a high incidence in the D region. *Cell 59*:171–183.

Reynolds, J. D., and Morris, B. (1983). The evolution and involution of Peyer's patches in fetal and postnatal sheep. *Eur. J. Immunol. 13*:627–635.

Reynolds, J. D., and Morris, B. (1984a). The effect of antigen on the development of Peyer's patches in sheep. *Eur. J. Immunol. 14*:1–6.

Reynolds, J. D., and Morris, B. (1984b). The emigration of lymphocytes from Peyer's patches in sheep. *Eur. J. Immunol. 14*:7–13.

Rich, A. R., Lewis, M. R., and Wintrobe, M. M. (1939). The activity of the lymphocyte in the body's reaction to foreign protein, as established by the identification of the acute splenic tumor cell. *Bull. Johns Hopkins Hosp. 65*:311–327.

Ritter, M. A., and Lebacq, A.-M. (1977). Embryonic bursa development in vitro. *Eur. J. Immunol.* 7:468–475.

Rolink, A., Kudo, A., Karasuyama, H., Kikuchi, Y., and Melchers, F. (1991). Long-term proliferating early pre-B cell lines and clones with the potential to develop to surface Ig-positive, mitogen reactive B cells in vitro and in vivo. *EMBO J.* 10:327–336.

Rosse, C. (1976). Small lymphocyte and transitional cell populations in the bone marrow: Their role in the mediation of immune and hemopoietic progenitor cell functions. *Int. Rev. Cytol.* 45:155–290.

Rosse, C., Cole, S. B., Appleton, C., Press, O. W., and Clagett, J. (1978). The relative importance of the bone marrow and spleen in the production and dissemination of B lymphocytes. *Cell. Immunol.* 37:254.

Russell, E. S. (1979). Hereditary anemias of the mouse: A review for geneticists. *Adv. Genet.* 20:357–457.

Ryser, J. E., and Vassalli, P. (1974). Mouse bone marrow lymphocytes and their differentiation. *J. Immunol.* 113:719–728.

Saga, Y., Tung, J. S., Shen, F. W., and Boyse, E. A. (1986). Sequences of Ly-5 cDNA: Isoform-related diversity of Ly-5 mRNA. *Proc. Natl. Acad. Sci. USA 83*:6940–6944.

Scheid, M. P., Landreth, K. S., Tung, J. S., and Kincade, P. W. (1982). Preferential but nonexclusive expression of macro-molecular antigens on B lineage cells. *Immunol. Rev. 69*:141–159.

Schrader, J. W. (1986). The panspecific hemopoietin of activated T lymphocytes (interleukin-3). *Annu. Rev. Immunol. 4*:205–230.

Segal, G. M., McCall, E., Stueve, T., and Bagby, Jr., G. C. (1987). Interleukin 1 stimulates endothelial cells to release multilineage human colony-stimulating activity. *J. Immunol. 138*:1772–1778.

Shortman, K., and Howard, M. (1979). Non-specific and specific responses of different B-cell subsets to antigen stimulation. In *B Lymphocytes in the Immune Response* (Cooper, M. D., Mosier, D., Scher, E., and Vitetta, E., eds.). New York: Elsevier/North Holland, pp. 97–106.

Simons, A. and Zharhary, D. (1989). The role of IL-4 in the generation of B lymphocytes in the bone marrow. *J Immunol. 143*:2540–2545.

Spangrude, G. J., Heimfeld, S., and Weissman, I. L. (1988). Purification and characterization of mouse hematopoietic stem cells. *Science 241*:58–62.

Stocker, J. M. (1977). Functional maturation of B cells in vitro. *Immunology 32*:275–281.

Sugiyama, H., Akira, S., Yoshida, N., Kishimoto, S., Yamamura, Y., Kincade, P., Honjo, T., and Kishimoto, T. (1982). Relationship between the rearrangement of immunoglobulin genes, the appearance of a B lymphocyte antigen, and immunoglobulin synthesis in murine pre-B cell lines. *J. Immunol. 128*:2793–2797.

Sugiyama, H., Akira, S., Kikutani, H., Kishimoto, S., Yamamura, Y., and Kishimoto, T. (1983). Functional V region formation during in vitro culture of a murine immature B precursor cell line. *Nature 303*:812–815.

Tavassoli, M., and Crosby, W. H. (1970). Bone marrow histogenesis: A comparison of fatty and red marrow. *Science 169*:291–293.

Tavassoli, M., and Yoffey, J. M. (1983). In *Bone Marrow Structure and Function*. New York: Alan R. Liss.

Teale, J. M., and Mandel, T. E. (1980). Ontogenetic development of B-lymphocyte function and tolerance susceptibility in vivo and an in vitro fetal organ culture system. *J. Exp. Med. 151*:429–445.

Thomas, M. L., Barclay, A. N., Gagnon, J., and Williams, A. F. (1985). Evidence from cDNA clones that the rat leukocyte-common antigen (T200) spans the lipid bilayer and contains a cytoplasmic domain of 80,000 M_r. *Cell 41*:83–93.

Thorens, B., Mermod, J.-J., and Vassalli, P. (1987). Phagocytosis and inflammatory stimuli induce GM-CSF mRNA in macrophages through posttranscriptional regulation. *Cell 48*:671–679.

Turpen, J. B., Cohen, N., Deparis, P., Jaylet, A., Tompkins, R., and Volpe, E. P. (1982). Ontogeny of amphibian hemopoietic cells. In *The Reticuloendothelial System,* vol. 3 (Cohen, N., and Sigel, M. M., eds.). New York: Plenum Press, pp. 569–587.

Velardi, A., and Cooper, M. A. (1984). An immunofluorescence analysis of the ontogeny of myeloid, T and B lineage cells in mouse hemopoietic tissues. *J. Immunol. 133*:672–677.

Visser, J.W.M., Bauman, J.G.J., Mulder, A. H., Eliason, J. F., and Deleeuw, A. M. (1984). Isolation of murine pluripotent hemopoietic stem cells. *J. Exp. Med. 159*:1576–1590.

Warner, N. L., Szenberg, A., and Burnet, F. M. (1962). The immunological role of different lymphoid organs in the chicken: I. Dissociation of immunological responsiveness. *Aust. J. Exp. Biol. Med. Sci. 40*:373–387.

Weber, W. T., and Mausner, R. (1977). Migration patterns of avian embryonic bone marrow cells and their differentiation to functional T and B cells. In *Avian Immunology* (Benedict, A. A., ed.). New York: Plenum Press, pp. 47–59.

Weidanz, W. P., Konietzko, D., and Lerman, S. P. (1971). The effect of combined chemical bursectomy on antibody formation at the cellular level. *J. Reticuloendothel. Soc. 9*:635.

Whitlock, C., Denis, K., Robertson, D., and Witte, O. (1985). In vitro analysis of murine B-cell development. *Annu. Rev. Immunol. 3*:213–235.

Whitlock, C. A., Robertson, D., and Witte, O. N. (1984). Murine B lymphopoiesis in long term culture. *J. Immunol. Methods 67*:353–369.

Whitlock, C. A., and Witte, O. N. (1982). Long term culture of B lymphocytes and their precursors from murine bone marrow. *Proc. Natl. Acad. Sci. USA 79*:3608–3612.

Witte, P. L., Kincade, P. W., and Vetvicka, V. (1986). Interculture variation and evolution of B lineage lymphocytes in long-term marrow culture. *Eur. J. Immunol. 16*:779–788.

Witte, P. L., Robinson, M., Henley, A., Low, M. G., Stiers, D. L., Perkins, S., Fleischman, R. A., and Kincade, P. W. (1987). Relationships between B-lineage lymphocytes and stromal cells in long term bone marrow cultures. *Eur. J. Immunol. 17*:1473–1484.

Wu, A. M., Till, J. E., Siminovitch, L., and McCulloch, E. A. (1968). Cytological evidence for a relationship between normal hematopoietic colony-forming cells and cells of the lymphoid system. *J. Exp. Med. 127*:455–463.

Yancopoulos, G. D., and Alt, F. W. (1986). Regulation of the assembly and expression of variable-region genes. *Annu. Rev. Immunol. 4*:339–368.

Yang, W. C., Miller, S. C., and Osmond, D. G. (1978). Maturation of bone marrow lymphocytes: II. Development of Fc and complement receptors and surface immunoglobulin studied by rosetting and radioautography. *J. Exp. Med. 148*:1251–1270.

Yoffey, J. M. (1931). Preliminary note on mammalian lymphoid tissue and its relation to the remainder of the haemopoietic system. *J. Anat. 65*:333–338.

Yoffey, J. M. (1933). The quantitative study of lymphocyte production. *J. Anat. 67*:250–262.

Yoffey, J. M. (1977). Lymphocyte loading of bone marrow sinusoids: A microcirculatory puzzle. *Adv. Microcirc. 7*:49–67.

Yoffey, J. M. (1981). Characterization of hematopoietic stem cells: The fundamental role of the transitional cell compartment. In *Advances in Morphology of Cells and Tissues.* New York: Alan R. Liss, pp. 201–212.

Zettergren, L. D. (1982). Ontogeny and distribution of cells in B lineage in the American leopard frog, *Rana pipiens. Dev. Comp. Immunol. 6*:311–320.

Zettergren, L. D., Chen, C., and Ewert, D. (1980). Pre-B and B cells in chick embryo urogenital tissues: A newly reported site for cells in early B lineages in endothermic vertebrates. *Am. Zool. 20*:793.

IV

MOLECULAR ASPECTS OF RECOGNITION AND REGULATION

12

Functional and Molecular Aspects of Antigen Processing and Presentation

JOHN W. SEMPLE,

TERRY L. DELOVITCH

The antigen receptors on B and T cells have both genetic and structural similarities, yet antigen recognition by T cells differs qualitatively from that of B cells. T cell receptors (TCRs) generally do not recognize native antigen. They recognize a degraded or processed form of the native antigen on the surface of an antigen presenting cell (APC) in association with molecules encoded by the major histocompatibility complex (MHC). B cell immunoglobulins (Igs), on the other hand, can recognize epitopes on the native antigen and/or on degraded forms of the antigen. This chapter is primarily concerned with the mechanisms of antigen processing and how APCs can generate immunogenic epitopes of proteins that are recognized by T cells.

DEVELOPMENTAL ASPECTS OF ANTIGEN PROCESSING AND PRESENTATION

Invertebrates

Little is known about the events of antigen processing by invertebrate organisms. To date, there is no evidence that molecules homologous to those encoded by the mammalian MHC exist within invertebrate species, which casts doubt on whether antigen (Ag) processing and presentation mechanisms are required in these species.

Engulfment reactions such as phagocytosis take place in virtually all the invertebrate phyla and probably represent the rudimentary beginnings of the antigen processing mechanisms seen in higher vertebrates. Phagocytic cells that resemble macrophages, the classical mammalian APC, are found in most invertebrate species. One of the earliest descriptions of phagocytosis was in the small mollusk *T. fimbria* by Ernst Haeckel (1862). Hemocytes from this animal can engulf particles of indigo dye and store them in cytoplasmic vacuoles in a manner similar to an amoeba. Eventually, enzymes secreted into the lumen of the vacuole degrade the dye with subsequent vacuole disappearance. This form of degradation can be observed in mammalian macrophages, although it usually correlates with a terminal degradation event that differs from antigen processing, as will be discussed.

Vertebrates

Within vertebrate phyla, a fairly uniform plan of organization is seen, and molecules associated with the immume system—such as those encoded by the MHC—are found in a variety of species. Mammalian MHC-like molecules are present in advanced amphibians (anurans), fishes (teleosts), reptiles, and birds and correlate with (the presence) of in vitro mixed lymphocyte culture (MLC) and allograft reactions (Roitt, 1989). The clawed toad *Xenopus* expresses both class I and class II MHC molecules that are polymorphic and of similar structure to mammalian MHC molecules (Horton, 1988). Its class II molecules are made up of an α and a β chain, each of which has a molecular mass of 30 to 35 kD. In addition, *Xenopus* contains helper and suppressor T cell subsets similar to those in mammals. These cells show MHC-restricted antigen recognition, and they can secrete T cell growth factors similar to mammalian interleukin 2 (IL-2) when stimulated with the T cell mitogen phytohemagglutinin (PHA) (Horton, 1988). The thymus of *Xenopus* contains both stromal and epithelial cells, which are probably involved in education of T cell precursors and may act as APC. Because of these apparent anatomical and functional similarities between *Xenopus* and mammals, it is reasonable to assume that antigen processing and presentation pathways exist in lower vertebrates for the generation of an immune response. However, there are few data available on this particular aspect of the developmental immunobiology of vertebrates. Most of the current data on antigen processing has been generated within mammalian systems, and it will be summarized here. It is conceivable, however, that any organism with a sufficiently developed immune system may depend, in part, on the processes discussed in this chapter. Table 12–1 summarizes the existing knowledge of antigen processing and presentation events during phylogenetic development.

ANTIGEN PROCESSING IN MAMMALIAN SYSTEMS: AN OVERVIEW

The generation of an immune response to a soluble protein antigen is initiated by the binding of antigen to an APC, which subsequently degrades (processes) the antigen into immunogenic peptides that are recognized by antigen-specific T cells. The activated T cells, in turn, secrete lymphokines and in this way stimulate various immune responses such as Ig production and cytotoxic lymphocyte generation. The basic requirements for cells to act as APCs are as follows: (a) expression of either class I or class II MHC mole-

Table 12–1.   Summary of Events Related to Antigen Processing
and Presentation within Different Phyla

Invertebrates	*Vertebrates*
Phagocytic cells that can engulf foreign material	Cell-mediated and humoral immunity in all phyla
Recognition mechanisms of nonself cell types in some phyla	Possess T and B lymphocytes and lymphoid tissues
	MHC molecules with related reactions are known to exist in teleosts and higher phyla
	Antigen processing events known to exist in all mammals

Adapted from Roitt et al. (1989).

cules on their surface; (b) existence of biochemical pathways that degrade antigen and; (c) ability to synthesize and secrete cytokines such as IL-1 (Unanue, 1984). These requirements are not mutually exclusive. The classical APC is the macrophage (Unanue, 1984), although a variety of other cell types that meet the above criteria also function as APC and may be important for the generation of an immune response in vivo. These APC types include dendritic cells, epithelial cells, monocytes, and B cells.

There have been numerous reports of a requirement for antigen processing by APC for antigen recognition by, and stimulation of, T cells. Shimonkevitz et al. (1983) demonstrated that native ovalbumin (OVA) could be presented to T cells by viable APC, but that the T cells failed to respond when the APC were prefixed with glutaraldehyde. The T cell response could be rescued, however, if cyanogen bromide (CNBr)-cleaved peptides of OVA were added to the fixed APC. This cell-free proteolysis bypassed the need for metabolic processing and, for the first time, directly demonstrated an antigen processing requirement for T cell recognition. Interestingly, another protein, myoglobin, seems to require only denaturation or unfolding in order to be presented to T cells (Berzofsky et al., 1988).

T lymphocytes do not generally recognize soluble native protein antigens. Ag processing is thus an essential mechanism to transform the native Ag into a structure that can bind to MHC molecules on the surface of an APC and be recognized by T cells. At this point, it is appropriate to distinguish between reactions within lysosomes that are terminally degradative and those that are related to antigen processing. Many of the lysosomal reactions degrade proteins into their constituent amino acids, whereas antigen processing mechanisms do not necessarily use these pathways and only partially degrade protein antigens. Antigen processing mechanisms are generally less destructive and lead to the production of unfolded forms of the protein or peptides. These mechanisms may lead to the exposure of normally unexposed internal hydrophobic regions of proteins that have a high ability to bind to MHC antigens. This has been observed for a variety of protein antigens, including myoglobin (Berzofsky et al., 1988), lysozyme (Allen et al., 1984), and fibrinogen (Allen, 1987). The nature of the processing events, together with the influence of the different MHC haplotypes, can lead to the expression of a hierarchy of antigenic determinants for T cell recognition (Berzofsky et al., 1988; Gammon et al., 1987), which will ultimately determine the orientation of the immune response to suppression or stimulation.

Processing Mechanisms and APC Types

In order for an antigen to be processed by an APC, it must first bind either specifically (through cell surface Ig, or sIg, on B cells) or nonspecifically to a component on the plasma membrane (Delovitch et al., 1989). Subsequently, a sequence of biochemical events has been proposed to occur within the APC, which leads to the production of a proteolytically processed immunogenic peptide (Delovitch et al., 1988a,b). This proteolysis may be mediated by enzymes at either the plasma membrane (Buus and Werdelin, 1986; Semple et al., 1989) or within the cell (Semple et al., 1989). Ultimately, the antigen peptide associates with MHC molecules through the antigen binding site, leading to the formation of a specific peptide/MHC complex that is then recognized by the T cell receptor. The particular peptide(s) of a given antigen that binds to MHC molecules determine the magnitude of the T cell stimulation and thus the net immune response.

The nature of the APC that processes an antigen may also control the generation of various antigenic peptides, thereby allowing for differential T cell stimulation. For example, antigen-specific B cells express sIg receptors that bind and endocytose antigen, making a greater concentration of antigen available for processing by a B cell (Lanzavecchia, 1987). This leads to effective B cell presentation of antigen at concentrations up to 1,000-fold lower than those required for presentation by dendritic cells or macrophages. As a result, an enhanced immune response may follow, particularly a secondary response due to a larger memory B cell population. Nonspecific binding of an antigen to APC is probably the major route of antigen processing, due to the greater number of antigen nonspecific APC types within the host. Furthermore, MHC class II–negative cells, such as fibroblasts, can also process antigens, which can then be presented by other MHC class II–positive cells (Roska and Lipsky, 1985). These mechanisms thus increase the number of different cell types that can potentially process antigen and influence the T cell repertoire.

The critical experiments now generally accepted as showing whether a particular peptide derived from a protein antigen binds to MHC and stimulates T cells to secrete IL-2 are based on presentation of the peptide by fixed APC or by planar membranes containing Ia molecules (Shimonkevitz et al., 1983; Watts et al., 1984). Neither fixed APC nor planar membranes have processing capabilities, and thus a peptide cannot be further degraded in these systems. T cells activated by antigen presented in these systems secrete interleukin 2 (IL-2), which can be measured by its ability to stimulate the growth of IL-2–dependent cell lines such as CTLL.

To determine the cellular mechanisms involved in the processing of an antigen, various agents that alter cellular metabolism have been used. For example, chloroquine and NH_4Cl are concentrated within lysosomes and raise lysosomal pH. APCs treated with chloroquine are incapable of presenting some soluble protein and bacterial antigens to T cells, and this has been shown to correlate with inhibition of antigen processing (Unanue, 1984). Unfortunately, most of these drugs have pleiotropic effects on cells and may affect many intracellular reactions. More recently, therefore, much research has focused on the identification of APC enzymes that are capable of degrading native antigens into immunogenic peptides. One such enzyme system that has received much attention is the cathepsin system. Cathepsins are a group of thiol (cathepsins A, B, and L) and acid (cathepsin D) proteases that are found primarily in lysosomes. Recent experiments by Berzofsky and his colleagues (Berzofsky et al., 1988; Takahashi et al., 1989) using various specific inhibitors of the different cathepsins, particularly cathepsin B, showed that such inhibitors abrogate the presentation of myoglobin peptides to T cells. These results suggest that a single enzyme may be sufficient to cleave a native protein into immunogenic peptides.

An understanding of the biochemical events of antigen processing that occur within APCs is critical to the identification of processed peptides that stimulate a physiological immune response. It is possible that modification of processing events can alter the immunogenicity of proteins in a manner that will facilitate the development of efficient vaccines. Our laboratory is primarily interested in the processing and presentation of insulin. Below, we will summarize our attempts to use insulin as a model antigen in an effort to further understand the pathways of antigen processing and presentation by an APC and the influence of APC on immune responsiveness. Most insulin-treated diabetics produce detectable anti-insulin antibodies, and some insulin-resistant patients produce high-titred anti-insulin antibodies that neutralize the metabolic effects of the hormone.

An understanding of the mechanisms that stimulate an immune response to insulin may also help elucidate some of the underlying mechanisms of autoimmune type I diabetes.

PROCESSING AND PRESENTATION OF INSULIN

Human insulin is a small irregular protein with a molecular mass of 5.75 kD. It is made up of a 21 amino acid A chain that is disulfide-linked through residues A7–B7 and A20–B19 to a 30 amino acid B chain. The A chain contains an intrachain disulfide loop between residues A6 and A11. This loop is present in one of two amphipathic regions of insulin (DeLisi and Berzofsky, 1985) and is the immunodominant structure that is recognized by T cells from all mammalian species that have been tested, except guinea pigs (DeLisi and Berzofsky, 1985; Naquet et al., 1987).

The mechanisms of insulin processing within non-APC types have been extensively studied due to the important role of insulin in the regulation of various metabolic pathways (Duckworth, 1988). Two catabolic pathways have been proposed for the processing of insulin by adipocytes and hepatocytes. The first involves the reduction of the disulfide bonds of insulin by the enzyme glutathione insulin transhydrogenase, which results in the separation of the A and B chains (Katzen and Setten, 1962; Varandani, 1972). The second and probably most important pathway is the direct proteolysis of insulin by the enzyme insulin protease (EC 3.4.22.11) (Duckworth and Kitabchi, 1981), which cleaves insulin between residues B16 and B17. Both mechanisms result in the production of intermediate forms of degraded insulin and leave the molecule susceptible to further attack by nonspecific proteases. It is not known at present whether these mechanisms could be responsible for the generation of immunogenic peptides.

We have previously demonstrated that insulin has to be processed before it can bind to class II MHC molecules (Phillips et al., 1986) and that various peptides produced in vitro can stimulate insulin-specific T cells to secrete IL-2 (Naquet et al., 1987). The protocol we used to study in situ processing of insulin by an APC is a combination and modification of techniques previously devised to study insulin processing by non-lymphoid cells (Caro et al., 1982; Hamel et al., 1986). Figure 12–1 presents a flow chart summary of the methods used to generate insulin peptides and then isolate them from three APC compartments termed extracellular, plasma-membrane associated, and intracellular. The general scheme involves the exposure of APC to physiological concentrations (10^{-9} M) of radiolabeled insulin at 37°C for varying lengths of time (0–4 hours) (Semple et al., 1989). The processed insulin peptides are then isolated from the APC compartments and analyzed by various biochemical techniques, including high performance liquid chromatographs (HPLC), determination of amino acid composition, and N-terminal sequencing (Semple et al., 1989).

Processing of Radiolabeled Human Insulin by B Cell APC

Iodinated proteins have been used extensively to study a variety of biochemical processes. When B cell APCs are incubated with ^{125}I-labeled human insulin (HI), a number of iodinated peptides are generated in each cellular compartment in a time-dependent manner (Semple et al., 1989). One HI peptide isolated from all three compartments is composed of residues A1–A14 disulfide-linked to residues B7–B26 (A1–A14/B7–B26) (Semple et

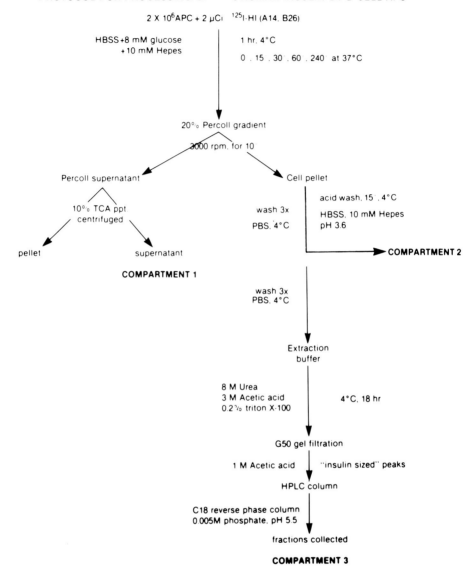

PROTOCOL FOR PROCESSING OF ^{125}I-HUMAN INSULIN BY B CELL APC

Figure 12–1.   Flow chart of the antigen-processing protocol used to produce insulin peptides in the three cellular APC compartments. Compartment 1, Extracellular compartment; Compartment 2, Plasma membrane-associated compartment; Compartment 3, Intracellular compartment.

al., 1989). The generation of this peptide in the extracellular compartment was due to release of an enzyme with characteristics similar to those of an insulin-specific degrading enzyme (Duckworth et al., 1988). Kinetic analyses revealed that this peptide appears at the plasma membrane within 15 minutes, then intracellularly by 30 minutes, and finally in the extracellular compartment by 60 minutes (Semple et al., 1989). The unlabeled peptide, when produced in sufficient quantity, can stimulate insulin-specific T cells to secrete IL-2. However, this peptide will not stimulate such T cells when it is presented by planar membranes or by fixed B cells, indicating that it requires further processing to be recognized by T cells (Semple et al., 1989). Nonetheless, these data indicate that insulin can be processed at various cellular sites within an APC and that processing at the plasma membrane precedes that which takes place intracellularly and extracellularly.

The above studies identified, for the first time, an immunogenic peptide produced by APC in situ. However, because an iodine molecule is large enough and sufficiently polar to change both the conformation and charge of insulin—and, hence, its natural biochemical activity—we sought to develop a radiolabeled insulin that was chemically identical to the unlabeled molecule. We reasoned that this would allow us to study the structure and immunogenicity of naturally processed insulin peptides produced by APC. Using recombinant HI that was biosynthetically labeled with ^{35}S- and ^{3}H-labeled amino acids in *E. coli* (Semple et al., 1988), we have been able to identify several distinct naturally processed peptides in various APC compartments. These physiologically processed peptides will allow us to more definitively map the biochemical pathways used by APC to process insulin into its immunogenic components.

Role of IDE in the Generation of Immunogenic Insulin Peptides

A third enzyme, termed insulin degrading enzyme (IDE), was recently described by Shii and Roth (1986) and is probably identical to the insulin protease described above (Duckworth and Kitabchi, 1981). It has been found in a variety of cell types, including red blood cells, hepatocytes, adipocytes, and lymphocytes. IDE is primarily a cytosolic enzyme, although a small proportion of IDE activity ($\leq 10\%$) is associated with the plasma membrane of lymphocytes (Yaso et al., 1987). Hamel et al. (1988) showed endosomal localization for this enzyme within hepatocytes. IDE is composed of a single polypeptide chain with a molecular mass of 110 kD, and it functions as a neutral thiol metalloproteinase with a pH optimum of 6–7, depending on cell type. The enzyme is relatively specific for insulin in that it cleaves insulin, glucagon, and insulin-like growth factor (IGF-II) but not proinsulin or IGF-I (Misbin and Almira, 1989). Sulfhydryl inhibitors, such as N-ethylmaleimide, inactivate IDE, but the activity of the enzyme is resistant to other protease inhibitors, such as leupeptin, pepstatin, and antipain. IDE has been shown to cleave insulin on the carboxy-terminal side of residues A11, A13, A14, and A17 of the A chain (Delovitch et al., 1989; Duckworth et al., 1987) and of residues B4, B9, B10, B12, B13, B14, B15, B16, B18, B20, B24, B25, B26, and B27 of the B chain (Delovitch et al., 1989; Duckworth et al., 1988). When they are injected into hepatoma cells, anti-IDE monoclonal antibodies can inhibit the degradation of radioiodinated insulin, indicating that the enzyme has a major role in intracellular insulin degradation (Shii and Roth, 1986). It has therefore been of interest to us to determine whether the enzyme has a primary role in the generation of insulin peptides that are immunogenic for T cells.

The major peptides of insulin produced by IDE digestion contain residues A1–A14 disulfide-linked to various B chain peptides such as B7–B26, B1–B9, or B5–B9 (Delovitch et al., 1989; Semple et al., 1989). All of these peptides contain the T cell immunogenic A-chain loop (A6–A11) which can activate T cells to secrete IL-2. Our aim has been to unravel the in situ pathways of insulin processing and presentation and to determine how

IDE Generated Peptides

1 B26-B30
2. B1-B4
3. B10-B13
4. B10-B14
5. A14-A17
6. B14-B16
7. B25-B30
8. A15-A21 / B13-B20
9. A1-A14 / B5-B9
10. A1-A14 / B1-B9
11. A12-A21 / B18-B26
12. A1-A14
13. A12-A20 / B19-B24
14. A16-A20 / B19-B27

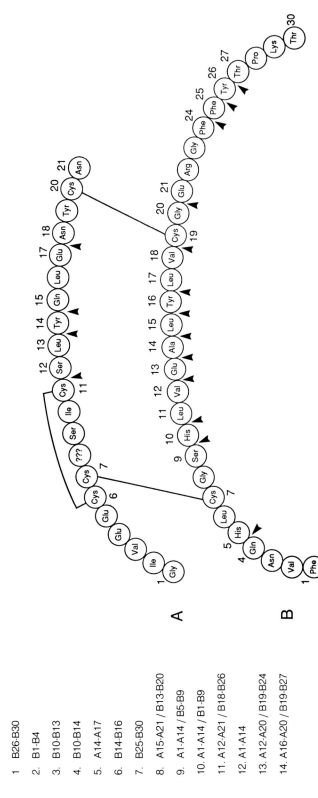

Figure 12–2.   Summary of the known cleavage sites of human insulin by IDE and the human insulin peptides purified by HPLC.

these pathways regulate T cell antigen recognition and antibody production by B lymphocytes (Delovitch et al., 1988b, 1989).

IDE can be readily purified from various cell types by immunoaffinity chromatography using monoclonal anti-IDE antibodies (not directed at the catalytic site of the enzyme) coupled to protein-A-Sepharose (Delovitch et al., 1989). IDE isolated in this manner retains its activity and can digest insulin into a variety of peptides that can be separated by HPLC. Figure 12-2 summarizes the IDE cleavage sites of HI and the peptides generated after an 18-hour digestion. The two peptides obtained in greatest yield comprise residues A1-A14/B1-B9 and A1-A14/B5-B9. Both peptides can stimulate insulin-specific T cells to secrete IL-2 when presented by unfixed B cell APC. However, the peptides fail to activate T cells when presented by prefixed APC, indicating that they require further processing. Similar results have been obtained for all insulin peptides generated to date, by a variety of methods (Naquet et al., 1987; Semple et al., 1989). Table 12-2 summarizes the immunological parameters of the various insulin peptides.

To further investigate whether IDE is primarily responsible for the generation of immunogenic insulin peptides by APC, experiments were performed using anti-IDE monoclonal antibodies and B cell APCs. When B cell APCs were pre-incubated with anti-IDE antibodies for 1 hour at 37°C and then exposed to radioiodinated HI, the generation of radiolabeled peptides was completely inhibited (Delovitch et al., 1989). This inhibition could be due to redistribution of the enzyme by the antibody, thus preventing contact between insulin and IDE. When they are present throughout the course of an in vitro APC assay, anti-IDE antibodies also inhibit the presentation of insulin by B cell APCs to T cells. These experiments suggest that IDE is a major enzyme within B cell APCs and is involved in processing of insulin into peptides that bind MHC class II molecules and acti-

Table 12-2. Summary of the Processing and Presentation Parameters of Various Insulins and their Peptides

Peptide (Source)	T Stimulation[a] (IL-2 secretion)		MHC Binding (Class II)
	Unfixed APC	Fixed APC	
A1-A14/B1-B16 (pork)	+[b]	−[c]	+[d]
A1-A14/B7-B15 (pork)	+	−	+
A1-A14/B7-B26 (human[e])	+	−	N/D[f]
A1-A14/B1-B9 (human)	+	−	+
A1-A14/B5-B9 (human)	+	−	N/D
A1-A14 (pork)	−	−	−
A chain (pork)	−	−	−
B chain (pork)	−	−	+
Insulin			
pork	+	−	−
human	+	−	N/D
beef	−	−	+

[a]Based on a T cell hybridoma that recognizes pork insulin in association with I-A^d.
[b]B cell APC containing I-A^d in its plasma membrane.
[c]No response detected at antigen concentrations as high as 100 μM
[d]Binding of I-A^d.
[e]Human and pork insulin peptides are identical; they differ only by one amino acid at residue B30.
[f]Not determined.

Table 12–3. Summary of the Monoclonal Anti-IDE Antibodies Used and Their Effects on Insulin Processing and Presentation

| Antibodies[b] | Inhibition Studies[a] | |
	Processing by B Cells[c]	Presentation by B cells to T cells[d]
VC6[e] (control antibody)	−	−
9B12 (non-catalytic site of human IDE)	+ + +	+ +
31H7 (catalytic site of human IDE)	+	+ +
28H1 (recognizes mouse and rat IDE)	+ +	+ + +

[a]Results of inhibition of processing experiments as determined by the ratio of control inhibition/specific inhibition $\times$ 100: (+), 25% inhibition; (+ +), 50% inhibition; (+ + +), 75% inhibition.
[b]From Shii and Roth (1986).
[c]Inhibition of production of radioiodinated HI peptides by pre-incubation with antibody.
[d]Inhibition by antibody of IL-2 production by insulin-specific T cell hybridomas.
[e]VC6 [anti-(anti-Ia.2)] (Phillips et al., 1984) was used as a control monoclonal antibody in all experiments.

vate T cells. Table 12–3 summarizes the effects of monoclonal anti-IDE antibodies on insulin processing and presentation to T cells.

SUMMARY

We have shown that the principal disulfide-linked insulin peptides produced by IDE can bind to class II MHC molecules but require further processing in order to stimulate T cells (Delovitch et al., 1989). It is therefore possible that IDE is responsible for early events in insulin processing (Fig. 12–3). Based on our analyses of the kinetics and cellular sites of insulin processing by B cells, and the T cell immunogenicity of insulin peptides, we have proposed a model for insulin processing into immungenic peptides by APC. Insulin binds to the APC and is initially cleaved by IDE either at the plasma membrane or intracellularly. This cleavage produces insulin intermediates, which can bind to MHC molecules if they are transported to the appropriate site. At this step, the insulin peptide assumes a particular conformation within the MHC antigen binding groove. This conformation depends not only on the cellular environment but also on the interaction of the peptide with the particular MHC molecule. If the MHC-associated peptide is susceptible to further proteolytic degradation, a new conformation would be generated, which would then be recognized by the T cell receptor, leading to stimulation.

MHC-directed processing of insulin is a possible explanation for our results obtained to date, and it may explain why certain mouse MHC haplotypes are either responders (H-2^d) or non-responders (H-2^k) to HI. It may also provide an explanation for the production of high titred anti-insulin antibodies by certain insulin-treated diabetic individuals. These possibilities are currently being studied to better understand how an immune response to insulin is generated.

Acknowledgments

We would like to thank Dr. Richard Roth, Department of Pharmacology, Stanford University, for generously supplying us with the monoclonal anti-IDE antibodies. We also extend special thanks to Janet Ellis and Edwin R. Speck for their superb technical assistance and to Drs. Alan H. Lazarus and Michael R. Christie for their helpful suggestions.

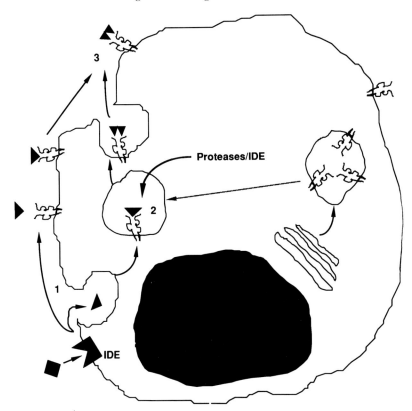

Figure 12–3. Proposed model of binding of insulin to an APC and the pathways of insulin processing by an APC. This model indicates that the class II MHC molecule may play a role in directing the processing of intermediate insulin peptides into minimal peptides available for T cell recognition. **Step 1:** Human insulin (diamonds) is partially digested by IDE, either at the plasma membrane or intracellularly. The intermediate products then bind to class II MHC molecules on the cell surface or in an endosome. **Step 2:** Proteases at the plasma membrane or within the endosome further degrade the intermediate insulin peptides bound to the MHC molecules. **Step 3:** The immunogenic minimal peptide-MHC complex is available on the plasma membrane for T cell recognition. Triangles refer to insulin peptides.

References

Allen, P. M. (1987). Antigen processing at the molecular level. *Immunol. Today* 8:270–273.

Allen, P. M., Strydom, D. J., and Unanue, E. R. (1984). Processing of lysozyme by macrophages: Identification of the determinant recognized by two T cell hybridomas. *Proc. Natl. Acad. Sci. USA 81*:2489–2493.

Berzofsky, J. A., Brett, S. J., Streicher, H. Z., and Takahashi, H. (1988). Antigen processing for presentation to T lymphocytes: Function, mechanisms, and implications for the T cell repertoire. *Immunol. Rev. 106*:5–31.

Buus, S., and Werdelin, O. (1986). Oligopeptide antigens of the angiotensin lineage compete for presentation by paraformaldehyde treated accessory cells to T cells. *J. Immunol. 136*:459–465.

Caro, J. F., Muller, G., and Glennon, J. A. (1982). Insulin processing by the liver. *J. Biol. Chem. 257*:8459–8466.

Cooper, E. L., Langlet, C., and Bierne, J., eds. (1987). *Developmental and Comparative Immunology.* New York: Alan R. Liss.

LeLisi, C., and Berzofsky, J. A. (1985). T-cell antigenic sites tend to be amphipathic structures. *Proc. Natl. Acad. Sci. USA* 82:7048–7052.

Delovitch, T. L., Lazarus, A. H., Phillips, M. L., and Semple, J. W. (1989). Antigen binding and processing by B cell APCs: Influence on T and B cell activation. *Cold Spring Harbor Symp. Quant. Biol.* 54:333–343.

Delovitch, T. L., Semple, J. W., Naquet, P., Bernard, N. F., Ellis, J., Champagne, P., and Phillips, M. L. (1988a). Pathways of processing of insulin by antigen presenting cells. *Immunol. Rev.* 106:195–222.

Delovitch, T. L., Semple, J. W., and Phillips, M. L. (1988b). Influence of antigen processing on immune responsiveness. *Immunol. Today* 9:216–218.

Draznin, B., Todd, W. W., Leitner, J. W., and Toothaker, D. R. (1981). Lysosomal and non-lysosomal pathways of intracellular insulin degradation in isolated rat hepatocytes. *Horm. Res.* 15:252–262.

Duckworth, W. C. (1988). Insulin degradation: Mechanisms, products and significance. *Endocrine Rev.* 9:319–345.

Duckworth, W. C., Hamel, F. G., Liepnicks, J. J., Peavy, D. E., Ryan, M. P., Hermodson, M. A., and Frank, B. A. (1987). Identification of A chain cleavage sites in intact insulin produced by insulin protease and isolated hepatocytes. *Biochem. Biophys. Res. Comm.* 147:615–621.

Duckworth, W. C., Hamel, F. G., Peavy, D. E., Liepnicks, J. J., Ryan, M. P., Hermodson, M. A., and Frank, B. A. (1988). Degradation products of insulin generated by hepatocytes and insulin protease. *J. Biol. Chem.* 263:1826–1833.

Duckworth, W. C., and Kitabchi, A. E. (1981). Insulin metabolism and degradation. *Endocrine Rev.* 2:210–233.

Gammon, G., Shastri, N., Cogswell, J., Wilbur, S., Sadegh-Nasseri, S., Krzych, U., Miller, A., and Sercarz, E. (1987). The choice of T cell epitopes utilized on a protein antigen depends on multiple factors distant from, as well as at, the determinant site. *Immunol. Rev.* 98:53–73.

Glieman, J., and Sonne, O. (1985). Uptake and degradation of insulin and of alpha$_2$-macroglobulin-trypsin complex in rat adipocytes: Evidence for different pathways. *Biochim. Biophys. Acta.* 845:124–130.

Haeckel, E. (1862). *Die Radiolarien* (Rhizopoda radiaria). Berlin: G. Reimer.

Hamel, F. G., Peavy, D. E., Ryan, M. P., and Duckworth, W. C. (1986). High performance liquid chromatographic analysis of insulin degradation by rat skeletal muscle insulin protease. *Endocrinology* 118:328–333.

Hamel, F. G., Posner, B. I., Bergeron, J.J.M., and Frank, B. A. (1988). Isolation of insulin degradation products from endosomes derived from intact rat liver. *J. Biol. Chem.* 263:6703–6708.

Horton, J. D. (1988). *Xenopus* and developmental immunobiology: A review. *Dev. Comp. Immunol.* 12:219–229.

Kaplan, D. R., Colca, J., and McDaniel, M. L. (1983). Insulin as a surface marker on isolated cells from rat pancreatic islets. *J. Cell. Biol.* 97:433–437.

Katzen, H. M., and Setten, Jr., D. (1962). Hepatic glutathione-insulin transhydrogenase. *Diabetes* 11:271–280.

Lanzavecchia, A. (1987). Antigen uptake and accumulation in antigen-specific B cells. *Immunol. Rev.* 99:39–51.

Lee, P., Matsueda, G. R., and Allen, P. M. (1988). T cell recognition of fibrinogen: A determinant on the A-chain does not require processing. *J. Immunol.* 140:1063–1068.

Misbin, R. I., and Almira, E. (1989). Degradation of insulin and insulin-like growth factors by enzyme purified from human erythrocytes: Comparison of degradation products. *Diabetes* 38:152–158.

Naquet, P., Ellis, J., Singh, B., Hodges, R. S., and Delovitch, T. L. (1987). Processing and presentation of insulin: I. Analysis of immunogenic peptides and processing requirements for insulin A loop-specific T cells. *J. Immunol. 139*:3955–3963.

Phillips, M. L., Harris, J. F., and Delovitch, T. L. (1984). Idiotypic analysis of anti-I-A^k monoclonal antibodies: I. Production and characterization of syngeneic anti-idiotypic mAb against anti-I-A^k Ab. *J. Immunol. 135*:2587–2594.

Phillips, M. L., Yip, C. C., Shevach, E. M., and Delovitch, T. L. (1986). Photoaffinity labelling demonstrates binding between Ia molecules and nominal antigen on antigen presenting cells. *Proc. Natl. Acad. Sci. USA 83*:5634–5638.

Roitt, I., Brostoff, J., and Male, J. (1989). *Immunology,* 2nd ed. London: Gower Medical.

Roska, A. K., and Lipsky, P. (1985). Dissection of the functions of antigen presenting cells in the induction of T cell activation. *J. Immunol. 135*:2953–2961.

Semple, J. W., Cockle, S. A., and Delovitch, T. L. (1988). Purification and characterization of radio-labelled biosynthetic human insulin from *E. coli:* Kinetics of processing by antigen presenting cells. *Mol. Immunol. 25*:1291–1298.

Semple, J. W., Ellis, J., and Delovitch, T. L. (1989). Processing and presentation of insulin: II. Evidence for intracellular, plasma membrane–associated and extracellular degradation of human insulin by antigen presenting B cells. *J. Immunol. 142*:4184–4193.

Shii, K., and Roth, R. A. (1986). Inhibition of insulin degradation by hepatoma cells after micro-injection of monoclonal antibodies to a specific cytosolic protease. *Proc. Natl. Acad. Sci. USA 83*:4147–4151.

Shimonkevitz, R., Kappler, J., Marrack, P., and Grey, H. M. (1983). Antigen recognition by H-2 restricted T-cells: I. Cell free antigen processing. *J. Exp. Med. 158*:303–316.

Takahashi, H., Cease, K. B., and Berzofsky, J. A. (1989). Identification of proteases that process distinct epitopes on the same protein. *J. Immunol. 142*:2221–2229.

Unanue, E. (1984). Antigen-presenting function of the macrophage. *Annu. Rev. Immunol. 2*:395–428.

Varandani, P. T. (1972). Insulin degradation: I. Purification and properties of glutathione-insulin transhydrogenase. *Proc. Natl. Acad. Sci. USA 69*:1681–1684.

Watts, T. H., Brian, A. A., Kappler, J., Marrack, P., and McConnell, H. M. (1984). Antigen presentation by supported planar membranes containing affinity purified I-A^d. *Proc. Natl. Acad. Sci. USA 81*:7564–7568.

Yaso, S., Yokono, K., Hari, J., Yonezowa, K., Shii, K., and Baba, S. (1987). Possible role of cell surface insulin degrading enzyme in cultured human lymphocytes. *Diabetologia 30*:27–32.

13

Ontogeny and Phylogeny of Complement

MATTEO ADINOLFI

Our understanding of the functions of the complement (C) system had its beginnings just before the turn of this century when it was shown that fresh serum had the ability to kill bacteria in the presence of antibodies (Bordet, 1909; Buchner, 1889; Ferrata, 1907; Pfeiffer and Issaeff, 1894). During the last decade, however, our appreciation of complement has changed substantially with the realization that, once activated, the proteins that make up the complement system perform several biological functions related to inflammation and thus play a major role in protection against infections by both specific and nonspecific mechanisms (Mayer, 1961; Müller-Eberhard, 1986; Rosen, 1974). Besides mediating cell lysis, the biological activities of complement include immune adherence, viral neutralization, chemotaxis, and release of lysosomal enzymes and histamine from basophils and mast cells. Other properties that have been attributed to some of the complement proteins are related to metabolism of bone, production of antibodies, and cell proliferation (Götze and Müller-Eberhard, 1976; Mayer, 1961; Müller-Eberhard, 1975, 1986; Porter and Reid, 1978; Rosen, 1974). Finally, several components of complement have been identified as "acute phase proteins" because their serum levels increase markedly during infection and inflammation (Adinolfi and Lehner, 1988; Lehner and Adinolfi, 1980).

In sera from either normal adults or newborn infants, the activation of complement, which is based on a cascade reaction, depends on the presence of all components in sufficient quantities to compensate for their rapid degradation following activation. Locally produced complement proteins are also essential for the triggering and maintenance of local inflammatory responses (Colten, 1976). It is not surprising, therefore, that the development of the complement system during fetal and perinatal life has been the subject of intense study and that, as a result of these studies, a clear pattern has slowly emerged about the timing and site of synthesis of many components of complement (Adinolfi, 1977, 1981; Colten, 1976; Rosen, 1974). In this chapter, the ontogeny and phylogeny of complement components will be reviewed after a brief account of their physicochemical and biological properties.

THE COMPONENTS OF COMPLEMENT AND THEIR MECHANISMS OF ACTIVATION

The complement system consists of about 30 plasma and cell membrane proteins, which, upon activation, interact with each other and acquire new biological properties. A unique characteristic of many components of complement is therefore their inherent ability to be transformed from soluble molecules to membrane constituents through the generation of fragments endowed with specific binding regions. In fact, cleavage of many complement proteins results in the formation of a fragment, or fragments, capable, for short periods of time, of binding to specific receptors on cells or bacteria or to other complement components (Götze and Müller-Eberhard, 1976; Müller-Eberhard, 1975, 1986; Porter and Reid, 1978; Rosen, 1974).

Some of the properties of the principal complement components and their levels in sera from normal adults are listed in Table 13–1. From an operational point of view, activation of the complement system proceeds along either of two pathways, which converge at the level of the cleavage of C3 by the classical and alternative "convertases" (Fig. 13–1).

Activation of the classical pathway is initiated when antigen-antibody (Ag-Ab) com-

Table 13–1.   Some Properties of Human Components of C and Their Levels
in Sera from Normal Adults

Protein	Levels in Serum (mg/100 ml)	Molecular Weight (kD)	Major Fragments	Chromosome Gene Location
Classic components				
C1q	18	410		1
C1r	5	83		12
C1s	5	83		12
C2	2.5	102	C2a, C2b	6
C4	43–64	206	C4a, C4b	6
Alternative pathway				
Properdin (P)	20	150		X
Factor B (B, C3 proactivator, C3PA)	20–22	93		6
Factor D (D, C3 proactivator convertase)	0.1–0.5	24		
C3	130	180	C3a, C3b, C3c, C3d	19
Terminal components				
C5	7	190	C5a, C5b	9
C6	6.5	120		5
C7	5.5	110		5
C8	8	151		1
C9	6	71		5
Regulatory proteins				
C1 inhibitor (C1-INH)	18	105		11
Factor I (C3b-INA)	3.5	100		4
C4 binding protein	25	500		1
Anaphylatoxin inactivator	4	300		
Factor H (β_1H)	13.3	150		1

CLASSICAL ACTIVATION   **ALTERNATIVE ACTIVATION**

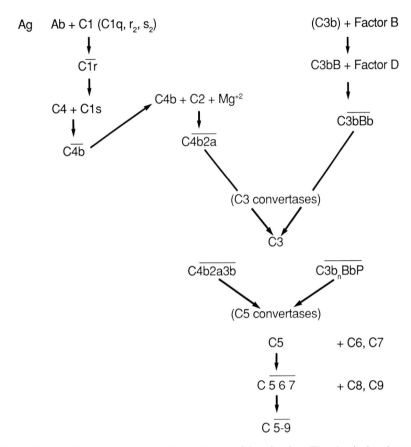

Figure 13–1.   Classical and alternative pathways of C activation. The classical pathway usually starts with the interaction of C1 with Ag-Ab complexes, whereas the alternative pathway is activated by fungal and bacterial cell wall polysaccharides. The C3 convertases formed by the two pathways cleave C3 and start the chain reaction involving the late components of C.

plexes interact with the first component of complement (Colten, 1976; Müller-Eberhard, 1986; Porter and Reid, 1978; Rosen, 1974). The activated components of complement are indicated by adding a bar above the symbol; thus $\overline{C1}$ designates the enzymatically active form of C1, and $\overline{C567}$ designates the activated complex of C5, C6, and C7 molecules. The fragments produced after cleavage are described by adding lowercase letters, as, for example, for the fragments of C3: C3a, C3b, C3c, and C3d. Most fragments have transient biological activity and then become inactive (e.g., iC3b). Human IgM and IgG antibodies, which undergo conformational changes following their interaction with specific antigens, are capable of initiating activation of complement via the classical pathway, although activation of the first component (C1) may also be induced by other factors such as C-reactive protein (Kaplan and Volanakis, 1974).

C1 complex ($C1qr_2s_2$) is present in serum as a calcium-dependent macromolecular structure made up of C1q, the recognition unit; and ($C1r,C1s)_2$, a catalytic subunit that comprises a tetrameric association of two serine protease proenzymes designated C1r and C1s (Carter et al., 1984; Müller-Eberhard and Lepow, 1965; Naff et al., 1964). Binding of C1q to immunoglobulin (Ig) leads to the conversion of C1r and C1s into active proteolytic enzymes. Activated C1s then interacts with C4 and C2, generating the C3 convertase. C1q molecules are made up of six globular "beads," each joined by a collagen-like connecting strand to a fibril-like central fraction. Electron micrographs of C1 complexes stabilized by chemical cross-linking indicate that the (C1r,C1s)$_2$ subunits fold into the center of the C1q molecule between the globular beads and the central bundle. From a functional point of view, the most important feature of the C1 complex is the location of (C1r,C1s)$_2$ in the center of C1q, allowing for close contact between the C1r and C1s catalytic sites.

cDNA clones for the C1r and C1s proenzymes have been isolated from a human liver library (Tosi et al., 1987), and analysis of their amino acid sequences has shown 40% homology between C1r and C1s (Carter et al., 1984). Both C1r and C1s genes map to the short arm (p13) of the human chromosome 13, and the genes are tightly linked (Tosi et al., 1987). C1r and C1s are both synthesized and secreted as single-chain polypeptides of 80 kD. When activated, C1r and C1s yield C-terminal catalytic chains of approximately 25 kD, which are disulfide-bound to N-terminal noncatalytic chains of 60 kD (Campbell et al., 1988).

In the classical pathway, C3 convertase is derived from C4 and C2 components. After proteolytic activation by the C1 complex, C4 is cleaved into two fragments: the larger, C4b, may bind directly to cell membranes. Alternatively, free C4b has been shown to enhance the activity of C1s in cleaving C2, its other substrate. This results in binding of a major component of C2 by cell surface receptors and formation of the $\overline{C42}$ complex. This complex, whether bound to cellular receptors or present in the fluid phase, acts as a C3 convertase and cleaves the third component of complement (C3). The genes coding for C2 and C4 in humans, as well as that which encodes Factor B, map to chromosome 6, between the MHC class I (HLA-A, B, and C) and class II (HLA-DR) loci (Fig. 13–2). The isolation of cDNA clones for these complement components has been achieved through screening of liver cDNA libraries with oligonucleotides synthesized on the basis of the known protein sequences (Campbell et al., 1988). These cDNA clones were then used to screen cosmid libraries of human genomic DNA to isolate clusters of overlapping

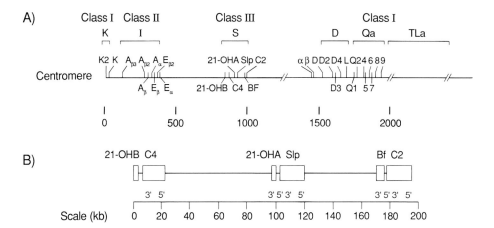

Figure 13–2. Molecular map of the human major histocompatibility complex (MHC). Class III encompasses the genes for C4, Factor B, and C2.

clones that contained the corresponding genes. These studies have revealed that C2 and Factor B genes are very closely linked. The two C4 loci (C4A and C4B) and the two loci for 21-hydroxylase (21OH), whose deficiency is responsible for congenital adrenal hyperplasia (CAH), map approximately 30 kb from the Factor B gene.

Comparison of C4A and C4B cDNA sequences has shown that the two isotypes are 99% homologous (Campbell et al., 1988). However, analysis of C4A and C4B amino acid sequences has provided the basis for understanding the serological (Giles, 1988) and functional (Belt et al., 1984; Law et al., 1984) differences between the two C4 isotypes. These two proteins, which adsorb onto the surface of red blood cells and can be detected as the blood group antigens Rodgers and Chido (Giles, 1988), are structurally similar to C3 and C5 and share homology with the proteinase inhibitor α_2-macroglobulin (Fig. 13–3). C4 is synthesized as a single chain protein and is subsequently processed to form a disulfide-linked three-chain (β, α, and γ) structure. This results in the removal of a signal peptide and four basic residues from between the β-α and the α-γ chain junctions (Bentley, 1988). C4, C2, and Factor B are highly polymorphic proteins. Although several cases of deficiency of C4 and C2 have been described, no individual with a complete deficiency of

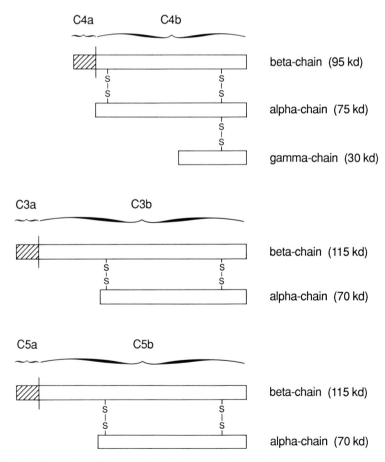

Figure 13–3.   C3, C4, and C5 are synthesized as single polypeptide proteins, which are then processed by proteolytic cleavage into subunits (α, β, and γ chains).

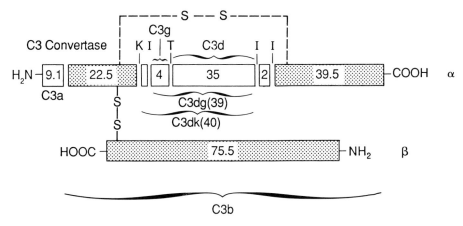

Figure 13–4. Cleavage of C3. The sequential degradation of native C3 to its biologically active fragments occurs in four steps. **Step 1** involves the cleavage of the α chain by the classical or alternative pathway convertases to form C3a and C3b. **Step 2** is mediated by Factor I in the presence of Factor H or the C3b receptor and results in formation of iC3b. The enzymes responsible for **step 3** cleavage to form C3d,g and C3c are not well defined. **Step 4** involves the cleavage of C3d,g by tryptic enzymes to form C3d and C3g. Numbers indicate molecular mass in kilodaltons (kD).

Factor B has been reported to date; this probably reflects the critical role of this component of complement. The complete sequence analysis of C2 and Factor B has demonstrated that they are closely related proteins, sharing 39% identity. Unlike other serine proteases, the N-terminal polypeptides of C2 (C2b) and Factor B (Ba) are not covalently linked by disulfide bonds to their respective catalytic chains (C2a and Bb) (Bentley, 1988).

C3 is synthesized as a single chain precursor (pro-C3) which is then processed by proteolytic cleavage into two subunits (Adinolfi et al., 1981; Müller-Eberhard, 1975; Porter and Reid, 1978; Sottzup-Jensen, et al., 1985). In its mature form, C3 is thus made up of two polypeptide chains with molecular mass of approximately 120 kD for the α chain and 75 kD for the β chain (Fig. 13–4). As a result of activation by either the classical C4,2 convertase or the alternative C3B(n)Bb convertase, the α chain is cleaved at several sites, yielding six fragments. This sequential degradation of C3 occurs in four distinct steps and starts with hydrolysis of the Arg-77–Ser-78 bond to generate C3a (9 kD) and C3b (186 kD). This proteolytic cleavage induces a rapid conformational change in C3b and leads to exposure of internal thioester bonds. C3b is then cleaved into two fragments that are still disulfide-bonded to the β chain. During the next step, C3b is cleaved at another site within the α chain, leading to the formation of C3d,g, which remains attached to the C-activating target, and of C3c, which is released in the fluid phase. The final stage of cleavage involves the splitting of C3d and C3g, mediated by plasmin, trypsin, and leukocyte elastase or cathepsin C.

The activation of C5 takes place by a mechanism similar to that of C3 (Bhakdi and Tranum-Jensen, 1984; Porter and Reid, 1978). C5 molecules are also formed by two polypeptide chains, with a molecular mass of 112 and 74 kD, respectively. The classical C5 convertase—C4b,2a,3b—like the alternative C5 convertase—C3b(n)Bb—cleaves the α chain at Arg-74-75 releasing C5a (11 kD). The larger fragment interacts with C6 to form C56 and promotes the assembly of the multimolecular C5–C9 complexes capable of damaging cell membranes. Recent studies have confirmed that these complexes, formed by C5b-8 monomers to which several molecules of C9 are bound, exhibit the morphology of a hollow protein channel (Bhakdi and Tranum-Jensen, 1984, 1991; Humphrey and Donsmashkin, 1969). When embedded within the lipid bilayer of a cell, aqueous trans-membrane pores are generated, which represent the major lesion caused by the activation of C and produce the lysis of single cells.

Activation of the alternative pathway is mediated by polysaccharides present on bac-terial and fungal cell walls or by aggregated IgA (Müller-Eberhard, 1975; Pillemer et al., 1954), and it involves the participation of Factor B, Factor D and native C3, which, in the presence of Mg^{2+}, generate the first alternative C3 convertase. When Factor B com-bines with C3b, it becomes susceptible to proteolysis by Factor D, which is a low molec-ular weight esterase, resistant to all protease inhibitors present in plasma, and is present in blood as a fully active enzyme. The alternative C3bBb convertase is a powerful C3 splitting enzyme that, upon binding, generates a receptor for properdin (P). Apparently P receptors reside in at least two closely spaced C3b molecules which, in combination with Bb and P fragments, result in the expression of C5 convertase (Kinoshita, 1991).

Activation of the C system is controlled at two basic levels (Müller-Eberhard, 1976). The first depends on the short biological half-life of most of the activated fragments; the cleaved products and complexes (e.g., $\overline{C567}$) are in fact unstable and decay rapidly. The second regulatory mechanism reflects the presence in sera of several inhibitors and inac-tivators that by combining with specific activated fragments block their functions or destroy them.

The regulatory proteins listed in Table 13–1 have been isolated and characterized. C1-inhibitor (C1-INH) combines stoichiometrically and inhibits the esterolytic activity of C1 either in fluid phase or bound to cells. C1-INH is polyspecific as it also inhibits the enzymatic activity of plasmin, kallikrein, lysosomal enzymes, and the activated Hage-man factor (Factor XIIa). Factor I (C3b-INA) binds to C3b in the fluid phase or on the cell surface and stops further activation of C3 convertase.

In addition to Factor I and Factor H, other serum proteins, such as the decay accel-erating factor (DAF) and the C4 binding protein (C4BP), act as regulatory proteins, as do membrane receptors for complement components (CR) (see next section).

cDNA clones for Factor H, DAF, C4BP, and the complement receptors CR1 and CR2 have been obtained (Bentley, 1988), and complete amino acid sequences are avail-able for all these proteins. Although they differ in molecular size, their structures are highly homologous and contain 60 amino acid repeating units termed SCR (short con-sensus repeats) (Reid et al., 1986).

Linkage analysis of allotypes of Factor H, C4BP, and CR1 indicates that these loci are closely clustered, while analysis of human-mouse cell hybrids and in situ hybridiza-tion have shown that they map to chromosome 1, as does the DAF gene (Bentley, 1988).

RECEPTORS FOR COMPONENTS OF COMPLEMENT

The capacity of a host to respond to bacterial infections also depends on the availability of tissue receptors for specific activated complement components. Several reviews on this topic have recently been published (Cooper et al., 1988; Fearon and Wong, 1983; Ross and Atkinson, 1985; Rother and Rother, 1986) so we will present only a brief account of this complex system (Table 13–2). Cellular receptors for C1q have been detected on neu-trophils, monocytes, null cells, and B lymphocytes using either the protein labeled with

Table 13–2.   C Receptors (CR) and Cells Expressing Them

Receptors	Specificity	Cells
C1q	C1a	Neutrophils, monocytes, null cells, and B lymphocytes
C3a	C3a (amino acids 72–77)	Mast cells
CR1	C3b, C4b, iC3b	Red cells, neutrophils, monocytes, macrophages, eosinophils, T and B lymphocytes, and glomerular podocytes
CR2	C3d, C3dg, iC3b, C3b	B lymphocytes
CR3	iC3b	as CR1 receptors
C5a	C5a	Mast cells, neutrophils, monocytes, macrophages
Factor H	Factor H	B lymphocytes, monocytes, neutrophils
DAF	C4b2a, C3bB4	Red cells, lymphocytes, platelets

[125]I or fluorescein, or red blood cells coated with C1q (Fearon and Wong, 1983; Ross and Atkinson, 1985; Tenner and Cooper, 1981). Analysis of binding of C1q labeled with ^{125}I suggests the presence of 150-kD receptors on the surface of granulocytes. The intact C1 molecular complex does not bind to mononuclear cells, suggesting that only the collagen-like tail region of C1q interacts with its specific receptor.

Four proteolytic products of C3 cleavage—C3b, iC3b, C3d,g, and C3d—have been shown to interact with three types of cellular receptors (Bianco et al., 1970; Dobson et al., 1981; Fearon and Wong, 1983; Ross and Atkinson, 1985). CR1 receptors are present on erythrocytes, neutrophils, eosinophils, monocytes, macrophages, B cells, some T cells, and mast cells, as well as on glomerular podocytes and bacteria. Natural killer cells seem to lack CR1 receptors. The presence of the receptors on different types of cells is best demonstrated using polyclonal and monoclonal antisera. Four codominantly expressed CR1 genes have been identified. The molecular size and gene frequency of each type are: A, 190 kD (0.83); B, 220 kD (0.16); C, 160 kD (0.01); D, 250 kD (<0.01) (Ross and Atkinson, 1985). The number of receptors expressed by erythrocytes is genetically regulated by two autosomal codominant alleles for high and low numbers of receptor molecules per cell. Thus, homozygotes for the "high" allele have 800,000 receptors per cell, whereas homozygotes for the "low" allele have 250,000 C1r receptors per cell (Fearon and Wong, 1983; Wilson et al., 1982). While only 12% of normal individuals are homozygous for the "low" allele, 53% of patients with systemic lupus erythematosus (SLE) are homozygous for this allele (Fearon and Wong, 1983; Miyakawa et al., 1981). This suggests that an inherited predisposition for the development of this disease may be associated with low or absent levels of CR1 receptors on glomerular podocytes (Kazatchrine et al., 1982). The number of receptors varies from one type of cell to another and in vitro it may be increased or decreased by changes in temperature (Fearon and Wong, 1983).

CR2 is present on human B lymphocytes as a 140-kD membrane glycoprotein, which, according to recent results, may also act as a receptor for Epstein-Barr virus (EBV) (Cooper et al., 1988). Isolated CR2 binds to the virus, and polyclonal anti-CR2 antisera may prevent EBV from binding and infecting B cells (Fearon and Wong, 1983; Ross and Atkinson, 1985). Biosynthetic approaches, as well as enzymatic investigations, have shown that CR2 is derived from a 115-kD precursor molecule that is later glycosylated to a 145-kD protein (Cooper et al., 1988). Weiss et al. (1986) isolated a partial cDNA clone for CR2 from a human B cell library and established that CR2 and CR1 are structurally related at the DNA level. At the protein level, the two receptors showed 60%–75% homology and contained an amino acid consensus repeat motif of 6–10 conserved residues (Bentley, 1988; Reid et al., 1986). This structural feature is found in six other C3 binding proteins, as well as in several unrelated proteins, including the interleukin receptors. CR1 and CR2 genes are located on the long arm of human chromosome 1 at band 3.2 (C1q32), suggesting that they were derived from a common ancestor by gene duplication. CR3, the iC3b receptor, is one of three related cell surface molecules, each made up of two noncovalently associated glycoprotein subunits, termed α and β. The α chains of these molecules are distinct but show sequence homology, while the β chains (95 kD) are identical. Patients who are deficient in this group of related molecules have been reported to suffer from recurrent infections (Fearon and Wong, 1983; Ross and Atkinson, 1985). Research on the structure and role of complement receptors is rapidly expanding, and the results of these studies will undoubtedly provide further insights into the functions of these molecules (Fearon and Wong, 1983; Ross and Atkinson, 1985).

DEVELOPMENT OF THE COMPLEMENT SYSTEM

Several approaches have been used to investigate the time and site of synthesis of complement components during fetal and perinatal life in humans and animals (Adinolfi, 1972, 1977; Colten, 1973, 1976). One of the first observations relating to the production of complement during fetal life was made by Hyde (1932). During the course of investigating a strain of guinea pigs with a recessively inherited deficiency of one component of complement, he demonstrated that heterozygous mothers could have offspring lacking the specific complement component. Furthermore, this protein did not cross the rodent placenta because the sera of the newborn guinea pigs—homozygous for the defect but born to heterozygote mothers—were not biologically active.

The first approach to investigate the levels of complement in normal human infants was based on comparison of total hemolytic activity in paired samples of cord and maternal blood (Arditi and Nigro, 1957; Traub, 1943; Wasserman and Alberts, 1940). These studies showed that complement activity was indeed present in cord blood, although the total lytic activity was only about half that detected in maternal samples.

In 1961, Fishel and Pearlman measured the serum levels of the first four components of complement (C1, C2, C4, and C3) using the so-called "R" reagents. These were samples of sera in which individual components of complement had been artificially inactivated. Using this method the relative amounts of the various components of complement were confirmed to be less in cord blood than in paired maternal samples. Within the limits of the techniques available at the time, it was also possible to establish that the reduced complement activity in cord sera was not due to the presence of anti-complementary factors.

During the last 15 years, the levels of the various complement components have been measured in fetal and newborn sera using specific antibodies, either in functional hemolytic assays or by immunoprecipitation techniques (Adinolfi and Beck, 1976; Adinolfi et al., 1968; Ballow et al., 1974; Fireman et al., 1969; Frank, 1979; Kohler, 1973; Sawyer et al., 1971). In some studies, the levels of the individual complement components were expressed as absolute concentrations, but in other investigations the amounts were reported as percentages of mean values in maternal sera or in sera from normal adults. Because the levels of some components of complement are elevated in maternal

Table 13-3. Mean Levels of Some Components of C in Newborn Sera[a]

Components of C (mg/ 100 ml or % of standard)	Normal Adult $\overline{X}$	Mother $\overline{X}$	Newborn $\overline{X}$	Levels in Newborn as Percentages of	
				Normal Adults	Maternal Sera
C1q (mg/100 ml)	18.5	16.4	12.4	67	75
C4	52.0	56.0	32.0	61	57
C4	48.1	53.3	29.9	62	56
C3		178.3	88.8		49
C3		139.3	75.0	54	
C3	130.0	143.4	54.4	42	38
C3	150.0	152.0	86.0	57	56
C5		11.9	5.8		49
C5	76.0	10.3	62.0	81	60
C7(%)	98.2		67.3	68	
C9(%)	93.0	176.6	16.5	18	9
Properdin (mg/100 ml)	19.3	23.1	156.0	81	67
Factor B (mg/100 ml)	14.4	27.1	11.4	79	42
Factor B (mg/100 ml)	33.0		17.0		51
Factor H	96.3	170.0	53.0	55	31
Factor I	105.0	140.0	60.0	57	43

[a]Some of the values have been recalculated from the original data (see Adinolfi 1977, 1981); values may differ from those listed in Table 13-1 due to the different "standard" used in early assays.

sera, figures expressed as percentages of mean values in maternal blood should be appropriately corrected. Table 13-3 lists the mean levels of most complement components in newborn samples.

C3 and C4 levels have been estimated in fetal sera using specific antibodies and immunoprecipitation techniques (Adinolfi et al., 1968; Fireman et al., 1969; Frank, 1979; Sawyer et al., 1971). These components are present in sera from human fetuses of more than 9 or 10 weeks gestational age, and their concentrations were found to increase with age. In cord sera the mean levels of C3 range from 38% to 56% of maternal values (Fig. 13-5). When expressed as percentages of the levels in sera from normal adults the mean amounts of C3 in cord sera vary from 42% to 71%. In newborns at term the mean levels of C4, expressed as the percentage of maternal mean values, range from 55% to 58% (Fig. 13-6), while slightly higher levels (60%–80%) were observed when concentration was expressed as the percentage of the mean values in sera from normal adults (Davis et al., 1979; Frank, 1979). The mean levels of C2 in cord sera are near 76% of the values detected in maternal samples (Adinolfi, 1981; Ballow et al., 1974; Sawyer et al., 1971).

C1q has been detected in sera from human fetuses of more than 20 weeks gestation and in all samples from newborns at term; the mean levels of C1q in cord blood are 65% of those in maternal samples, but the levels increase rapidly during the first few days after delivery and reach normal adult levels within a few weeks (Adinolfi, 1972; Frank, 1979; Mellbye et al., 1971). Samples from human fetuses of more than 5 weeks gestational age contain C5, and the mean levels of this complement component are near 50% of the values in maternal blood (Adinolfi, 1972; Frank, 1979; Fireman et al., 1969). The late complement components—C6, C7, C8, and C9—have all been detected in sera from normal newborns (Adinolfi, 1977; Adinolfi and Beck, 1976). The mean level of C7 in cord blood

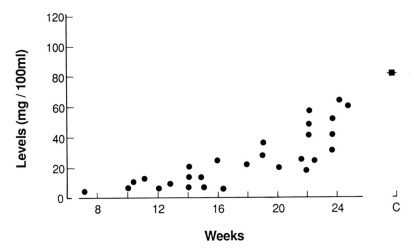

Figure 13–5.   Individual levels of C3 (mg/100 ml) in sera from human fetuses (ages indicated in weeks), and mean level of C3 in cord blood of infants born at term. (Modified from refs. 14,15.)

is near 70% of that in sera from normal individuals (Fig. 13–7), whereas the mean concentrations of C6 and C8 seem to be about 50% of the values detected in maternal samples.

In cord blood, levels of C9 are occasionally low and values near 50% of the mean levels in normal sera are rarely observed; this complement component has been detected only in human fetuses over 18 weeks old (Adinolfi and Beck, 1976). Factor B, Factor D, and properdin have all been detected in cord blood. Factor B has been detected in fetuses of over 16 weeks gestation, and its mean levels in newborn sera range from 50% to 70% of the values in sera from normal adults (Adinolfi and Beck, 1976; Stossell et al., 1973). It has been shown that the mean level of properdin in cord sera ranges between 53% to

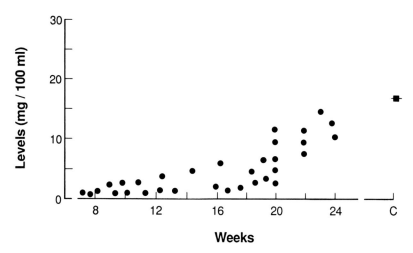

Figure 13–6.   Individual levels of C4 (mg/100 ml) in sera from human fetuses, and mean level of C4 in cord blood of infants born at term. (Modified from refs. 14,15.)

71% of the mean adult values. The concentration of properdin increases rapidly during the first months of life, and adult values are reached by about 1 year of age.

Regulatory proteins have also been detected in fetal and cord blood; C1-INH has been detected in fetuses as early as 4 weeks gestational age (Gitlin and Biasucci, 1969). Sera from fetuses more than 14 weeks old contain both Factor I (Fig. 13–8) and Factor H. The mean levels of these components of complement in cord blood are near 60% of the values detected in sera from normal adults (Adinolfi et al., 1981). As mentioned previously, the levels of complement components in cord blood have often been compared with those in maternal sera. Some complement components are present in higher concentrations during pregnancy than in normal adults; for example, the ratios of maternal to normal adult serum levels of C3 and C4 are near 1.8 and 1.9. Higher levels of C5, C9, and Factor B have also been detected in maternal sera than in sera from normal adults. In contrast, the levels of C1q and C2 in maternal sera do not seem to differ from those in normal adults (Adinolfi, 1981).

Due to the different techniques employed, some discrepancies have been noticed with regard to the estimation of complement components in maternal sera. For example, Mak (1978) detected levels of C3 and C4 in maternal sera that were 48% and 77% higher than the mean values in normal adults. Less dramatic increases have been noticed by other investigators. Mak used functional tests for his assays and found mean values of C3 and C4 in normal adults of 85% and 74%, respectively, of a standard control made up of pooled sera from 20 normal individuals; this inevitably affected his results. It is of interest that some of the components of complement that behave as "acute phase proteins" may be elevated in maternal sera. This is particularly true for C9 and Factor B, which, together with other acute phase proteins, are increased in sera during gestation (Adinolfi, 1981). The biological significance of the increased mean levels of Factor H and C3b-INA in sera from pregnant women is less clear due to the fact that these two complement components seem to play an important role in reducing the activity of the alternative pathway of complement.

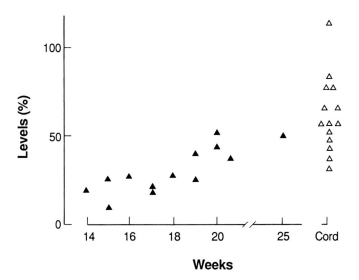

Figure 13–7. Levels of C7 in fetal and cord sera, expressed as percentages of the mean level in sera from normal adults (C7 = 5.5 mg/100 ml). (Modified from ref. 58.)

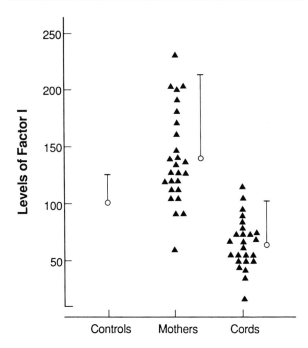

Figure 13–8.   Serum levels of Factor I in a group of normal adults, maternal sera, and cord samples, expressed as percentages of the mean value in adults (4 mg/100 ml). Means ± 2 SD are also indicated.

SITE OF SYNTHESIS OF COMPONENTS
OF COMPLEMENT DURING FETAL LIFE

Of the various approaches that have been used to study the site of synthesis of complement components during development, the most important is based on incubation of fetal tissues in media containing labeled amino acids. The cultured fluids are then dialyzed to remove excess labeled amino acids and then assayed by immunoelectrophoresis using specific antibodies. This is followed by autoradiography of the immunoplates (Adinolfi et al., 1968; Permutter and Colten, 1986; Thorbecke et al., 1965). The results of these tests should be critically evaluated, however, as protein interactions may produce false results. When possible, it is important to confirm the in vitro synthesis of complement components by demonstrating that they are present in the culture fluid as biologically active proteins.

The in vitro culture technique has the advantage that, in addition to providing information about the time of onset of synthesis, it identifies the tissues that are involved in the production of specific complement components. Newly synthesized C3 and C4 have been detected in culture supernatants of liver from human fetuses of more than 8 weeks gestation (Table 13–4). The de novo synthesis of these complement components has often been confirmed by demonstrating that they are hemolytically active. Synthesis of C3 and C4 has also been observed when human peritoneal cells and alveolar cells are cultured in vitro; synthesis of these components of complement was detected in tissues of fetuses more than 14 weeks old (Adinolfi and Beck, 1976; Permutter and Colten, 1986).

Table 13-4. Site of Synthesis of Human Components of C
During Fetal Life

Component		Age (weeks)[a]
C1	Intestinal epithelium; macrophages	19
C1q	Spleen; fibroblasts	14
C2	Macrophages (liver)	8
C4	Macrophages (liver)	8
C3	Liver cells	8
C5	Liver	9
	Spleen; liver	8–14
C6	Unknown, probably liver	—
C7	Liver (?)	14
C8	Probably liver	—
C9	Liver	12–18
C1-INH	Liver; macrophage	4

[a]Early detection using in vitro cultures. For references see Adinolfi (1972, 1977, 1981).

There is general agreement that in both humans and experimental animals, C4 and C2 are synthesized mainly by macrophages, and detection of these two complement components in liver cultures was attributed to the presence of macrophages in fetal liver tissues (Permutter and Colten, 1986).

In the past there has been some controversy about the site of synthesis of C1. Early studies by Colten et al. (1966) suggested that C1 was synthesized primarily, if not exclusively, in the small and large intestine in humans and guinea pigs. Almost simultaneously, Thorbecke et al. (1965) described the production of C1q by human and monkey liver, spleen, bone marrow, and lung cells, as well as by peritoneal or lung macrophages. The controversy about the site of synthesis of C1 was solved when it was shown that intestinal epithelial and mesenchymal cells synthesized the three subcomponents of C1 in quantities much higher than any other tissues, including monocytes (Permutter and Colten, 1986).

In human fetuses, most C1 macromolecules are produced by the columnar cells of the small intestine and colon; this has been confirmed by the demonstration that a human embryonal cell line derived from intestine (MA177) produces large quantities of these molecules in vitro. Early reports that C1q was synthesized mainly by lymphoid cells (Day et al., 1969) have not been confirmed, and there is no evidence at the present time that lymphocytes produce these complement components during fetal life. Details about the site and onset of synthesis of other complement components are still incomplete, but there is general agreement that C5 is produced mainly by liver cells from fetuses of more than 8 weeks gestational age (Adinolfi, 1981; Colten, 1973); synthesis of C5 was also shown using spleen, colon, lung, and peritoneal cells from human fetuses (Adinolfi et al., 1968). Evidence that spleen cells produce C5 has been provided by the demonstration that transplantation of bone marrow and spleen cells from normal mice into C5-deficient animals leads to the appearance of C5 in the recipients' serum (Phillips et al., 1969).

Fetal liver seems to be the site of synthesis of C7 and C9, although the types of cells that produce these two complement components have not yet been identified (Adinolfi, 1981). It has been suggested that C7 and C9 are synthesized by hepatocytes. This is consistent with the observation that rat hepatoma cells produce C9 together with albumin and C3, but do not synthesize C1 or C4 (Adinolfi, 1981).

With the exception of C1-INH, there is no information about the sites of synthesis of the regulatory components of complement during fetal life. Newly synthesized C1-INH was detected by Gitlin and Biasucci (1969) in culture supernatants of liver from a 4-week-old human fetus. This early synthesis of C1-INH by fetal liver was later confirmed, and the newly synthesized protein was shown to be biologically active (Campbell et al., 1986). It is of great interest that the rate of synthesis of this complement component in tissues from an 11-week-old fetus seemed to be similar to that observed in tissues from normal adults. In the adult, C1-INH is also produced by mononuclear phagocytes.

GENETICS OF COMPLEMENT COMPONENTS AND THEIR DEVELOPMENT

Many components of complement show a surprisingly high degree of genetic polymorphism. Several complement genes have now been cloned, and their chromosomal localization has been established (Campbell et al., 1986). The detection of different complement phenotypes in paired maternal and cord samples has often been exploited to investigate the ontogeny of these proteins. For example, Propp and Alper (1968) showed that discordant C3 phenotypes could be detected in pairs of maternal and cord sera. Different genetic variants of C4 and C6 were also detected between mothers and offspring (Alper et al., 1975; Bach et al., 1971), and discordant phenotypes of Factor H were observed in several pairs of maternal and cord samples.

Deficiencies of complement have also been used to confirm the fetal origin of some components. As already mentioned, this approach was first used to demonstrate the synthesis of complement in fetal guinea pigs (Hyde, 1932). Later Ruddy et al. (1970) were able to demonstrate synthesis of C2 in infants born to mothers with C2 deficiency.

ONTOGENY OF COMPLEMENT RECEPTORS

The ontogeny of complement receptors is an interesting area of research, which so far has only been partially explored. In most strains of mice investigated (BALB/c, C57Bl/6, C3H/He, DBA/2) a substantial time interval was observed between the appearance of B lymphocytes with surface Ig and those bearing C3 receptors. In BALB/c mice, for example, C3 receptor-bearing lymphocytes were detected beginning only 2 weeks after birth (2.6%); at 4 and 6 weeks the percentages of B lymphocytes expressing these receptors were 13.5% and 25.2%, respectively. AKR mice are an exception because between 16% and 28% of B cells express C3 receptors (CR3) at 1 to 2 weeks of age. The analysis of complement receptor expression in (AKR) F_1 mice and $F_1 \times$ parental backcross progeny suggests that the "high" or "low" expression of these molecules is an inherited trait (Rosenberg and Parish, 1977).

Early studies on the ontogeny of human CR3 suggested that they were acquired later in ontogeny, confirming the results in mice. B lymphocytes from human fetuses were claimed not to bind C3b-coated erythrocytes (Gupta et al., 1976); cells of the myeloid lineage were also found to acquire C3b receptors relatively late during their differentiation (Ross et al., 1976).

In contrast to these early findings, however, more recent investigations using polyclonal antibodies against CR3 showed that about 27% of pre-B cells from fetal bone marrow express this receptor, as compared to 30% of pre-B cells from normal adult bone marrow (Tedder et al., 1983). Pre-B cells were identified as cytoplasmic μ^+, but surface Ig^-, cells; when they were separated into large (12–18 μm) and small (6–12 μm) cell population, about 15% of the large pre-B cells were $CR3^+$ in both fetal and adult bone marrow. Because large pre-B cells give rise to small pre-B cells, it appears that CR3 expression increases with maturation. The great majority of IgM^+ B cells in fetal and newborn tissue were found to express CR3. In fact, about 60%–80% of fetal bone marrow IgM^+ immature B cells were found to react with anti-CR3, while the more mature IgM^+,D^+ cells were 90% $CR3^+$. In contrast, plasma cells from adult bone marrow and spleen rarely express CR3 (IgM^+ = 3%–4%; IgG^+ = 2.3%–2.8%; IgA^+ = 0.7%–0.9%). In view of these findings and those from similar studies that involved cells of myelomonocytic lineage, Tedder et al. (1983) suggested that CR3 expression increases from large pre-B cells (15%) to small pre-B cells (35%–60%) and to mature B-cells (99%), and then declines on plasma cells, which are only rarely CR3-positive (<2%). The expression of CR3 increases on cells of the myelomonocytic lineage with maturation.

Recently, the expression of CR1 and CR3 during fetal and perinatal life was investigated using monoclonal antibody (mAb) staining of cells before and after stimulation with formyl-methionyl-leucylphenylalanine (FMLP) or leukotriene B4 (LTB4) (Adinolfi et al., 1988). Higher levels of expression of the CR1 and CR3 molecules were detected on monocytes and neutrophils from cord blood samples than from adult controls using untreated cells or cells previously stimulated with the chemotactic factors (Figs. 13–9, 13–10). CR1 and CR3 molecules were also detected on peripheral blood monocytes and neutrophils obtained from fetuses of more than 14 weeks gestation and on subpopulations of cells from fetal bone marrow, spleen, and thymus (Adinolfi et al., 1988).

There are still uncertainties about the maturation of CR2 during fetal life and the development of B cells in adults. According to Cooper et al. (1988), CR2 molecules appear on the surface of immature B cells at the stage of early synthesis of μ (IgM) chains, while they are absent on activated B cells. Further investigations are needed to clarify the relationship between the expression of CR2 molecules and maturation of B cells, however.

PHYLOGENY OF COMPLEMENT

Studies of the evolution of complement have revealed that complex systems of interacting proteins with biological and physicochemical properties similar to those in human sera are also present in many of the mammalian vertebrate species investigated so far. In many species, including guinea pigs, mice, rabbits, sheep, and cattle, a large number of proteins have been identified, which are involved in both the classical and alternative pathways of complement activation (Brown et al., 1974; Day et al., 1975; Gigli and Austen, 1971; Rosen, 1974). Both pathways generate C5 convertase and lead to the same lytic stage of complement activity. In many instances, components of complement from one species can interact with those of other species, although with variable efficiency.

Extensive investigations on the site of synthesis of the various components of complement have been performed using in vitro techniques in primates, guinea pigs, mice, and rats (Colten, 1976; Rother et al., 1968; Siboo and Vas, 1965; Stecher and Thorbecke, 1967). The results are in good agreement with studies carried out in humans and have confirmed that the majority of complement proteins are produced by hepatocytes, mono-

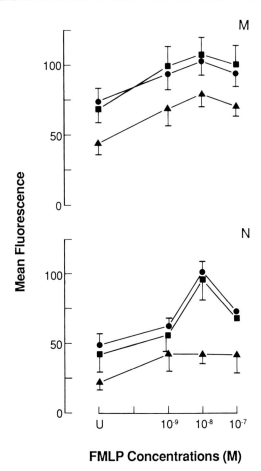

Figure 13–9.   Expression and enhancement of CR1 molecules on monocytes (M) and neutrophils (N) from normal adults (triangles), maternal (squares), and cord (circles) peripheral blood samples. The results are reported as mean values ($1 \pm SE$) from five samples in each group. The expression of CR1 was estimated on unstimulated (U) cells and after incubation with 10^{-9}, 10^{-8}, and 10^{-7} M FMLP solutions.

cytes, and macrophages. Recent studies have also demonstrated that several components of complement in mammals are polymorphic (Whitehouse, 1988) and that their genes are often clustered, thus forming families of structurally and functionally related proteins (Bentley, 1988). As in man, deficiencies of complement components have been identified in several species, including guinea pigs, rabbits, and mice (Day et al., 1975; Frank et al., 1971; Rother and Rother, 1961; Tachibana and Rosenberg, 1966). They provide useful models to investigate the effects of and possible therapies for complement abnormalities.

Relatively little information is available about the complement system in ecto-thermic vertebrates (Koppenheffer, 1987). Early studies by Gigli and Austen (1971) and Rosen (1974) suggested that vertebrate complement systems shared a common evolu-tionary origin but that differences existed between the mammalian and ectothermic ver-tebrate complement systems. Lamprey serum has been found to contain a factor that can lyse rabbit erythrocytes in the presence of Mg^{2+} but in the absence of antibodies. The lytic activity of lamprey sera is abolished by treatment with zymosan (Fujii and Murakawa, 1981; Noguchi et al., 1981). These and other observations in invertebrates have led to the suggestion that during the course of evolution, the primary role of complement was to promote phagocytosis rather than to mediate cell lysis and that the alternative pathway, involving properdin and Factor B, represents the earliest form of complement activity (Nonaka, 1985).

At least six components of complement have been identified in shark sera (Jensen et al., 1981). The first component of complement (C1n) was detected in the nurse shark, *Ginglymostoma cirratum,* and was found to be activated by shark antibodies to interact with guinea pig C4 to C9 proteins, but to be resistant to EDTA inactivation (Ross and Jensen, 1973). Intermediate and terminal components of complement in shark sera were also identified and were found to interact with mammalian complement proteins (Jensen et al., 1981). Studies intended to discriminate between classical and alternative pathways revealed instead that sera from rainbow trout, catfish, carp, and tuna contain complement components that are able to participate in both pathways of complement activation (Giclas et al., 1981; Kaastrup and Kock, 1983; Nonaka et al., 1981). These findings show

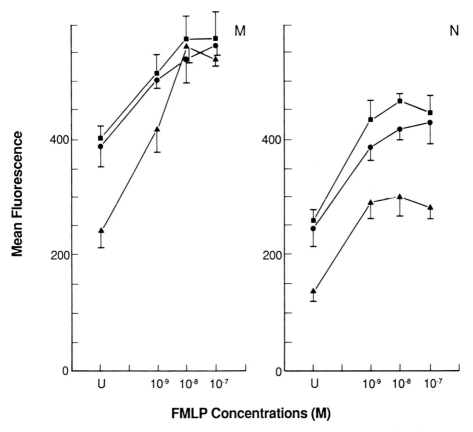

Figure 13–10.   Expression and enhancement of CR3 molecules on monocytes (M) and neutrophils (N) from normal adults (triangles), maternal (squares), and cord (circles) blood samples before (U) and after incubation with 10^{-9}, 10^{-8}, and 10^{-7} M FMLP solutions. Data are expressed as mean values ($1 \pm$ SE) from five samples in each group.

that the evolutionary origins of complement proteins date back at least to teleost fishes and have been highly conservative.

Less is known about complement-like systems in invertebrates. There is evidence that invertebrates such as *Asterias forbesi* (starfish) contain a protein that can cleave purified human C3 and is claimed to have C3-like proactivator activity (Day et al., 1970). Complement-like activity was detected in sea urchin coelomic fluid (Berthuessen, 1983). This activity, which was detected using rabbit red cells, could be destroyed by heating the fluid to 37°C for 30 minutes or by reducing the Ca^{2+} concentration. The physicochemical and biological properties of this factor remain uncertain, but preliminary studies suggest that it may represent a primordial form of an alternative pathway complement component. Further investigation of the phylogeny of complement at a molecular level should be rewarding and may help clarify the intricate and complex relationships between the complement system and other families of plasma proteins involved in coagulation and the response to injury and inflammation.

CONCLUSION

The complement system plays an essential role in immune mechanisms of defense against infection by preparing bacteria for phagocytosis and lysis (Cooper, 1991). Once activated, many components of complement perform other important functions, particularly as mediators of local tissue inflammation. In fact, many have been shown to act as acute phase reactants, and their levels in sera may reflect the course of many infections and inflammatory diseases (Lehner and Adinolfi, 1980).

With the exception of C9, most proteins that form the complement system are synthesized beginning early in fetal life and are present in sera from term infants at levels of over 50% of normal adult values. During the first months after birth, serum levels of complement components increase rapidly, and in the majority of cases normal adult amounts are reached within the first year of life.

There is general agreement that complement components are present in normal adults at concentrations in excess of those required for protection against infection. This conclusion is based mainly on the observation that heterozygotes for single components of complement, who have low levels of the specific proteins, are not usually more susceptible to infections than are normal individuals, and that some deficiencies involving multiple complement components are well tolerated. This suggests that the partially reduced complement activity in cord and infant sera is still adequate to provide protection against most infections that may occur during perinatal life.

In fact, the most frequent cause of infections during the early postnatal period is probably associated with a transient hypogammaglobulinemia resulting from the slow catabolism of maternally derived IgG together with insufficient endogenous antibody synthesis by the newborn. However, in children with recurrent infections, measurement of serum levels of complement components and functional activity are indicated in order to exclude the possibility that the infection results from a specific complement deficiency (Morgan and Walport, 1991).

The most common complement defect is due to C1-INH deficiency, where heterozygotes manifest the disease hereditary angioedema. In the great majority of affected individuals, absolute levels of C1-INH are reduced to less than 30%; in the remaining patients, normal amounts of a dysfunctional protein are present, which cross-reacts with immune sera raised against C1-INH.

Estimation of selected complement components that function as acute phase reactants may be useful to monitor the course of disease in children or adults with infections or chronic inflammation (Adinolfi, 1982; Drew and Arrojave, 1981). For example, the presence of C3 split products in sera has been claimed to be a useful tool in the diagnosis of bacterial infections in the newborn (Drew and Arrojave, 1981), while high levels of C9 and Factor B have been detected in sera of neonates suspected of being infected with rubella virus in utero or in infants with proven *E. coli* infections. High serum levels of C9, as well as other acute phase proteins, have been detected in sera from patients with Crohn's disease (Adinolfi, 1982).

Still, little is known about the possibility that recurrent infections or diseases in children may be associated with deficiencies of complement receptors; however, preliminary results suggest that abnormal expression of complement receptors is often associated with chronic recurrent infections in childhood. As soon as techniques using monoclonal antibodies for detection and characterization of complement receptors become more clinically available, it should be possible to fully evaluate the possible role of these structures in the predisposition to selected disorders.

Acknowledgment

Original work by the author cited in this chapter was financially supported by the Spastics Society. I am grateful to Adrienne Knight for typing and editing the manuscript.

References

Adinolfi, M. (1972). Ontogeny of components of complement and lysozyme. *Ciba Found. Symp.,* pp. 65–81.

Adinolfi, M. (1977). Human complement: Onset and site of synthesis during fetal life. *Am. J. Dis. Child. 131*:1015–1023.

Adinolfi, M. (1981). Ontogeny of complement, lysozyme and lactoferrin in man. In *Immunological Aspects of Infections in the Fetus and Newborn.* (Lambert, H. P., and Woods, C.B.S., eds.). London: Academic Press, pp. 19–52.

Adinolfi, M. (1982). Human complement C9 and Factor B in the diagnosis of infections in infants. *Acta. Paediatr. Scand. 71*:845–846.

Adinolfi, M., and Beck, S. (1976). Human complement—C7 and C9—in fetal and newborn sera. *Arch. Dis. Child. 50*:562–564.

Adinolfi, M., Cheetham, M., Lee, T., and Rodin, A. (1988). Ontogeny of human complement receptors CR1 and CR3: Expression of these molecules on monocytes and neutrophils from maternal, newborn and fetal samples. *Eur. J. Immunol. 18*:565–569.

Adinolfi, M., Dobson, N., and Bradwell, A. R. (1981). Synthesis of human complement, β1H, and C3b-INA during fetal life. *Acta. Paediatr. Scand. 70*:705–710.

Adinolfi, M., Gardner, B., and Wood, C.B.S. (1968). Ontogenesis of two components of human complement. β1E and β1C-1A globulins. *Nature 219*:189–191.

Adinolfi, M., and Lehner, T. (1988). C9 and Factor B as acute phase proteins and their diagnostic and prognostic value in disease. *Exp. Clin. Immunogenet. 5*:123–132.

Alper, C. A., Hobart, M. J., and Lachmann, P. J. (1975). Polymorphism of the sixth component of complement. In *Isoelectric Focusing* (Arbuth, J. P., and Beeley, J. A., eds.). London: Butterworth, pp. 306–312.

Arditi, E., and Nigro, N. (1957). Ricerche sul comportamento serico nell'immaturo. *Minerva. Pediatr. 9*:921–928.

Bach, S., Ruddy, S., and MacLaren, A. J. (1971). Electrophoretic polymorphism of the fourth component of human complement in paired maternal and fetal plasma. *Immunology* 21:869–878.

Ballow, M., Fang, F., Good, R. A., and Day, N. K. (1974). Developmental aspects of complement components in the newborn. *Clin. Exp. Immunol.* 18:259–266.

Belt, K. T., Carroll, M. C., and Porter, R. R. (1984). The structural basis of the multiple forms of human complement component C4. *Cell* 36:907–914.

Bentley, D. R. (1988). Structural superfamilies of the complement system. *Exp. Clin. Immunogenet.* 5:69–80.

Berthuessen, K. (1983). Complement-like activity in sea urchin coelomic fluid. *Dev. Comp. Immunol.* 7:21–31.

Bhakdi, S., and Tranum-Jensen, J. (1984). Mechanism of complement autolysis and the concept of channel-forming proteins. *Phil. Trans. R. Soc. Lond. Ser. B.* 306:311–324.

Bhakdi, S., and Tranum-Jensen, J. (1991). Complement lysis: A hole in a hole. *Immunol. Today* 12:318–320.

Bianco, C., Patrick, R., and Nussenzweig, V. (1970). A population of lymphocytes bearing a membrane receptor for antigen-antibody-complement complexes: I. Separation and characterization. *J. Exp. Med.* 132:702–720.

Bordet, J. (1909). Recherches sur la destruction extracellulaire des bacteries (Annales de l'Institut Pasteur, June 1895). Cited in *Studies in Immunology* (Bordet, J., ed.). London: Chapman and Hall, p. 81.

Buchner, H. (1889). Über die bakterientodende Wirkung des zellenfreien Blutserums. *Zentralbl. Bacteriol. Parasitenk. Infektionskrank. Hyg.* 6:561–565.

Campbell, C., Walner Smith, J. A., and Adinolfi, M. (1982). Acute phase proteins in chronic inflammatory bowel disease in childhood. *J. Pediatr. Gastroenterol.* 1:193–200.

Campbell, R. D., Carrol, M. C., and Porter, R. R. (1986). The molecular genetics of complement. *Adv. Immunol.* 38:203–244.

Campbell, R. D., Dunham, I., and Sargent, C. A. (1988). Molecular mapping of the HLA-linked complement genes and the RCA linkage group. *Exp. Clin. Immunogenet.* 5:81–98.

Carter, P. E., Dunbar, B., and Fothergill, J. E. (1984). Structure and activity of C1r and C1s. *Phil. Trans. R. Soc. Lond. Ser. B.* 306:293–299.

Colten, H. R. (1973). Biosynthesis of the fifth component of complement (C5) by human fetal tissues. *Clin Immunol. Immunopathol.* 1:346–352.

Colten, H. R. (1976). Biosynthesis of complement. *Adv. Immunol.* 22:67–118.

Colten, H. R., Borsos, T., and Rapp, H. J. (1966). In vitro synthesis of the first component of complement by guinea pig small intestine. *Proc. Natl. Acad. Sci. USA* 56:1158–1163.

Cooper, N. R. (1991). Complement evasion strategies of microorganisms. *Immunol. Today* 12:327–331.

Cooper, N. R., Moore, M. D., and Nemerow, G. R. (1988). Immunology of CR2, the lymphocyte receptor for Epstein-Barr virus and the C3d complement fragment. *Annu. Rev. Immunol.* 6:85–113.

Davis, C. A., Vallota, F. H., and Forristral, J. (1979). Serum complement levels in infancy: Age-related changes. *Pediatr. Res.* 13:1043–1046.

Day, N. K., Geiger, H., and Good, R. A. (1975). Complement. In *Molecular Pathology* (Good, R. A., Day, S. B., and Yunis, J. J., eds.). Springfield, Ill.: C. C. Thomas, pp. 115–160.

Day, N. K., Gewurz, H., Johannsen, R., Finstad, J., and Good, R. A. (1970). Complement and complement-like activity in lower vertebrates and invertebrates. *J. Exp. Med.* 132:941–950.

Day, N. K., Pickering, R. J., Gewurz, H., and Good, R. A. (1969). Ontogenetic development of the complement system. *Immunology* 16:319–326.

Dobson, N. J., Lambris, J. D., and Ross, G. D. (1981). Characteristics of isolated erythrocyte complement receptor type one (CR1, C4b-C3b receptor) and CR1 specific antibodies. *J. Immunol.* 126:693–698.

Drew, J. H., and Arrojave, C. M. (1981). Complement activation: Use in the diagnosis of infection in newborn infants. *Acta. Paediatr. Scand. 70*:255–256.

Fearon, D. T., and Wong, W. W. (1983). Complement ligand-receptor interactions that mediate biological responses. *Annu. Rev. Immunol. 1*:243–271.

Ferrata, A. (1907). Die Unwirksamkeit der komplexen Hamolysine in salzfreien Lösungen and ihre Ursache. *Berliner. Klin. Wochenschr. 44*:366.

Farries, T. C., and Atkinson, Y. P. (1991). Evolution of the Complement System. *Immunol. Today 12*:295–300.

Fireman, P., Zuchowski, D. A., and Taylor, P. M. (1969). Development of human complement system. *J. Immunol. 103*:25–31.

Fishel, C. W., and Pearlman, D. S. (1961). Complement components of paired mother-cord sera. *Proc. Soc. Exp. Biol. Med. 107*:695–699.

Frank, M. M. (1979). The complement system in host defense and inflammation. *Rev. Infect. Dis. 1*:483–501.

Frank, M. A., May, J., Gaither, T., and Ellman, L. (1971). In vitro studies of complement function in sera of C4-deficient guinea pig. *J. Exp. Med. 134*:176–187.

Fujii, T., and Murakawa, S. (1981). Immunity in the lamprey: III. Occurrence of the complement-like activity. *Dev. Comp. Immunol. 5*:251–259.

Gagnon, J. (1984). Structure and activation of complement. *Phil. Trans. R. Soc. Lond. Ser. B. 305*:301–309.

Giclas, P. C., Morrison, D. C., Curry, B. J., Laurs, R. M., and Ulevitch, R. Y. (1981). The complement system of the albacore tuna *Thunnus alalunga. Dev. Comp. Immunol. 5*:437–447.

Gigli, I., and Austen, K. F. (1971). Phylogeny and functions of the complement system. *Annu. Rev. Microbiol. 25*:309–332.

Giles, C. M. (1988). Antigenic determinants of human C4, Rodgers and Chido. *Exp. Clin. Immunogenet. 5*:99–114.

Gitlin, D., and Biasucci, A. (1969). Development of γG, γA, γM, β1C, β1A, C1 esterase inhibitor, ceruloplasmin, transferrin, hemopexin, haptoglobin, fibrinogen, plasminogen, α1-anti-trypsin, orosomucoid, β-lipoprotein, β_2-macroglobulin and prealbumin in the human conceptus. *J. Clin. Invest. 48*:1433–1446.

Götze, O., and H. J. Müller-Eberhard (1976). Alternative pathway of complement activation. *Adv. Immunol. 24*:1–35.

Gupta, S., Pahwa, R., O'Reilley, R., Good, R. A., and Siegal, F. P. (1976). Ontogeny of lymphocyte subpopulations in human fetal liver. *Proc. Natl. Acad. Sci. USA 73*:919–922.

Humphrey, J. H., and Donsmashkin, R. R. (1969). The lesions in cell membranes caused by complement. *Adv. Immunol. 11*:75–115.

Hyde, R. R. (1932). The complement deficient guinea pig: A study of an inheritable factor in immunity. *Am. J. Hyg. 15*:824–836.

Jensen, J. A., Festa, E., Smith, D. S., and Cayer, M. (1981). The complement system of the nurse shark: Haemolytic and comparative characteristics. *Science 214*:566–568.

Kaastrup, P., and Kock, C. (1983). Complement in the carp fish. *Dev. Comp. Immunol. 7*:781–783.

Kaplan, M. H., and Volanakis, J. E. (1974). Interaction of C-reactive protein complexes with the complement system: I. Consumption of human complement associated with the reaction of C-reactive protein with pneumococcal C-polysaccharide and with the choline phosphatides, lecithin and sphingomyelin. *J. Immunol. 112*:2135–2147.

Kazatchrine, M. D., Fearon, D. T., Appay, M. D., Kandet, C., and Bariety, Y. (1982). Immunohistochemical study of the human glomerular C3b receptor in normal kidney and in seventy-five cases of renal diseases. *J. Clin. Invest. 69*:900–912.

Kinoshita, T., (1991). Biology of complement: The overture. *Immunol. Today 12*:291–295.

Kohler, P. E. (1973). Maturation of the human complement system: I. Onset time and sites of fetal C1q, C4, C3 and C5 synthesis. *J. Clin. Invest. 52*:671–677.

Koppenheffer, T. L. (1987). Serum complement systems of ectothermic vertebrates. *Dev. Comp. Immunol.* *11*:279–286.

Lachmann, P. J. (1973). Complement. In *Defense and Recognition* (Porter, R. R. ed.). London: Butterworth, pp. 361–397.

Lachmann, P. J. (1979). Complement. In *The Antigens,* Vol. 5 (Sela, M., ed.). New York: Academic Press, pp. 283–353.

Law, S. K., Dodds, A. W., and Porter, R. R. (1984). A comparison of the properties of the two classes, C4A and C4B, of the human complement components C4. *EMBO J. 3*:1819–1823.

Lehner, T., and Adinolfi, M. (1980). Acute phase proteins C9, Factor B and lysozyme in recurrent oral ulceration and Behçet's syndrome. *J. Clin. Pathol. 33*:269–275.

Mak, L. W. (1978). The complement profile in relation to the "reactor" state: A study in the immediate post-partum period. *Clin. Exp. Immunol. 31*:419–425.

Mayer, M. M. (1961). Complement and complement fixation. In *Experimental Immunochemistry,* 2nd ed. (Kabat, E. A., ed.). Springfield, Ill.: C. C. Thomas, pp. 133–240.

Mellbye, O. J., Natvig, J. B., and Krarstein, B. (1971). Presence of IgG subclasses and C1q in human cord sera. In *Protides of Biological Fluids,* Vol. 18 (Peeters, H., ed.). Oxford: Pergamon, pp. 127–131.

Miyakawa, Y., Yamada, A., Kosaka, K., Tsuda, F., and Mayumi, M. (1981). Defective immune-adherence (C3b) receptor on erythrocytes from patients with systemic lupus erythematosus. *Lancet ii*:493–497.

Morgan, B. P., and Walport, M. Y. (1991). Complement deficiency and disease. *Immunol. Today 12*:301–305.

Müller-Eberhard, H. J. (1975). The complement system. In *The Plasma Proteins,* vol. 1. (Putnam, F. W., ed.). New York: Academic Press, pp. 394–432.

Müller-Eberhard, H. J. (1986). The membrane attack complex of complement. *Annu. Rev. Immunol. 4*:503–528.

Müller-Eberhard, H. J., and Lepow, I. H. (1965). C1 esterase effect on activity and physicochemical properties of the fourth component of complement. *J. Exp. Med. 121*:819–833.

Naff, G. B., Pensky, J., and Lepow, I. H. (1964). The macromolecular nature of the first component of human complement. *J. Exp. Med. 119*:593–613.

Noguchi, A., Hanafusa, K., and Ohnishi, K. (1981). Properdin-like pathway in the serum of lamprey *(Lampeter japonica).* In *Aspects of Developmental and Comparative Immunology* (Solomon, B., ed.). London: Pergamon, pp. 487–495.

Nonaka, M. (1985). Evolution of the complement system. *Dev. Comp. Immunol. 9*:377–387.

Nonaka, M., Yamaguchi, N., Natasuume-Sakki, S., and Takashi, M. (1981). The complement system of rainbow trout *(Salmo sairdueri):* I. Identification of the serum lytic system homologous to mammalian complement. *J. Immunol. 126*:1489–1494.

Osler, A. G., and Sandberg, A. L. (1973). Alternate complement pathways. *Prog. Allergy 17*:51–92.

Permutter, D. H., and Colten, H. R. (1986). Molecular immunology of complement biosynthesis. *Annu. Rev. Immunol. 4*:231–251.

Pfeiffer, R., and Issaeff, R. (1894). Über die specifische Bedeutung der Choleraimmunität. *Z. Hyg. Infectionskrank. 17*:355–400.

Phillips, M. E., Rother, V. A., Rother, K. O., and Thorbecke, G. J. (1969). Studies on the serum proteins of chimeras: III. Detection of donor type C5 in allogenic and congenic post-irradiation chimeras. *Immunology 17*:315–321.

Pillemer, L., Blum, L., and Lepow, I. H. (1954). The properdin system and immunity: I. Demonstration and isolation of a new serum protein, properdin, and its role in immune phenomena. *Science 120*:279–285.

Porter, R. R., and Reid, K.B.M. (1978). The biochemistry of complement. *Nature 275*:699–704.

Propp, R. P., and Alper, C. A. (1968). C3 synthesis in the human fetus and lack of transplacental passage. *Science 162*:672–673.

Reid, K.B.M., Bentley, D. R., and Campbell, R. D. (1986). Complement system proteins which interact with C3b or C4b. *Immunol. Today* 7:230–234.

Rosen, F. S. (1974). Complement: Ontogeny and phylogeny. *Transplant Proc.* 6:47–50.

Rosenberg, Y. J., and Parish, C. R. (1977). Ontogeny of the antibody forming cell lines in mice: IV. Appearance of cells bearing Fc receptors, complement receptors and surface immunoglobulin. *J. Immunol.* 118:612–617.

Ross, G. D., and Atkinson, J. P. (1985). Complement receptor structure and function. *Immunol. Today* 6:115–119.

Ross, G. D., Jarowski, C. I., Rabellino, E. M., and Winchester, R. J. (1976). The sequential appearance of Ia-like antigens and two complement receptors during the maturation of human neutrophils. *J. Exp. Med.* 147:730–744.

Ross, G. D., and Jensen, J. A. (1973). The first component (C1n) of the complement system of nurse shark *(Ginglymostoma cirratum). J. Immunol.* 110:911–918.

Rother, K., and Rother, U. (1986). Biological functions of the complement system. *Prog. Allergy* 39:24–100.

Rother, U., and Rother, K. (1961). Über einen angerboren komplement Defect bei Kaninchen. *Immunitätsforsch. Exp. Ther.* 121:224–231.

Rother, U., Thorbecke, G. J., and Stecher-Levin, V. J. (1968). Formation of C3 by rabbit liver tissue in vitro. *Immunology* 14:649–655.

Ruddy, S., Klemperer, M. R., and Rosen, F. S. (1970). Hereditary deficiency of the second component of complement (C2) in man: Correlation of C2 haemolytic activity with immunochemical measurements of C2 protein. *Immunology* 18:943–954.

Sawyer, M. K., Forman, M. J., Kuplic, L., and Stiehm, E. R. (1971). Developmental aspects of the human complement system. *Biol. Neonate* 19:148–162.

Shin, H. S., Pickering, J., and Mayer, M. M. (1971). The fifth component of the guinea pig complement system: II. Mechanism of SAC12435b formation and C5 consumption by EAC1423. *J. Immunol.* 106:473–479.

Siboo, R., and Vas, S. I. (1965). Studies on in vitro antibody production. III. Production of complement. *Can. J. Microbiol.* 11:415–422.

Sottzup-Jensen, L., Stepanik, T. M., Kristensen, T. et al. (1985). Common evolutionary origin of α_2-macroglobulin and complement C3 and C4. *Proc. Natl. Acad. Sci. USA* 82:9–13.

Stecher, V. J., and Thorbecke, G. J. (1967). Sites of synthesis of serum proteins: I. Serum proteins produced by macrophages in vitro. *J. Immunol.* 99:643–652.

Stossel, T. P., Alper, C. A., and Rosen, F. S. (1973). Opsonic activity in the newborn: Role of properdin. *Pediatrics* 52:134–137.

Tachibana, D. K., and Rosenberg, L. T. (1966). Fetal synthesis of Hc[1], a component of mouse complement. *J. Immunol.* 97:213–215.

Tedder, T. F., Fearon, D. T., Garland, G. L., and Cooper, M. D. (1983). Expression of C3b receptors on human B cells and myelomonocytic cells but not natural killer cells. *J. Immunol.* 130:1668–1673.

Tenner, A. J., and Cooper, N. R. (1981). Identification of types of cells in human peripheral blood that bind C1q. *J. Immunol.* 126:1174–1179.

Thorbecke, G. J., Hochwald, G. M., van Furth, R., Müller-Eberhard, H. J. and Jacobsen, E. R. (1965). Problems in determining the sites of synthesis of complement reactions. *Ciba Found. Symp.*, pp. 99–114.

Tosi, M., Duponche, C., Meo, T., and Julier, C. (1987). Complete cDNA sequence of human complement C1s and close physical linkage of the homologous genes C1s and C1r. *Biochemistry* 26:8516–8528.

Traub, M. (1943). The complement activity of the serum of healthy persons, mothers and newborn infants. *J. Pathol. Bacteriol.* 55:447–552.

Wasserman, P., and Alberts, E. (1940). Complement titre of blood of the newborn. *Proc. Soc. Exp. Biol. Med.* 45:563–564.

Weiss, J. J., Fearon, D. T., Klickstein, L. B., Wong, W. W., Richards, S. A., de Bruyn, A., Kops, A., Smith, J. A., and Weis, J. H. (1986). Identification of a partial cDNA clone for the C3d/Epstein-Barr virus receptor of human B lymphocytes: Homology with the receptor for fragments C3b and C4b of the third and fourth components of complement. *Proc. Natl. Acad. Sci. USA 83*:5639–5643.

Whitehouse, D. B. (1988). Genetic polymorphisms of animal complement components. *Exp. Clin. Immunogenet. 5*:69–80.

Wilson, J. G., Wong, W. W., Schur, P. H., and Fearon, D. T. (1982). Mode of inheritance of decreased C3b receptors on erythrocytes of patients with systemic lupus erythematosus. *N. Engl. J. Med. 307*:981–986.

14

Origin and Function of Cytokines Involved in the Development of Immune Responses

JAMES D. WATSON, ROGER J. BOOTH,

KATHRYN E. CROSIER

Cytokines are polypeptides that regulate the development and function of cells that make up the hematopoietic and immune systems. All cytokines discovered so far have been shown to be pleiotropic, exerting activity on more than one cell type, making it clear that the regulatory mechanisms influencing development and function of cells in these systems are closely interwoven. We focus here on the development of those cells that are involved in the initiation of immune responses, thymus-derived (T) lymphocytes and antigen-presenting cells (APCs), which are believed to be predominantly macrophages and dendritic cells. The immune system is normally activated by the entry of pathogens into the body, with subsequent processing of pathogenic antigens within APCs. The cellular processes that result from T lymphocytes binding these processed antigens lead to the synthesis of a wide range of cytokines, which then form a communication network that not only stimulates lymphocyte development and function but also influences the processes involved in cell differentiation and renewal throughout the hematopoietic system. We discuss here the cell biology of T lymphocyte development, the cellular mechanisms underlying antigen processing, the activation of helper T lymphocytes, and the properties of cytokines produced as a result of the initiation of immune responses. The function of cytokines in the developmental biology of the immune system is today a fascinating area of study. The promise of the future is that such regulatory molecules can be harnessed selectively to manipulate immune and hematopoietic function.

ORIGIN OF THYMUS-DERIVED LYMPHOCYTES

During T cell ontogeny, a large number of cells of diverse phenotypes arise from a common progenitor (Kingston et al., 1985). The events that take place during T cell development are poorly defined, but they occur mainly in the thymus and result in the production of a number of mature subpopulations that subsequently migrate to peripheral lymphoid organs. Four classes of T cell have been recognized on the basis of their cell surface phenotypes and the assays in which they function. They are cytotoxic (or killer)

Table 14–1.   Surface Phenotypes of T Lymphocyte
Subpopulations

Surface Marker	T_H	T_{DTH}	T_K	T_S
Antigen receptor (TCR $\alpha\beta$)	+	+	+	+
CD3	+	+	+	+
CD4	+	+	–	–
CD8	–	–	+	+
Class II MHC determinants	–	–	–	+?

T cells (T_K), helper T cells (T_H), suppressor T cells (T_S), and delayed-type hypersensitivity cells (T_{DTH}) which can be distinguished from T_H only on the basis of their activity rather than their phenotype (Table 14–1).

T cell development begins in the hematopoietic stem cell pool and is followed by migration of early progenitors to the thymus for further maturation. The thymus supports a large population of dividing lymphocytes, with an estimated one-third replacement of these cells daily. The vast majority of the cells that originate in the thymus also die there, and only a very small proportion go on to become members of the four effector subsets and leave the thymus. Two major problems await resolution: How is T cell division regulated in the thymus? What controls development of the effector T classes?

Relatively little is known of the intrathymic events that lead to the production of mature, non-self-reactive T lymphocytes (Scollay et al., 1984a,b). The thymus produces vast numbers of lymphocytes each day (10^8 in mice) but exports only a minor subset to the periphery (about 10^6 cells per day). Thymus lymphocyte development is a continual process, and the time spent in each maturational phase is relatively short. In the thymus of a mature mouse, early thymocytes, which constitute less than 3% of the cells, are referred to as double-negative cells because they bear neither the CD4 nor the CD8 molecules characteristic of mature T cells. Within the thymus, double-negative cells proliferate, rearrange, and express genes that give rise to the antigen binding receptor and in a few days they lead to the predominant thymic cell type—the cortisone-sensitive, double-positive cells ($CD4^+ CD8^+$), located in the cortex. More than 50% of these cells bear surface antigen binding receptors at much lower concentrations than those found on peripheral T cells. The mature thymic lymphocytes are single-positive cells, bearing either CD4 or CD8. It is not yet known whether these cells proceed from an intermediate double-positive precursor or whether they are produced directly from single-positive progenitors. They are cortisone-resistant, express high levels of antigen binding receptors, and appear to be localized in the medulla.

The thymic anlage of mice is colonized by cells of hematopoietic origin at about day 11 of embryonic development. Proliferation and maturation of these cells within the fetal thymus then occurs very rapidly over a period of 7 days, resulting in the establishment of distinct subpopulations of thymocytes and immunocompetence. The ontogeny of fetal thymocytes has also been studied by following the appearance of cell surface antigens and the rearrangement and subsequent expression of genes that constitute the antigen binding receptor of T cells (Born et al., 1985; Haars et al., 1983; Owen and Jenkinson, 1984). These events can also be conveniently followed in vitro in fetal thymic lobes maintained in organ culture where the temporal sequence of T cell development is similar to that observed in vivo (Skinner et al., 1987).

The non-lymphoid architecture of thymic lobes consists of a number of different stromal cell types, which include epithelial, dendritic, and monocytic cells. These non-lymphoid cells are essential for the continued development of lymphoid components in vitro. The removal of thymocyte precursors from this organized stromal environment prevents further thymocyte proliferation and development in cell culture. The mechanisms by which stromal cells influence thymocyte maturation remain to be elucidated. In addition, there is extensive innervation of the thymus from a branch of the vagus nerve, although the impact of the neuroendocrine system on thymic development is largely unknown.

Effector T Lymphocytes

The cells of each effector class that leave the thymus are mature, antigen-sensitive cells that migrate to peripheral lymphoid organs. There they require interaction with specific antigen as the stimulus to drive cell division and subsequent differentiation to terminal effector function. The various T cell classes differ in the way they recognize antigen. For T_H this is reflected in a need for antigen to be presented on the cell surface bound to class II major histocompatibility complex (MHC) gene products, whereas T_K cells respond to antigen bound to class I MHC molecules (Benacerraf and Germain, 1978; Doherty et al., 1976). However, antigen-specific receptors on T_H and T_K cells are structurally similar to one another, and thus cell surface molecules other than these receptors must influence the antigen specificity of T cells. CD8 on T_K cells is involved in the recognition of class I MHC-encoded molecules and associated antigens, while CD4 is involved in the binding of T_H to antigens associated with class II MHC molecules (Fig. 14–1).

Antigen Processing and Molecules Encoded within the MHC

The three-dimensional structure of a class I MHC molecule (human HLA-A2) has now been described. What is of great interest is the finding that class I MHC molecules possess a site that binds peptide fragments in such a way that they can be presented to T cells (Bjorkman et al., 1987a,b). These peptide fragments are the products of processed antigen. Although the structure of the analogous binding site in class II MHC molecules has not yet been determined, an overall similarity is likely because of the similarity of the domain structure and genomic organization of class I and II molecules, the similarity in their C-terminal domain sequences, and the fact that T cells restricted to class I or class II antigens use the same $\alpha\beta$ heterodimeric receptor structure (Brown et al., 1988).

One of the most interesting aspects of immune recognition concerns the nature of antigen processing events and the peptides that bind to class I or II MHC molecules. The antigen binding receptor of T cells appears to recognize antigen only in the form of degraded peptide fragments noncovalently associated with class I or II MHC molecules on the surface of cells (Babbit et al., 1985; Townsend et al., 1986a,b). Thus, foreign antigens must first be processed inside antigen presenting cells, and then the processed fragments are directed to sites on class I or class II MHC molecules if they are to be successfully displayed on the cell surface for T cell recognition. A dichotomy appears to exist in the manner in which foreign antigens are processed intracellularly. For peptides that bind to class II MHC molecules, antigen processing appears to involve endocytosis and deg-

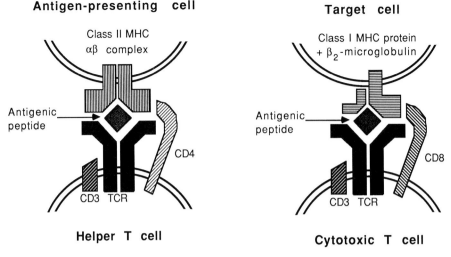

Figure 14–1.   Recognition of antigen by T lymphocytes. Although helper T lymphocytes and cytotoxic (killer) T lymphocytes use the same α and β polypeptide chains in their antigen receptors, they each recognize antigens associated with different surface structures on target cells. Helper cells recognize antigenic peptides associated with class II MHC proteins on antigen presenting cells, whereas cytotoxic T cells recognize antigens associated with class I MHC molecules. The $\alpha\beta$ receptors on both helper and cytotoxic T cells are associated with a complex of polypeptides called CD3, which is responsible for transmitting to the inside of the cell the information that the receptor has bound antigen. Also associated with the TCR/CD3 complex is CD4 in the case of helper cells and CD8 on cytotoxic cells. The CD4 and CD8 molecules are probably involved in recognition of non-polymorphic MHC determinants on target cells.

radation of antigens within acidic lysosomes. In contrast, peptides that associate with class I MHC molecules appear to be derived from endogenous protein synthesis, such as viral proteins synthesized within virus-infected cells (Fig. 14–2).

Because class I and class II MHC molecules are distributed on various somatic cells, it seems most likely that these cells possess the same intrinsic capacity to process proteins as do classical APCs. For example, fibroblasts transfected with class II genes are capable of presenting antigen to T lymphocytes (Germain and Malissen, 1986). Thus the phenomenon of antigen processing may actually reflect a fundamental intracellular activity present in virtually all cells of the body.

For many years, macrophages were regarded as the only APCs. It is now clear, however, that other cell types can also act as APCs, and, because of their tissue distribution, they may be important in certain physiological situtations. In mice, skin Langerhans cells, liver Kupffer cells, spleen dendritic cells, various epithelial cells, and B (bursal or bone marrow–derived) cells have all been shown to be capable of presenting antigen to T lymphocytes (Schuler and Steinman, 1985). The striking feature that all APCs have in common is their ability to express class II MHC molecules on their surface.

Although the first step in the activation of T_H cells is antigen processing, the events that follow are not so evident. It is not clear whether recognition of the class II MHC molecules expressed on the APC is all that is needed to activate T_H cells in the induction process, or whether another active event is required. For example, when macrophages are used as APCs, they appear to release interleukin 1 (IL-1) during the antigen-binding process by T cells. IL-1 released from macrophages appears to function as a signal to convert T_H cells to a cytokine-synthesizing state and to stimulate them to express surface receptors for some of the cytokines that they produce—an autocrine growth response. In order that macrophages secrete IL-1, it would be reasonable to assume that they can somehow sense that the antigen binding process by T cells has occurred, but the mechanism behind this "sensing" is not known. Further, the requirement for IL-1 in the T_H activation process may not be absolute because other cell types that function as APCs, such as dendritic, epidermal, and B cells, do not readily secrete IL-1. Also, glutaraldehyde fixation of APCs, as a means of preventing active metabolic processing events from taking place, has been shown in a number of situations to result in effective antigen presentation, implying that IL-1 release may not be obligatory.

The Cytokine Cascade and Different Subclasses of T_H Cells

A number of cytokines are simultaneously synthesized and secreted by individual T_H cells following their recognition of antigen on antigen presenting cells. In mice, T_H cell clones can be subdivided into two categories based on the spectrum of cytokines that they produce following activation (Mosmann and Coffman, 1987). $T_H 1$ cells produce interleukin

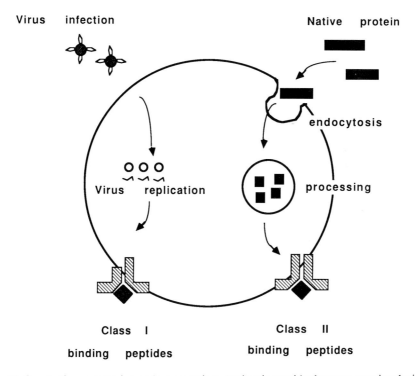

Figure 14–2. Antigen processing and presentation. Antigenic peptides become associated with either class I or class II MHC molecules on the surface of cells depending on the manner in which they are processed within the cell. Endogenous antigens, such as those produced intracellularly during an infection, associate with class I MHC. Exogenous antigens taken up by endocytosis are processed and presented preferentially and associate with class II MHC structures.

2 (IL-2), interferon γ (IFN-γ), and lymphotoxin, while T_H2 cells produce interleukin 4 (IL-4), interleukin 5 (IL-5), and interleukin 6 (IL-6). Both cell types secrete interleukin 3 (IL-3) and the granulocyte-macrophage colony stimulating factor (GM-CSF). There is, however, some controversy over the validity of this classification of T_H cells, particularly with reference to human T_H cells. The major dilemma faced, then, concerns the role of various cytokines released by T_H cells. Some of these, such as IL-2, IL-4, and IL-6, have as their major targets B lymphocytes and T lymphocytes—the cell types needed for cell-mediated and humoral immunity. However, other cytokines, such as IL-3, GM-CSF, and IL-5, act on progenitors of cells of various hematopoietic lineages. Clearly, the initiation of immune responses is not required for maintenance of steady-state hematopoiesis. Athymic mice, for example, lack T_H cells but have normal hematopoietic development and function.

CLONING OF CYTOKINE GENES AND THEIR RECEPTORS

Because T_H cells secrete a number of different cytokines, crude preparations usually exert a range of biological activities, and it is difficult to determine whether effects are due to one molecule or to a number of molecules. Attempts to purify cytokines from crude preparations by classical biochemical techniques led to the realization that these mediators have very high specific activities (typically between 10^{-11} and 10^{-13} M) and are heterogeneous with respect to both size and charge. Some of the heterogeneity can be attributed to variable glycosylation, but the question of whether each activity is the property of a single protein or a family of similar molecules is difficult to resolve formally by purely biochemical means. The development of recombinant DNA technology offered an alternative approach to this problem, and in 1982 the first successful molecular cloning of a DNA copy (cDNA) of the mRNA sequence coding for human IFN-γ was completed. Since then, IL-1 to IL-7, GM-CSF, the granulocyte colony stimulating factor (G-CSF), and the macrophage colony stimulating factor (M-CSF) have all been cloned, and the answers to many of the questions that have puzzled biochemists have begun to emerge. With the aid of these cDNA clones, it is possible to probe the chromosomal DNA to study the genes themselves and the ways by which their expression is regulated. What has already become clear is that all the cytokine genes that have been studied in this way are present only as single copies in the genome, and so any biochemical heterogeneity in the products of these genes must result from post-transcriptional and/or post-translational modifications of the mRNA or protein, respectively. There is no evidence for the re-arrangement of genetic elements coding for cytokines during the development of mature cytokine-secreting cells, as has been found to occur, for example, in the immunoglobulin (Ig) genes during B lymphocyte ontogeny. The genes for IL-1 to IL-7, IFN-γ, GM-CSF, G-CSF, and M-CSF appear to have the same architecture in DNA in essentially all cells of the body.

Cloning Strategies

The cloning of nucleotide sequences coding for cytokines generally begins with the construction of a library of cDNA clones synthesized from from the mRNA of cytokine-secreting cells. This library is then screened to identify cytokine-coding sequences in one of three ways. (1) If a partial amino acid sequence of the cytokine is known, then a series of radioactive nucleotide probes likely to code for part of the peptide can be synthesized and used to probe the library for complementary sequences by colony hybridization. Any positive clones that are detected by this procedure are used to select, by hybridization, complementary mRNA sequences, which are then translated in *Xenopus* oocytes or in a cell-free system to yield proteins that can be tested for cytokine activity in a bioassay. (2) When no protein sequence information is available, cDNA clones containing cytokine sequences can only be identified by using a bioassay or precipitation with a specific antibody following translation of hybrid-selected mRNA. (3) An alternative procedure, which was used originally to identify murine IL-2 and murine IL-3 cDNA clones, involved the construction of cDNA libraries in a plasmid vector that could be expressed in eukaryotic cells. Clones that contained functional cytokine cDNA could then be identified by transfecting recombinant plasmids into a monkey cell line and subsequently testing for the production of cytokine.

A summary resulting from gene cloning studies currently available is as follows.

Interleukin 1

The cDNA for IL-1 was cloned from a lipopolysaccharide (LPS)-stimulated human macrophage library and revealed the presence of two distinct, but distantly related, IL-1 molecules, termed IL-1 α and IL-1 β. Amino acid homology between human IL-1 α and human IL-1 β was shown to be 26%, although nucleotide homology was higher at 45%, raising the possibility that a gene duplication event led to the existence of the two IL-1 molecules. The nucleotide sequences predicted primary translation products of 271 (M_r 30,606) and 269 amino acids (M_r 30,749) for IL-1 α and IL-1 β, respectively, although biological activity was shown to reside in the C-terminal 159 and 153 amino acids. This finding was consistent with findings that had shown IL-1 activities to have M_r's of approximately 30,000 and 17,000 and suggested that higher molecular weight precursors were proteolytically cleaved to yield active forms of M_r 17,500. The primary translation product of IL-1 α, but not IL-1 β, is biologically active, although it is possible that this phenomenon represents more efficient cleavage of IL-1 α to the smaller, active form. Neither IL-1 α nor IL-1 β contains a hydrophobic sequence associated with secreted proteins, raising the possibility that these molecules are not actively secreted but, rather, may be released from cells that have been structurally damaged. This remains a major concern in the postulation that IL-1 secretion by APCs is obligatory for the activation of T_H cells. The gene for human IL-1 β has been assigned to chromosome 2q14.

Interleukin 2

Human IL-2 was cloned from a cDNA library prepared from Jurkat cells and coded for a polypeptide of 153 amino acids with a predicted signal sequence of 20 amino acids and a calculated M_r of 15,421 for the mature protein. This agreed with the observed M_r of 15,000 for native IL-2 purified from Jurkat cell culture supernatant. This material is not glycosylated, in contrast to IL-2 from human tonsil cells that displays heterogeneity due to variable glycosylation. The gene for human IL-2 is located on chromosome 4q.

Interleukin 3

Murine IL-3 cDNA clones were isolated from the cell line WEHI-3B, but the search for an equivalent human IL-3 molecule, using both cellular and molecular approaches, proved difficult. Through studies attempting to identify a human sequence by hybridization with a mouse cDNA, and by comparing the sequences of murine and rat IL-3, it became evident that the IL-3 gene was not highly conserved between species. Using a mammalian cell expression cloning system, a cDNA clone was identified that encoded a novel hematopoietic growth factor activity produced by the gibbon lymphosarcoma cell line UCD-144-MLA. This activity caused proliferation of human chronic granulocytic leukemic cells in the presence of a neutralizing antibody to GM-CSF. The cDNA sequence had low but significant homology (29% amino acid, 45% nucleotide) with murine IL-3, and it was used to probe a human fetal liver genomic library to isolate the human IL-3 gene. The structure of the gene for IL-3 is similar in all species studied, containing five exons and four introns of similar respective sizes. In humans, the gene maps to the long arm of chromosome 5. The IL-3 moleculae is processed at the amino-terminus to yield a mature polypeptide. As with IL-2, it is variability in glycosylation that results in the heterogeneous size of molecular species (M_r 14,000–30,000) observed in native IL-3.

Interleukin 4

IL-4 was initially described as a co-stimulant, along with anti-IgM antibodies, of B cell proliferation and secretion of antibody. It was subsequently shown to induce resting B cells to increase expression of class II MHC molecules, to enhance the secretion of IgE and IgG1 by LPS-stimulated B cells, and to stimulate the proliferation of several IL-2– and IL-3–dependent cell lines. That these multiple activities were mediated by a single molecule was confirmed by the cloning and expression of a cDNA that encoded murine IL-4. The DNA sequence codes for a polypeptide of 140 amino acids with a hydrophobic leader sequence of 20 residues and a predicted M_r of 14,000 for the mature protein. The observed M_r (20,000) of the purified native molecule is accounted for by three potential N-linked glycosylation sites.

Interleukin 5

IL-5 is a T cell–derived cytokine that was shown to act as an eosinophil differentiation factor for murine and human cells, and to possess B cell growth factor activity for murine cells. Cloning of cDNAs for IL-5 revealed the structure of human IL-5 to consist of 134 amino acids and to share 70% amino acid homology with the murine molecule. As with other cytokines, post-translational modification with carbohydrate is likely. There are two potential N-glycosylation sites. The gene is present in single copy, contains four exons and three introns, and was mapped to human chromosome 5q23.3 to 5q32.

Interleukin 6

Another example in which different biological activities were subsequently shown to be due to a common protein is IL-6. Previously known as B cell stimulatory factor-2 (BSF-

2), IL-6 is inducible in lymphocytes stimulated with mitogens, and it was shown to be involved in the final pathways of B cell maturation. The cloned cDNA containing 184 amino acids showed sequence homology with G-CSF and was subsequently demonstrated to be identical to a 26-kD protein from fibroblasts, to interferon β_2 (IFN-β_2), to the hybridoma/plasmacytoma growth factor, to a novel T cell–activating factor, and to a murine-active GM-CSF. The human IL-6 gene contained five exons and showed marked similarity in gene structure to G-CSF. Although the conservation in amino acid sequence between murine and human IL-6 is only 41%, there is some species cross-reactivity, which suggests that even though the gene sequence has diverged during speciation, the tertiary structure has probably been maintained. The gene for IL-6 is located on human chromosome 7.

Interleukin 7

IL-7 was purified from the culture supernatant of a murine-adherent bone marrow stromal cell line that had been infected with the SV40 virus. The purified molecule had an M_r of 25,000 and stimulated the proliferation of precursor cells of the B-lymphoid lineage derived from long-term bone marrow cultures. A cDNA clone was isolated using direct expression in COS-7 cells. IL-7 contains a leader sequence of 25 amino acids, followed by 129 amino acids, resulting in a predicted M_r of 14,000. There are two potential N-linked glycosylation sites and six cysteine residues, which are probably involved in intramolecular disulfide bonds.

Colony Stimulating Factors

GM-CSF was initially characterized in media conditioned by lung tissue from mice injected with endotoxin. A cDNA encoding murine GM-CSF was subsequently cloned from the T cell line LB3, and a human GM-CSF cDNA was isolated from the T-lymphoblast cell line Mo, which carries a human T-lymphotropic virus II. Homologies between murine and human GM-CSFs were 60% at the amino acid and 70% at the nucleotide level. The GM-CSF molecules of both species contain four cysteine residues, suggesting the presence of two internal disulfide bridges to retain structural integrity. Purification to homogeneity of the natural and recombinant proteins yielded identical molecules with sizes ranging between 14,000 and 35,000 depending on the extent of glycosylation at the two potential N-glycosylation sites. Restriction enzyme analysis of DNA from the Mo cell line and from human liver suggested that GM-CSF is encoded by a single gene. The gene for human GM-CSF contains four exons and has been localized to chromosome 5q21-32.

G-CSF was initially described as an activity present in endotoxin-treated mouse lung-conditioned medium and in human placental-conditioned medium. Human G-CSF was purified to apparent homogeneity from the squamous carcinoma cell line CHU-2 and from the bladder carcinoma cell line 5637. Following NH$_2$-terminal sequencing of the purified protein, oligonucleotide probes were constructed and used to probe cDNA libraries constructed from the CHU-2 line and the 5637 line. The two clones were identical apart from the insertion of three additional amino acids in the CHU-2–derived G-CSF between residues 35 and 36 of the shorter protein (derived from the 5637 line). This difference has been shown to have arisen through the use of alternative splicing of the primary transcript of the gene. The CHU-2 line possesses alternative donor splice sites at the 5' end of intron 2, and these sites are nine nucleotides apart. G-CSF has no potential sites for N glycosylation but is O-glycosylated, resulting in less heterogeneity in molecular size than exists for IL-3 or GM-CSF. The M_r of natural G-CSF purified from the 5637 line was 18,500 and of recombinant G-CSF cloned from the same source and expressed in *Escherichia coli* was 18,800, as determined by sodium dodecyl sulfate-polyacrylamide gel electrophoresis (SDS-PAGE), under reducing conditions. These values agree with a predicted M_4 of 18,671 deduced from the amino acid sequence. The gene for G-CSF is made up of five exons and four introns, is present in the human genome as a single copy, and is located on human chromosome 17.

M-CSF, also termed colony stimulating factor 1 (CSF-1), was purified from human urine and shown to be a glycoprotein of M_r 47,000 to 76,000 consisting of two identical subunits, of which 14,500 in each subunit was polypeptide. Oligonucleotide probes prepared from the amino-terminal sequence were used to screen a human genomic library, and then one of the clones so isolated was used to screen a cDNA library constructed from 1.5 to 2 kb mRNA, which was extracted from the pancreatic carcinoma cell line MIA PaCa following induction with phorbol ester. Using hybridization studies of M-CSF–specific mRNA from a number of cellular sources, the predominant transcript was found to be 4 kb in size, and a corresponding cDNA clone was isolated from a human SV40-transformed trophoblast cell line. There is only one copy of the M-CSF gene, and the different mRNA species result from alternative splicing of the primary transcript with an extra 894 nucleotides present in the protein-coding region and a longer 3' non-coding region in the extended sequence. The insertion of 894 residues preserves the reading frame within the protein-coding region, and the two transcripts encode polypeptides of 224 and 552 amino acids, respectively. These polypeptides are then processed to result in related but distinct forms of M-CSF. The processed polypeptides contain 145 and 223 residues with expected M_r's of 26,000 and 16,000, respectively. N glycosylation and association to form dimers result in M-CSF molecules of M_r's 40,000–50,000 and 70,000–90,000. The unprocessed polypeptides share a common carboxy-terminal sequence of 75 amino acids, which contains a hydrophobic region of 23 residues followed by a sequence of charged amino acids. This structure is consistent with that proposed for transmembrane insertion, and it suggests that mature M-CSF may be generated by proteolytic cleavage from the cell membrane. However, structural cDNA mutants that lack this region also express and secrete M-CSF from transfected COS-1 cells in an apparently normal manner. The human gene for M-CSF has been assigned to chromosome 1p13–21.

Interferon γ

IFN-γ is an inducible cytokine produced by activated T cells. A cDNA encoding human IFN-γ has been isolated, contains 146 amino acids, and does not show significant homology with either interferon α or interferon β. Native IFN-γ can be resolved into two components of M_r's 20,000 and 25,000, which represent glycosylation either at residue 28 only or at both residues 28 and 100. IFN-γ is encoded by a single gene, which is situated on human chromosome 12q24.1.

Cytokine Receptors

Cytokines interact with target cells through the binding to specific receptors expressed on the cell surface. These receptors are expressed on progenitor and mature cells of the immune and hematopoietic systems, and the binding of cytokines initiates a range of responses in target cells that includes proliferation, differentiation, and activation of effector functions. Binding experiments established that for each cytokine there exists a corresponding receptor. The binding of growth factor to receptor occurs with high affinity (10^{-9} to 10^{-12} M), and binding studies have shown that the number of cytokine receptors on responsive cells is relatively low. This is particularly so for IL-1, IL-3, GM-CSF, G-CSF, erythropoietin, and tumor necrosis factor (TNF), for which individual cells express 10 to 5,000 receptors, whereas for M-CSF the numbers are somewhat higher at 20,000 to 75,000 receptors per cell. The biological effects of hematopoietic growth factors may be achieved with low levels of receptor occupancy.

Despite the difficulties imposed by the relatively low numbers of receptors on cells, cDNAs encoding several of the growth factor receptors have been isolated. IL-1 α and IL-1 β bind to the same cell surface receptor, which has recently been characterized from mouse T cells and shown to be a member of the Ig superfamily. A further member of the Ig superfamily was identified following isolation of a cDNA for the human IL-6 receptor. Complementary DNAs encoding the α chain of the human IL-2 receptor have been isolated, a cDNA encoding the murine IL-4 receptor has been identified, and the receptor for erythropoietin has recently been cloned.

While there have been major advances in the application of molecular biology to the isolation of genes that encode hematopoietic growth factors and their receptors, one of the most elusive areas in modern biology remains the identity of the intracellular events that occur following ligand–receptor interactions. There is evidence that the receptor for M-CSF possesses tyrosine kinase activity and is related to the c-*fms* proto-oncogene product. Because proteins with tyrosine kinase activity are encoded by a number of viral oncogenes and their cellular homologues, this activity is considered to be important in the mitogenic function of a number of cytokine receptors. Analyses of the amino acid sequences of the cytoplasmic portions of the IL-1 and IL-6 receptors have not revealed domains typical of either protein tyrosine kinases or protein tyrosine kinase acceptor sites, and the cytoplasmic tail of the IL-2 receptor is too small (13 amino acids) to encode such an activity. There is evidence accumulating, however, that phosphorylation of growth factor receptors does occur following the binding of ligand, and that this may involve receptor-associated kinase activities. This has been demonstrated for the IL-3 receptor, where a tyrosine residue is phosphorylated on the p140 protein, and recently for the IL-1 receptor.

REGULATION OF CYTOKINE GENE EXPRESSION

Several important issues in hematopoietic and lymphoid cell development can now be approached through an understanding of the control of gene expression. These include the cell specificity of cytokine production, the inducibility of the system in response to environmental stress, and the means by which the composition of cells within and between individual lineages is maintained. The cell types that are essential to intiate immune responses, T_H cells and macrophages, are clearly major sources of cytokines. Further, activation of these cells by antigenic stimuli, or by events that mimic antigenic stimuli, remains the most potent means of inducing cytokine synthesis. Thus far there is only evidence for constitutive production of M-CSF. Normal activated T cells have been shown to express high levels of mRNA for GM-CSF and low levels of IL-3 mRNA; however, mRNA transcripts for G-CSF and M-CSF have not been detected in these cells.

Monocytes that have been activated with IFN-γ, endotoxin, or phorbol esters express high levels of mRNA for G-CSF and M-CSF but not for GM-CSF. Primary human bone marrow stromal cells and other stromal elements, including endothelial cells and fibroblasts, can be induced by IL-1 to express mRNA for GM-CSF and G-CSF. These findings have suggested that although M-CSF, G-CSF, and GM-CSF play a role in hematopoietic cell production in times of stress such as infection, there is no clear role for these molecules in steady-state hematopoiesis.

The genes for IL-3 and GM-CSF are in close physical linkage in both murine and human genomes. In humans, the IL-3 and GM-CSF genes are tandemly arrayed, separated by 9 kb, on chromosome 5q. Because of the proximity of the genes and their similarities in both gene structure and biological functions, IL-3 and GM-CSF may have evolved from a common ancestral gene. Located on the long arm of chromosome 5 are also the genes for IL-5, the M-CSF receptor (c-*fms*), the endothelial cell growth factor, and the receptor for platelet-derived growth factor. The presence of this cluster of growth factor and growth factor receptor genes raises the question of whether their juxtaposition may be important for coordination or control of their expression.

The most highly conserved sequences of the murine and human genes for GM-CSF are in the promoter regions and in the 3' noncoding regions of the mRNA. Sequences in the promoter region are also shared by other cytokine genes, suggesting that these sequences may be important in the regulation of gene expression. A promoter region decanucleotide consensus sequence consisting to 5'GRGRTTYCAY3' (R = A or G; Y = C or T) was identified for the murine IL-2, IL-3, and GM-CSF genes and for the human IL-2 gene. This sequence also occurs in two regions upstream of the transcription initiation site of the human GM-CSF gene. Furthermore, when a 660-nucleotide fragment, 5' from the initiation codon of the human GM-CSF gene, was linked to the marker chloramphenicol acetyltransferase (CAT) gene, and the recombinant constructs were transfected into a human T cell leukemia virus type 1 (HTLV-1)–infected T lymphoblast cell line, the 5' flanking sequence directed increased expression of the CAT gene after cell activation with phytohemagglutinin and phorbol 12-myristate 13-acetate. This suggested that the 660-nucleotide fragment contained *cis*-acting regulatory elements, and it appeared to be cell-specific in that no CAT activity was detected in stimulated, Epstein-Barr virus–transformed B cells infected with the construct. Nuclear proteins that bind to two cytokine-specific sequences, CK-1 (5' GAGATTCCAC 3') and CK-2 (5' TCAGGTA 3'), were identified, and these sequences spanned nucleotides -95 to -86 from the transcription initiation site of the GM-CSF gene. The presence of these inducible nuclear proteins was specific for cells that produced GM-CSF. CK-1 sequences are also evident in the IL-2, IL-3, GM-CSF, and G-CSF genes (human and murine), whereas CK-2 sequences are found only in the IL-3 and GM-CSF genes of both species. It is possible that the extended, conserved sequence of CK-1 and CK-2 may be important in the coordinate expression of IL-3 and GM-CSF from T cells.

CYTOKINES AND DEVELOPMENT

The term cytokine is usually used to describe a subset of hormone-like polypeptides with immunoregulatory effects as distinct from growth factors that modulate proliferation and differentiation in non-immunological cell types. However, the distinction between cytokines and growth factors is becoming increasingly blurred as the two types of regulators have many similarities, and many cytokines act on cells outside the immune system. Table 14–2 summarizes some of the known effects of the cytokines. It can be seen that most of them are multifunctional.

Analysis of the role of cytokines in development is complicated by a number of factors. Within the hematopoietic and immune systems the production and development of a particular cell lineage can be stimulated by more than one cytokine, and the control of

Table 14–2. Cytokines and Their Properties

Cytokine	Molecular Weight (M_r)	Source	Activity
Interleukin 1	17,000	Monocytes Antigen presenting cells	Second signal for T cell activation Activation of various inflammatory responses Induction of neuropeptides Neurotransmission in hypothalamus
Interleukin 2	21,000	Helper T cells	Proliferation of activated T cells B cell activation NK cell activation Induction of neuropeptides in pituitary
Interleukin 3	15–30,000	Helper T cells	Proliferation of cells from many hemopoietic lineages
Interleukin 4	16–20,000	Helper T cells	Proliferation of activated B cells Modulation of CSF activity Proliferation of activated T cells
Interleukin 5	46,000	Helper T cells	Proliferation of eosinophil precursors Functional activation of mature eosinophils Enhanced IgA synthesis by activated B cells
Interleukin 6	19–21,000	Helper T cells	B cell differentiation Plasmacytoma growth factor T lymphocyte activation Hepatocyte stimulation Anti-viral activity Induction of ACTH production in pituitary
Interleukin 7	25,000	Bone marrow Stromal cells	Growth of pre-B cells Growth of early thymocytes
Interferon γ	20–25,000	Helper T cells	Anti-viral activity Macrophage activation Enhanced expression of class II MHC determinants on APC and activated B cells Enhancement of NK activity
GM-CSF	14–35,000	Helper T cells	Granulocyte and macrophage growth and differentiation
G-CSF	20,000	Monocytes	Granulocyte growth and differentiation
M-CSF	70–90,000, 45–50,000	Fibroblasts	Monocyte/macrophage growth and differentiation

blood cell formation is characterized by apparent redundancy of the factors involved (Metcalf, 1989). The effects of a particular cytokine on a single cell type are not invariant, but are critically dependent on the background set by other signaling molecules. For example, IL-4 acting alone has no proliferative effects on normal murine hematopoietic progenitor cells, but it enhances colony formation stimulated by G-CSF, M-CSF, or erythropoietin. In contrast, IL-4 inhibits colony formation stimulated by IL-3.

Along with the complications of additive, synergistic, and inhibitory effects of cytokines on one another, the analysis of cytokine biology is greatly complicated by the ability of one factor to induce the synthesis of others. The resultant cascade of mediator release may either amplify the effect of the original stimulus or lead to feedback inhibition of the primary cytokine. The multitude of molecular signals to which cells are continually exposed can therefore be viewed as a signaling language in which, like all languages, the information content of any individual component depends on its context (Sporn and Roberts, 1988).

Cytokines and Intrathymic Development

Although IL-2, IL-4, and IL-7 are all T cell growth hormones, it is not known whether any of these cytokines is produced in the thymus or regulates cell proliferation in this organ. Interestingly, IL-2 receptor–bearing cells can be identified in the thymus. In the murine thymus, a small population of blast cells, which do not express CD4 or CD8 surface antigens, have receptors for IL-2. These CD4$^-$ CD8$^-$ blasts are the logical precursors of the CD4$^+$ and CD8$^+$ lymphoblasts associated with functional T_H or T_K activity. However, the role of IL-2 receptors in cell differentiation in the thymus is not clear, and so far IL-2 has been shown not to be needed during thymocyte maturation.

IL-7 has been shown to stimulate proliferation of early (day 12–14) murine fetal thymocytes in each of three culture systems: single cell suspension cultures; conventional fetal thymus organ cultures (FTOC), in which intact thymic lobes were cultured at the gas-fluid culture interface; and lobe submersion cultures (LSC), in which thymic lobes were cultured totally immersed in medium. In suspension cultures, IL-7 and IL-2 each induced a DNA synthetic response in a short-term assay, but neither cytokine supported continued cell growth. In FTOC, the addition of exogenous IL-7 resulted in a two-fold increase in cell number over that which normally developed in FTOC during a 7-day period. However, the most striking effects of IL-7 were noted in LSC, a system in which thymocyte growth was totally dependent on the addition of exogenous cytokine. Cells proliferated for a period of approximately 2 weeks in IL-7, and cell viability could be maintained even longer. A high percentage of cells recovered after 7–14 days from IL-7–supplemented LSC resembled the earliest detectable fetal thymocytes by expressing Pgp-1, by the lack of CD4, CD8, and CD3 surface antigens, and by exhibiting an increased expression of the IL-2 receptor. These results suggest that IL-7 may favor the growth of cells that appear early in the T cell lineage and may also act as the stimulus to induce the expression of IL-2 receptors.

Cytokines and Effector T Cell Functions

In the peripheral lymphoid organs, interaction of T cells with antigen leads to the increased expression of receptors for IL-2 and a consequent proliferation in response to

IL-2. IL-6 also appears to prepare T cells for responses to other cytokines, and IL-6-activated cells also respond well to IL-4 or IL-7. Although IL-2 is known primarily as a growth hormone, it also stimulates T cells to secrete a variety of other cytokines. There are now many reports of IL-2 inducing the secretion by T cells of IFN-γ, which itself has a variety of effects on immune function. First, IFN-γ induces the expression of class II MHC molecules on macrophages, which is important for processes that regulate the activation of helper T cells. Second, it enhances the susceptibiltiy of virus-infected cells to cytotoxic T cells. Third, it enhances natural killer activity and stimulates the appearance of cytolytic T cells. Each of the responses elicited by IFN-γ serves to enhance immunity through cell-mediated responses. The other group of mediators released upon stimulation of T cells by IL-2 are those activities that have B cell growth activity. In particular, IL-2 appears to stimulate the release of IL-4, IL-5, and IL-6 from T cells.

B Cell Maturation

The induction of antibody synthesis requires an interaction between T_H cells and B cells. The helper function of T cells can be attributed to their production of cytokines that stimulate B cell division and regulate the isotype of antibody produced (Kishimoto et al., 1984). However, precise cytokine function in B cell activation and antibody production is difficult to understand because of the diverse multiple effects that have been recorded.

The function of IL-2 in T cell growth, together with the apparent symmetry between the processes responsible for induction of antigen-sensitive B and T cells, has resulted in an extensive search for a growth factor similar to IL-2, but with specificity for B cells. The difficulty in this search lies in distinguishing between (a) the development of B-lineage cells, i.e., processes analogous to those that are confined to the thymus for T lymphocytes, and (b) the induction of mature antigen-sensitive B cells, i.e., a population comparable to peripheral T lymphocytes. Although there are several reports of the establishment of continuous B cell lines, no B cell–specific regulator analogous to IL-2 for T cells has emerged. The use of B cell stimulators such as LPS, dextran sulfate, or anti-immunoglobulin reagents led to the discovery of the cytokines, including IL-4, IL-5, and IL-6, which act synergistically with these stimulators.

IL-4 was first shown to act as a co-stimulator with anti-immunoglobulin and to cause proliferation of resting B cells. In addition, IL-4 induced the expression of class II MHC molecules on B cells. IL-4 also induces isotype switching and production of IgG1 and IgE by activated B cells. The increase in class II MHC molecules on B cells induced by IL-4 may lead to more effective presentation of antigen to T cells, thereby increasing the efficiency of T–B cell collaboration. However, the multiple effects of IL-4 on T cells, mast cells, and B cells make it difficult to focus on a primary role for this cytokine.

IL-5 was initially described as a growth factor for a B cell hybridoma, BCL_1, and it has subsequently been shown to induce IgM, IgG, and IgA secretion from activated B cells. Again, as IL-5 stimulates the differentiation of eosinophils within bone marrow, and appears to induce the development of cytolytic T cells, its effects are clearly not confined to B cells.

IL-6 was identified by its ability to induce antibody production by preactivated human B cells; this was followed by the demonstration that IL-6 enhanced the growth of B cell hybridomas and plasmacytomas. IL-6 also possesses T cell–activating properties, but its primary physiological role is not known.

Cytokines and Macrophage Development

Throughout life, macrophages, which are derived from bone marrow, can differentiate into several diverse subpopulations of cells. These include blood monocytes, alveolar macrophages, inflammatory macrophages such as peritioneal exudate cells (PEC), Kupffer cells, dendritic cells, glial cells, tissue histiocytes, and Langerhans cells. All these cells appear to function as effective APCs. Activated macrophages are also a source of the cytokines IL-1, GM-CSF, G-CSF, M-CSF, and IL-6. Although much has been achieved in the molecular analysis of regulators that stimulate macrophage growth and differentiation in culture, the physiological roles of IL-3, M-CSF, GM-CSF, and G-CSF in these processes are not nearly so clear. Macrophages and other APC types to which they give rise are important to the immune system, and there are aspects of their role in antigen presentation that should be stressed. The expression of class II MHC molecules by macrophages is not constitutive. Because of the pivotal role of class II MHC molecules in T_H cell activation, the regulation of their expression on APCs is a critical step in the control of immune responsiveness. Although there may also be regulation of the class I MHC expression, this appears to be a matter of degree of expression rather than presence or absence.

To maintain the basal level of class II molecules expressed in tissue macrophages, there appears to be a need for continuous differentiation of stem cells to macrophages. The extent of class II molecule expression can be regulated in two ways: (1) the interaction of class II positive macrophages with some agents that trigger phagocytosis maintains class II molecule expression and retards the transition from positive to negative expression, and (2) IFN-γ produced by antigen-stimulated T cells preferentially induces class II molecule expression in previously immature macrophages. The production of IFN-γ by T cells takes place after activation of the T_H cells by antigen-bearing, class II positive macrophages—a clear example of the symbiotic relationship between these two cell types. The same molecule, IFN-γ, released by T cells stimulates both class II expression and cytotoxic T cell activity. Studies show that macrophages fixed after having processed antigen can stimulate the growth of T cell clones, which suggests that secreted products, such as IL-1, may not be required for stimulation of recently activated T cells. Thus, a requirement for IL-1 production by an antigen presenting cell might be limited to the induction of growth in memory or perhaps naive T cells. Alternatively, membrane-associated IL-1 may be able to substitute for secreted IL-1 under certain conditions.

Cytokines and Phylogeny

Much of the work on cytokines to date has concentrated on mammalian species and has demonstrated a considerable degree of sequence conservation and functional cross-reactivities among the molecules derived from different members of this family. The questions of whether cytokines or cytokine-like molecules exist in lower order vertebrates and invertebrates, and whether some conclusions can be drawn about evolutionary or phylogenetic significance are beginning to be explored. An IL-1–like protein was found in the coelomic fluid and coelomocytes of an echinoderm, the starfish, *Asterias forbesi* (Beck and Habicht, 1986), and, more recently, IL-1–like activity was discovered in eight North American species of tunicates (Beck et al., 1989). Tunicate IL-1 has physical and biochemical properties that are strikingly similar to both mammalian and echinoderm IL-1, and its activity in the thymocyte proliferation assay can be inhibited by rabbit antibodies against human IL-1, which indicates considerable functional as well as structural homology. For example, Hamby and colleagues (1986) isolated IL-1–like activity from fish and reported that fish lymphocytes can respond to human IL-1. These data suggest that IL-1 is an ancient and important molecule, and the fact that it was conserved throughout the invertebrates and the vertebrates attests to its importance in the evolution of host defenses, and perhaps in evolution itself.

Of the other cytokines, IL-2–like activity was found in supernatants of mitogen-activated lymphocytes from amphibians *(Xenopus)* and fish *(Cyprinus)*. Human, gibbon, and rat (but not mouse) IL-2 were able to stimulate carp lymphocytes, but, interestingly, no cross-reactivity between *Xenopus* and human or murine IL-2 was observed (Watkins and Cohen, 1985). It is conceivable, therefore, that IL-2 has been less conserved throughout evolution than has IL-1. With the availability of gene probes for all of the murine and human cytokines, it is now possible to begin to investigate the phylogeny of cytokines in a more systematic way and to address the question of the significance of these molecules in evolution and development.

CONCLUSION

Over the last 10 years, much has been learned about the hematopoietic and lymphoid cell lineages and the factors that can influence their development. A number of cytokines have been identified that appear to be intimately associated with development and differentiation in cell culture, but the question of their physiological importance has yet to be answered (Clark and Kamen, 1987). Studies using panels of monoclonal antibodies and flow cytometry have yielded extensive information about the phenotypic complexity and the relationships among many of the cell types within the hematopoietic lineages. With this information, together with cytokines available in large quantities from cloned genes, it should be possible to probe the fine structure of the hematopoietic and immune systems and to discover their physiological role and potential immunotherapeutic applications.

Acknowledgments

The authors acknowledge the support of the Health Research Council of New Zealand, Auckland Division; the Cancer Society of New Zealand Inc.; the Auckland Medical Research Foundation; the Wellcome Trust Foundation, London; and the Leukaemia and Blood Foundation of New Zealand.

References

Babbitt, B. P., Allen, P. M., Matsueda, G., Haber, E., and Unanue, E. R. (1985). The binding of immunogenic peptides to Ia histocompatibility molecules. *Nature 317*:359–361.

Beck, G., and Habicht, G. S. (1986). Isolation and characterization of a primitive IL-1-like protein from an invertebrate, *Asterias forbesi. Proc. Natl. Acad. Sci. USA 83*:7429–7433.

Beck, G., Vasta, G. R., Marchalonis, J. J., and Habicht, G. S. (1989). Characterization of interleukin 1 activity in tunicates. *Comp. Biochem. Physiol. 928*:93–98.

Benacerraf, B., and Germain R. N. (1978). The immune response genes of the major histocompatibility complex. *Immunol. Rev. 38*:70–119.

Bjorkman, P. J., Saper, M. A., Samraoui, B., Bennett, W. S., Strominger, J. L., and Wiley, D. C. (1987a). Structure of the human class I histocompatibility antigen, HLA-2. *Nature 329*:506–512.

Bjorkman, P. J., Saper, M. A., Samraoui, B., Bennett, W. S., Strominger, J. L., and Wiley D. C.

(1987b). The foreign antigen binding site and T cell recognition regions of class I histocompatibility antigens. *Nature 329*:512–518.

Born, W., Yague, J., Palmer, E., Kappler, J., and Marrack, P. (1985). Rearrangement of T-cell receptor β-chain genes during T-cell development. *Proc. Natl. Acad. Sci. USA 82*:2925–2929.

Brown J. H., Jardetzky, T., Saper, M. A., Samraoui, B., Bjorkman, P. J., and Wiley, D. C. (1988). A hypothetical model of the foreign antigen binding site of class II histocompatibility molecule. *Nature 332*:845–850.

Clark, S. L., and Kamen, R. (1987). The human hematopoietic colony-stimulating factors. *Science 236*:1229–1237.

Doherty, P. C., Blanden, R. C., and Zinkernagel, R. M. (1976). Specificity of virus-immune effector T cells for H-2K or H-2D compatible interactions: Implications for H-antigen diversity. *Immunol. Rev. 29*:89–124.

Germain, R. N., and Malissen, B. (1986). Analysis of the expression of class-II major histocompatibility complex–encoded molecules by DNA-mediated gene transfer. *Annu. Rev. Immunol. 4*:281–315.

Haars, R., Kronenberg, M., Gallatin, W. M., Weissman, I. L., Owen, F. L., and Hood, L. (1986). Rearrangement and expression of T cell antigen receptor and γ genes during thymic development. *J. Exp. Med. 164*:1–24.

Hamby, B., Huggins, E., Lachman, L., Dinarello, C., and Sigel, M. (1986). Fish lymphocytes respond to human IL-1. *Lymphokine Res. 5*:157–162.

Kingston, R., Jenkinson, E. J., and Owen, J.J.T. (1985). A single stem cell can recolonize an embryonic thymus, producing phenotypically distinct T-cell populations. *Nature 317*:811–813.

Kishimoto, T., Yoshizaki, K., Kimoto, M., Okada, M., Kuritani, T., Kikutani, H., Shimizu, K., Nakagawa, T., Nakagawa, N., Miki, Y., Kishi, H., Fukunaga K, Yoshikubo, T., and Tada, T. (1984). B cell growth and differentiation factors and mechanism of B cell activation. *Immunol. Rev. 78*:97–118.

Metcalf, D. (1989). The molecular control of cell division, differentiation commitment and maturation in haemopoietic cells. *Nature 339*:27–30.

Mosmann, T. R., and Coffman, R. L. (1987). Two types of mouse helper T-cell clone: Implications for immune regulation. *Immunol. Today 8*:223–227.

Owen, J.J.T., and Jenkinson E. J. (1984). Early events in T lymphocyte genesis in the fetal thymus. *Am. J. Anat. 170*:301–310.

Schuler, G., and Steinman, R. M. (1985). Murine epidermal Langerhans cells mature into potent immunostimulatory dendritic cells in vitro. *J. Exp. Med. 161*:526–546.

Scollay, R., Bartlett, P., and Shortman, K. (1984a). T cell development in the adult murine thymus: Changes in the expression of the surface antigens Ly2, L3T4 and B2A2 during development from early precursor cells to emigrants. *Immunol. Rev. 82*:79–103.

Scollay, R., Wilson, A., and Shortman, K. (1984b). Thymus cell migration: Analysis of thymic emigrants with markers that distinguish medullary thymocytes from peripheral T cells. *J. Immunol. 132*:1089–1094.

Skinner, M., Le Gros, G. S., Marbrook, J., and Watson, J. D. (1987). Development of fetal thymocytes in organ cultures: Effect of interleukin 2. *J. Exp. Med. 165*:1481–1493.

Sporn, M. B., and Roberts, A. B. (1988). Peptide growth factors are multifunctional. *Nature 332*:217–219.

Townsend, A.R.M., Basten, J., Gould, K., and Brownlee, G. G. (1986a). Cytotoxic T lymphocytes recognize influenza haemagglutinin that lacks a signal sequence. *Nature 324*:575–577.

Townsend, A.R.M., Rothbard, J., Gotch, F. M., Bahadur, G., Wraith, D., and McMichael, A. J. (1986b). The epitopes of influenza nucleoprotein recognized by cytotoxic T lymphocytes can be defined by short synthetic peptides. *Cell 44*:959–968.

Watkins, D., and Cohen, N. (1985). The phylogeny of interleukin-2. *Dev. Comp. Immunol. 9*:819–824.

15

Functional and Molecular Aspects of T Cell Activation

GORDON B. MILLS

The mechanisms by which interaction of lymphocytes with foreign antigen initiates cell activation are only beginning to be identified. An understanding of the interplay of signals that determines how a particular foreign insult leads to the appropriate cellular response has been even more elusive. The difficulties in elucidating regulatory events in the immune response are further compounded when one considers the requirements for maturation and differentiation of thymus-derived (T) and bone marrow-bursal–derived (B) cells. Although it appears that the transmembrane signals generated in mature and immature cells are qualitatively similar, their physiologic roles in determining cellular repertoire, self recognition, and tolerance in immature cells are unknown.

SIGNAL TRANSDUCTION

Background

Cells interact with their environment either through direct cell–cell contact or through the action of intercellular mediators such as steroid and polypeptide hormones and neurotransmitters. Steroid hormones permeate the cell membrane and interact directly with intracellular mediators. Although steroid hormones influence immune function both directly or indirectly, little is known about the role of this pathway in the normal immune response. Therefore, steroid hormones will not be considered further in this discussion.

Polypeptide hormones, neurotransmitters, and cell–cell contact appear to use similar signal transduction mechanisms. In each case, a ligand—either a soluble factor or a cell surface molecule—binds to a specific cell surface receptor. If the affinity of the complex formed—either through high-affinity binding to one receptor, as in the case of lymphokines, or through low-affinity binding by a group of receptors, as in the case of antigen presentation and other cell–cell interactions—is sufficient, the conformation of the receptor changes. This conformational change is translated across the membrane, and it subsequently activates intracellular enzymes, which then generate second messengers. These second messengers, in turn, activate a cascade of enzymes that transduce the signal across the cytosol to the nucleus. The existence of the cascade allows for discrete sites of crosstalk between signals transduced through several activated cell surface receptors, which "fine-tune" the message.

Although a variety of intracellular second messengers are generated after activation of cell surface receptors, most pathways activate protein kinases (Hanks et al., 1988; Hunter, 1987; Hunter and Cooper, 1985; Yarden and Ullrich, 1988). Phosphorylation of enzymes can either increase or decrease their activity (Hunter, 1987; Yarden and Ullrich, 1988). G proteins and their regulators, which control the activity of adenyl and guanyl cyclase as well as phospholipase C (PLC), are substrates for phosphorylation (Ellis et al., 1990; Margolis et al., 1989; Meisenhelder et al., 1989; Molloy et al., 1989; O'Brien et al., 1989). Similarly, phosphorylation of both the substrates and the products of PLC alters their availability and ability to act as second messengers (Berridge, 1987; Kikkawa and Nishizuka, 1986; Nishizuka 1988). The affinity or availability of a variety of DNA binding proteins is also regulated by their level of phosphorylation or by the phosphorylation of binding proteins (Alexander and Cantrell, 1989; Cyert and Thorner, 1989; Hunter, 1989; Saksela et al., 1989). In addition, cell cycle progression requires the coordinated expression and phosphorylation and dephosphorylation of a number of intracellular kinases (Cyert and Thorner, 1989; Hunter, 1989). Therefore, the net phosphorylation state, which is the consequence of the interplay between kinases and phosphatases, of a variety of molecules has the potential to underlie most of the mechanisms involved in signal transduction.

Serine, threonine, and tyrosine are the major protein kinase substrates, accounting for approximately 95%, 5%, and 0.02% of the total, respectively (Hunter, 1989). Despite being quantitatively minor, tyrosine phosphorylation appears to play a particularly important role in signal transduction that regulates cell growth. Several growth factor receptors, including those for epidermal growth factor (EGF), platelet-derived growth factor (PDGF), and macrophage colony stimulating factor (M-CSF, also called CSF-1) contain intrinsic tyrosine kinase domains (Yarden and Ullrich, 1988). Others, such as the insulin receptor and CD4 or CD8, interact either covalently or noncovalently with proteins that possess tyrosine kinase activity (Rudd et al., 1988; Veillette et al., 1989a,b; White and Khan, 1986; Yarden and Ullrich, 1988). Furthermore, a majority of identified oncogenes are tyrosine kinases (Hunter, 1987; Hunter and Cooper, 1985; Hanks et al., 1988; Yarden and Ullrich, 1988). Inactivation of the tyrosine kinase activity through site-directed mutagenesis, injection of anti-phosphotyrosine antibodies, or incubation with specific tyrosine kinase inhibitors blocks the mitogenic and transforming activities of these genes (Glenney et al., 1988; Hunter, 1987; Levitski and Schlessenger, 1989; Yarden and Ullrich, 1988). In addition, increasing the levels of phosphorylated tyrosine residues by blocking phosphatase activity can increase cell proliferation and can confer a transformed phenotype on some cell lineages (Klarland, 1985).

Receptor Complexes

Several polypeptide hormone receptors, such as those for EGF, PDGF, and M-CSF, comprise a ligand recognition domain, a transmembrane segment, and an intracellular tyrosine kinase domain (Yarden and Ullrich, 1988). In others, such as the insulin receptor, the ligand binding molecule is associated with a kinase through a disulfide linkage (White and Khan, 1986). This allows conformational changes in the extracellular portion of the molecule to directly alter the catalytic domain. However, even in these cases, it appears that activation of kinase activity does not occur as a consequence of direct alteration in the kinase domain secondary to ligand binding. Rather, it appears that ligand binding results in the formation of oligomeric complexes (Bishayee et al., 1989; Carraway et al., 1989; Honegger et al., 1989; Northwood and Davis, 1989; Obermaier-Kusser et al., 1989). Cross-phosphorylation of the apposed intracellular domains has then been proposed to allow access of the kinase domain to the intracellular substrates involved in signal transduction.

Other receptors function as part of a complex containing a transmembrane molecule with an extracellular ligand binding domain and several molecules involved in signal transduction. Several of the signal transducing molecules, including some of the tyrosine kinase oncogenes and G proteins, are myristoylated, ensuring localization to the cell membrane (Gilman and Casey, 1988; Hanks et al., 1988; Hunter, 1987; Hunter and Cooper, 1985; Neer and Clapham, 1988). In other signal transducing systems, such as the B and T cell antigen-receptors, transmembrane proteins are integral components of the activation complex (Campbell and Cambier, 1990; MacDonald and Nabholz, 1986; Weiss et al., 1986; Weissman et al., 1988).

In receptors that function as a complex, the signal transduction molecule may be free to move in the membrane and to associate with several different cell surface ligand binding molecules (Gilman and Casey, 1988; Neer and Clapham, 1988). Variable associations between ligand binding molecules and signal transduction molecules may allow a single ligand binding molecule to interact with more than one signal transduction mechanism, either in the same or in different cells. For example, a single β adrenergic receptor can interact with adenyl cyclase or with PLC as its second messenger–generating system (Gilman and Casey, 1988; Neer and Clapham, 1988). Conversely, this could allow several distinct receptor species to interact with and regulate the same signal transducing pathway. For example, in the immune system, the T cell antigen receptor (TCR), Thy-1, CD2, and CD23 all appear to signal through interaction with the CD3 complex (Breitmeyer et al., 1987; Guimezanes et al., 1988; Gunter et al., 1987).

Many of the receptors that are specific for the immune system have separate polypeptides mediating ligand binding and signal transduction. Both TCR and surface immunoglobulin (sIg) have short intracellular chains and have no obvious mechanism for signal transduction. Therefore, the TCR and sIg likely signal through their associations with signal transduction complexes (Campbell and Cambier, 1990; Hombach et al., 1988; MacDonald and Nabholz, 1986; Sakaguchi et al., 1988; Samelson et al., 1990; Weiss et al., 1986; Weissman et al., 1988). The lymphokine receptors can be divided into two broad families, which have similarity either to Ig (Aguet et al, 1988; Curtis et al., 1989; Kumar et al., 1989; Sims et al., 1988; Yamasaki et al., 1988) or to a family of receptors that includes those for IL-2, IL-3, IL-4, GM-CSF, EPO, prolactin, and the growth hormone (Bazan, 1989; D'Andrea et al., 1989; Hatakeyama et al., 1989a,b; Waldman, 1989; Mosley et al., 1989). However, both types of lymphokine receptors (with the exception of one of the two IL-2 binding molecules) have larger intracellular domains than either the sIg or TCR has. These intracellular domains may interact with signal transduction molecules by as yet unidentified mechanisms. The demonstration that the cell surface adhesion molecules CD4 and CD8 associate with and activate the relatively T cell–specific tyrosine kinase (LCK) provides a potential signal transduction mechanism for at least one class of lymphocyte-associated receptor (Rudd et al., 1988; Hurley and Setton, 1989; Veillette et al., 1989a,b). The most extreme example of these mechanisms in the immune system is seen in the interleukin 6 (IL-6) receptor, in which the extracellular portion of the IL-6 receptor interacts with a transmembrane signal transducing molecule (Yamasaki et al., 1988). The transmembrane portion of the IL-6 binding molecule appears to be functionally irrelevant other than as an anchor in the cell membrane (Taga et al., 1989).

Because cross-linking of sIg or TCR by physiological ligands, mitogenic lectins, or antibodies appears to be obligatory for cellular activation, it is intriguing to speculate that oligomerization of the receptor complex may play a role in signal transduction. In the case of EGF, PDGF, and M-CSF receptors, oligomerization mediated by components of the extracellular portion of the receptor is associated with activation and may be necessary for signal transduction. In addition, the EGF receptor phosphorylates the NEU receptor when they are apposed, suggesting that this may be a general mechanism for receptor interaction. Following activation of T lymphocytes, the TCR and adhesion molecules co-localize to the pole of the T cell that interacts with the antigen presenting cell (APC) (Kupfer et al., 1987; Rivas et al., 1988). This would bring the TCR into close approximation with adhesion molecules (Kupfer et al., 1987; Makgoba et al., 1989; Rivas et al., 1988), and interaction between these molecules could then lead to signal transduction.

LYMPHOID ACTIVATION

Resting T or B cells display a restricted number and pattern of cell surface receptors (Dinarello, 1986; Greene and Leonard, 1986; MacDonald and Nabholz, 1986; Mills, 1989; Paetkau and Mills, 1989; Smith, 1988; Weiss et al., 1986; Weissman et al., 1988). Following productive activation of the antigen receptor, a number of different lymphokine receptors are expressed on the cell surface. The interaction of these newly expressed receptors with their specific ligands leads to proliferation and differentiation of the responding cells (Mills, 1989; Mills et al., 1988a,b, 1991; Smith, 1988; Waldmann, 1989). Therefore, activation of a particular antigen receptor determines the specificity of the response. The intracellular machinery of the cell, determined by its lineage and state of differentiation, establishes the type of response generated. The array of lymphokine receptors expressed, together with the pattern of lymphokines produced, regulates the magnitude of the response. The duration of the response is determined by the kinetics of local lymphokine production combined with the array of lymphokine receptors on the cell surface.

Activation of lymphocytes normally occurs within a cognate recognition complex made up of antigen presenting cells (APCs) and helper (T_H) and effector T lymphocytes. The interaction between TCR and antigen in the context of MHC has relatively low affinity. The complex is stabilized by the presence of a number of adhesion molecules, such as CD2, CD4, CD8, intercellular adhesion molecule (I-CAM) and lymphocyte function antigen (LFA) on the surface of the responding and stimulating cells; the adhesion molecules do not function solely to stabilize the complex but appear to be directly involved in signal transduction (Bierer et al., 1988; Bockenstedt et al., 1988; Boyd et al., 1988; Haynes et al., 1989; Janeway et al., 1989; Makgoba et al., 1989; Marlin and Springer, 1987; Moretta et al., 1989; St-Pierre et al., 1989). This may result in amplification of the signal transduced through the TCR or may communicate a negative signal in the absence of TCR ligation.

Within this activation complex, the concentrations of lymphokines are sufficient for them to bind to their high-affinity receptors. In some cases, lymphokines are secreted specifically at the pole of the cell involved in the complex, thus further increasing the local concentrations (Poo et al., 1988). Because lymphokines bind to their receptors and are then endocytosed and degraded, most of the secreted lymphokine never leaves the activation complex (Fujii et al., 1986; Greene and Leonard, 1986; Matsushima et al., 1986; Mills, 1989; Paetkau and Mills, 1989; Smith, 1988). Dilution in interstitial fluid ensures that lymphokine concentrations decrease rapidly as the distance from the complex is increased. Furthermore, systemic free lymphokine concentrations are limited by the presence of binding proteins, which in the case of IL-2 and IL-4 can be secreted forms of the receptor itself, together with rapid degradation and excretion (Borth and Lugar, 1989; Mills, 1989; Mills et al., 1991; Mosley et al., 1989; Paetkau and Mills, 1989; Rubin et al., 1985).

Both receptor expression and lymphokine production appear to be coordinately regulated. For example, the T_H1 subclass of helper T cells in mice produces an array of lymphokines that is ideally suited for generation of antibody responses, whereas the T_H2 subclass produces an array that would preferentially stimulate cytotoxic T lymphocyte responses. The presence of common upstream regulatory regions in the genes, coupled with expression of DNA binding proteins, results in specific patterns of mRNA synthesis for lymphokines and their receptors (Paetkau and Mills, 1989; Taniguichi, 1988). The presence of common downstream regions in lymphokine mRNAs, which regulate their sensitivity to degradative enzymes such as RNAase H, and the potential for regulated expression of these enzymes further enhance the specificity of protein production (Paetkau and Mills, 1989; Taniguichi, 1989).

In most cases, lymphokine functions are pleiotropic and are not restricted to cells involved in immune responses. Receptors for IL-1 appear to be present on almost all cell types, suggesting that this lymphokine functions in a number of cellular responses (Dinarello, 1986). Even IL-2, which is among the most functionally restricted of the lymphokines, stimulates B cells, T_H cells, T effector cells, natural killer (NK) cells, lymphokine activated killer (LAK) cells, endothelial cells, and macrophages. As well, lymphokines can induce either proliferation or differentiation of cells within a single lineage. The end response is determined by the intracellular machinery of the cell, together with responses to signals transduced through lymphokine receptors, adhesion molecules, and antigen receptors.

Despite the fact that many lymphokines have similar tertiary structures and that their receptors contain similar binding regions, lymphokines generally interact only with their specific receptors and do not bind to receptors for other lymphokines (Brandhuber et al., 1987; Cohen et al., 1986; Smith, 1988). However, IL-3 and granulocyte macrophage colony stimulating factor (GM-CSF) appear to bind to a common receptor, as well as to specific individual receptors (Park et al., 1989). Nevertheless, despite binding to different receptors, several lymphokines appear to exert similar effects on cells. For example, cloned T lymphocyte lines can proliferate in response to either IL-2 or IL-4 or to IL-2, IL-3, or GM-CSF (Kupper et al., 1987; Morla et al., 1988; Paetkau and Mills, 1989), indicating that the same cell can express and respond to signals transduced through a variety of receptors. Similarly, transfection of IL-2 or EGF receptors, which activate tyrosine kinases (Mills, 1989; Mills et al., 1990, Paetkau and Mills, 1989; Yarden and Ullrich, 1988) into IL-3–dependent cells renders those cells responsive to IL-2 or EGF, respectively (Hatakeyama et al., 1989a; Pierce et al, 1988). Because several of the lymphokine receptors belong to a gene superfamily (based on sequence comparison), it is likely that they induce similar or at least overlapping intracellular signals. This is supported by the observation of increases in tyrosine phosphorylation following activation of the IL-2, IL-

3, IL-4, G-CSF, GM-CSF, and interferon γ receptors in several cell types, and by the ability to bypass requirements for lymphokines by transfecting cells with oncogenes with tyrosine kinase activity (Evans et al., 1990; Mills, 1989; Mills et al., 1990; Mills et al., 1991b; Morla et al., 1988; Paetkau and Mills, 1989; Pierce et al., 1985; Saltzman et al., 1988; Sorensen et al., 1989; Wheeler et al., 1987). Therefore, many of the mechanisms of action of lymphokines are redundant, allowing for the generation of similar responses through a variety of mechanisms.

Activation of cell surface receptors can lead to additive, synergistic, or antagonistic effects, depending on the cell type and on the response assayed. Although activation of the TCR is required to induce optimal expession of most lymphokine receptors, concurrent engagement of lymphokine receptors and TCR decreases the ultimate response, probably through attenuation of transmembrane signaling (Jenkins et al., 1987; Mills, 1989; Otten et al., 1986, 1987). Similarly, IL-2 and IL-4 synergize in the proliferative response of some T lymphoid cell lines (Paetkau and Mills, 1989), whereas IL-4 can inhibit the effect of IL-2 on B cell activation and proliferation (Karray et al., 1988; Tigges et al., 1989). Therefore, receptor activation should not be considered in isolation, and the ultimate response represents the crosstalk between signals transduced through an array of receptors on the cell surface. In addition, a given biochemical event should not be considered to be solely stimulatory or negative. The final effect will reflect the character, magnitude, and duration of the signals, together with the lineage and differentiation state of the cell.

SIGNAL TRANSDUCTION WITHIN THE IMMUNE SYSTEM

General Considerations

Activation of PLC, along with adenyl and guanyl cyclase, was reported following interaction of a variety of lymphoid cell receptors with their ligands (Gelfand et al., 1987a,b; Hadden and Coffey, 1982; MacDonald and Nabholz, 1986; Mills et al., 1985; Weiss et al., 1986). Each of these pathways leads to activation of specific intracellular kinases. In addition, independent of activation via these pathways, activation of T lymphocytes through the TCR or through lymphokine receptors leads to phosphorylation of a variety of intracellular substrates (Mills et al., 1991).

Within the immune system, it is not clear how the conformational changes that follow ligand binding activate intracellular enzymes. In addition, little is known about how activation signals transit the cytosol and alter gene expression in the nucleus. Within the nucleus, availability of DNA binding proteins or their post-translational modification regulates the synthesis of new mRNA and, eventually, cell cycle progression and differentiation. Studies with antisense constructs indicate that function of both the MYC and MYB nuclear proto-oncogenes is required for proliferation of T lymphocytes (Gewirtz et al., 1989; Heikkila et al., 1989). Transient MYC expression is a consequence of activation of both TCR and lymphokine receptors (Broome et al., 1987; Mills et al., 1988a,b). However, prolonged activation of cell surface receptors and signaling pathways is required to induce cell proliferation (Gelfand et al., 1987a,b; Kumagai et al., 1987, 1988; Mills et al., 1988). MYC expression returns to baseline levels before commitment to proliferation (Kumagai et al., 1987, 1988) indicating that, although it may be required for cell proliferation, MYC expression is not sufficient to induce cell proliferation. However, abnormal expression of an activated *myc* gene can bypass the requirement for growth factors in T cell proliferation (Rapp et al., 1985). MYB expression is a later event and may be the consequence of activation of lymphokine receptors rather than a direct consequence of perturbation of the TCR (Pauza, 1987; Stern and Smith, 1986). Neither MYC nor MYB appears to be required for early events transduced by the TCR complex, due to the fact that lymphokines, lymphokine receptors, and the transferrin receptor are expressed normally in cells in which MYC or MYB protein production is blocked (Gewirtz et al., 1989; Heikkila et al., 1989). However, these cells do not synthesize DNA, suggesting that MYC and MYB may be required for exit from G_1 and entry into the S phase (Gewirtz et al., 1989; Heikkila et al., 1989).

Activation of the TCR is not sufficient to induce cell proliferation or differentiation (Cantrell and Smith, 1984; MacDonald and Nabholz, 1986; Mills et al., 1988a; Smith, 1988; Weiss et al., 1986). Rather it induces the expression of lymphokine receptors on the cell surface and the production of lymphokines. It is the interaction of lymphokines with their receptors that induces differentiation and proliferation. By analogy with well-defined systems in fibroblasts, activation of T lymphocytes renders them competent to respond to lymphokine progression factors. As such, it would be expected that the signals induced by the TCR, which lead to competence, would be different, in nature or at least in kinetics, from those induced by activation of lymphokine receptors. This indeed seems to be the case.

T Cell Receptors

Phospholipase C

Within the immune system, PLC represents the best-characterized system of cell activation through TCR and sIg. PLC hydrolyzes specific membrane phospholipids (phosphatidylinositols) to produce diacylglycerol (DAG) and inositol phosphates (IP) (Berridge, 1987; Kikkawa and Nishizuka, 1986; Nishizuka, 1988). DAG activates several members of the phospholipid and calcium-dependent protein kinase C (PKC) family (Nishizuka, 1988). PKC, a serine/threonine kinase, is important in cell growth regulation, as suggested by the action of the tumor promoters that bind to and activate PKC and by transformation of cells by overexpression of PKC (Berridge, 1987; Kikkawa and Nishizuka, 1986; Nishizuka, 1988). In addition, arachidonic acid side chains are present in the majority of membrane lipids hydrolyzed by PLC. Arachidonic acid released from DAG can activate some forms of PKC and can also provide a substrate for eicosanoid and prostaglandin production (Nishizuka, 1988).

One of the inositol phosphates, inositol trisphosphate (IP_3), releases calcium from intracellular stores, and either IP_3 or inositol tetrakisphosphate (IP_4), which is derived from IP_3, may play a role in opening membrane calcium channels (Gardiner, 1989). Increases in cytosolic free calcium ($[Ca^{2+}]_i$) activate the calcium/calmodulin-dependent kinases and increase the affinity of some forms of PKC for DAG, thus increasing kinase activity (Nishizuka, 1988).

Following activation of the TCR and other T cell surface molecules, such as CD2, CD4, CD23, and Thy-1, increased PLC activity has been demonstrated through measurements of DAG production, IP hydrolysis, phosphatidylinositol turnover, $[Ca^{2+}]_i$, PKC translocation, and PKC activity (Alcover et al., 1988; Gelfand et al., 1987a,b; Gunter et al., 1987; MacDonald and Nabholz, 1986; Moretta et al., 1987; Weiss et al., 1986; Weissman et al., 1988). Increases in $[Ca^{2+}]_i$, in turn, lead to membrane hyperpolarization, and activation of PKC can increase Na^+/H^+ exchange through the Na^+/H^+ antiport, leading to cytosolic alkalinization (Gelfand et al., 1987a,b, 1988; Mills et al., 1986, 1990a,b).

CD2, CD23, and Thy-1 appear to increase PLC activity through interaction with CD3 (Alcover et al., 1988; Gunter et al., 1987; Moretta et al., 1987). This suggests that CD3 is functionally linked to PLC and that each of these surface molecules interacts with PLC through CD3. Because T cells lacking η/ζ heterodimers but containing ζ homodimers do not signal through PLC (Mercep et al., 1988; Stanley et al., 1989), it is possible that η is associated with the pathway that leads to PLC activation. However, as discussed below, tyrosine phosphorylation appears to be integral to the pathway that leads to PLC activation, and cells lacking η have normal tyrosine kinase activity (Mercep et al., 1988; Stanley et al., 1990). Intriguingly, although CD3 appears to be required for Thy-1–mediated increases in PLC activity, it is not required for Thy-1–induced increases in tyrosine phosphorylation (Hsi et al., 1989). Although cross-linking of CD2, CD23, and Thy-1 can induce processes such as tyrosine phosphorylation without activating PLC or inducing changes in $[Ca^{2+}]_i$, these are not sufficient to induce IL-2 production (Alcover et al., 1988; Gunter et al., 1987; Moretta et al., 1987). This supports the hypothesis (see below) that under many circumstances increases in $[Ca^{2+}]_i$ are required for IL-2 production and thus for normal T cell proliferation.

The mechanism by which perturbation of the TCR or sIg leads to activation of PLC is unclear. In B lymphocytes, interference with G proteins (by incubation of either permeabilized cells with G protein inhibitors or of intact cells with bacterial toxins that inhibit G proteins) blocks activation of PLC, suggesting that a G protein is involved in the activation of PLC. However, the role of G proteins in T lymphocytes is more controversial. Stimulation of both sIg and the TCR lead to increased tyrosine phosphorylation of components of the G activating protein (GAP) complex (Puil et al., 1990), suggesting a possible mechanism for G protein activation. Tyrosine phosphorylation precedes detectable increases in PLC activity and $[Ca^{2+}]_i$, suggesting that tyrosine phosphorylation may be an upstream event, which leads to activation of PLC (June et al., 1990). This is supported by observations from other systems that PLC is itself a substrate for tyrosine phosphorylation (Margolis et al., 1989; Meisenhelder et al., 1989; Wahl et al., 1989). In addition, tyrphostins, which are potent specific inhibitors of tyrosine phosphorylation (Levitski and Schlessenger, 1989; Stanley et al., 1990), block both sIg- and TCR-mediated PLC activation and increases in $[Ca^{2+}]_i$ (Padeh et al., 1991). Therefore, it seems likely that stimulation of antigen receptors on lymphocytes leads to activation of tyrosine kinases, which, in turn, results in increased PLC activity.

Proliferation of T and B lymphocytes can be induced by incubation with phorbol esters, which activate PKC, and by cation ionophores, which increase $[Ca^{2+}]_i$ (Delia et al., 1984; Finkel et al., 1987; Havran et al., 1987; Roifman et al., 1986; Weiss et al., 1987). This has been interpreted as suggesting that activation of PLC with production of DAG and IP_3 is sufficient for T cell proliferation. However, it must be emphasized that phorbol esters are poorly metabolized by cells, resulting in prolonged and marked activation of PKC (Berridge, 1987; Nishizuka, 1988). In contrast, DAG production is transient and DAG is rapidly metabolized. In many cases, membrane permeant DAG analogues substitute poorly, if at all, for phorbol esters in cell activation (Ebanks et al., 1989; Kumagai et al., 1988). Therefore, it is unlikely that phorbol esters mimic the effects of TCR activation on PKC in these systems.

Protein Kinase C

As discussed above, PKC translocates to the cell membrane following activation of the TCR, and its activity is increased. The role of PKC activity in T cell proliferation is somewhat controversial. In general, studies in which PKC has been depleted or inhibited suggest that PKC activation may be required for propagation of activation signals through the TCR and, as well, that it plays a role in feedback inhibition and signal termination (Mills et al., 1988a,b, 1989a,b 1990b, 1991b; Valge et al., 1988).

Studies with H7, a putative PKC inhibitor, indicate that it blocks T cell proliferation and killing activity. However, H7 is not specific for PKC and its toxic effects can kill lymphocytes at levels of H7 that are 10 times lower than those required to inhibit PKC in intact T lymphocytes (Mills et al., 1988a). Sphingosine and staurosporine, which are effective PKC inhibitors, can augment IL-2 production mediated through the TCR at concentrations that block PKC in intact T lymphocytes (Mills et al., 1989a,b, 1990b). This suggests that in some T cells PKC may function as a feedback inhibitor that limits IL-2 production.

PKC can be depleted from cells by prolonged incubation with high concentrations of phorbol esters. In at least one cell line this depletion markedly inhibits IL-2 production mediated by the TCR (Valge et al., 1988). However, in other cell lines similar treatment augments TCR-induced IL-2 production (Mills et al., 1989a,b, 1990b). Therefore, in at least some circumstances PKC activation is not obligatory for IL-2 secretion.

Acute activation of PKC in murine lymphocytes completely inhibits TCR-mediated activation of PLC and increases in $[Ca^{2+}]_i$ (Mills et al., 1989). Several potential mechanisms contribute to this effect. Activation of PKC leads to serine phosphorylation of the ϵ chain of the TCR, to increased internalization of the TCR complex, and, perhaps, to a decrease in signal transduction. PKC can also directly phosphorylate PLC, leading to decreased activity. PKC also phosphorylates and inactivates some G proteins that may play a role in signal transduction. Similarly, PKC activates enzymes that increase the metabolism of inositol phosphates and may play a role in signal termination. In support of this hypothesis, depleting PKC from LBRM 331A5 cells increases ligand-induced increases in $[Ca^{2+}]_i$, IP levels, and phosphatidylinositol hydrolysis and it results in augmented IL-2 production (Mills et al., 1989a,b). Therefore, it is likely that transient activation of PKC following activation of the TCR plays a dual role in signal transmission and in signal termination.

Cytosolic Calcium

Increases in $[Ca_2^+]_i$ have been shown to be required for cell proliferation and lymphokine production in response to activation of the TCR by anti-CD3 antibodies and some mitogens (Gelfand et al., 1987a,b; Mills et al., 1985; Weiss et al., 1986). However, some mitogens and antibodies can induce IL-2 production in the absence of changes in $[Ca^{2+}]_i$ (Gelfand et al., 1985). In addition, the requirement for changes in $[Ca^{2+}]_i$ can be bypassed in many cases by activation of PKC with phorbol esters (Gelfand et al., 1985, 1987a,b). In some cells, activation of PKC abrogates changes in $[Ca^{2+}]_i$ mediated through the TCR as it augments IL-2 production (Mills et al., 1989a,b). Therefore, increases in $[Ca^{2+}]_i$ are not an obligatory step in lymphokine production.

Several T cell lines have been identified in which ligation of the TCR does not lead to activation of PLC or increases in $[Ca^{2+}]_i$ (Mercep et al., 1988; Stanley et al., 1989, 1990). In at least some of these lines, this nonetheless leads to substantial production of IL-2. Activation of the TCR in these cells activates the Na^+/H^+ antiport and induces tyrosine phosphorylation (Hsi et al., 1989; Mercep et al., 1988; Stanley et al., 1989, 1990), indicating that at least some aspects of signal transduction are intact. This also suggests that both of these processes do not occur solely as a secondary effect of activation of PLC.

Source of Calcium

Increases in $[Ca^{2+}]_i$ can occur either through release of calcium from intracellular stores or by influx of calcium from extracellular sources. In lymphocytes, both processes occur following perturbation of the TCR (Gardner, 1989; Gelfand et al., 1987a,b). Release from intracellular stores appears to be a consequence of IP_3 interacting with specific endoplasmic reticulum components called calcisomes. The mechanism leading to the opening of calcium channels is less clear; however, patch clamp studies suggest that IP_3 or perhaps IP_4 may be responsible for opening transmembrane channels (Gardner, 1989). Studies in which changes in $[Ca^{2+}]_i$ due to release from intracellular stores or uptake of extracellular calcium are blocked by intracellular or extracellular calcium chelators suggest that transmembrane uptake of calcium is required for T cell activation (Gelfand et al., 1989).

Tyrosine Phosphorylation

In addition to activation of PKC and calcium/calmodulin-dependent kinases, perturbation of the TCR leads to increases in tyrosine phosphorylation of T lymphocyte proteins, including the ζ chain of the TCR (Baniyash et al., 1988; Patel et al., 1987; Samelson et al., 1986a,b) and components of the GAP activation complex (Puil et al., 1990), as well as a number of as yet unidentified proteins (Hsi et al., 1989; Mills, 1989; Mills et al., 1990a,b; Stanley et al., 1990). Because none of the members of the TCR complex contain intrinsic kinase activity, this must occur through activation of an intracellular tyrosine kinase or inactivation of a tyrosine phosphatase. The increased tyrosine phosphorylation is not likely a consequence of PLC activity, as it is observed in cells in which PLC is not activated by cross-linking the TCR (Hsi et al., 1989; Mercep et al., 1988; Stanley et al., 1990). In addition, it occurs before changes in cytosolic calcium and IP production (June et al., 1990), and it is not mimicked by activation of protein kinase C in the presence or absence of increases in $[Ca^{2+}]_i$ (Hsi et al., 1989; Mercep et al., 1988; Stanley et al., 1990). In contrast, increases in $[Ca^{2+}]_i$ appear to inhibit or reverse tyrosine phosphorylation mediated through the TCR, possibly as a consequence of activation of tyrosine phosphatases (Mills, 1989). Similarly, increases in cyclic adenosine monophosphate (cAMP) decrease anti-TCR–induced tyrosine phosphorylation, at least of ζ (Patel et al., 1987), suggesting that adenyl cyclase and tyrosine kinases cross-talk. This may also explain in part the negative effects of cAMP on IL-2 production and T cell proliferation (Hadden and Coffey, 1982; Johnson et al., 1988; Mary et al., 1987).

Several observations may explain the mechanisms that lead to tyrosine phosphorylation. Cross-linking of CD4 or CD8 leads directly to activation of the relatively T cell–specific tyrosine kinase, LCK (Rudd et al., 1988; Veillette et al., 1989a,b). Ability to bind LCK is required for CD8 function, suggesting that this is an important physiological interaction (Zamoyska et al., 1989). In addition, LCK is a substrate for CD45 tyrosine phosphatase activity (Ostergaard et al., 1989). CD45 tyrosine phosphatase activity appears to be required for activation through the TCR due to the fact that production of both antigen-mediated and mitogen-mediated IL-2 is inhibited in CD45-negative T cell mutants (Koretzky et al., 1991). Dephosphorylation of tyrosine 505 of LCK increases its tyrosine kinase activity, and the ζ chain of the TCR is a substrate for LCK (Ostergaard et al., 1989). LCK levels, as well as its state of phosphorylation on serine and threonine residues, are also regulated by T cell activation (Marth et al., 1987, 1989). In lymphocytes of lpr mice, which overexpress the tyrosine kinase FYN, ζ is constitutively tyrosine phosphorylated (Samelson et al., 1986a,b). This is associated with inhibition of activation of PLC following perturbation of the TCR (Scholz et al., 1988), suggesting that ζ tyrosine phosphorylation dissociates the TCR from the signal transduction pathway that activates PLC.

Signal Transduction Across the Membrane

None of the identified components of the TCR complex or of lymphokine receptors contain regions that are associated with signal transduction based on studies in other systems. Therefore, the mechanisms by which perturbation of these receptors leads to activation of intracellular events is unknown. In addition, although changes in phosphorylation of a variety of substrates, particularly the receptors themselves, have been identified, none of these has been linked to the propagation of the signal across the cytosol. Phosphorylation of the receptors may uncouple receptors from signal transduction, as in the case of tyrosine phosphorylation of the ζ chain of the CD3 complex or of dissociation of LCK from CD8 following activation of PKC.

In several systems, a family of G proteins link surface receptors to intracellular second messenger–generating systems, such as adenyl and guanyl cyclase. PLC, and specific membrane ion channels. These G proteins are composed of a trimeric complex. Following activation of receptors, guanine triphosphate (GTP) binds to the α subunit and dissociates it from the other components of the complex. The α subunit in the GTP-bound form is responsible for signal transduction. The signal is terminated by hydrolysis of GTP to guanine disphosphate (GDP) by intrinsic GTPase activity of the α subunit. The role of G proteins in signal transduction has been probed (1) with pertussis and cholera toxins, which dissociate the G protein complexes from receptor activation; (2) with mastoparan, which is a direct G protein activator; (3) with aluminum fluoride, which substitutes for the γ phosphate and activates G proteins; and (4) with non-hydrolyzable GTP and GDP analogues in permeabilized cells.

Considerable controversy still exists as to the role of G proteins in signal transduction in the immune system. In permeabilized B lymphocytes, non-hydrolyzable GDP analogues inhibit activation of PLC mediated by cross-linking of sIg, providing convincing evidence that G proteins are involved in signal transduction from sIg to PLC. This is supported by the ability of bacterial toxins, which inactivate G proteins in intact cells, to inhibit B cell function. In some but not all T lymphoid cells, pertussis and cholera toxins inhibit TCR-mediated activation and lymphokine secretion (Stanley et al., 1989). However the failure of these reagents to inhibit lymphokine secretion, despite their ability to adenosine diphosphate (ADP)-ribosylate G proteins in at least some T lymphocytes, suggests that cholera and pertussis toxin–sensitive G proteins are not obligatory components of the activation pathway (Stanley et al., 1989). Direct activation of G proteins with AlF_4^- increases activity of PLC and receptor phosphorylation, suggesting that lymphocytes contain G proteins that are linked to PLC (O'Shea et al., 1987). However, this does not conclusively establish that TCR signal tranduction is mediated through a G protein. Similarly, activation or inhibition of G proteins in permeabilized lymphocytes can lead to tyrosine phosphorylation (G. B. Mills and S. Grinstein, unpublished), suggesting that G proteins regulate tyrosine phosphatases or kinases. However, this does not demonstrate that this is the normal pathway for activation of tyrosine kinases by the TCR.

Adhesion Molecules

It is clear that adhesion molecules participate in the formation of activation complexes. In particular, adhesion complexes form readily in the absence of TCR interaction, and antibodies to adhesion molecules can decrease both the formation of adhesion complexes and the proliferative responses (Makgoba et al., 1989). It also appears that adhesion molecules can transmit signals directly. As indicated above, cross-linking of CD4 or CD8 increases the tyrosine kinase activity of LCK. In addition, cross-linking of adhesion molecules can alter signals transmitted by antibodies against the TCR complex without affecting their ability to bind.

Therefore, activation of T cells by antigen appears to require cross-linking of both TCR and adhesion molecules. It is attractive to speculate that signals transmitted through adhesion molecules in the absence of activation of the TCR impart a negative signal to the cell, rendering it transiently or perhaps permanently refractory to activation through the TCR. Supporting this is the observation that antibodies to CD4, CD8, CD11, and CD18 (α and β chain of LFA-1) can inhibit or stimulate adhesion-independent anti-CD3–induced T cell proliferation.

Crosstalk

In resting T cells, TCR, CD4, CD8, and CD45 are not in close proximity on the cell surface (Kupfer et al., 1987; Rivas et al., 1988; Rojo et al., 1989). This prevents potentially inappropriate interactions between these cell surface adhesion and signaling molecules. However, these molecules become more closely associated after activation. This occurs by at least two mechanisms. Cell surface adhesion molecules co-cap at the site of contact between antigen presenting cells and T lymphocytes (Kupfer et al., 1987). In addition, following activation of the TCR by cross-linking, the TCR complex and CD4 or CD8 can be co-immunoprecipitated, suggesting that a noncovalent interaction occurs between these molecules (Rivas et al., 1988; Rojo et al., 1989).

Cross-linking of adhesion molecules (CD4, CD8, and CD45) concurrently with the TCR alters both biochemical and proliferative responses. Cross-linking of CD4 and CD8 with the TCR increases the proliferative response. Intriguingly, increases in $[Ca^{2+}]_i$ and T cell proliferation were both inhibited when CD45 was cross-linked into complexes with CD3, CD2, or CD28 on T cells. Cross-linking CD45 and CD4 augmented the observed change in $[Ca^{2+}]_i$.

Taken together, these results suggest that activation of the TCR leads to activation of CD45. Activation of the tyrosine kinase, FYN, which appears to exist in a T cell–specific form, is one possible mechanism. CD45 then dephosphorylates LCK, which results in an increase in its tyrosine kinase activity. This may transmit an activation signal through phosphorylation of other intracellular substrates, but it also transmits an inhibitory signal through phosphorylation of ζ and dissociation of the TCR from activation of PLC. In addition, it appears that appropriate temporal-spatial localization and activation of TCR and adhesion molecules is required for modulation of signal transduction and signal termination.

Synthesis

Two potential activation cascades, each of which is sufficient for TCR activation and feedback inhibition, may be present in cells. The first is mediated through activation of PLC and its consequences. The second is activation of tyrosine kinases and phosphatases. It is likely that these two cascades do not exist in isolation but, rather, work together to regulate and fine tune the ultimate response—in this case, production of lymphokines and expression of lymphokine receptors. Preliminary evidence indeed suggests that this is true, in that phosphorylation of CD8 by PKC results in its dissociation from LCK. Furthermore, because inhibiting tyrosine phosphorylation prevents activation of PLC, tyrosine phosphorylation and PLC activation may represent sequential components of the same cascade.

Lymphokines

Because, as suggested above, most lymphokines function as progression signals, it seemed likely that their receptors would be coupled to an array of intracellular activation pathways or substrates different from those of the antigen receptors. This indeed appears to be the case. In contrast to antigen receptors, lymphokine receptors do not activate phosphatidylinositol-specific PLC or increase $[Ca^{2+}]_i$. Several reports suggest that lymphokines, in particular IL-1 and IL-2, do activate PKC, perhaps through production of DAG as a consequence of activation of phosphatidylcholine or phosphatidylethanolamine-specific phospholipases. Nevertheless, IL-1 and IL-2 activity is not prevented by blocking PKC activity or depleting cells of PKC, and IL-1 and IL-2 effects are seen in cells that constitutively lack functional PKC (Mills et al., 1988a,b, 1989a,b, 1990a,b).

Granulocyte CSF (G-CSF), granulocyte-monocyte CSF (GM-CSF), IL-2, IL-3, IL-4, and interferon γ have all been reported to increase tyrosine phosphorylation of specific intracellular substrates (Evans et al., 1990; Farrar and Ferris, 1989; Ferris et al., 1989a,b; Mills, 1989; Mills et al., 1990a,b, 1991a,b; Morla et al., 1988; Paetkau and Mills, 1989; Pierce et al., 1985; Saltzman et al., 1988; Sorensen et al., 1989; Wheeler et al., 1987). In cells that can be stimulated by several of these ligands, both overlapping and distinct patterns of tyrosine phosphorylation can be observed (Ferris et al., 1989a; Morla et al., 1988; N. Zhang and G. B. Mills, unpublished). In addition, the TCR and lymphokines induce markedly different patterns of tyrosine phosphorylation (Mills, 1989; Mills et al., 1990a; Paetkau and Mills, 1989; Saltzman et al., 1990). This suggests that the lymphokines activate different kinases or phosphatases. Alternatively, different substrates may associate with the receptor complex and thus be available to the receptor associated kinase for phosphorylation. At least in the case of IL-2, interrupting tyrosine phosphorylation also blocks IL-2-induced proliferation (Zhang et al., 1992), suggesting that IL-2–induced tyrosine phosphorylation is an integral event for IL-2–induced cell proliferation. IL-2 and IL-3 both increase tyrosine phosphorylation of their receptors (Asao et al., 1990; Mills, 1989;

Mills et al., 1990a,b; Sharon et al., 1989; Sorensen et al., 1989). IL-2 also increases phosphorylation of members of the GAP complex (Puil et al., 1990), the 85 kD regulatory subunit of phosphatidylinositol-3 kinase (Augustine et al., 1991; Merida et al., 1991) and the LCK and RAF kinases (Bruce et al., 1991; Hatekayama et al., 1991; Horak et al., 1991), two intracellular kinases implicated in signal transduction (Morrison et al., 1989).

A serine-rich domain of the intracellular domain of the IL-2 receptor β chain (IL-2R β) has been demonstrated to be required for IL-2 induced proliferation (Fung et al., 1991; Hatekayama et al., 1990a) and association with intracellular tyrosine kinases (Fung et al., 1991). A similar domain of the erythropoietin receptor is required for erythropoietin-induced cellular proliferation and tyrosine phosphorylation emphasizing the importance of this domain (Muira et al., 1991). Surprisingly, the LCK tyrosine kinase appears to associate with an acidic domain of the IL-2R β chain (Hatekayama et al., 1991) that is not essential for cellular proliferation (Hatekayama et al., 1990a). The FYN tyrosine kinase has also been reported to be activated following stimulation of the IL-2 receptor (Augustine et al., 1991); however, mimicking the effects of IL-2 on activation of FYN does not bypass the requirement for IL-2 to induce cell proliferation (Augustine et al., 1991). In addition, IL-2 can induce proliferation of cells that lack LCK and/or FYN (Mills et al., 1992), and IL-2 inhibits proliferation of IL-2R-transfected MT-1 cells (Tsudo et al., 1989) that lack FYN (G. B. Mills, unpublished), eliminating the possibility that these kinases are obligatory for IL-2–induced cellular proliferation. In addition, IL-2 induces proliferation of cells that lack FGR, LTK, FYN, and LCK (Mills et al., 1992). Therefore, either the system is redundant and several different kinases can mediate IL-2–induced signaling, or a different kinase is responsible for IL-2–induced cellular proliferation.

A role for G proteins in lymphokine function is supported by evidence that IL-1 and IL-2 increase GTP binding and hydrolysis in T lymphocyte membranes (O'Neill et al., 1990; Evans et al., 1987) and that IL-2, IL-3, GM-SCF but not IL-4 (Satoh et al., 1991), and activation of the TCR increase the amount of GTP associated with RAS (Downward et al., 1990). In general lymphokine function is not inhibited by cholera or pertussis toxin, suggesting that cholera or pertussis toxin–sensitive G proteins are not required for signal transduction. However, Mizel has suggested that IL-1 signals cells through increases in cAMP mediated through a pertussis toxin–sensitive G protein (Chedid et al., 1989; Shirakawa et al., 1988). In contrast, pertussis toxin does not inhibit IL-1 augmentation of IL-2 production in LBRM331A5 cells despite ADP-ribosylation of G proteins, suggesting that at least in some systems IL-1 signaling is independent of G proteins (Mills et al., 1990b). Furthermore, although IL-1 augments IL-2 production, cAMP inhibits IL-2–induced proliferation (Mary et al., 1987), further questioning the role of cAMP in IL-1–induced IL-2 production.

Several other candidate systems have been proposed for IL-2–induced signal transduction, including hydrolysis of glycosylphosphatidylinositols, nuclear prolactin, and induction of ornithine decarboxylase (Clevenger et al., 1991; Eardley and Koshland, 1991; Legraverend et al., 1989; Merida et al., 1990). However, neither their role in cell proliferation nor the mechanisms leading to their activation have been ascertained.

Following ligand binding, many growth factor receptors are internalized at an increased rate. This could act as a signal termination mechanism, resulting in fewer receptors on the cell surface. However, in some cases receptor internalization is required for function of the receptor. Clearly this is the case for the transferrin receptor, which internalizes transferrin and iron. This may also be the case in other systems. For example, the insulin and EGF receptor kinases remain active following internalization and phosphorylate specific substrates within the cell (Cohen and Fava, 1985; Klein et al., 1987). Therefore, receptor phosphorylation and internalization likely plays a role in signal termination and perhaps in propagation of the activation signal. Most lymphokines and their receptors are internalized at increased rates following ligand binding (Fujii et al., 1986; Greene and Leonard, 1986; Matsushima et al., 1986; Smith, 1988). However, internalization of the IL-2 receptor is not sufficient to lead to cell proliferation (Hatekayama et al., 1989). Although the function of internalized IL-2 or receptors is not clear, at least in one system, interfering with internalization prevents cell proliferation (Kumar et al., 1987). Phosphorylation of receptors leads to increased rates of internalization of the receptors, and, at least in the case of the IL-2 receptor, blocking tyrosine phosphorylation prevents ligand-induced internalization (Zhang and Mills, unpublished).

Therefore, it appears that cytokine receptors signal cells through the activation of intracellular kinases and that, similar to the TCR, surface immunoglobulin, the nerve growth factor receptor, and CD4 and CD8, this is through association with intracellular kinases. The kinases stimulated by cytokines are beginning to be characterized; however, the mechanism(s) by which they are regulated remain to be determined. Nevertheless, our understanding of the biochemical processes leading to T cell proliferation is improving rapidly.

Crosstalk

Concurrent activation of the TCR and lymphokine receptors can result in decreased responsiveness, as measured by signal transduction and by cell proliferation (Jenkins et al., 1987; Mills, 1989; Otten et al., 1986, 1987). Several potential mechanisms are available to explain this process. Tyrosine kinase activity mediated through the IL-2 receptor could induce phosphorylation of LCK, which could, in turn, phosphorylate ζ and dissociate the TCR from PLC. Cyclic nucleotides regulate serine/threonine kinases and decrease signaling by PLC. PLC increases production of DAG and IP, which, in turn increase cytosolic calcium. Cytosolic calcium and DAG can increase the activity of PKC, which feeds back on the PLC pathway and on cyclic nucleotide production. Prostaglandins increase cytosolic calcium, which can interact with each of the preceding pathways. However, although several potential pathways have been described, the actual physiologic mechanisms are unclear.

SIGNAL TRANSDUCTION IN THE THYMUS

As discussed above, functional T cell responses occur as a consequence of perturbation of the TCR concomitantly with the activation of signals through adhesion molecules such as CD45, CD4, and CD8. Crosstalk between these signals is important, and altering the signals mediated through one of these receptors may result in an aberrant response. Furthermore, activation of adhesion molecules in the absence of cross-linking of the TCR or before cross-linking of the TCR can alter responses. Similarly, signals transduced through the TCR and lymphokine receptors may interact to modulate the message that reaches the nucleus. Therefore, functional activation of mature lymphocytes occurs as the result of a highly regulated interplay between transmembrane signals transduced through an array of receptors on the cell surface. These signals cross-talk both to regulate the activation response and also to terminate the activation signal.

The TCR, adhesion molecules, and lymphokines and their receptors appear to play an important role in intrathymic proliferation, differentiation, and selection of T lymphocytes (Ramsdell and Fowlkes, 1989; Takei, 1988). However, in contrast to peripheral blood lymphocytes and mature thymocytes, immature thymocytes do not express the complete array of surface signal transduction molecules but, rather, pass through stages of differentiation in which only partial arrays are expressed (discussed in detail elsewhere in this book). For example signal transduction may be very different in cells that do not express CD4 or CD8 than in cells that express CD4 or CD8 or in cells that express CD45 but not TCR or CD4 or CD8. Because these adhesion molecules themselves signal lymphocytes and modulate signals delivered following activation of the TCR, it is likely that different signals will reach the nucleus of immature and mature lymphocytes even after interaction with the same extracellular ligands. This may play a role in the high rate of proliferation and death of thymocytes, as well as in selecting the receptor repertoire of thymocytes.

Activation of PLC

Phorbol esters and cation ionophores induce proliferation of human and murine thymocytes, similar to their effects on mature lymphocytes (Delia et al., 1984; Finkel et al., 1987; Havran et al., 1987; Roifman et al., 1986; Weiss et al., 1987). However, only mature thymocytes appear to be responsive to these mediators (Delia et al., 1984; Finkel et al., 1987; Havran et al., 1987; Weiss et al., 1987). Immature, $CD4^+8^+$ thymocytes do not proliferate following activation with ionophores and phorbol esters. This suggests that only relatively mature thymocytes can be induced to proliferate in response to activation of PKC and increases in $[Ca^{2+}]_i$. The failure of immature thymocytes to proliferate may be explained, in part, by calcium-mediated programmed cell death (see below).

Stimulation of the TCR, Thy-1, or CD2 on thymocytes, as in mature T cells, results in increased PLC activity (Finkel et al., 1987; Roifman et al., 1986; Taylor et al., 1984, 1988; Tsien et al., 1982; Weiss et al., 1987). This then results in increases in $[Ca^{2+}]_i$ and PKC activation through the release of IP_3 and DAG. Changes in $[Ca^{2+}]_i$ in immature thymocytes have either been reported to be identical to those in mature thymocytes and lymphocytes (Roifman et al., 1986; Weiss et al, 1987) or to be lower than those in mature cells, due to decreased influx of extracellular calcium (Finkel et al., 1987; Havran et al., 1987). However, although cross-linking of the TCR activates PLC in both mature and immature thymocytes, it is not sufficient to induce proliferation, particularly of immature thymocytes (Finkel et al., 1987; Havran et al., 1987; Roifman et al., 1986; Tsoukas et al., 1987; Weiss et al., 1987). This appears to be due to an inability of TCR activation to induce IL-2 production and IL-2 receptor expression (Boyer et al., 1989; Havran et al., 1987; Mills et al., 1976, Paetkau et al., 1976; Roifman et al., 1986; Tsoukas et al., 1987).

In mature T cells, activation of PLC with subsequent increases in $[Ca^{2+}]_i$ and activation of PKC likely plays a major, if not obligatory, role in inducing proliferation of resting T cells. However, in immature thymocytes, increases in $[Ca^{2+}]_i$ lead to programmed cell death through apoptosis (DNA fragmentation). This may explain, in part, the high rate of intrathymic cell death. This cell death can be induced through activation of the TCR (McConkey et al., 1989a) or by increasing $[Ca^{2+}]_i$ with ionophores (Kizaki et al., 1989). Prolonged increases in $[Ca^{2+}]_i$ appear to be required to result in cell death (McConkey et al., 1988a,b). This is somewhat surprising in that anti-CD3–induced changes in $[Ca^{2+}]_i$ appear to be shorter lived in immature thymocytes than in mature thymocytes or lymphocytes, due to lower influxes of extracellular calcium (Finkel et al., 1987; Havran et al.,1987).

Cation ionophores, or anti-CD3, do not induce rapid cell death or DNA fragmentation in resting mature lymphocytes (McConkey et al., 1989a). However, in some T cell lines, activation by these agents is associated with decreased proliferative responses, which may be a consequence of programmed cell death (Mercep et al., 1988; Stanley et al., 1990). However, in mature T cells, ligand-induced cell death may occur through pathways that do not involve increases in $[Ca^{2+}]_i$ due to the fact that ligand-induced apoptosis occurs in cells in which the TCR is not coupled to PLC and thus is not associated with increased $[Ca^{2+}]_i$ (Mercep et al., 1988).

PKC activation has been variably associated with both increased (Kizaki et al., 1989) and decreased (McConkey et al., 1989b) rates of calcium-induced apoptosis. IL-1 appears to inhibit calcium-induced apoptosis, purportedly through production of DAG and activation of PKC (McConkey et al., 1989a,b). This is somewhat unexpected in that activation of the TCR in thymocytes leads to activation of PLC (Finkel et al, 1987; Roifman et al, 1986; Taylor et al., 1984; Tsien et al., 1982; Weiss et al., 1987), which increases DAG production in parallel with IP_3 production and $[Ca^{2+}]_i$ (Berridge, 1987). Therefore, stimulation of the TCR on immature thymocytes should lead to activation of PKC. Perhaps the magnitude or the duration of TCR-induced PKC activation is not sufficient to block calcium-induced apoptosis, or IL-1 and phorbol esters may prevent apoptosis through a non-PKC–mediated process.

It appears that activation of the TCR on immature thymocytes in the absence of other signals leads to cell death. In the presence of appropriate accessory signals, such as IL-1 or perhaps other signals that activate PKC, cell death may not occur.

Tyrosine Phosphorylation

Early reports indicated that mitogenic stimulation of thymocytes leads to tyrosine phosphorylation of lipomodulin (Hirata et al., 1984), suggesting that the TCR in thymocytes also activates a tyrosine kinase. More recent reports have concentrated on LCK and the ζ chain of the TCR complex, which are known to be tyrosine phosphorylated following activation of mature lymphocytes.

LCK is detectable in murine day-15 thymocytes, and thus it precedes expression of either CD4 or CD8 (Veillette et al., 1989b). Following expression of CD4 and CD8, LCK can be shown to associate with CD4 and CD8 in day 19 double-positive thymocytes. In postnatal thymocytes, LCK also associates with CD4 and CD8 in both double- and single-positive subsets (Veillette et al., 1989b).

Similar to more mature cells, cross-linking of CD4 and CD8 on immature CD4$^+$8$^+$ thymocytes or on single-positive thymocytes results in an increase in LCK tyrosine kinase activity (Veillette et al., 1989a,b), suggesting that CD4 and CD8 are functionally linked to LCK. Similar to the situation in mature T cells, cross-linking of CD3 or Thy-1 in thymocytes does not alter tyrosine phosphorylation of LCK (Veillette et al., 1989a,b).

Immediately following isolation of thymocytes, ζ appears to be tyrosine phosphorylated in both single-positive and double-positive thymocytes, perhaps as a consequence of in vivo thymocyte activation (Nakayama et al., 1989). Following in vitro incubation, ζ becomes dephosphorylated (Nakayama et al., 1989). Two different groups tested the ability of cross-linking of CD3, CD4, or CD8 to induce tyrosine phosphorylation of ζ in vitro. One group suggested that cross-linking of all these surface antigens, as well as cell–

cell interactions, is sufficient to induce tyrosine phosphorylation of ζ (Nakayama et al., 1989), whereas the other group failed to observe any changes in tyrosine phosphorylation of ζ after cross-linking (Veillette et al., 1989b). These differences may stem from different in vitro incubation periods before activation by cross-linking. Dephosphorylation of ζ induced by in vitro culture may have allowed detection of ligand-induced rephosphorylation of ζ in the experiments of Nakayama et al. (1989). Cross-linking of CD3 increased tyrosine phosphorylation of 72- and 120-kD proteins (Veillette et al., 1989b) and also of ζ in immature $CD4^+8^+$ thymocytes (Nakayama et al., 1989), indicating that tyrosine kinases also associate with the TCR beginning early in ontogeny. Tyrosine kinases are therefore present in immature T cells and respond to activation of cell surface antigens. The tyrosine phosphorylation pattern in thymocytes (Veillette et al., 1989b) appears to be quite different from that in more mature cells (Hsi et al., 1989; Mills, 1989; Mills et al., 1990a,b; Paetkau and Mills, 1989; Stanley et al., 1990; Veillette et al., 1989a). This may be a consequence of association of different intracellular substrates with the kinase. Alternatively, it may represent activation of different patterns of intracellular tyrosine kinases and phosphatases in mature and immature cells.

TCR Expression

Early in ontogeny, γ/δ may be the predominant TCR. Signaling through a TCR complex composed of γ/δ on mature cells appears to be indistinguishable from signaling through α/β TCR complexes (Bismuth et al., 1988; Krangel et al., 1987; Panatelo et al., 1987a,b). However, signal transduction through γ/δ on immature thymocytes has not been well characterized as yet.

The relative amounts of α, β, γ (Martinez-Valdez et al., 1988a,b), and perhaps δ expression in thymocytes appear to be regulated, at least in part, through changes in intracellular $[Ca^{2+}]_i$ and activation of PKC. Increases in $[Ca^{2+}]_i$ increase γ expression, whereas activation pf PKC antagonizes this effect (Martinez-Valdez et al., 1988a). Activation of PKC increases the levels of α and β mRNA, whereas increases in $[Ca^{2+}]_i$ decrease α and β mRNA (Martinez-Valdez et al., 1988b). This suggests that a switch from $[Ca^{2+}]_i$-mediated signaling to PKC-dominant signaling during maturation could play a role in the increased α/β expression in thymocytes. This would be in agreement with the results described above for sensitivity to $[Ca^{2+}]_i$-mediated apoptosis, in that more mature thymocytes or lymphocytes may have constitutive PKC activation, which protects them from apoptosis.

Special Cases

Although signal transduction has not been studied extensively in thymocyte subsets, it is intriguing to hypothesize what the consequence of expression of partial receptor arrays might be.

$CD3^+$, $CD4^-$, $CD8^-$

In mature lymphocytes and in single- or double-positive thymocytes, approximately 50% of LCK is tightly associated with CD4 or CD8 (Veillette et al., 1989a,b). Because CD4 and CD8 do not associate to any significant degree with the TCR in resting lymphocytes (Kupfer et al., 1987; Rivas et al., 1988; Rojo et al., 1989), LCK is spatially separated from TCR. This may function to limit LCK-mediated phosphorylation of the ζ chain of the TCR in resting lymphocytes. In CD4$^-$8$^-$ lymphocytes, LCK may be free to move along the cell membrane and, in more mature thymocytes, to phosphorylate intracellular substrates including the ζ chain of the TCR. Because ζ phosphorylation appears to uncouple the TCR from activation of PLC (Samelson et al., 1986a,b; Sholz et al., 1988), an inability to signal through the TCR may result in these cells.

lpr Mice

In *lpr* mice, thymic differentiation appears to be blocked at the stage of a 16-day thymus and thymocytes are primarily CD4$^-$8$^-$, Thy-1 dull (Altman et al., 1981; Katagiri et al., 1987; Rosenberg et al., 1985; Wofsky et al., 1981). Thymocytes or thymocyte cell lines from these mice overexpress the tyrosine kinase FYN (Permutter, 1989), as well as MYB, which is required for T cell proliferation (Gewirtz et al., 1989), and RAF (Mustelin and Altman, 1989; Rosenberg et al., 1985), a serine/threonine kinase that is a substrate for a variety of tyrosine kinases (Morrison et al., 1989), perhaps including the TCR- and IL-2–activated kinases. Although *lpr* thymocytes proliferate rapidly in vivo, a marked decrease in mitogen and anti-CD3 proliferation occurs in vitro, likely as a consequence of decreased IL-2 production and IL-2 receptor expression (Altman et al., 1981; Katagiri et al., 1987; Wofsky et al., 1981). Similarly, signal transduction through the TCR is abnormal in *lpr* thymocytes with decreased ligand-induced IP production and changes in [Ca^{2+}]$_i$ (Sholz et al., 1988). Because increases in [Ca^{2+}]$_i$ are normally required for IL-2 production in resting lymphocytes (Gelfand et al., 1987a,b; Mills et al., 1985) and because IL-2 increases expression of its own receptor (Reem et al., 1985), the decreased IL-2 production and IL-2 receptor expression seen in activated thymocytes from *lpr* mice may be the result of aberrant signal transduction through the TCR, leading to decreased IP production and changes in [Ca^{2+}]$_i$.

Thymocytes from *lpr* mice show constitutive phosphorylation of the ζ chain of the TCR (Samelson et al., 1986a,b). Because ζ is a substrate for LCK (Barber et al., 1989; Veillette et al., 1989a), the increased ζ phosphorylation may be a consequence of LCK activity. In lpr thymocytes and in normal CD4$^-$8$^-$ thymocytes, LCK is not restrained through association with CD4 or CD8 and may have access to ζ. Access of LCK to ζ may play a role in the constitutive ζ phosphorylation observed in lpr thymocytes, as well as in thymocytes from normal mice (Nakayama et al., 1989; Samelson et al., 1986a,b).

The consequence of ζ phosphorylation is not known. However, mature T cells that lack ζ do not signal normally through the TCR (Sussman et al., 1988). Lpr thymocytes, in which ζ is tyrosine phosphorylated, also signal poorly through the TCR. Therefore, it is possible that tyrosine phosphorylation of ζ also dissociates the TCR complex from its signal transduction machinery. This may act as a feedback inhibition mechanism or as a mechanism for crosstalk with lymphokine receptors in mature lymphocytes.

CD45

Expression of the CD45 tyrosine phosphatase precedes expression of the TCR and of CD4 and CD8 (Thomas and Lefrancois, 1988). This may result in changes in the levels of tyrosine phosphorylation of various proteins. In CD45$^+$, TCR$^-$, CD4$^-$, CD8$^-$ thymocytes,

CD45 cross-linking results in marked increases in IL-2 and IL-2 receptor mRNA synthesis, suggesting that CD45 and its ligand may be the dominant regulatory elements in these cells (Deans et al., 1989). In addition, several forms of CD45 derived by alternative splicing are expressed on thymocytes, and this may also result in differential regulation of CD45 tyrosine phosphatase activity (Thomas and Lefrancois, 1988). Finally, because LCK is a substrate for CD45 and because tyrosine dephosphorylation is known to regulate the activity of the LCK tyrosine kinase (Mustelin et al., 1989), the differences in CD45 expression between immature thymocytes and mature lymphocytes are likely to lead to changes in LCK activity. Because the ζ chain of the TCR complex is a substrate for LCK (Veillette et al., 1989a) and because ζ phosphorylation is probably associated with the decreased ability to transmit signals through the TCR (Scholz et al., 1988), CD45 may also alter the activation of thymocytes that express different CD45 splice variants in the presence of the TCR complex.

$CD2^+$, $CD3^-$

Interestingly, in light of the failure of anti-CD3 to induce IL-2 production, IL-2 receptor expression or proliferation in immature $CD3^+$, $CD4^+$, $CD8^+$ thymocytes (Boyer et al., 1989; Havran et al., 1987; Roifman et al., 1986; Tsoukas et al., 1987), anti-CD2 antibodies induce all of these processes in even more immature T lymphocytes prior to CD3, CD4, and CD8 expression (Fox et al., 1985; Reem et al., 1987). This is surprising in that CD2 is thought to signal through its association with CD3 in mature lymphocytes.

$IL-2R^+$, $CD3^-$

IL-2 appears to play an obligatory role in thymocyte maturation in vivo (Jenkins et al., 1987; Tentori et al., 1988), perhaps by activating cells that express the IL-2 receptor in the absence of the TCR complex and produce IL-2 in an autocrine manner (Pearse et al., 1989; Toribio et al., 1989). This may correlate with the ability of IL-2 to induce proliferation of $CD3^-$, $CD4^-$, $CD8^-$ thymocytes (Fox et al., 1985; Reem et al., 1987). Because activation through the TCR decreases the ability to signal through the IL-2 receptor, activation of the IL-2 receptor in thymocytes in the absence of the TCR may have quite different functional consequences than activation of the IL-2 receptor in mature T cells.

SUMMARY

The role of the intracellular biochemical changes that occur as a consequence of receptor activation on mature lymphocytes are only beginning to be understood. Clearly, multiple activation pathways are involved, and these pathways interact to result in the highly regulated proliferation and differentiation of lymphocytes. Full understanding of these pathways is further hindered by the likelihood that the array of signals transmitted and their functional consequences will likely differ in each lymphocyte subtype. In addition, depending on the particular foreign insult and the environment of the responding cell, differences in the receptors stimulated and the duration of their activation will also likely occur. Although some inroads are being made into these questions in mature cells that express a well-defined series of surface receptors, little progress has been made in understanding the functional consequences of expression of partial or different arrays of receptors in thymocytes and in B cell precursors.

Acknowledgments

I would like to thank Drs. Puil, Pawson, Arami, Greene, and Roifman and Nan Zhang for sharing unpublished data. This work was supported by the Medical Research Council of Canada and the National Cancer Institute of Canada. G. Mills is a Medical Research Council of Canada Scientist and a McLaughlin Scientist.

References

Aguet M., Dembic, Z., and Merlin G. (1988). Molecular cloning and expression of the human interferon-γ receptor. *Cell 55*:273–280.

Alcover, A., Alberini, C., Acuto, O., Clayton, L. K., Transy, C., Spagnoli, G. C., Moingeon, P., Lopez, P., and Reinherz, E. L. (1988). Interdependence of CD3-Ti and CD2 activation pathways in human T lymphocytes. *EMBO J. 7*:1973–1980.

Alexander, D. R., and Cantrell D. A. (1989). Kinases and phosphatases in T-cell activation. *Immunol. Today 10*:200–203.

Altman, A., Theofilopoulos, A. N., Weiner, R., Katz, D. H., and Dixon, F. J. (1981). Analysis of T cell function in autoimmune murine strains: Defects in production of and responsiveness to interleukin 2. *J. Exp. Med. 154*:791–798.

Asao, H., Takeshita, T., Nakamura, M., Nagata, K., and Sugamura, K. (1990). Interleukin 2 (IL-2)–induced tyrosine phosphorylation of IL-2 receptor p75. *J. Exp. Med. 171*:637–644.

Augustine, J., Sutor, S., and Abraham, R. (1991). Interleukin 2- and polyoma middle T antigen-induced modification of phosphatidylinositol 3-kinase activity in activated T lymphocytes. *Mol. Cell. Biol. 11*:4431–4440.

Baniyash, M., Garcia-Morales, P., Luong, E., Samelson, L. E., and Klausner, R. D. (1988). The T cell antigen receptor zeta chain is tyrosine phosphorylated upon activation. *J. Biol. Chem. 263*:18225–18229.

Barber, E. K., Dasgupta, J. D., Schlossman, S. F., Trevillyan, J. M., and Rudd, C. E. (1989). The CD4 and CD8 antigens are coupled to a protein-tyrosine kinase (p56[lck]) that phosphorylates the CD3 complex. *Proc. Natl. Acad. Sci. USA 86*:3277–3281.

Bazan, J. F. (1989). A novel family of growth factor receptors: A common binding domain in the growth hormone, prolactin, erythropoietin and IL-6 receptors, and the p75 IL-2 receptor β-chain. *Biochem. Biophys. Res. Comm. 164*:788–795.

Berridge, M. J. (1987). Inositol trisphosphate and diacylglycerol: Two interacting second messengers. *Annu. Rev. Biochem. 56*:159–167.

Bierer, B. E., Peterson, A., Barbosa, J., Seed, B., and Burakoff, S. J. (1988). Expression of the T-cell surface molecule CD2 and an epitope-loss CD2 mutant to define the role of lymphocyte function-associated antigen 3 (LFA-3) in T-cell activation. *Proc. Natl. Acad. Sci. USA 85*:1194–1199.

Bishayee, S., Majumdar, S., Khire, J., and Das, M. (1989). Ligand-induced dimerization of the platelet-derived growth factor receptor: Monomer-dimer interconversion occurs independent of receptor phosphorylation. *J. Biol. Chem. 264*:11699–11705.

Bismuth, G., Faure, F., Theodorou, I., Debre, P., Hercend, T. (1988). Triggering of the phosphoinositide transduction pathway by a monoclonal antibody specific for the human γ/δ T cell receptor. *Eur. J. Immunol. 18*:1135–1141.

Bockenstedt, L. K., Goldsmith, M. A., Dustin, M., Olive, D., Springer, T. A., and Weiss, A. (1988). The CD2 ligand LFA-3 activates T cells but depends on the expression and function of the antigen receptor. *J. Immunol. 141*:1904–1911.

Borth, W., and Luger, T. A. (1989). Identification of α₂-macroglobulin as a cytokine binding plasma protein: Binding of interleukin-1γ to "F" α₂-macroglobulin. *J. Biol. Chem. 264*:5818–5824.

Boyd, A. W., Wawryk, S. O., Burns, G. F., and Fecondo, J. V. (1988). Intercellular adhesion molecule 1 (I-CAM-1) has a central role in cell–cell contact-mediated immune mechanisms. *Proc. Natl. Acad. Sci. USA 85*:3095–3101.

Boyer, P. D., Diamond, R. A., and Rothenberg, E. V. (1989). Changes in inducibility of IL-2 receptor α-chain and T cell-receptor expression during thymocyte differentiation in the mouse. *J. Immunol.* *142*:4121–4126.

Brandhuber, B. J., Boone, T., Kenney, W. C., and McKay, D. B. (1987). Three-dimensional structure of interleukin-2. *Science 238*:707–710.

Breitmeyer, J. B., Daley, J. F., Levine, H. B., and Schlossman, S. F. (1987). The T11 (CD2) molecule is functionally linked to the T3/Ti T cell receptor in the majority of T cells. *J. Immunol. 139*:2899–2905.

Broome, H. E., Reed, J. C., Godillot, E. P., and Hoover, R. G. (1987). Differential promoter utilization by the c-*myc* gene in mitogen- and interleukin-2–stimulated human lymphocytes. *Mol. Cell Biol. 7*:2988–2993.

Bruce, T., Rapp R., App, H., Greene, I., Dobashi, K., and Reed, J. (1991). Interleukin 2 (IL-2) induces tyrosine phosphorylation and activation of p72-74 raf-1 kinase in T lymphocytes. *Proc. Natl. Acad. Sci. USA 88*:1227–1231.

Campbell, K. S., and Cambier, J. C. (1990). B lymphocyte antigen receptors (mIg) are non-covalently associated with a disulfide linked, inducibly phosphorylated glycoprotein complex. *EMBO J. 9*:37–44.

Cantrell, D. A., and Smith, K. A. (1984). The interleukin-2 T-cell system: A new cell growth model. *Science 224*:1312–1316.

Carraway, K. L., III, Koland, J. G., and Cerione, R. A. (1989). Visualization of epidermal growth factor (EGF) receptor aggregation in plasma membranes by fluorescence resonance energy transfer. Correlation of receptor activation with aggregation. *J. Biol. Chem. 264*:8699–8704.

Chedid, M., Shirakawa, F., Naylor, P., and Mizel, S. B. (1989). Signal transduction pathway for IL-1: Involvement of a pertussis toxin–sensitive GTP-binding protein in the activation of adenylate cyclase. *J. Immunol. 142*:4301–4306.

Clevenger, C., Altmann, S., and Prystowsky, M. (1991). Requirement of nuclear prolactin for interleukin-2–stimulated proliferation of T lymphocytes. *Science 253*:77–79.

Cohen, S., and Fava, R. A. (1985). Internalization of functional epidermal growth factor:receptor/kinase complexes in A-431 cells. *J. Biol. Chem. 260*:12351.

Cohen, F. E., Kosen, P. A., Kuntz, I. D., Epstein, L. B., Ciardelli, T. L., and Smith, K. A., (1986). Structure-activity studies of interleukin-2. *Science 234*:349–352.

Curtis, B. M., Gallis, B., Overell, R. W., McMahan, C. J., DeRoos, P., Ireland, R., Eisenman, J., Dower, S. K., and Sims, J. E. (1989). T-cell interleukin 1 receptor cDNA expressed in Chinese hamster ovary cells regulates functional responses to interleukin 1. *Proc. Natl. Acad. Sci. USA 86*:3045–3054.

Cyert, M. S., and Thorner, J. (1989). Putting it on and taking it off: Phosphoprotein phosphatase involvement in cell cycle regulation. *Cell 57*:891–892.

D'Andrea, A. D., Fasman, G. D., and Lodish, H. F. (1989). Erythropoietin receptor and interleukin-2 receptor β chain: A new receptor family. *Cell 58*:1023–1024.

Deans, J. P., Shaw, J., Pearse, M. J., and Pilarski, L. M. (1989). CD45R as a primary signal transducer stimulating IL-2 and IL-2R mRNA synthesis by CD3⁻4⁻8⁻ thymocytes. *J. Immunol. 143*:2425–3431.

Delia, D., Greaves, M., Villa, S., and DeBraud, F. (1984). Characterization of the response of human thymocytes and blood lymphocytes to the synergistic mitogenicity of 12-0-tetradecanoyl-phorbol-13-acetate (TPA)-ionomycin. *Eur. J. Immunol. 14*:720–726.

Dinarello, C. A. (1986). Interleukin 1. *Rev. Infect. Dis. 6*:51–62.

Downward, J., Graves, J., Warne, P., Rayter, S., and Cantrell, D. (1990). Stimulation of p21 RAS upon T-cell activation. *Nature 346*:719–723.

Eardley, D., and Koshland, M. (1991). Glycosylphosphatidylinositol: A candidate system for interleukin-2 signal transduction. *Science 251*:78–85.

Ebanks, R., Roifman, C., Mellors, A., and Mills G. B. (1989). The diacylglycerol analog, 1,2-sn-dioctanoylglycerol, induced increases in cytosolic free Ca^{2+} and cytosolic acidification through a protein kinase C–independent process. *Biochem. J. 258*:689–698.

Ellis, C., Moran, M., McCormick, F., and Pawson, T. (1990). Phosphorylation of GAP and GAP-associated proteins by transforming and mitogenic tyrosine kinases. *Nature 343*:377–381.

Evans, S. W., Beckner, S. K., and Farrar, W. L. (1987). Stimulation of specific GTP binding and hydrolysis activities in lymphocyte membrane by interleukin-2. *Nature 325*:166–168.

Evans, J.P.M., Mire-Sluis, A. R., Hoffbrand, A. V., and Wickremasinghe, R. G. (1990). Binding of G-CSF, GM-CSF, tumor necrosis factor-α and γ-interferon to cell surface receptors on human myeloid leukemia cells triggers rapid tyrosine and serine phosphorylation of a 75-kD protein. *Blood 75*:88–93.

Farrar, W. L., Evans, S. W., Rapp, U. R., and Cleveland, J. L. (1987). Effects of antiproliferative cyclic AMP on interleukin 2–stimulated gene expression. *J. Immunol. 139*:2075–2080.

Farrar, W. L., and Ferris, D. K., (1989). Two-dimensional analysis of interleukin 2–regulated tyrosine kinase activation mediated by the p70-75 β subunit of the interleukin 2 receptor. *J. Biol. Chem. 264*:12562–12567.

Ferris, D. K., Willette-Brown, J., Linnekin, D., and Farrar, W. L. (1989a). Comparative analysis of IL-2 and IL-3 induced tyrosine phosphorylation. *Lymphokine Res. 8*:215–224.

Ferris, D. K., Willette-Brown, J., Ortaldo, J. R., and Farrar, W. L. (1989b). IL-2 regulation of tyrosine kinase activity is mediated through the p70-75 β-subunit of the IL-2 receptor. *J. Immunol. 143*:870–876.

Finkel, T., McDuffie, M., Kappler, J. W., Marrack, P., Cambier, J. C. (1987). Both immature and mature T cells mobilize Ca^{2+} in response to antigen receptor crosslinking. *Nature 330*:179–182.

Fox, D. A., Hussey, R. E., Fitzgerald, K. A., Bensussan, A., Daley, J. F., Schlossman, S. F., and Reinherz, E. L. (1985). Activation of human thymocytes via the 50 kD T11 sheep erythrocyte binding protein induces the expression on interleukin 2 receptors on both T3$^+$ and T3$^-$ populations. *J. Immunol. 134*:330–338.

Fujii, M., Sugamura, K., Sano, M., Nakai, M., Sugita, K., and Hinuma, Y. (1986). High-affinity receptor-mediated internalization and degradation of interleukin 2 in human T cells. *J. Exp. Med. 163*:550–560.

Fung, M. R., Scearce, R. M., Hoffman, J. A., Peffer, N. J., Hammes, S. R., Hosking, J. B., Schmandt, R., Kuziel, W. A., Haynes, B. F., Mills, G. B., and Greene, W. C. (1991). Tyrosine and serine/threonine kinases associate with the functional β subunit of the human interleukin-2 receptor. *J. Immunol. 147*:1253–1260.

Gardner, P. (1989). Calcium and T lymphocyte activation. *Cell 59*:15–16.

Gelfand, E. W., Cheung, R. K., Mills, G. B., and Grinstein, S. (1985). Mitogens trigger a calcium-independent signal for proliferation in phorbol ester–treated lymphocytes. *Nature 315*:419–420.

Gelfand, E. W., Cheung, R. K., Mills, G. B., and Grinstein, S. (1987a). Regulatory role of membrane potential in the response of human T lymphocytes to phytohemagglutinin. *J. Immunol. 138*:527–531.

Gelfand, E. W., Cheung, R. K., Mills, G. B., and Grinstein S. (1988). Uptake of extracellular Ca^{2+} and not recruitment from internal stores is essential for T-lymphocyte proliferation. *Eur. J. Immunol. 18*:917–922.

Gelfand, E. W., Mills, G. B., Cheung, R. K., Lee, J.W.W., and Grinstein, S. (1987b). Transmembrane ions fluxes during activation of human T lymphocytes. Role of Ca^{2+}, Na^+/H^+ exchange and phospholipid turnover. *Immunol. Rev. 95*:59–87.

Gewirtz, A. M., Anfossi, G., Venturelli, D., Vapreda, S., Sims, R., and Calabretta, B. (1989). G1/S transition in normal human T-lymphocytes requires the nuclear protein encoded by c-*myb*. *Science 245*:180–183.

Gilman, A. G., and Casey, P. J. (1988). G protein involvement in receptor-effector coupling. *J. Biol. Chem. 263*:2577–2582.

Glenney, Jr., J. R., Chen, W. S., Lazar, C. S., Walton, G. M., Zokas, L. M., Rosenfeld, M. G., and Gill, G. N. (1988). Ligand-induced endocytosis of the EGF receptor is blocked by mutational inactivation and by microinjection of anti-phosphotyrosine antibodies. *Cell 52*:675–681.

Greene, W. C., and Leonard, W. J. (1986). The human interleukin 2 receptor. *Annu. Rev. Immunol.* 4:69–77.

Guimezanes, A., Buferne, M., Pont, S., Pierres, M., and Schmitt-Verhulst, A. M. (1988). Interactions between the Thy-1 and T-cell antigen receptor pathways in the activation of cytotoxic T cells: Evidence from synergistic effects, loss variants, and anti-CD8 antibody-mediated inhibition. *Cell. Immunol.* 113:435–442.

Gunter, K. C., Germain, R. N., Kroczek, R. A., Saito, T., Yokoyama, W. M., Chan, C., Weiss, A., and Shevach, E. M. (1987). Thy-1-mediated T-cell activation requires co-expression of CD3/Ti complex. *Nature* 326:505–507.

Hadden, J. W., and Coffey, R. G. (1982). Cyclic nucleotides in mitogen-induced lymphocyte proliferation. *Immunol. Today* 3:299–301.

Hanks, S. K., Quinn, A. M., and Hunter, T. (1988). The protein kinase family: Conserved features and deduced phylogeny of the catalytic domains. *Science* 241:42–49.

Hatkeyama, M., Kono, T., Kobayashi, N., Kawaharar, A., Levin, D., Perlmutter, R., and Taniguchi, T. (1991). Interaction of the IL-2 receptor with src-family kinase p56LCK: identification of a novel intermolecular association. *Science* 252:1523–1528.

Hatakeyama, M., Mori, H., Doi, T., and Taniguchi, T. (1989a). A restricted cytoplasmic region of IL-2 receptor γ chain is essential for growth signal transduction but not for ligand binding and internalization. *Cell* 59:837–843.

Hatakeyama, M., Tsudo, M., Minamoto, S., Kono, T., Doi, T., Miyata, T., Miyasaka, M., and Taniguchi, T. (1989b). Interleukin-2 receptor β chain gene: Generation of three receptor forms by cloned human α and β chain cDNA's. *Science* 244:551–556.

Havran, W. L., Poenie, M., Kimura, J., Tsien, R., Weiss, A., and Allison, J. P. (1987). Expression and function of the CD3-antigen receptor on murine CD4[+]8[+] thymocytes. *Nature* 330:170–172.

Haynes, B. F., Telen, M. J., Hale, L. P., and Denning, S. M. (1989). CD44-A molecule involved in leukocyte adherence and T-cell activation. *Immunol. Today* 10:423–424.

Heikkila, R., Schwab, G., Wickstrom, E., Loke, S. L., Pluznik, D. H., Watt, R., and Neckers, L. M. (1989). A c-myc antisense oligodeoxynucleotide inhibits entry into S phase but not progress from G_0 to G_1. *Nature* 328:445–447.

Hirata, F., Matsuda, K., Notsu, Y., Hattori, T., and Carmine, R. D. (1984). Phosphorylation at a tyrosine residue of lipomodulin in mitogen-stimulated murine thymocytes. *Proc. Natl. Acad. Sci. USA* 81:4717–1423.

Hombach, J., Leclercq, L., Radbruch, A., Rajewsky, K., and Reth, M. (1988). A novel 34-kD protein co-isolated with the IgM molecule in surface IgM–expressing cells. *EMBO J.* 7:3451–3459.

Honegger, A. M., Kris, R. M., Ullrich, A., and Schlessinger, J. (1989). Evidence that autophosphorylation of solubilized receptors for epidermal growth factor is mediated by intermolecular cross-phosphorylation. *Proc. Natl. Acad. Sci. USA* 86:925–931.

Horak, I. D., Gress, R., Lucas, P., Horak, E., Waldmann, T., and Bolen, J. (1991). T-lymphocyte interleukin 2–dependent tyrosine protein kinase signal transduction involves the activation of p56lck. *Proc. Natl. Acad. Sci. USA* 88:1996–2000.

Hsi, E., Siegel, N., Minami, E., Luong E., and Klausner, R. (1989). T cell activation induces rapid tyrosine phosphorylation of a limited number of substrates. *J. Biol. Chem.* 264:10863–10870.

Hunter, T. (1987). A thousand and one protein kinases. *Cell* 50:823–824.

Hunter, T. (1989). Protein-tyrosine phosphatases: The other side of the coin. *Cell* 58:1013–1014.

Hunter, T., and Cooper, J. A. (1985). Protein-tyrosine kinases. *Annu. Rev. Biochem.* 54:897–910.

Hurley, T. R., and Sefton, B. M. (1989). Analysis of the activity and phosphorylation of the *lck* protein in lymphoid cells. *Oncogene* 4:265–270.

Janeway, C. A., Rojo, J., Saizawa, K., Dianzani, U., Portoles, P., Tite, J., Haque, S., and Jones, B. (1989). The co-receptor function of murine CD41. *Immunol. Rev.* 109:77–92.

Jenkins, M. K., Pardoll, D. M., Mizuguchi, J., Chused, T. M., and Schwartz, R. H. (1987). Molecular events in the induction of a nonresponsive state in interleukin 2–producing helper T-lymphocyte clones. *Proc. Natl. Acad. Sci. USA 84*:5409–5413.

Johnson, K. W., Davis, B. H., and Smith, K. A. (1988). cAMP antagonizes interleukin 2–promoted T-cell cycle progression at a discrete point in early G_1. *Proc. Natl. Acad. Sci. USA 85*:6072–6079.

June, C. H., Fletcher, M. C., Ledbetter, J., and Samelson, L. E. (1990). Increases in tyrosine phosphorylation are detectable before phospholipase C activation after T cell receptor stimulation. *J. Immunol. 144*:1591–1599.

Karray, S., DeFrance, T., Merle-Beral, H., Banchereau, J., Debre, P., and Galanaud, P. (1988). Interleukin 4 counteracts the interleukin 2–induced proliferation of monoclonal B cells. *J. Exp. Med. 168*:85–92.

Katagiri, K., Katagiri, T., Eisenberg, R. A., Ting, J., and Cohen, P. L. (1987). Interleukin 2 responses of *lpr* and normal L3T4⁻/Lyt-2⁻ T cells induced by TPA plus A23187. *J. Immunol. 138*:149–153.

Kikkawa, U., and Nishizuka, Y. (1986). The role of protein kinase C in transmembrane signalling. *Annu. Rev. Cell. Biol. 2*:149–162.

Kizaki, H., Tadakuma, T., Odaka, C. Muramatsu, J., and Ishimura, Y. (1989). Activation of a suicide process of thymocytes through DNA fragmentation by calcium ionophores and phorbol esters. *J. Immunol. 143*:1790–1796.

Klarlund, J. K. (1985). Transformation of cells by an inhibitor of phosphatases acting on phosphotyrosine in proteins. *Cell 41*:707–710.

Klein, H. H., Friedenberg, Matthaei, S. and Olefsky. J. M. (1987). Insulin receptor kinase following internalization in isolated rat adipocytes. *J. Biol. Chem. 262*:10557–10562.

Koretzky, G., Pincus, J., Schultz, T., and Weiss A. (1991). Tyrosine phosphatase CD45 is required for T-cell antigen receptor and CD2-mediated activation of a protein tyrosine kinase and interleukin 2 production. *Proc. Natl. Acad. Sci. USA 88*:2037–2041.

Krangel, M. S., Bierer, B. E., Devlin, P., Clabby, M., Strominger, J. L., McLean, J., and Brenner, M. B. (1987). T3 glycoprotein is functional although structurally distinct on human T-cell receptor γ T lymphocytes. *Proc. Natl. Acad. Sci USA 84*:3817–3822.

Kumagai, N., Benedict, S., Mills, G. B., and Gelfand, E. W. (1987). Requirements for simultaneous presence of phorbol esters and calcium ionophores in the expression of human T lymphocyte proliferation related genes. *J. Immunol. 139*:1393–1399.

Kumagai, N., Benedict, S., Mills, G. B., and Gelfand, E. W. (1988). Induction of competence and progression signals in human T lymphocytes by phorbol esters and calcium ionophores. *J. Cell. Physiol. 137*:329–336.

Kumar, A., Moreau, J.-L., Gibert, M., and Theze, J. (1987). Internalization of interleukin 2 (IL-2) by high affinity IL-2 receptors is required for the growth of IL-2–dependent T cell lines. *J. Immunol. 139*:3680–3687.

Kumar, C. S., Muthukumaran, G., Frost, L. J., Noe, M., Ahn, Y. H., Mariano, T. M., and Pestka, S. (1989). Molecular characterization of the murine interferon γ receptor cDNA. *J. Biol. Chem. 264*:17939–17942.

Kupfer, A., Singer, S. J., Janeway, C. A., and Swain, S. L. (1987). Coclustering of CD4 (L3T4) molecule with the T-cell receptor is induced by specific direct interaction of helper T cells and antigen-presenting cells. *Proc. Natl. Acad. Sci USA 84*:5888–5893.

Kupper, T., Flood, P., Coleman, D., and Horowitz, M. (1987). Growth of an interleukin 2/interleukin 4–dependent T cell line induced by granulocyte-macrophage colony-stimulating factor (GM-CSF). *J. Immunol. 138*:4288–4292.

Legraverend, C., Potter, A., Holta, E., Alitalo, K., and Andersson, C. (1989). Interleukin-2 induces a rapid increase in ornithine decarboxylase mRNA in a cloned murine T lymphocyte cell line. *Exp. Cell. Res. 181*:273–281.

Levitski, A., and Schlessinger, J. (1989). Tyrphostins inhibit epidermal growth factor receptor tyrosine kinase activity in living cells and EGF-stimulated cell proliferation. *J. Biol. Chem.* *264*:14503–14512.

MacDonald, H. R., and Nabholz M. (1986). T-cell activation. *Annu. Rev. Cell Biol.* *2*:231–239.

Makgoba, M. W., Sanders, M. E., and Shaw, S. (1989). The CD2–LFA-3 and LFA-1–ICAM pathways: Relevance to T-cell recognition. *Immunol. Today* *10*:417–419.

Margolis, B., Rhee, S. G., Felder, S., Mervic, M., Lyall, R., Levitzki, A., Ullrich, A., Zilberstein, A., and Schlessinger, J. (1989). EGF induces tyrosine phsophorylation of phospholipase C-II: A potential mechanism for EGF receptor signaling. *Cell* *57*:1101–1107.

Marlin, S. D., and Springer, T. A. (1987). Purified intercellular adhesion molecule-1 (ICAM-1) is a ligand for lymphocyte function-associated antigen 1 (LFA-1). *Cell* *51*:813–817.

Marth, J. D., Lewis, D. B., Cooke, M. P., Mellins, E. D., Gearn, M. E., Samelson, L. E., Wilson, C. B., Miller, A. D., and Perlmutter, R. M. (1989). Lymphocyte activation provokes modification of a lymphocyte-specific tyrosine kinase (P56*lck*). *J. Immunol.* *142*:2430–2435.

Marth, J. D., Lewis, D. B., Wilson, C. B., Gearn, M. E., Krebs, E. G., and Perlmutter, R. M. (1987). Regulation of pp56*lck* during T-cell activation: Functional implications for the src-like protein tyrosine kinases. *EMBO J.* *6*:2727–2734.

Martinez-Valdez, H., Doherty, P. J., Thompson, E., Benedict, S. H., Gelfand, E. W., and Cohen, A. (1988a). Antagonistic effects of calcium ionophores and phorbol esters on T cell receptor mRNA levels in human thymocytes. *J. Immunol.* *140*:361–365.

Martinez-Valdez, H., Thompson, E., and Cohen, A. (1988b). Phorbol esters and calcium ionophores differentially regulate the transcription of γ-T-cell antigen receptor gene in human thymocytes. *J. Biol. Chem.* *263*:4043–4048.

Mary, D., Aussel, C. Ferrua, B., and Fehlmann, M. (1987). Regulation of interleukin 2 synthesis by cAMP in human T cells. *J. Immunol.* *139*:1179–1184.

Matsushima, K., Yodoi, J., Tagaya, T., and Oppenheim, J. J. (1986). Down-regulation of interleukin 1 (IL 1) receptor expression by IL 1 and fate of internalized ^{125}I-labelled IL 1B in a human large granular lymphocyte cell line. *J. Immunol.* *137*:3183–3188.

McConkey, D. J., Hartzell, P., Jondal, M., and Orrenius, S. (1989a). Inhibition of DNA fragmentation in thymocytes and isolated thymocyte nuclei by agents that stimulate protein kinase C. *J. Biol. Chem.* *264*:13399–13405.

McConkey, D. J., Hartzell, P., Amador-Pérez, J. F., Orrenius, S., and Jondal, M. (1989b). Calcium-dependent killing of immature thymocytes by stimulation via the CD3/T cell receptor complex. *J. Immunol.* *143*:1801–1810.

Meisenhelder, J., Suh, P. G., Rhee, S. G., and Hunter, T. (1989). Phospholipase C-γ is a substrate for the PDGF and EGF receptor protein-tyrosine kinases in vivo and in vitro. *Cell* *57*:1109–1114.

Mercep, M., Bonifacino, J. S., Garcia-Morales, P., Samelson, L. E., Klausner, R. D., and Ashwell, J. D. (1988). T cell CD3-zeta,eta heterodimer expression and coupling to phosphoinositide hydrolysis. *Science* *242*:571–573.

Merida, I., Diez, E., and Gaulton, N. (1991). Interleukin 2 binding activates a tyrosine phosphorylated phosphatidylinositol-3 kinase. *J. Immunol.* *147*:2202–2207.

Merida, I., Pratt, J., and Gaulton, G. (1990). Regulation of interleukin 2–dependent growth responses by glycophosphatidylinositol molecules. *Proc. Natl. Acad. Sci. USA* *87*:9421–9425.

Mills, G. B. (1989). Activation of lymphocytes by lymphokines. *Curr. Top. Membr. Transport* *35*:495–535.

Mills, G. B., Arima, N., May, C., Schmandt, R., Miyamoto, N., Greene, W. (1992). Neither LCK nor FYN are obligatory for IL-2 signal transduction in HTLVI-infected T cells. *International Immunology*: in press.

Mills, G. B., Benedict, S., Mellors, A., Grinstein, S., and Gelfand E. W. (1988a). Transmembrane signalling by interleukin 2. In: *Interleukin 2* (Smith, K. A., ed.). San Diego: Academic Press, pp. 113–135.

Mills, G. B., Cheung, R. K., Cragoe, E. J., Grinstein, S., and Gelfand, E. W. (1986). Activation of the Na^+/H^+ antiport is not required for proliferation of human T lymphocytes. *J. Immunol.* *136*:1150–1154.

Mills, G. B., Cheung, R. K., Grinstein, S., and Gelfand, E. W. (1985). Increase in cytosolic free calcium concentration is an intracellular messenger for the production of interleukin 2 but not for the expression of the interleukin 2 receptor. *J. Immunol. 134*:1640–1643.

Mills, G. B., Girard, P., Grinstein, G., and Gelfand, E. W. (1988b). Interleukin 2 induces proliferation of T cell mutants lacking protein kinase C. *Cell 55*:91–100.

Mills, G. B., May, C., Hill, M., Ebanks, R. Mellors, A., and Gelfand E. W. (1989a). Physiologic activation of protein kinase C limits interleukin 2 secretion. *J. Immunol. 142*:1995–2003.

Mills, G. B., May, C., Hill, M., and Gelfand, E. W. (1989b). Role of protein kinase C in interleukin 1, anti-T3 and mitogenic lectin induced interleukin 2 secretion. *J. Cell. Physiol. 141*:310–317.

Mills, G. B., May, C., McGill, M., Fung, M., Baker, M., Sutherland, R., and Greene, W. C. (1990a). Interleukin 2 induced tyrosine phosphorylation: $IL2R\beta$ is tyrosine phosphorylated. *J. Biol. Chem. 265*:3561–3567.

Mills, G. B., Stanley J., Stewart, D., Mellors, A., and Gelfand, E. W. (1990b). Interrelationship between signals transduced by phytohemagglutinin and interleukin 1. *J. Cell Physiol. 142*:539–551.

Mills G. B., Zhang N., Schmandt, R., Fung, M., Greene, W., Mellors, A., and Hogg, D. (1991b). Transmembrane signalling by interleukin 2. *Biochem. Trans. 115*:277–285.

Molloy, C. J., Bottaro, D. P., Fleming, T. P., Marshall, M. S., Gibbs, J. B., and Aaronson, S. A. (1989). PDGF induction of tyrosine phosphorylation of GTPase activating protein. *Nature 342*:711–714.

Moretta, A., Ciccone, E., Pantaleo, G., Tambussi, G., Bottino, C., Melioli, G., Mingari, M. C., and Moretta, L. (1989). Surface molecules involved in the activation and regulation of T or natural killer lymphocytes in humans. *Immunol. Rev. 111*:145–150.

Moretta, A., Poggi, A., Olive, D., Bottino, C., Fortis, C., Pantaleo, G., and Moretta, L. (1987). Selection and characterization of T-cell variants lacking molecules involved in T-cell activation (T3, T-cell receptor, T44, and T11): Analysis of the functional relationship among different pathways of activation. *Proc. Natl. Acad. Sci. USA 84*:1654–1660.

Morla, A. O., Schreurs, J., Miyajima, A., and Wang, J. J. (1988). Hematopoietic growth factors activate the tyrosine phosphorylation of distinct sets of proteins in interleukin-3–dependent murine cell lines. *Mol. Cell. Biol. 8*:2214–2218.

Morrison, D. K., Kaplan, D. R., Escobedo, J. A., Rapp, U. R., Roberts, T. M., and Williams, L. T. (1989). Direct activation of the serine/threonine kinase activity of Raf-1 through tyrosine phosphorylation by the PDGF β-receptor. *Cell 58*:649–657.

Mosley, B., Beckmann, M. P., March, C. J., Idzerda, R. L., Gimpel, S. D., VandenBos, T., Friend, D., Alpert, A., Anderson, D., Jackson, J., Wignall, J. M., Smith, C., Gallis, B., Sims, J. E., Urdal, D., Widmer, M. B., Cosman, D., and Park, L. S. (1989). The murine interleukin-4 receptor: Molecular cloning and characterization of secreted and membrane bound forms. *Cell 59*:335–341.

Muira, O., D'Andrea, A., Kabat, D., and Ihle, J. (1991). Induction of tyrosine phosphorylation by the erythropoietin receptor correlates with mitogenesis. *Mol. Cell. Biol. 11*:4895–4902.

Mustelin, T., and Altman, A. (1989). Do CD4 and CD8 control T-cell activation via a specific tyrosine protein kinase? *Immunol. Today 10*:189–191.

Mustelin, T., Coggeshall, K. M., and Altman, A. (1989). Rapid activation of the T-cell tyrosine protein kinase $pp56^{lck}$ by the CD45 phosphotyrosine phosphatase. *Proc. Natl. Acad. Sci. USA 86*:6302–6306.

Nakayama, T., Singer, A., Hsi, E. D., and Samelson, L. E. (1989). Intrathymic signalling in immature $CD4^+CD8^+$ thymocytes results in tyrosine phosphorylation of the T-cell receptor zeta chain. *Nature 341*:651–653.

Neer, E. J., and Clapham, D. E. (1988). Roles of G protein subunits in transmembrane signalling. *Nature* 333:129–132.

Nishizuka, Y. (1988). The molecular heterogeneity of protein kinase C and its implications for cellular regulation. *Nature* 334:661–665.

Northwood, I. C., and Davis, R. J. (1989). Protein kinase C inhibition of the epidermal growth factor receptor tyrosine protein kinase activity is independent of the oligomeric state of the receptor. *J. Biol. Chem.* 264:5746–5751.

Obermaier-Kusser, B., White, M. F., Pongratz, D. E., Su, Z., Ermel, B., Muhlbacher, C., and Haring, H. U. (1989). A defective intramolecular autoactivation cascade may cause the reduced kinase activity of the skeletal muscle insulin receptor from patients with noninsulin-dependent diabetes mellitus. *J. Biol. Chem.* 264:9497–9503.

O'Brien, R., Houslay, M. D., Brindle, N.P.J., Milligan, G., Whittaker, J., and Siddle, K. (1989). Binding to GDP-agarose identifies a novel 60kDa substrate for the insulin receptor tyrosyl kinase in mouse NIH-3T3 cells expressing high concentrations of the human insulin receptor. *Biochem. Biophys. Res. Commun.* 158:743–749.

O'Neill, L.A.J., Bird, T. A., Gearing, J. H., and Saklatvala, J. (1990). Interleukin-1 signal transduction: Increased GTP binding and hydrolysis in membranes of a murine thymoma line. *J. Biol. Chem.* 265:3146–3152.

O'Shea, J. J., Urdahl, K. B., Luong, H. T., Chused, T. M., Samelson, L. E., and Klausner, R. D. (1987). Aluminum fluoride induces phosphatidylinositol turnover, elevation of cytoplasmic free calcium, and phosphorylation of the T cell antigen receptor in murine T cells. *J. Immunol.* 139:3463–3470.

Ostergaard, H. L., Shackelford, D. A., Hurley, T. R., Johnson, P., Hyman, R. Sefton, B. M., and Trowbridge, I. S. (1989). Expression of CD45 alters phosphorylation of the lck-encoded tyrosine protein kinase in murine lymphoma T-cell lines. *Proc. Natl. Acad. Sci. USA* 86:8959–8963.

Otten, G., Herold, K. C., and Fitch, F. W. (1987). Interleukin 2 inhibits antigen stimulated lymphokine synthesis in helper T cells by inhibiting calcium-dependent signalling. *J. Immunol.* 139:1348–1357.

Otten, G., Wilde, D. B., Prystowsky, M. B., Olshan, J. S., Rabin, H., Henderson, L. E., and Fitch, F. W. (1986). Cloned helper T lymphocytes exposed to interleukin 2 become unresponsive to antigen and concanavalin A but not to calcium ionophore and phorbol ester. *Eur. J. Immunol.* 16:217–225.

Padeh, S., Levitzki, A., Mills, G. B., and Roifman, C. M. (1991). Activation of phospholipase C in B cells is dependent on tyrosine phosphorylation. *J. Clin. Invest.* 87:1114.

Paetkau, V., and Mills, G. B. (1989). Cytokines and the mechanisms of lymphocyte activation. *Immunol. Allergy Clin. N. Am.* 9:21–44.

Pantaleo, G., Ferrini, S., Zocchi, M. R., Bottino, C., Biassoni, R., Moretta, L., Moretta, A. (1987a). Analysis of signal transducing mechanisms in CD3+ CD4− CD8− cells expressing the putative T cell receptor γ gene product. *J. Immunol.* 139:3580–3585.

Pantaleo, G., Olive, D., Poggi, A., Pozzan, T., Moretta, L., and Moretta, A. (1987b). Antibody-induced modulation of the CD3/T cell receptor complex causes T cell refractoriness by inhibiting the early metabolic steps involved in T cell activation. *J. Exp. Med.* 166:619–629.

Park, L. S., Friend, D., Price, V., Anderson, D., Singer, J., Prickett, K. S., and Urdal, D. L. (1989). Heterogeneity in human interleukin-3 receptors: A subclass that binds human granulocyte/macrophage colony stimulating factor. *J. Biol. Chem.* 264:5420–5429.

Patel, M., Samelson, L. E., and Klausner, R. D. (1987). Multiple kinases and signal transduction. *J. Biol. Chem.* 262:5831–5837.

Pauza, C. D. (1987). Regulation of human T-lymphocyte gene expression by interleukin 2: Immediate-response genes include the proto-oncogene c-myb. *Mol. Cell. Biol.* 7:342–348.

Pearse, M., Wu, L., Egerton, M., Wilson, A., Shortman, K., and Scollay, R. (1989). A murine early thymocyte developmental sequence is marked by transient expression of the interleukin 2 receptor. *Proc. Natl. Acad. Sci. USA 86*:1614–1620.

Perlmutter, R. M. (1989). T cell signaling. *Science 245*:344–346.

Pierce, J. H., Di Fiore, P. P., Aaronson, S. A., Potter, M., Pumphrey, J., Scott, A., and Ihle, J. N. (1985). Neoplastic transformation of mast cells by Abelson-MuLV: Abrogation of IL-3 dependence by a nonautocrine mechanism. *Cell 41*:685–693.

Pierce, J. H., Ruggiero, M., Fleming, T. P., DiFiore, P. P., Greenberger, J. S., Varticovski, L., Schlessinger, J., Rovera, G., and Aaronson, S. A. (1988). Signal transduction through the EGF receptor transfected in IL-3–dependent hematopoietic cells. *Science 239*:628–630.

Poo, W.-J., Conrad, L., and Janeway, Jr., C. A. (1988). Receptor-directed focusing of lymphokine release by helper T cells. *Nature 332*:378–380.

Puil, L., Mills, G. B., and Pawson A. (1990). Signal transduction through the TCR-CD3 complex IL-2 receptor results in the phosphorylation of GAP and GAP-associated proteins. *Blood 76*(Suppl. 1):217a.

Ramsdell, F., and Fowlkes, B. J. (1989). Engagement of CD4 and CD8 accessory molecules is required for T cell maturation. *J. Immunol. 143*:1467–1471.

Rapp, U. R., Cleveland, J. L., Brightman, K., Scott, A., and Ihle, J. N. (1985). Abrogation of IL-3 and IL-2 dependence by recombinant murine retroviruses expressing v-*myc* oncogenes. *Nature 317*:434–438.

Reem, G. H., Carding, S., and Reinherz, E. L. (1987). Lymphokine synthesis is induced in human thymocytes by activation of the CD2 (T11) pathway. *J. Immunol. 139*:130–135.

Reem, G. H., Yeh, N.-H., Urdal, D. L., Kilian, P. L., and Farrar, J. J. (1985). Induction and up-regulation of high-affinity interleukin 2 receptors on thymocytes and T cells. *Proc. Natl. Acad. Sci. USA 82*:8663–8670.

Rivas, A., Takada, S., Koide, J., Sonderstrup-McDevitt, G., and Engleman, E. G. (1988). CD4 molecules are associated with the antigen receptor complex on activated but not resting T cells. *J. Immunol. 140*:2912–2917.

Roifman, C., Mills, G. B., Cheung, R. K., and Gelfand, E. W. (1986). Mitogenic response of human thymocytes: Identification of functional Ca^{2+}-dependent and independent signals. *Clin. Exp. Immunol. 66*:139–149.

Rojo, J. M., Saizawa, K., and Janeway, C. A. (1989). Physical association of CD4 and the T-cell receptor can be induced by anti-T-cell receptor antibodies. *Proc. Natl. Acad. Sci. USA 86*:3311–3316.

Rosenberg, Y. J., Malek, T. R., Schaeffer, D. E., Santoro, T. J., Mark, G. E., Steinberg, A. D., and Mountz, J. D. (1985). Unusual expression of IL 2 receptors and both the c-*myb* and c-*raf* oncogenes in T cell lines and clones derived from autoimmune MRL-lpr/lpr mice. *J. Immunol. 134*:3120–3124.

Rubin, L. A., Kurman, C. C., Fritz, M. E., Biddison, W. E., Boutin, B., Yarchoan, R., and Nelson, D. L. (1985). Soluble interleukin 2 receptors are released from activated human lymphoid cells in vitro. *J. Immunol. 135*:3172–3177.

Rudd, C. D., Trevillyan, J. M., Dasgupta, J. D., Wong, L. L., and Schlossman, S. F. (1988). The CD4 receptor is complexed in detergent lysates to a protein-tyrosine kinase (pp58) from human T lymphocytes. *Proc. Natl. Acad. Sci. USA 85*:5190–5196.

Sakaguchi, N., Kashiwamura, S., Kimoto, M., Thalmann, P., and Melchers, F. (1988). B lymphocyte lineage-restricted expression of *mb-1*, a gene with CD3-like structural properties. *EMBO J. 7*:3457–3461.

Saksela, K., Makela, T. P., Evan, G., and Alitalo, K. (1989). Rapid phosphorylation of the L-*myc* protein induced by phorbol ester tumor promoters and serum. *EMBO J. 8*:149–156.

Saltzman, E. M., Thom, R. R., and Casnellie, J. E. (1988). Activation of a tyrosine protein kinase is an early event in the stimulation of T lymphocytes by interleukin-2. *J. Biol. Chem. 263*:6956–6959.

Saltzman, M., White, K., and Casnellie, E. J. (1990). Stimulation of the antigen and interleukin-2 receptor on T lymphocytes activates distinct tyrosine protein kinases. *J. Biol. Chem.* 265:10138–10142.

Samelson, L., Davidson, W. F., Morse, H. C., and Klausner R. D. (1986a). Abnormal tyrosine phosphorylation on T-cell receptor in lymphoproliferative disorders. *Nature* 324:674–677.

Samelson, L. E., Patel, M. D., Weissman, A. M., Harford, J. B., and Klausner, R. D. (1986b). Antigen activation of murine T cells induces tyrosine phosphorylation of a polypeptide associated with the T cell antigen receptor. *Cell* 46:1083–1088.

Samelson, L., Phillips, A., Luong, E., and Klausner R. (1990). Association of the FYN tyrosine kinase with the T-cell antigen receptor. *Proc. Natl. Acad. Sci. USA* 87:4358–4362.

Satoh, T., Nafafuku, M., Miyajima, A., and Kaziro, Y. (1991). Involvement of ras p21 protein in signal-transduction pathways from interleukin 2, interleukin 3, and granulocyte/macrophage colony-stimulating factor, but not from interleukin 4. *Proc. Natl. Acad. Sci. USA* 88:3314–3318.

Scholz, W., Isakov, N., Mally, M. I., Theofilopoulos, A. N., and Altman, A. (1988). Lpr T cell hyporesponsiveness to mitogens linked to deficient receptor-stimulated phosphoinositide hydrolysis. *J. Biol. Chem.* 263:3626–3631.

Sharon, M., Gnarra, J., and Leonard, W. (1989). The β chain of IL-2 receptor (p70) is tyrosine-phosphorylated on YT and HUT-102B2 cells. *J. Immunol.* 143:2530–2537.

Shirakawa, F., Yamashita, U., Chedid, M., and Mizel, S. B. (1988). Cyclic AMP—an intracellular second messenger for interleukin 1. *Proc. Natl. Acad. Sci. USA* 85:8201–8205.

Sims, J. E., March, C. J., Cosman, D., Widmer, M. B., MacDonald, H. R., McMahan, C. J., Grubin, C. E., Wignall, J. M., Jackson, J. L., Call, S. M., Friend, D., Alpert, A. R., Gillis, S., Urdal, D. L., and Dower, S. K. (1988). cDNA expression cloning of the IL-1 receptor, a member of the immunoglobulin superfamily. *Science* 241:585–589.

Smith, K. A. (1988). Interleukin 2: Inception, impact, and implications. *Science* 240:1169–1172.

Sorensen, P., Mui, A.L.F., and Krystal, G. (1989). Interleukin-3 stimulates the tyrosine phosphorylation of the 140-kilodalton interleukin-3 receptor. *J. Biol. Chem.* 264:19253–19259.

Stanley, J. B., Gorczynski, R. M., Delovitch, T., and Mills G. B. (1989). IL 2 secretion is pertussis toxin sensitive in a T lymphocyte hybridoma. *J. Immunol.* 142:3546–3552.

Stanley, J., Huang, C.-K., Love, J., Gorczynski, R., and Mills, G. B. (1990). Tyrosine phosphorylation is an obligatory event in IL2 production. *J. Immunol.* 145:2189–2198.

Stern, J. B., and Smith, K. A. (1986). Interleukin-2 induction of T-cell G_1 progression and c-*myb* expression. *Science* 233:203–206.

St-Pierre, Y., Nabavi, N., Ghogawala, Z., Glimcher, L. H., and Watts, T. H. (1989). A functional role for signal transduction via the cytoplasmic domains of MHC class II proteins. *J. Immunol.* 143:808–813.

Sussman, J. J., Bonifacino, J. S., Lippincott-Schwartz, J., Weissman, A. M., Saito, T., Klausner, R. D., and Ashwell, J. D. (1988). Failure to synthesize the T cell CD3-ζ chain: Structure and function of a partial T cell receptor complex. *Cell* 52:85–93.

Taga, T., Hibi, M., Hirata, Y., Yamasaki, K., Yasukawa, K., Matsuda, T., Hirano, T., and Kishimoto, T. (1989). Interleukin-6 triggers the association of its receptor with a possible signal transducer, gp130. *Cell* 58:573–580.

Takei, F. (1988). IL-2 and IL-4 stimulate different subpopulations of double-negative thymocytes. *J. Immunol.* 141:1114–1119.

Taniguchi, T. (1988). Regulation of cytokine gene expression. *Annu. Rev. Immunol.* 6:439–450.

Taylor, M. V., Hesketh, T. R., and Metcalfe, J. C. (1988). Phosphoinositide metabolism and the calcium response to concanavalin A in S49 T-lymphoma cells. A comparison with thymocytes. *Biochem. J.* 249:847–852.

Taylor, M. V., Metcalfe, J. C., Hesketh, T. R., Smith, G. A., and Moore, J. P. (1984). Mitogens increase phosphorylation of phosphoinositides in thymocytes. *Nature* 312:462–464.

Tentori, L., Longo, D. L., Zuniga-Pflucker, J. C., Wing, C., and Kruisbeek, A. M. (1988). Essential role of the interleukin 2 receptor pathway in thymocyte maturation in vivo. *J. Exp. Med.* *168*:1741–1745.

Thomas, M. L., and Lefrancois, L. (1988). Differential expression of the leucocyte-common antigen family. *Immunol. Today* *9*:320–323.

Tigges, M. A., Casey, L. S., and Koshland, M. E. (1989). Mechanisms of interleukin-2 signaling: Mediation of different outcomes by a single receptor and transduction pathway. *Science* *243*:781–784.

Toribio, M. L., De la Hera, A., Marcos, M.A.R., Márquez, C., and Martínez-A., C. (1989). Activation of the interleukin 2 pathway precedes CD3-T cell receptor expression in thymic development: Differential growth requirements of early and mature intrathymic subpopulations. *Eur. J. Immunol.* *19*:9–15.

Tsien, R. Y., Pozzan, T., Rink, T. J. (1982). T-cell mitogens cause early changes in cytoplasmic free Ca^{2+} and membrane potential in lymphocytes. *Nature* *295*:68–70.

Tsoukas, C. D., Landgraf, B., Bentin, J., Lamberti, J. F., Carson, D. A., and Vaughan, J. H. (1987). Structural and functional characteristics of the CD3 (T3) molecular complex on human thymocytes. *J. Immunol.* *138*:3885–3891.

Tsudo, M., Karasuyama, H., Kitamura, F., Nagasaka, Y., Tanaka, T., and Miyasaka, M. (1989). Reconstitution of a functional IL-2 receptor by the β-chain cDNA: A newly acquired receptor transduces negative signal. *J. Immunol.* *143*:4039–4044.

Valge, V. E., Wong, J.G.P., Datlof, B. M., Sinskey, A. J., and Rao, A. (1988). Protein kinase C is required for responses to T cell receptor ligands but not to interleukin-2 in T cells. *Cell* *55*:101–112.

Veillette, A., Bookman, M. A., Horak, E. M., Samelson, L. E., and Bolen, J. B. (1989a). Signal transduction through the CD4 receptor involves the activation of the internal membrane tyrosine-protein kinase p56*lck*. *Nature* *338*:257–259.

Veillette, A., Zuniga-Pflucker, C., Bolen, J. B., and Kruisbeek, A. M. (1989b). Engagement of CD4 and CD8 expressed on immature thymocytes induces activation of intracellular tyrosine phosphorylation pathways. *J. Exp. Med.* *170*:1671–1680.

Wahl, M. I., Olashaw, N. E., Nishie, S., Rhee, S. G., and Pledger, W. J. (1989). Platelet-derived growth factor induces rapid and sustained tyrosine phosphorylation of phospholipase C-γ in quiescent BALB/c 3T3 cells. *Mol. Cell Biol.* *9*:2934–2938.

Waldmann, T. A. (1989). The multi-subunit interleukin-2 receptor 1,2. *Annu. Rev. Biochem.* *58*:875–911.

Weiss, A., Dazin, P. F., Shields, R., Fu, S. M., and Lanier, L. L. (1987). Functional competency of T cell antigen receptors in human thymus. *J. Immunol.* *139*:3245–3249.

Weiss, A., Imboden, J., Hardy, K., Manger, B., Terhorst, C., and Stobo, J. (1986). The role of the T3/antigen receptor complex in T-cell activation. *Annu. Rev. Immunol.* *4*:593–602.

Weissman, A. M., Baniyash, M., Hou, D., Samelson, L. E., Burgess, W. H., and Klausner, R. D. (1988). Molecular cloning of the zeta chain of the T cell antigen receptor. *Science* *239*:1018–1020.

Wheeler, E. F., Askew, D., May, S., Ihle, J. N., and Sher, C. (1987). The v-*fms* oncogene induces factor-independent growth transformation of the interleukin-3 dependent myeloid cell line FDPC-P1. *Mol. Cell Biol.* *7*:1673–1679.

White, M. F., and Kahn, R. C. (1986). The insulin receptor and tyrosine phosphorylation. *Enzymes* *27*:247–258.

Wofsky, D., Murphy, E. D., Roths, J. B., Daupinee, M. J., Kipper, S. B., and Talal, N. (1981). Deficient interleukin 2 activity in MRL/LPR and C57BL/6J mice bearing the *lpr* gene. *J. Exp. Med.* *154*:1671–1679.

Yamasaki, K., Taga, T., Hirata, Y., Yawata, H., Kawanishi, Y., Seed, B., Taniguchi, T., Hirano, T., and Kishimoto, T. (1988). Cloning and expression of the human interleukin-6 (BSF-2/IFNB 2) receptor. *Science* *241*:825–827.

Yarden, Y., and Ullrich, A. (1988). Growth factor receptor tyrosine kinases. *Annu. Rev. Biochem.* 57:443–457.

Zamoyska, T., Deham, P., Gorman, S. D., Hoegen, P. V., Bolent, J. B., Veillette, A., and Parnes, J. R. (1989). Inability of CD8δ' polypeptides to associate with p56^{lck} correlates with impaired function in vitro and lack of expression in vivo. *Nature 342*:278–280.

Zhang, N., Huang, C.-K., Chung, A., and Mills, G. B. (1992). Tyrosine phosphorylation is an obligatory event for IL2-induced cell cycle progression. Submitted.

16

Activation and Inactivation in Regulation of the B Cell Response

ERWIN DIENER, RATI FOTEDAR,

ANIMESH SINHA

For many years the B lymphocyte has been the main focus of attention for those interested in the mechanisms that control cellular responses to external stimuli. From descriptions of antigen binding to cell surface immunoglobulin (Ig) receptors, measurements of the proliferative activity of cells, as reflected by incorporation of radionucleotides into DNA, and of the population kinetics of antibody-secreting cells, immunologists have developed a fairly accurate picture of discernible physiologic states of B cells as they respond to external stimuli. For a long time, the definition of these stimuli was highly controversial, particularly after it was shown that an immunocompetent B cell must not only bind antigen but also interact with non-B cell–derived co-stimulators to undergo a full immune response, including antibody secretion. After the discovery of antigen-specific helper T cells, it appeared reasonable to postulate that T cells might cooperate with B cells through secretion of antigen-specific helper factors. Although numerous reports of these factors have been published, molecular analysis has so far failed to provide evidence of their biochemical nature, including the existence of secretory forms of T cell antigen receptors. Thus, contemporary opinion favors a direct cooperative interaction via an antigen bridge between antigen-specific T cells and B cells as part of the initial triggering stimulus. Following this contact, antigen nonspecific lymphokines secreted by T cells and acting at short range on B cells may contribute additional signals. For reasons of experimental convenience, in-depth analysis of B cell growth and differentiation has, for the most part, been carried out by using a variety of specific and nonspecific ligands, including T cell–independent (TI) antigens, polyclonal B cell activators, and antibodies directed at antigen receptors, in place of antigen-specific helper T cells.

The published work concerning the types of cytokines required for B cell activation has given rise to many interpretations, some of which are directly contradictory. Some of these contradictions no doubt reflect alternative pathways for B cell activation. It is also clear, however, that the lack of a clear understanding of B cell activation mechanisms reflects the range of experimental models used by different investigators. Many of these models failed to take sufficient account of B cell heterogeneity, of the possible contamination of B cells with other cell types, and of the stimulatory artifacts that arise from foreign serum proteins present in culture media. In addition, the use of mycoplasma-infected cell lines in reconstitution experiments, along with poorly defined cytokine preparations, has also contributed to this problem. Nevertheless, it now appears that several separate signals mediate functionally distinct steps in the B cell response, including activation from the resting G_0 to the G_1 state, followed by proliferation and differentiation into antibody-secreting effector cells (Leanderson et al., 1987; Melchers and Lernhardt, 1985; Zubler and Kanagawa, 1982). It has been known for some time that T cell– and macrophage-derived factors are required for activated B lymphocytes to undergo a full immune response. In recent years these factors have been identified as various lymphokines and monokines, many of which are now available in pure form through the use of recombinant DNA technology. While it is clear that B cell activation to the G_1 phase of the cell cycle is required for subsequent proliferation and maturation, it has been difficult to assign signals provided by specific cytokines to particular stages of B cell development (discussed by Leanderson et al., 1987).

During the course of an immune response, most B lymphocytes undergo blastogenesis and then mitosis before antibody synthesis and secretion (Diener et al., 1973, 1976; Russell and Diener, 1970). Nevertheless, as early as 1970 it was shown that within the first few days after in vitro or in vivo stimulation with a TI antigen, up to 30% of the cells making antibody to that antigen had differentiated from small B lymphocytes into antibody-secreting cells without undergoing mitosis (Russell and Diener, 1970). Other investigators have shown that B lymphocytes treated with polyclonal activators and then cultured with inhibitors of DNA synthesis still matured into antibody-secreting cells (Andersson and Melchers, 1974). B lymphocytes specific for certain antigens undergo proliferation without antibody production when stimulated in vitro with the polyclonal B cell activator lipopolysaccharide (LPS) in the absence of T cells and macrophages (Inazawa et al., 1985; Jaworski et al., 1982). Work by other groups has shown that treatment with anti-IgM antibody suppresses LPS-induced antibody production by B lymphocytes without affecting their proliferative response (Leanderson and Forni, 1984). These observations suggest that proliferation and maturation of B cells are controlled by two separate signals. Ever since the first description of non-antigen–specific T cell–derived lymphokines by Schimpl and Wecker (1972, 1975), attempts have been made to relate antigen- and lymphokine-mediated signals to distinct stages of the B cell response and thus to construct a minimal model of the underlying mechanisms. The availability of more sophisticated methods to probe the metabolic states of B cells following activation has greatly enhanced our understanding of B cell signaling mechanisms.

Studies of the molecular basis of transmembrane signaling by B cell mitogens have helped shed light on the role of various second messengers derived from products of phospho-inositide breakdown. These mediators lead to translocation and activation of protein kinase C, and the biological response of the cell is then thought to be mediated by phosphorylation-dependent regulatory events (reviewed by Cambier et al., 1987). To further understand B cell activation, one needs to elucidate the molecular mechanisms that regulate B cell division (DNA replication) and B cell maturation (inducible gene expression). The availability of cell-free systems to study DNA replication, chromatin assembly, and inducible tissue-specific transcription provides a powerful experimental approach for this work.

While an understanding of the molecular biology of the B cell response to external stimuli should allow us to define the intracellular pathways that lead from signal transduction to selective gene activation, delineation of the cellular interactions involved in B cell stimulation provides a bridge from the single cell to the organismal levels. The work we shall describe below, on mechanisms of B cell triggering, has been directed to develop a minimal model for cytokine-dependent responses in B cells, induced by T-independent antigens, polyclonal activators, and anti-Ig antibodies. Cell proliferation and antibody secretion were used to assess B cell responses, as we were able to assign distinct externally derived signals to these two parameters.

Those who prefer to describe the immune system in the language of Newtonian formalism, as linear progressions of causes and effects, extrapolate the dialectic definition of tolerance as the conceptual antithesis of immunity to the binding event between cell surface receptors and antigen. Consequently, one has to look for circumstances in which antigen recognition is translated by the cell into either activation or inactivation. Theories traditionally known as "clonal deletion and abortion hypotheses" (Bretscher and Cohn, 1970; Lederberg, 1959) are attempts to formulate these circumstances. Earlier work from this laboratory has contributed extensively to define the conditions under which an antigen induces unresponsiveness rather than an immune response in B lymphocytes. It was shown that the extent to which cell surface receptors were cross-linked by an appropriate ligand represented a critical parameter in determining whether immunocompetent B cells were negatively or positively regulated (reviewed by Diener and Feldmann, 1972). Unfortunately, as later experiments suggested, this parameter applied only to certain antigens but not to others that induced equivalent degrees of receptor cross-linking (reviewed by Diener and Waters, 1986). Thus, parameters other than receptor cross-linking alone must be involved in mechanisms of self tolerance, a conclusion that makes the clonal abortion hypothesis, as formulated by Lederberg (1959), particularly attractive. According to this model, the immature, antigen receptor–bearing lymphocyte can uniquely be rendered unresponsive after engagement of its receptors with ligands of sufficiently high avidity. From this, it follows that the distinction between tolerogenic and non-tolerogenic antigens that is observed when dealing with mature immunocompetent B cells should not apply to immature B cells. Later in this chapter we shall discuss experimental evidence that fails to support this conclusion.

ACTIVATION OF B LYMPHOCYTES

Type of Ligand Used to Initiate B Cell Response

Early experimental attempts to define the mechanisms that control B lymphocyte responses were based on the assumption that such mechanisms should be unaffected by either the type of ligand used to initiate the response or the cell surface acceptor sites recognized by a particular ligand. For this reason, the anti-Ig antibody has been generally accepted as a polyclonal analogue that can substitute for antigen. It has now been unequivocally established that ligand-mediated cross-linking of surface Ig (sIg) receptors on B lymphocytes per se evokes biochemical alterations such as hydrolysis of phosphatidylinositol 4, 5 bisphosphate (Cambier et al., 1987; Klaus et al., 1987), which is believed to be a part of the triggering pathway. Under the co-stimulatory influence of the T cell–derived B cell growth factor IL-4, anti-Ig antibody–stimulated B cells undergo DNA synthesis and proliferation (Noma et al., 1986).

Cross-linking of cell surface membrane structures other than antigen receptors on B lymphocytes may also lead to B cell activation, the classical example being the polyclonal B cell activation induced by LPS (Möller, 1975). For a long time, most investigators in the field considered the LPS-induced antibody response to be independent of T cells and macrophages, although responses to other T cell–independent (TI) polyclonal activators, such as polymerized flagellin, were known to be macrophage-dependent (Lee et al., 1976). Subsequent work by other investigators suggested the possibility that LPS induces macrophages to secrete T cell–activating factors, which, in turn, stimulate B cells (Hoffman et al., 1979). This possibility was confirmed in work that showed that the IgM response induced by LPS was indeed dependent on T cells (Jaworski et al., 1982) (Table 16–1). However, only the maturation of proliferating B lymphocytes was T cell–dependent, and B cell growth was readily induced by LPS in the absence of T cell–derived factors. Thus, T cell–depleted spleen cells stimulated in vitro with LPS gave a proliferative response in the absence of the B cell growth factor, IL-4 (formerly called BCGF). In marked contrast, however, the proliferative response of anti-Ig–stimualted spleen cells was entirely dependent on IL-4 (Inazawa et al., 1985) (Fig. 16–1).

The above experiments on the role of T cells in B cell maturation were carried out with B cells that had been affinity enriched by rosetting with chicken red blood cells (CRBC) and then stimulated with LPS to proliferate and to generate CRBC-specific antibodies. Surprisingly, when similar experiments were carried out with trinitrophenyl (TNP)-affinity–enriched B cells or with B cells that had been affinity-enriched for fluorescein (FLU) (M. Inazawa, unpublished data), the maturation of LPS-stimulated proliferating B cells was shown to be entirely independent of T lymphocytes or of secreted products from the EL-4 thymoma line (Inazawa et al., 1985) (Fig. 16–2). This difference in T cell requirement for CRBC- and TNP-specific B lymphocytes to mature into AFC held true even when mixtures of CRBC- and TNP-specific B cells were cultured with LPS, thus excluding the possibility that one of the B cell populations was contaminated with T cells. These observations may explain why earlier reports on mechanisms controlling B cell responses have failed to show that B cell proliferation and maturation are controlled by different lymphokines. Because both CRBC and hapten affinity-purified B lymphocytes used in our work were small cells with a sedimentation velocity of <5 mm/hour, it is possible that TNP- and FLU-specific B cells, and perhaps the majority of B lymphocytes, which by physical criteria are considered to be in a resting state, may nevertheless be antigen-experienced and hence do not require maturation-promoting signals. Given this uncertainty, the use of affinity-purified TNP- or FLU-specific B cell populations in poly-

Table 16–1. Anti–Thy-1.2 Treatment Abrogates T Cell Participation in LPS-Mediated IgM Response[a]

Contents of Culture			
Rosette-Forming B Cells	Non–Rosette-Forming B Cells	LPS	AFC/Culture ($\bar{x} \pm SE$)
C'	C'	+	1,244 ± 84
α Thy-1.2 + C'	C'	+	1,080 ± 103
C'	α Thy-1.2 + C'	+	160 ± 68
α Thy-1.2 + C'	α Thy-1.2 + C'	+	75 ± 32

[a]Unprimed spleen cells were treated with complement (C') or anti–Thy-1.2 antibody and C' and subsequently subjected to antigen affinity purification with chicken red blood cells (CRBC) to obtain an enriched population of CRBC-reactive, CRBC-rosette-forming B cells (RFC). The non–rosette-forming cells (NRFC) as a source of T cells and macrophages served to reconstitute the LPS response if the T cells were left intact. CRBC-specific antibody-forming cells (AFC) as a response to in vitro stimulation by LPS were measured in the hemolytic plaque assay 4 days after onset of culture.
Jaworski et al., 1982. *J. Exp. Med.* 155:248.

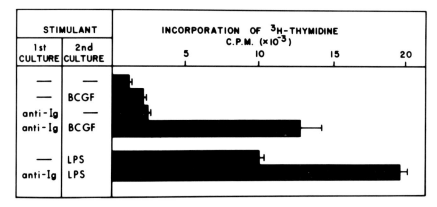

A

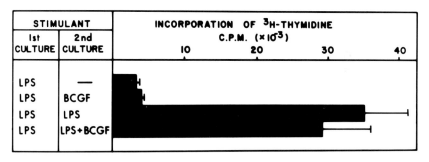

B

Figure 16–1.   (**A**) Effect of IL-4 on the rate of ^{3}H-thymidine incorporation by B lymphocytes following activation with anti-Ig antibody. T cell–deficient spleen cells were precultured for 48 hours with or without anti-Ig antibody, washed, recultured for 24 hours with or without IL-4 or LPS, and assayed for thymidine incorporation. All cultures were in serum-free medium. (**B**) IL-4 has no effect on the rate of ^{3}H-thymidine incorporation by B lymphocytes following activation with LPS. T cell–deficient spleen cells were precultured for 48 hours with LPS, washed, recultured for 72 hours with IL-4 or LPS or both, and assayed for thymidine incorporation. All cultures were in serum-free medium. Bars represent the mean $\pm$ SE. (Inazawa et al., 1985. *Lymphokine Res.* 4:343.)

clonal antibody responses as an experimental readout appears unsuitable for the definition of function-specific lymphokine activities.

Role of Macrophages in B Cell Response

Macrophages have also been extensively studied, chiefly for their role as antigen presenting cells (APCs) for T lymphocytes, particularly the L3T4 (CD4) helper T cell subset (Sprent et al., 1986). Cellular and molecular analysis has firmly established that the T cell receptor complex recognizes antigen only after it has been processed by macrophages and then re-expressed as peptidic fragments in physical association with class II (Ia) molecules

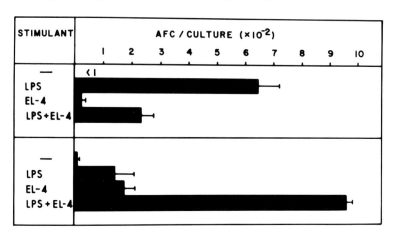

Figure 16–2.   Requirement for maturation factors in LPS-induced antibody-forming cell (AFC) response by antigen affinity-enriched B cells depends on the antigen with which B cells were selected: 2×10^3 affinity-enriched B cells were cultured for 4 days in serum-free medium in the presence of stimulants as indicated. *Upper panel:* B cells enriched for TNP. *Lower panel:* B cells enriched for CRBC. Bars represent the mean ± SE. (Inazawa et al., 1985. *Lymphokine Res.* 4:343.)

of the major histocompability (MHC) complex (Babbitt et al., 1985; Buus et al., 1986, 1987; Unanue, 1984). An additional function had to be assigned to macrophages, however, after it was shown that they also appeared to play some role in IgM responses to TI antigens (Lee et al., 1976). This suggested the existence of macrophage-mediated control mechanisms that were auxiliary to cognate macrophage–T cell interactions. Strong additional evidence in support of this conclusion came from subsequent experiments in which we tried to reconstitute responses in macrophage-depleted spleen cell cultures with macrophages from either the Ia⁻ P388D1 or the Ia⁺ P388AD cell line; we reasoned that presentation of a T cell–dependent antigen in immunogenic form to T lymphocytes could only be mediated by Ia⁺ macrophages. Thus, if only Ia-associated antigen presentation by macrophages was involved in the IgM response to these antigens, Ia⁻ macrophages would not be expected to restore immunocompetence. However, reconstitution of immunocompetence was observed with macrophages from the Ia⁻ cell line. These data were further strengthened by experiments in which the response of macrophage-depleted spleen cells to TI antigens was restored by splenic macrophages whose ability to process and present antigen to T cells had been blocked by treatment with the lysosomotropic agent chloroquine (Sinha et al., 1987).

Monokine Required in B Cell Responses to TI-1 and TI-2 Antigens

Using antigen affinity-enriched B lymphocyte populations cultured in vitro under conditions in which T cells and macrophages were limiting, we were able to substitute macrophage-culture supernatants for macrophages. The reconstituting agent in these supernatants was concluded to represent either some trophic factor or, more interestingly, a monokine with specific effects on activated B lymphocytes. We felt that IL-1 was among the monokines likely to reconstitute immunocompetence in macrophage-depleted spleen cell cultures. This prediction was based on experiments in which the Ia^+ P388AD.2 and the Ia^- P388D1 cell lines, both of which secrete IL-1, were shown to restore responses to the TI-1 antigen TNP-Brucella (TNP-Ba) and the TI-2 antigen TNP-Ficoll (TNP-F). In contrast, the Ia^+ P388NA.10 cell line, which does not produce IL-1, failed to reconstitute these responses. Indeed, adherent cells could be entirely replaced by recombinant IL-1 in IgM responses to the TI-1 antigens TNP-Ba and TNP-LPS and to the TI-2 antigens TNP-F and TNP-Dextran. These experiments strongly support the conclusion that IL-1 is the only macrophage-derived cytokine required for the generation of optimal levels of IgM antibodies to TI antigens (Sinha et al., 1987) (Fig. 16–3).

Role of IL-2 and IL-1 in the Maturation of LPS-Stimulated B Lymphocytes

Most known cytokines whose biological activity was originally defined for a single cell type were later found to have effects on a number of other cells. For example, the monokine IL-1, which was originally shown to contribute to T cell responses, was also found

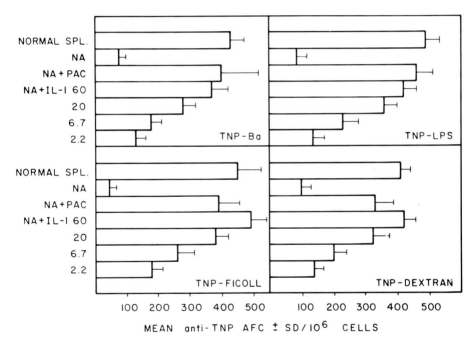

Figure 16–3. Replacement of adherent (A) cells by recombinant IL-1. CBA/CaJ normal or A cell–deficient spleen cells were cultured with TNP-Ba, NP-LPS, TNP-Ficoll, or TNP-Dextran. A cell–deficient spleen cells were reconstituted with 10^5 CBA/CaJ A cells (adherent peritoneal cells) or the indicated concentrations (U/ml) of recombinant murine IL-1. Cultures were assayed for TNP-specific antibody-forming cells (AFC) after 4 days. A cell–deficient NA spleen cells cultured with IL-1 (60, 20, 6.7, or 2.2 U/ml) in the absence of antigen yielded <10 AFC/culture. (Sinha et al., 1987. *J. Immunol. 138*:4143.)

to cause B cell proliferation, when used in conjunction with anti-Ig antibody and IL-4 (Chiplunkar et al., 1986; Howard et al., 1983). IL-1 was also shown to act as a cofactor in B cell proliferation and maturation induced by anti-Ig antibody and Dextran sulfate (Chiplunkar et al., 1986), by antigen alone (Pike and Nossal, 1985), or by antigen and IL-2 (Hoffmann et al., 1987; Leibson et al., 1982). The demonstration on B lymphocytes of IL-2 receptors with similar affinity and surface kinetics as those expressed on T cells (Zubler et al., 1987) further underscores the physiological significance of this lymphokine for B cells. Although earlier work suggested that IL-2 selectively affects B cell maturation, it did not formally exclude the possibility that it does so indirectly, via T cells (Parker, 1982). With the development of tissue culture conditions that allowed us to study antigen affinity-enriched B lymphocytes at low densities of a few thousand cells per culture, we were able to assess the roles of IL-2 and IL-1 on B cell proliferation and maturation. When CRBC affinity-enriched B lymphocytes (of which >99% were small sized, IgM$^+$ cells containing less than 0.6% $\pm$ 0.2% Thy-1.2$^+$ cells and 0.9% $\pm$ 0.5% esterase positive monocyte/macrophages) were cultured at 6×10^3 cells per culture in the presence of LPS, they underwent proliferation only (Tables 16–2 and 16–3). In the presence of LPS and supernatant of the IL-2–containing EL-4 thymoma line, however, these cells were triggered not only to proliferate but also to mature into CRBC-specific IgM antibody-forming cells (AFC) (Table 16–3). Importantly, the ability of the EL-4 supernatant to cause LPS-stimulated, proliferating B cells to mature into AFC could also be achieved with recombinant IL-2. B cells exposed to IL-2 alone underwent neither proliferation nor maturation, analogous to B cells exposed to EL-4 supernatant alone. In light of our previous results concerning replacement of macrophages by IL-1 in B cell responses to TI anigens, we looked for a combined effect of IL-2 and IL-1 on LPS-stimulated B cells. In the presence of the two cytokines, LPS-stimulated B cells were induced to mature into AFC at a much higher rate than in the presence of IL-2 alone. This marked synergism between IL-2 and IL-1 critically depends on the initial, LPS-induced B cell proliferation (Fotedar and Diener, 1988). The proliferative response by LPS-stimulated B cells was only marginally affected by these cytokines (Table 16–2).

Table 16–2.   Effect of Recombinant IL-2 and IL-1 on
Proliferation of Antigen Affinity-Enriched B Cells

Stimuli[a]	^3H-Thymidine Incorporation[b] (cpm/culture) $\pm$ SD	
	Experiment 1	Experiment 2
None	229 $\pm$ 36	281 $\pm$ 80
LPS	13,959 $\pm$ 765	15,329 $\pm$ 1,878
IL-1	409 $\pm$ 192	293 $\pm$ 36
IL-2	378 $\pm$ 82	255 $\pm$ 53
IL-1 + IL-2	369 $\pm$ 122	306 $\pm$ 66
LPS + IL-1	23,337 $\pm$ 1,291	17,764 $\pm$ 1,289
LPS + IL-2	28,052 $\pm$ 1,051	22,618 $\pm$ 767
LPS + IL-2 + IL-1	29,054 $\pm$ 1,416	23,724 $\pm$ 2,796

[a]Chicken red blood cell antigen-enriched B cells were cultured in the presence of LPS, 500 units/ml of IL-2, and 400 units/ml of IL-1.
[b]Cultures were pulsed on day 4 with ^{3}H-thymidine.
Fotedar et al., 1988. *Lymphokine Res.* 7:393.

Table 16–3. Effect of Recombinant IL-2 and IL-1 on
Maturation of Proliferating B Cells

	AFC/Culture[b] ± SE	
Stimuli[a]	*Experiment 1*	*Experiment 2*
None	<1	<1
LPS	34 ± 18	20 ± 14
EL-4 Sup	<1	ND[c]
LPS + EL-4 Sup	572 ± 40	ND
IL-1	<1	<1
IL-2 (500)	<1	<1
IL-2 (50)	<1	<1
L-2 + IL-1	<1	<1
LPS + IL-1	52 ± 23	40 ± 14
LPS + IL-2 (500)	312 ± 81	292 ± 95
LPS + IL-2 (50)	280 ± 59	315 ± 90
LPS + IL-2 (500) + IL-1	1,058 ± 207	1,744 ± 349
LPS + IL-2 (50) + IL-1	436 ± 48	1,040 ± 259

[a]Chicken red blood cell antigen-enriched B cells were cultured in the presence of LPS,
50 and 500 units/ml of IL-2, and 400 units/ml of IL-1.
[b]Cultures were assayed on 4.5 days for the number of antibody-forming cells (AFC).
[c]Not done.
Fotedar et al., 1988. *Lymphokine Res. 7*:393.

INACTIVATION OF B LYMPHOCYTES

Theoretical Considerations

Historically, research on B cell activation has paralleled attempts to understand antigen-specific inactivation, also called B cell tolerance or unresponsiveness. Fundamental to the interpretation of experimental data on B cell tolerance has been the idea that under certain critical conditions, engagement of B cell surface immunoglobulin (sIg) receptors by appropriate ligands, such as antigen or anti-Ig antibodies, may cause the cell to become refractory to subsequent immunogenic challenge. As postulated by different competing theories, the conditions deemed critical to the induction of unresponsiveness were thought to reflect quantitative parameters of the ligand or, alternatively, the physiological state of the B cells. In either case, B cell unresponsiveness was seen to arise from the interaction of B cell receptors with their ligands, without the need for cellular or humoral mediators. Thus the various versions of clonal deletion and abortion hypotheses (Bretscher and Cohn, 1970; Lederberg, 1959) distanced themselves from theories that postulated the involvement of suppressor T cells as the chief mechanism in control of tolerance to "self." Among the most challenging studies in recent years on B cell tolerance, including its putative role in tolerance to self antigens, have been those from Nossal's laboratory (Nossal and Pike, 1975, 1980). In essence, these studies aimed at testing the conditions under which immature, surface Ig$^+$ B lymphocytes were rendered unresponsive after engagement of their antigen receptors with appropriately selected ligands, chiefly hapten conjugates of mammalian IgG and anti-μ chain antibody. The basic observations by Nossal and his colleagues and by Metcalf and Klinman (1976) may be interpreted as confirming Lederberg's (1959) clonal abortion hypothesis. According to this theory, a potentially self-reactive, immature but Ig^+ B lymphocyte may receive an abortive signal on recognition of self antigen. A problem that is not satisfactorily addressed by this theory concerns the possibility of accidental abortion of B cell clones whose recognition specificity is directed at antigens other than self antigens. Because B cells mature in the bone marrow throughout life, and not just during fetal ontogeny of the immune system, they must at all times be considered accessible to pathogen-derived antigens. Clonal abortion of potentially immunoprotective B cell clones under these circumstances could be fatal. One could also raise the argument that early abortion of low-affinity, potentially self-reactive, B cells whose antigen receptors happened to cross-react with pathogen-derived antigens, could equally be life-threatening.

Experimental Facts That Disagree with Clonal Abortion and Deletion Models

In disagreement with the clonal abortion/deletion hypothesis was the observation that selected antigens, which were administered transplacentally to the murine fetus and shown to distribute throughout the fetal body, including hematopoietic tissue, failed to cause clonal abortion of B cells in the offspring (reviewed by Diener and Waters, 1986). Significantly, the only exception occurred with human IgG, which has traditionally been the tolerogen of choice in the work of many investigators. The fallacy of generalizing interpretations of data obtained with mammalian IgG as tolerogen is perhaps best illustrated by experiments designed to compare the mechanisms by which trinitrophenylated human gammaglobulin (HGG) and bovine serum albumin (BSA) induce tolerance in immature B lymphocytes of the neonatal mouse. Thus clonal analysis of responsiveness to the TI-1 antigen TNP-LPS of B lymphocytes from systemically tolerant animals that had been neonatally treated with either TNP-HGG or TNP-BSA clearly showed that only TNP-HGG had caused the functional deletion of TNP-specific clones (Waters and Diener, 1983). In sharp contrast, B cells from TNP-BSA–treated mice were capable of generating the full affinity range of TNP-specific IgM antibody in response to TNP-LPS (Fig. 16–4). At the systemic level, unresponsiveness in TNP-BSA–treated mice was shown to be maintained by BSA-specific suppressor T cells (Fig. 16–5). The only conclusion to be drawn from this observation is that some difference in the molecular structures of IgG and BSA determines whether tolerance at the systemic level is controlled by clonal abortion or active suppression of helper T cells. Quite possibly, the potential of IgG to cause clonal abortion of B cells resides in the Fc portion of the molecule (Waldschmidt et al., 1983).

The notion of the relationship between an antigen's molecular structure and its potential to cause clonal abortion of B cells received further support from yet another set of experiments. In these, trinitrophenylated methylcellulose (TNP-MC) was shown to be a potent hapten-specific tolerogen for adult murine spleen cells in vivo and in vitro. Interestingly, the hapten conjugate of oxidized MC (TNP-MCO) was not tolerogenic, although the two antigenic forms had the same molecular weight. Furthermore, TNP-MCO had the same or even a slightly higher binding avidity for TNP-specific B cells than TNP-MC, which excludes the possibility that lack of tolerogenicity results from loss of binding avidity. The observation that the difference in tolerogenic potential of the two molecular forms of the antigen is seen both on mature and on immature B cells of the neonate (von Borstel et al., 1983) is incompatible with the clonal abortion hypothesis. Evidently, if the clonal abortion hypothesis has any validity in explaining tolerance to "self," it must

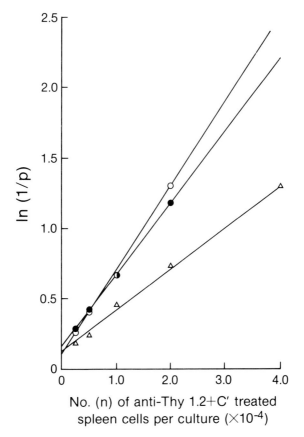

Figure 16–4. TNP-specific precursor frequencies in spleen cells from normal 12–14-week (solid circles) and from age- and sex-matched TNP $_{10}$HGG-treated (open triangles) or TNP$_{10}$BSA-treated (open circles) CBA/CaJ mice depleted of T cells and co-cultured in limiting numbers with 3.5×10^6 syngeneic thymocyte fillers. TNP-LPS was added at 1 µg/ml. The abscissa represents the number of cells/culture well ($\times 10^{-4}$); the ordinate represents the natural logarithm of the inverse fraction (p) of wells negative for AFC. Very low frequencies [$1n(1/p) < 0.01$] were obtained from wells that contained thymocytes alone. (Waters et al., 1983. *Eur. J. Immunol.* *13*:928.)

exclude an antigen's molecular properties other than the binding avidity for immature B lymphocytes as having any effect on the tolerogenic potency.

Finally, selected tolerogens, such as polymerized flagellin (POL), were shown to render B cells refractory to immunogenic challenge without causing them to become physically deleted. Thus, the first evidence that tolerance may not be due to the physical, clonal deletion of B cells derived from studies in which B cells rendered unresponsive through exposure to high doses of POL in vitro were shown to have the antigen still bound to their surface receptors. Indeed these cells were unable to "clean" the antigen off their surface membrane. In contrast, B cells that had been exposed to low, immunogenic concentrations of antigen cleared it from their surface membrane through receptor capping, endocytosis, or shedding. Interestingly, exposure of tolerant cells to colchicine not only led to modulation of the antigen, but also the generation of new antigen receptors. In spite

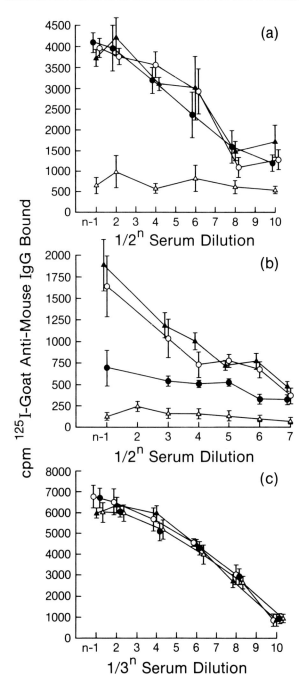

Figure 16–5. Antigen-specific suppression in spleens of $TNP_{10}BSA$-treated mice (panels **b** and **c**) and its absence in TNP_{10} HGG-treated mice (panel **a**). Treated mice were from groups receiving 1–2 mg of antigen twice weekly since birth. Each point on the ordinate represents mean anti-TNP–specific serum antibody titers $\pm$ SE of a group of five mice as determined by radioimmunoassay. (**a**) $TNP_{10}HGG$ challenge: (solid triangles) 5×10^7 normal spleen; (open circles) 15×10^7 normal spleen; (open triangles) 10×10^7 $TNP_{10}HGG$-tolerant spleen; (solid

of this, these cells remained refractory to immunogenic doses of the same antigen. Surprisingly, however, their unresponsive state could be reversed by the addition of adherent, macrophage-like cells (Diener et al., 1976). In subsequent studies, Nossal and Pike (1980) described similar observations in vivo and called this type of B cell unresponsiveness "clonal anergy."

DISCUSSION

On reviewing the large number of articles on B lymphocyte activation that have been published within recent years, one is impressed, and bewildered at the same time, by the pleiotropic effects of the steadily increasing collection of cytokines on many cell types. Because most immunologists are reductionists at heart, they have contributed to the current phenomenology over B cell reactivity to a great variety of extrinsic and intrinsic stimuli, without much concern for any potential artifacts derived from the investigative practice of arbitrarily applying boundary conditions to a complex system. Thus in the light of dynamic theories of complex systems, of which the immune system undoubtedly represents a prime example, it is impossible at the present time, for epistemological reasons, to arrive at an accurate flow diagram of mechanisms of B cell activation or a diagram defining the functional specificity and temporal sequence of action of cooperating cell-external stimuli. In this respect, we are faced with problems akin to those that plague interpretations of the possible immunoregulatory significance of idiotype networks. The imposition of boundaries on a biological phenomenon is greatly aided by introducing a multiplicity of investigative techniques or even different disciplines to its analysis (Dyke, 1988). This approach, when extended to the B cell response by concurrently analyzing both intercellular parameters and the much more accurately definable intracellular molecular dynamics, is likely to yield a physiologically more meaningful functional definition of the stimuli that are external and internal to the immune system than has been the case to date.

In our opinion the quasi-philosophical comments made above also apply to our understanding of self tolerance. In light of the experimental evidence we can no longer consider clonal abortion of B lymphocytes as the key mechanism controlling auto-antibody responses. This conclusion derives further support from numerous observations confirming the existence of potentially auto-reactive B cells as part of a normally functioning immune system (Yung et al., 1973). Under appropriate conditions, such cells may even become activated (Unlig et al., 1985). Of particular interest in this respect is the work by Borel and his collaborators who provided strong evidence that the lack of antibodies directed at the complement component C5 in normal mice is due to the presence of regulatory T cells (Harris et al., 1982).

---

circles) 5×10^7 normal spleen + 10×10^7 $TNP_{10}HGG$-tolerant spleen. (b) $TNP_{10}BSA$ challenge: (solid triangles) 5×10^7 normal spleen; (open circles) 10×10^7 normal spleen; (open triangles) 5×10^7 $TNP_{10}BSA$-tolerant spleen; (solid circles) 5×10^7 normal spleen + 5×10^7 $TNP_{10}BSA$-tolerant spleen. (c) $NIP_{20}HGG$ challenge: (solid triangles) 5×10^7 normal spleen; (open circles) 15×10^7 normal spleen; (open triangles) 10×10^7 TNP_{10} BSA-tolerant spleen; (solid circles) 5×10^7 normal spleen + 10×10^7 $TNP_{10}BSA$-tolerant spleen. (Waters et al., 1983. *Eur. J. Immunol.* *13*:928.)

Other strong evidence against clonal abortion and deletion as a mechanism of self tolerance in B cells stems from experiments on mice that were rendered transgenic for neo-self antigen egg-lysozyme and for a high-affinity lysozyme-specific antibody. Analysis of these mice demonstrated the presence of B lymphocytes that expressed the transgene-encoded Ig receptor. Interestingly, these cells showed a marked reduction from normal control levels of surface IgM but not IgD and remained refractory to conventional antigenic challenge (Goodnow et al., 1988). Whether the anergic state (Nossal and Pike, 1980) of these B cells can be reversed by mitogens or through viral transformation (Unlig et al., 1985) remains to be answered. One is reminded of a suggestion first made by Allison (Allison et al., 1971), that clonal deletion may only occur at the level of potentially self-reactive T cells rather than in B cells.

References

Allison, A. C., Denman, A. M., and Barnes, R. D. (1971). Cooperating and controlling functions of thymus-derived lymphocytes in relation to autoimmunity. *Lancet 2*:135–140.

Andersson, J., and Melchers, F. (1974). Maturation of mitogen-activated bone marrow–derived lymphocytes in the absence of proliferation. *Eur. J. Immunol. 4*:533–539.

Babbitt, D. P., Allen, P. M., Matsueda, G., Haber, E., and Unanue, E. R. (1985). Binding of immunogenic peptides to Ia histocompatibility molecules. *Nature 317*:359–361.

Bretscher, P. A., and Cohn, M. (1970). A theory of self-nonself discrimination. *Science 169*:1042–1049.

Buus, S., Colon, S., Smith, C., Freed, J., Miles, C., and Grey, H. (1986). The interaction between a "processed" ovalbumin peptide and Ia. *Proc. Natl. Acad. Sci. USA 83*:3968–3971.

Buus, S., Sette, A., Colon, A., Miles, C., and Grey, H. (1987). The relation between major histocompatibility complex (MHC) restriction and the capacity of Ia to bind immunogenic peptides. *Science 235*:1353–1358.

Cambier, J. C., Justement, L. B., Newell, M. K., Chen, Z. Z., Harris, L. K., Sandoval, V. M., Klemsz, M. J., and Ransom, J. T. (1987). Transmembrane signals and intracellular "second messengers" in the regulation of quiescent B lymphocyte activation. *Immunol. Rev. 95*:37–57.

Chiplunkar, S., Langhorne, J., and Kaufmann, S.H.E. (1986). Stimulation of B cell growth and differentiation by murine recombinant interleukin 1. *J. Immunol. 137*:3748–3752.

Diener, E., and Feldmann, M. (1972). Relationship between antigen and antibody-induced suppression of immunity. *Transplant. Rev. 8*:76–103.

Diener, E., Kraft, N., and Armstrong, W. D. (1973). Antigen recognition: I. Immunological significance of antigen binding cells. *Cell. Immunol. 6*:80–86.

Diener, E., Kraft, N., Lee, K.-C., and Shiozawa, C. (1976). Antigen recognition: IV. Discrimination by antigen-binding immunocompetent B cells between immunity and tolerance is determined by adherent cells. *J. Exp. Med. 143*:805–821.

Diener, E., and Waters, C. A. (1986). Immunological quiescence towards self: Rethinking the paradigm of clonal abortion. In *Paradoxes in Immunology.* Boca Raton, Fla.: CRC Press, pp. 27–40.

Dyke, C. (1988). *The Evolutionary Dynamics of Complex Systems.* New York: Oxford University Press.

Fotedar, R., and Diener, E. (1988). The role of recombinant IL-2 and IL-1 in murine B cell differentiation. *Lymphokine Res. 7*:393–402.

Goodnow, C. C., Crosbie, J., Adelstein, S., Lavoie, T. B., Smith-Gill, S. J., Brink, R. A., Pritchard-Briscoe, H., Wotherspoon, J. S., Loblay, R. H., Raphael, K., Trent, R. J., and Basten, A. (1988). Altered immunoglobulin expression and functional silencing of self-reactive B lymphocytes in transgenic mice. *Nature 334*:676–682.

Harris, D. H., Cairnes, L., Rosen, F. S., and Borel, Y. (1982). A natural model of immunologic tolerance. Tolerance to murine C5 is mediated by T cells, and antigen is required to maintain unresponsiveness. *J. Exp. Med.* 156:567–584.

Hoffmann, M. K., Gilbert, K. M., and Hirst, J. A. (1987). An essential role for interleukin 1 and a dual function for interleukin 2 in the immune response of murine B lymphocytes to sheep erythrocytes. *J. Mol. Cell Immunol.* 3:29–36.

Hoffmann, M. K., Koenig, S., Mittler, R. S., Oettgen, H. F., Ralph, P., Galanos, C., and Hammerling, U. (1979). Macrophage factor controlling differentiation of B cells. *J. Immunol.* 122:497–502.

Howard, M., Mizel, S. B., Lachman, L., Ansel, J., Johnson, B., and Paul, W. E. (1983). The effect of interleukin-1 on anti-immunoglobulin–induced B cell proliferation. *J. Exp. Med.* 157:1529–1543.

Inazawa, M., Shiozawa, C., Fotedar, R., and Diener, E. (1985). How relevant are growth and maturation factors to the B lymphocyte response induced by LPS? *Lymphokine Res.* 4:343–350.

Jaworski, M. A., Shiozawa, C., and Diener, E. (1982). Triggering of affinity-enriched B cells. Analysis of B cell stimulation by antigen-specific helper factor or lipopolysaccharide: I. Dissection into proliferative and differentiative signals. *J. Exp. Med.* 155:248–263.

Klaus, G.G.B., Bijsterbosch, M. K., O'Garra, A., Harnett, M. M., and Rigley, K. P. (1987). Receptor signaling and crosstalk in B lymphocytes. *Immunol. Rev.* 99:19–38.

Leanderson, T., Andersson, J., and Rajasekar, R. (1987). Clonal selection in B cell growth and differentiation. *Immunol. Rev.* 99:53–69.

Leanderson, T., and Forni, L. (1984). Effects of μ-specific antibodies on B cell growth and maturation. *Eur. J. Immunol.* 14:1016–1021.

Lederberg, J. (1959). Genes and antibodies. *Science* 129:1649–1653.

Lee, K.-C., Shiozawa, C., Shaw, A., and Diener, E. (1976). Requirement for accessory cells in the antibody response to T cell–indepedent antigens in vitro. *Eur. J. Immunol.* 6:63–68.

Leibson, H. J., Marrack, P., and Kappler, J. (1982). B cell helper factors: II. Synergy among three helper factors in the response of T cell– and macrophage-depleted B cells. *J. Immunol.* 129:1398–1402.

Melchers, F., and Lernhardt, W. (1985). Three restriction points in the cell cycle of activated murine B lymphocytes. *Proc. Natl. Acad. Sci. USA* 82:7681–7685.

Metcalf, E. S., and Klinman, N. R. (1976). In vitro tolerance induction of neonatal murine B cells. *J. Exp. Med.* 143:1327–1340.

Möller, G. (1975). One nonspecific signal triggers B lymphocytes. *Transplant. Rev.* 23:126–137.

Noma, Y., Sideras, P., Naito, T., Bergstedt-Lindquist, S., Azuma, C., Severinson, E., Tanabe, T., Kinashi, T., Matsuda, F., Yaoita, Y., and Honjo, T. (1986). Cloning of cDNA encoding the murine IgG$_1$ induction factor by a novel strategy using SPG promoter. *Nature* 319:640–646.

Nossal, G.J.V., and Pike, B. L. (1975). Evidence for the clonal abortion theory of B lymphocyte tolerance. *J. Exp. Med.* 141:904–917.

Nossal, G.J.V., and Pike, B. L. (1980). Clonal anergy: Persistence in tolerant mice of antigen-binding B lymphocytes incapable of responding to antigen or mitogen. *Proc. Natl. Acad. Sci. USA* 77:1602–1606.

Parker, D. C. (1982). Separable helper factors support B cell proliferation and maturation to Ig secretion. *J. Immunol.* 129:469–474.

Pike, B. L., and Nossal, G.J.V. (1985). Interleukin 1 can act as a B cell growth and differentiation factor. *Proc. Natl. Acad. Sci. USA* 82:8153–8157.

Russell, P. J., and Diener, E. (1970). The early antibody-forming response to *Salmonella* antigens; a study of morphology and kinetics in vivo and in vitro. *Immunology* 19:651–667.

Schimpl, A., and Wecker, E. (1972). Replacement of T cell function by a T cell product. *Nature* 237:15–17.

Schimpl, A., and Wecker, E. (1975). A third signal in B cell activation given by TRF. *Transplant. Rev.* 23:176–188.

Sinha, A. A., Guidos, C., Lee, K.-C., and Diener, E. (1987). Functions of accessory cells in B cell responses to thymus-independent antigens. *J. Immunol. 138*:4143–4149.

Sprent, J., Schaefer, M., Lo, D., and Korngold, R. (1986). Functions of purified L3T4[+] and Lyt-2[+] cells in vitro and in vivo. *Immunol. Rev. 91*:195–218.

Unanue, E. R. (1984). Antigen-presenting function of the macrophage. *Annu. Rev. Immunol. 2*:395–428.

Unlig, H., Rutter, G., and Derrick, R. (1985). Self-reactive B lymphocytes detected in young adults, children and newborns after in vitro infection with Epstein-Barr virus. *Clin. Exp. Immunol. 62*:75–84.

von Borstel, R. C., Diner, U. E., Waters, C. A., and Diener, E. (1983). Tolerance induction during ontogeny: III. Carrier recognition by the immature and adult immune system determines tolerogenicity of hapten-carrier conjugates. *Cell. Immunol. 81*:229–242.

Waldschmidt, T. J., Borel, Y., and Vitetta, E. S. (1983). The use of haptenated immunoglobulins to induce B cell tolerance in vitro. The roles of hapten density and the Fc portion of the immunoglobulin carrier. *J. Immunol. 131*:2204–2209.

Waters, C. A., and Diener, E. (1983). Tolerance induction during ontogeny: II. Distinct unresponsive states in immature mice question the generality of clonal abortion. *Eur. J. Immunol. 13*:928–935.

Yung, L. L., Diener, E., McPherson, T. A., Barton, M. A., and Hyde, H. A. (1973). Antigen-binding lymphocytes in normal man and guinea pig to human encephalitogenic protein. *J. Immunol. 110*:1383–1387.

Zubler, R. H., and Kanagawa, O. (1982). Requirement for three signals in B cell responses: II. Analysis of antigen- and Ia-restricted T helper cell–B cell interaction. *J. Exp. Med. 156*:415–429.

Zubler, R. H., Werner-Favree, C., Wen, L., Sekita, K. I., and Straub, C. (1987). Theoretical and practical aspects of B cell activation: Murine and human systems. *Immunol. Rev. 99*:281–299.

V

DEVELOPMENT OF THE IMMUNE SYSTEM IN HEALTH AND DISEASE

17

Immunological Manipulation and Immunopharmacology of the Thymosins

KAREN K. OATES,

ALLAN L. GOLDSTEIN

The thymus is an endocrine organ that synthesizes many different proteins having unique chemical structures and hormonal activities. The thymus gland and its hormones are responsible for the development and function of the cellular (T-cell–dependent) compartment of the immune system (Fig. 17–1). As early as 1935 Gregoire demonstrated that regeneration of the cortical regions in irradiated thymus occurred only if circulating lymphocytes were allowed to reach the thymus. If cells were prevented from reaching this area, the thymic cortex remained devoid of lymphocytes. In the early 1960s it was shown that neonatally thymectomized animals exhibited severe defects of primary and secondary immunity with the development of a wasting syndrome characterized by retarded growth, poor eating, lethargy, increased susceptibility to infection, and high mortality (Miller and Good, 1961). These animals primarily exhibited defects in T cell numbers and functions as manifested by loss of ability to reject allografts and decreased delayed hypersensitivity responses. When neonatally thymectomized animals were given implants of thymus, the animals were protected from the wasting syndrome (Fig. 17–2). The severity of wasting is much less in adult-thymectomized rodents, presumably reflecting, at least in part, the long (6–9 month) life span of established T cell lymphocyte populations.

In 1963, Osoba and Miller demonstrated that implantation of cell-impermeable diffusion chambers containing thymus tissue into neonatally thymectomized animals could prevent wasting, reverse lymph node atrophy, and establish immunologic competence. The chamber prevented direct contact between the implanted thymus and circulating lymphocytes of the host. These experiments strongly suggested that the reconstitutive effects of the thymus are the result of humoral rather than cellular products. Beginning in the late 1960s and continuing up to the present, many efforts have been made to purify thymic extracts and to identify the "humoral" compounds produced in the thymus that are responsible for restoring immunocompetence. In 1966, Goldstein et al. described the isolation of a partially purified thymic preparation called thymosin, which re-established (to various degrees) immunocompetence in immunosuppressed animals. This prepara-

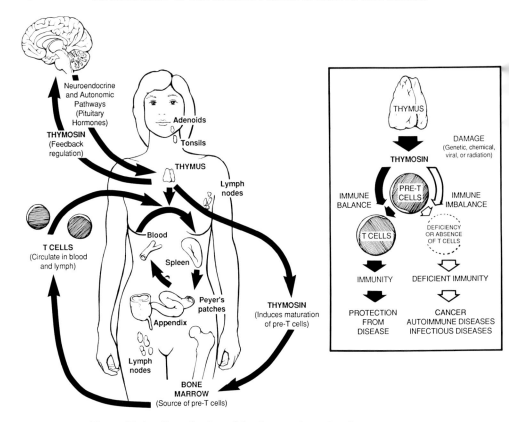

Figure 17–1. Organization of the thymus-dependent immune system.

tion enhanced lymphocytopoiesis in vivo and prevented development of the post-thy-mectomy wasting syndrome. Subsequent studies have led to the isolation and character-ization of additional factors, which have been shown to induce expression of T cell differentiation markers and to increase lymphokine production (Table 17–1).

The general protocol for isolating thymosin fraction 5 (TF5) (which contains a group of biologically active thymic peptides) from calf thymus was initially described by Hooper et al. (1975). In brief, thymus tissue is homogenized then centrifuged at 14,000 g, and the supernatant is heated to 80°C to produce a large amount of white flocculent precipitate. The supernatant that remains after the heating step is precipitated with acetone and ammonium sulfate (25%–50%), and the ammonium sulfate precipitate is ultrafiltered through an Amicon DC-2 hollow fiber system and then desalted on a Sephadex G-25 column. The resulting mixture of polypeptides is designated thymosin fraction 5. Precip-itation with ammonium sulfate (50%–90%) has been used in an alternative procedure for the preparation of a fraction designated thymosin fraction 5A, which is used for purifi-cation of large quantities of the peptide, thymosin β_4. There is substantial evidence that TF5 has effects on all the major compartments of the lymphoid system, including the bone marrow, thymus, and peripheral lymphoid tissue (Low et al., 1979). TF5 contains at least three classes of molecules: (1) thymic hormones produced predominantly by thy-mic epithelial cells; (2) lymphokine-like molecules isolated from thymocytes; and (3) other peptides and cellular constituents, such as thymosin β_1 peptide (ubiquitin) (Low and Goldstein, 1979) and trace amounts of polyamines (Folkers and Shiek, 1984). At least 40–50 unique peptides that can be separated by isoelectric focusing are found in TF5. They are typically heat stable, acidic polypeptides with molecular mass ranging from 1 to 15 kD.

Several of the thymic peptides have been purified to homogeneity and sequenced. A summary of the biochemical and biological properties of most of the well-characterized thymic preparations is presented in Table 17–1. A number of thymic hormones have been reported to restore varying degrees of T cell–dependent immunity in immunodeficient animals. Among the functions restored are allograft rejection, mixed lymphocyte reactions, graft vs. host reactions, and the induction of T lymphocyte markers.

Nomenclature for the thymosins is based on the order of their isolation and the isoelectric point (pI) of the peptide, as determined using a standardized isoelectric focusing technique for TF5 (Goldstein et al., 1977). Polypeptides having pI less than 5.0 are termed α thymosins; β thymosins have pI's between 5.0 and 7.0, while γ thymosins have pI's greater than 7.0. Consecutive subscript numbers are used to identify the individual polypeptides as they are isolated and become chemically and biologically characterized. To date, eight polypeptides from the α fraction and six from the β fraction of TF5 have been characterized.

The mechanisms of action of the thymic hormones are currently being investigated. Pharmacologic studies, along with direct measurements of cyclic nucleotides and cytosolic-free calcium concentrations, indicate that thymic hormones mediate their effects on lymphocyte activation and maturation through a second messenger mechanism. One of the earliest detectable effects of TF5 on thymocytes is an increase in intracellular cyclic GMP (cGMP) levels (Naylor and Goldstein, 1979). Stimulation of cGMP levels can be

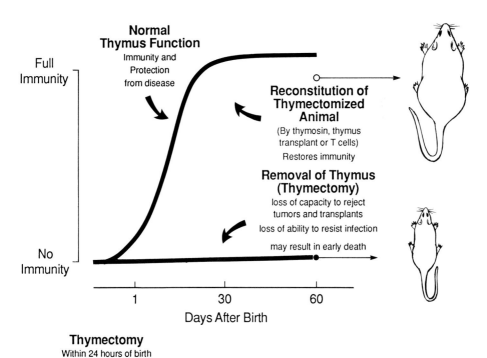

Figure 17–2. Effects of thymectomy in mice. Neonatal thymectomy prevents normal development of an immune response. Thymus grafts, T cells, and thymic hormones can restore the immune function.

Table 17-1.   Properties of Well-Characterized Partially Purified Prohormones, Thymosin
Preparations, and Purified Peptides

Name of Preparation	Principal Reference	Chemical Properties	Biological Effects
Thymosin fraction 5 (TF5)	Hooper et al., 1975	Family of heat-stable, acid polypeptide 1–15 kD	Induces T cell differentiation and enhances immunological function in animal models and humans; increases ACTH, B-endorphin, and glucocorticoid release; stimulates production of MIF, TCGF (IL-2), β and γ interferon, CSF and other lymphokines
Thymosin α 1 (Tα_1)	Low and Goldstein, 1979	Polypeptide of 28 residues, 3.108 kD pI 4.2, sequence determined	Induces enhancement of MIF, interferon, and lymphotoxin production; modulates TdT activity; increases viral, fungal, and tumor immunity; increases IL-2 in aging mice; amplifies T cell immunity in humans and prolongs survival in patients with lung cancer who are receiving radiotherapy
des (25-28) Tα_1	Caldarella et al., 1983	Polypeptide of 24 residues identical to TA1 positions 1–24	No biological activity reported
Prothymosin α	Haritos et al., 1985a	113 amino acids, TA1 at N-terminal position, 30 AA sequenced, pI. 3.55, 13.5 kD	Similar biological activity to Tα_1 in protecting mice against opportunistic infections with *Candida albicans*
Thymosin α_7	Low and Goldstein (unpublished)	Acid polypeptide, 2 kD, pI 3.5	In vitro enhancement of suppressor T cells; expression of Lyt-1,2,3-positive cells
Thymosin α_{11}	Caldarella et al., 1983	Polypeptide of 35 residues, N-terminal 28 residues identical to Tα_1	Similar biological activity to Tα_1 in protecting mice against opportunistic infections with *C. albicans*
Thymosin β_3	Low and Goldstein (unpublished)	Polypeptide of 49 residues, 5.7 kD, N-terminal 43	Similar biological activities to thymosin β_4

seen as early as 1 minute after exposure of thymocytes to TF5 in vitro. Maximum effects are observed after 5–10-minute exposure. TF5 causes an influx of calcium into thymocytes, suggesting that the increases in cGMP are calcium dependent. TF5 has not been shown to increase cAMP levels. Based on studies of cellular subpopulations, the increase in cGMP appears to mediate maturation and activation of more mature or committed lymphocytes in the thymus and spleen. It is not known whether the calcium flux acts

Table 17–1.   *(Continued)*

Name of Preparation	Principal Reference	Chemical Properties	Biological Effects
		residues, identical to thymosin β_4	
Thymosin β_4	Low and Goldstein, 1982	Polypeptide of 43 residues, 4.963 kD, pI 5.1, sequence determined	Induces TdT in vivo and in vitro in bone marrow cells from normal and athymic mice; in vivo induction of TdT in thymocytes of immunosuppressed mice; stimulates release of LH-RH (LRF); enhances thymocyte allogenic MLR
Thymosin β_7	Low and Goldstein, 1979	Polypeptide of 42 residues, partial homology to thymosin β_4	No biological activity reported
Thymosin β_8	Hannappel et al., 1982	Polypeptide of 39 residues, sequence determined, partial homology to thymosin β_4	No biological activity reported
Thymosin β_9	Hannappel et al., 1982	Polypeptide of 41 residues, 80% sequence homology to thymosin β_4	No biological activity reported
Thymosin β_{10}	Erickson-Viitanen et al., 1983b	Polypeptide of 42 residues, sequence determined	Stimulates thymocyte MLR
Thymosin β_{10} Arg	Ruggieri et al., 1983	Polypeptide of 42 residues, sequence determined, 75% homology with thymosin β_4 (Arg substituted for Ser at position 39)	No biological activity reported
Thymosin β_{11}	Erickson-Viitanen and Horeker, 1984	Polypeptide of 42 residues, 75% sequence homology with thymosin β_4	No biological activity reported

directly on guanylcyclase or via the phosphodiesterase system, arachidonic acid metabolism, or a calcium binding protein such as calmodulin. TF5 and the well-studied purified hormone thymosin α_1 (Tα_1) have both been shown to stimulate production of prostaglandin E_2 (PGE$_2$) in immature lymphocytes from the spleens of adult thymectomized mice, and from thymocytes or thymus of normal mice (Garaci et al., 1983). A long-acting synthetic analogue of PGE$_2$ (6, 16-dimethyl PGE$_2$-methyl ester) was able to partially

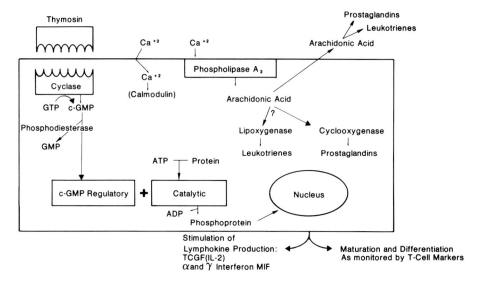

Figure 17–3.   Proposed mechanisms for activation of T cells by thymosin.

restore thymic function in adult thymectomized mice, thus mimicking the action of thy-mic hormones and suggesting that prostaglandin synthesis may be involved in the early activation of T cells (Thurman et al., 1979). TF5 and $T\alpha_1$ have also been shown to stim-ulate production of PGE_2 by immature lymphocytes. A composite of these proposed mechanisms is presented in Figure 17–3.

THE ALPHA THYMOSINS

Thymosin α_1 ($T\alpha_1$) was the first of the α thymosins to be described, and it was isolated by ion exchange chromatography with CM cellulose and DEAE cellulose, followed by Sephadex G-75 gel filtration (Low and Goldstein, 1979). $T\alpha_1$ has now been synthesized by solution and solid phase procedures (Wong and Merrifield, 1980) and has also been produced in its desacetyl form using recombinant DNA techniques in *E. coli* (Wetzel et al., 1980). $T\alpha_1$ appears to be produced in the thymus as a longer precursor peptide having a molecular mass of approximately 16 kD. Haritos et al. (1985a,b) isolated from rat thy-mus a polypetide of approximately 113 amino acid residues, which contains the complete $T\alpha_1$ sequence at its amino-terminus (Fig. 17–4). This molecule has been termed prothy-mosin α. It has a molecular mass of approximately 12 kD and a pI of 3.55; available data suggest that prothymosin α is the precursor molecule from which $T\alpha_1$ is derived.

$T\alpha_1$, like many of the other purified peptides, appears to have selective sites of action. $T\alpha_1$ induces helper T cells and the expression of T cell phenotypic markers such as θ and Lyt-1,2,3$^+$ (Goldstein et al., 1981) not only in immature precursor cells but also in thy-mocytes and peripheral blood lymphocytes (PBL) (Hu et al., 1980). These observations are consistent with its localization within the thymus as well as its presence in the systemic circulation. In peripheral blood lymphocytes from mice, guinea pigs, and humans, $T\alpha_1$ causes increased secretion of lymphokines such as MIF (migration inhibition factor) (Thurman et al., 1981) and the α and β interferons (Huang et al., 1981). More recently, Sztein et al. (1986) found that $T\alpha_1$ can increase interleukin 2 (IL-2) receptor expression induced by the mitogenic lectin, phytohemagglutinin (PHA). These latter studies may shed new light on the mechanisms by which thymic hormones modulate immune functions. We postulate that the thymosins may function to increase the number of lymphokine receptors in the presence of mitogens (and/or antigens) and thus allow T cells to respond more efficiently to lymphokine-mediated signals.

Many of the individual thymosin preparations show a considerable overlap in their ability to modulate the functions of particular T cell subsets. In all cases, the thymosins enhance immune functions in immunologically incompetent animals. None of the isolated peptides, however, expresses the complete spectrum of biological activities included within TF5.

$T\alpha_1$ has been detected in relatively high concentrations within the central nervous system, particularly the subcortical nuclei including the hypothalamus and the pituitary gland (Hall et al., 1982). When micropunch sections of discrete brain regions were homogenized and assayed for $T\alpha_1$-like reactivity by radioimmunoassay, highest activity

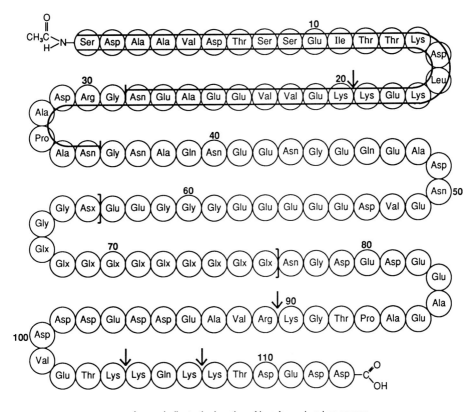

Arrows indicate the location of Lys-Arg or Lys-Lys groups, which are potential sites for proteolytic processing.

Figure 17–4.   Primary amino acid sequence of $T\alpha_1$, $T\alpha_{11}$, and prothymosin α.

was found in the median eminence (6.33 ± 0.24 pg/µg protein) and the arcuate nucleus (6.53 ± 0.54 pg/µp protein), whereas the hypothalamic and diencephalon nuclei had significantly lower levels of immunoreactive $T\alpha_1$. The concentration of $T\alpha_1$ was even higher in the pituitary gland, particularly in the neuro-intermediate posterior lobe (14.0 ± 1.8 pg/µg protein) and the anterior lobe (3.9 ± 0.4 pg/µg protein). It is unlikely that these peptides reach the pituitary by diffusion from the blood because the density of immunoreactive $T\alpha_1$ in the brain and pituitary is unaltered after whole-body perfusion with phosphate buffered saline (PBS) (Palaszymski et al., 1983). Certain neuroblastoma cell lines express appreciable amounts of immunoreactive $T\alpha_1$ (T. Moody, unpublished, and J. McGillis and J. McClure, unpublished). Due to the high density of endogenous $T\alpha_1$-like peptides in the brain and pituitary, we postulate that they play an important neuroendocrine regulatory role. Other evidence for effects on the central nervous system (CNS) is seen when $T\alpha_1$ is injected into the cerebroventrical system of chronically cannulated mice; this results in a significant rise in serum corticosterone. The release of corticosterone does not occur when cultured adrenal fasciculata cells are incubated with $T\alpha_1$ (Vahouny et al., 1983). These latter studies suggest that the corticogenic effects of $T\alpha_1$ occur at the level of the CNS or pituitary. Increased levels of corticosteroids cause changes in immunity such as decreased antibody titers (Van Dijk et al., 1976). To date, our studies suggest an important link between the endocrine functions of the thymus and the brain. $T\alpha_1$ and other thymic peptides not only appear to directly augment T cell numbers and function but also function as important immunoregulatory signals that may act via the CNS (Fig. 17–5). Thymosin peptides appear to modulate and suppress neuroendocrine processes that may directly or indirectly influence immune functions.

Recent evidence has shown that $T\alpha_1$ enhances mitogen-induced IL-2 production in vivo but not in vitro (Zatz et al., 1984), whereas studies with TF5 demonstrated enhanced mitogen-induced IL-2 production in vitro. In other words, $T\alpha_1$, a component of TF5, does not lead to enhanced IL-2 production by mitogen-stimulated PBL in vitro, although several investigators have reported enhanced in vivo production. Using anti-$T\alpha_1$ monoclonal antibodies, we studied the interaction of $T\alpha_1$ with other components of TF5 (Oates et al., 1984). Anti-$T\alpha_1$ monoclonal antibodies blocked a portion of the increased IL-2 production induced by TF5, suggesting an important role for $T\alpha_1$ in regulation of IL-2 secretion. To date, however, we have not been able to identify a single component within TF5 that mediates the IL-2 response. Our hypothesis is that $T\alpha_1$ may be one factor in a cascade of thymic peptide interactions required for IL-2 augmentation by TF5. In 1985 we showed that TF5 and aspirin both augment IL-2 production by normal PBL in the presence of mitogen and that their effects are additive (Zatz et al., 1985), suggesting that they act by distinct mechanisms. Aspirin inhibits production of prostaglandins by macrophages (presumably by inhibition of the cyclooxygenase pathway) thus leading to enhanced IL-2 production. We therefore conclude that TF5 and $T\alpha_1$ act as modulators of immunological responsiveness as measured by IL-2 production, although their effects are not mediated through the cyclooxygenase pathway.

Haritos et al. (1985a,b) reported the isolation and characterization of the prohormone prothymosin α, which is composed of 113 amino acids and which cross-reacts with $T\alpha_1$ in the $T\alpha_1$ radioimmunoassay. These same investigators published the primary amino acid sequence of prothymosin α and showed that the first 28 amino-terminal residues are identical to those of $T\alpha_1$. Prothymosin α is present in highest concentrations in the thymus and spleen, and it has been shown to protect mice against opportunistic infections, confirming that its biological activity is similar to that of $T\alpha_1$. Another peptide,

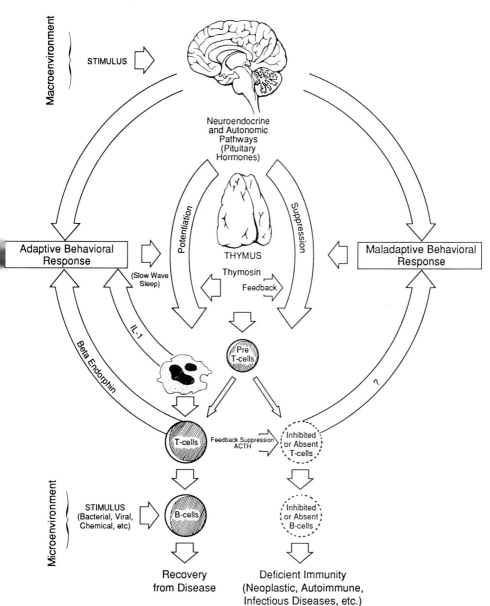

Figure 17–5. The thymus–brain connection: Thymosins can modulate neuroendocrine responses.

parathymosin α, has been identified and shown to have partial structural homology with prothymosin α. Parathymosin α, as isolated from rat thymus, contains approximately 105 amino acid residues, of which the first 30 amino-terminal residues show 43% homology with Tα₁ and prothymosin α (Haritos et al., 1985b). Parathymosin α has been shown to inhibit the immune-enhancing effects of prothymosin α. The role of parathymosin α as a competitive suppressor of immune function is still unclear. Another thymic peptide, thymosin α_{11} (Tα_{11}), was initially isolated from TF5 by HPLC (Caldarella et al., 1983) and was found to be identical to Tα_1 but for the addition of seven C-terminal amino acids. It is not yet known if these additional amino acid residues confer other biological activity on this molecule.

THE BETA THYMOSINS

To date six thymosins have been isolated from the β region of thymosin fraction 5 and 5A. The β thymosins show a high degree of structural conservation (Fig. 17–6). Although molecular size varies among the different β thymosins, there are 29 invariant residues, suggesting divergence from a common parental gene. Phylogenetic distribution studies of the β thymosins have shown that thymosin β_4 is the most widely distributed member of this group as it is found in all vertebrate classes except bony fish, which contain a variant termed thymosin β_{11} (Erickson-Viitanen et al., 1983a,b). Other variant forms of thymosin β_4 (Van Dijk et al., 1976) and thymosin β_{10} (Ruggieri et al., 1983) have also been isolated.

Thymosin β_3 and β_4 are the most extensively characterized members of the β thymosin family, and their biological activities are summarized in Table 17–1. Although Tα_1 appears to act primarily on mature T-cells, thymocytes, and PBL, thymosins β_3 and β_4 appear to act at earlier stages in T cell differentiation. Thymosin β_4 has been found to induce TdT expression in TdT-negative murine bone marrow cells in vivo and in vitro in a dose-dependent manner (Pazmino et al., 1978). Thymosin β_4 also increases terminal deoxynucleotidyl transferase (TdT) activity in thymocytes from hydrocortisone-treated mice, while thymosin β_3 inhibits macrophage migration in an antigen-independent fashion (Rebar et al., 1981).

Thymosin β_4 has also been found to stimulate secretion of luteinizing hormone-releasing factor (LRF) from the medial basal hypothalamic nuclei of female rats (Rebar

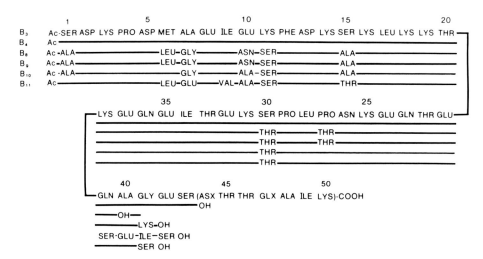

Figure 17–6.   Primary amino acid sequence of the β thymosins.

et al., 1981), while luteinizing hormone (LH) was released from pituitary glands perfused in sequence with hypothalami. However, thymosin β_4 itself did not stimulate release of LH from perfused pituitaries. These data provide the first evidence of a direct effect of the endocrine thymus on the hypothalamus and suggest a potential role for thymic peptides in reproductive function.

PHYLOGENETIC DISTRIBUTION OF THYMOSIN ALPHA

$T\alpha_1$ is present in the circulation of mammals (human, mouse, rat, cat, dog, monkey, baboon) and has been localized to the epithelial cells of the thymic subcapsular cortex and medulla (Haynes et al., 1983). The presence of immunoreactive $T\alpha_1$ has been reported in body fluids or culture supernatant from fish, invertebrates, and bacteria using a specific radioimmunoassay (Oates et al., 1988).

Recent data suggest that certain hormones and other biologically active mediators may have had their phylogenetic and ontogenetic origins as local tissue factors (LeRoith et al., 1980). A subset of these messenger molecules subsequently developed into hormones and neurotransmitters, in parallel with increased developmental and morphological compartmentation. This would account for the biological effects of many vertebrate hormones in unicellular organisms. It has also been shown that a $T\alpha_1$-like peptide is present in body fluids and blood of a wide range of organisms from annelida (earthworms, 15,000 pg/ml) to primates (humans, 670 pg/ml). Given this background, we looked for the presence of $T\alpha_1$ in prokaryotic unicellular organisms (Oates et al., 1988). $T\alpha_1$ was extracted from these organisms and separated further by preparative reverse-phase HPLC followed by a specific radioimmunoassay. Table 17–2 summarizes the concentrations of $T\alpha_1$-like material in bacterial and protozoan homogenates before and after extraction from a C18 octadecylsilane bonded Sep-Pak column (Waters Assoc.).

The explanation for the presence of a $T\alpha_1$-like peptide in unicellular organisms is not clear. The presence of $T\alpha_1$ in non-thymic sites (spleen and brain) and in blood in humans suggests an endocrine and/or autocrine role. Our studies indicate that $T\alpha_1$ is highly conserved during evolution (Fig. 17–7) and should therefore be added to the growing list of mammalian hormones that are also present in lower forms of life, including bacteria and plants. The occurrence of a $T\alpha_1$-like peptide in unicellular organisms suggests that it may play an autocrine or a primitive neuromodulating role in these cells.

Table 17–2. Thymosin α_1–Like Peptide in Some Unicellular Organisms

	Before Extraction (pg/mg protein)	After Extraction (pg/mg protein)	Culture Supernatants (pg/ml)
Bacterium			
M. phlei	208	182	2,209
M. tuberculosis	4,750	3,860	1,121
Protozoan			
Tetrehymena pyriformis	23,923	16,000	not tested
Uninoculated media	undetected	undetected	N/A

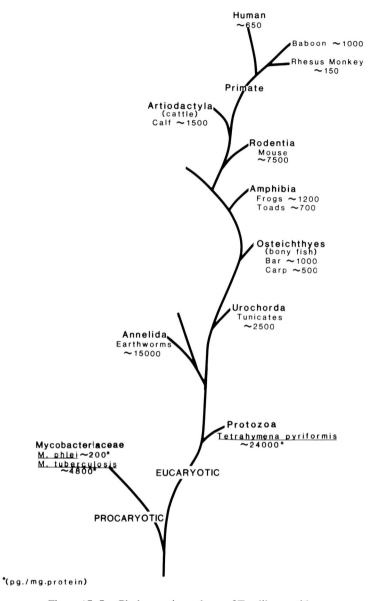

Figure 17–7. Phylogenetic analyses of $T\alpha_1$-like peptides.

THERAPEUTIC APPLICATIONS OF THYMOSIN

The success of thymic hormones in reconstituting immunodeficient animals has suggested myriad applications for the thymosins in the treatment of human diseases. There has been no significant toxicity of TF5 or $T\alpha_1$ in any clinical study to date, although an occasional patient may develop an allergic reaction to TF5. Immunodeficient patients who have had the greatest response to thymosin are those who suffer from primary thymic immunodeficiencies such as DiGeorge syndrome or thymic aplasia (Barrett et al., 1980).

Three patients with ataxia telangiectasia (a disorder of both T and B cell immunity) have also been treated with thymosin for periods of 6 to 12 months with variable responses (Wara et al. 1985). Prior to thymosin therapy, all of these patients had experienced numerous severe systemic infections, with infected eczema, low T cell numbers, and decreased PBL proliferative responses. After 6 months of therapy, all three patients showed increased total numbers of T cells, and one had an elevated PBL response to PHA. Two patients had normalization of mixed lymphocyte responses. All of these are parameters for assessing the cellular arm of the immune system. All three patients improved clinically and had fewer infections. Thymosin has also been used successfully in the treatment of hypothymic or dysthymic disorders, although it does not appear to be effective in the treatment of primary immune deficiencies due to stem cell defects, as in severe combined immunodeficiency disease (SCID). TF5 or $T\alpha_1$ may be of potential use in the treatment of secondary immunodeficiency diseases, including acquired immuno-deficiency syndrome (AIDS), autoimmune diseases, infectious diseases, immune sup-pression secondary to chemotherapy or surgery, and senesence of immunity with aging: all of these are currently under active investigation.

In 1980, the National Cancer Institutes (NCI) initiated, through its biological response modifier (BRM) program, a series of phase I and phase II clinical trials using TF5 and $T\alpha_1$. The most interesting and potentially important results were obtained in trials in patients with lung cancer. Thymosin was found to be effective in inducing clinical responses in immunosuppressed patients with lung cancer (Schulof et al., 1984). In a 3-year phase II randomized double-blind trial at the George Washington University Med-ical Center it was demonstrated that patients with non-oat cell lung cancer treated with a loading dose regimen of $T\alpha_1$ showed normalization of T cell function compared to a placebo-treated group ($p = .04$) and that patients treated with $T\alpha_1$ on a twice-weekly schedule maintained normal helper T cell percentages ($p = .04$) (Schulof et al., 1985). $T\alpha_1$ treatment was also associated with significant improvements in relapse-free intervals and overall survival. The beneficial effects of $T\alpha_1$ were more pronounced in patients with small tumor burden. The positive results obtained in this phase II study indicated that phase III trials were needed with larger patient populations to definitively establish the effect of $T\alpha_1$ treatment in lung cancer. Such a trial is currently in progress under the aus-pices of the Radiation Therapy Oncology Group (RTOC) and the Mid-Atlantic Oncology Program (MAOP) cooperative groups of the NCI.

Clinical trials of TF5 or $T\alpha_1$ in patients with primary and secondary immunodefi-ciency diseases have been in progress since 1974. Several of these studies have found that TF5 and $T\alpha_1$ have immunorestorative effects in patients with impaired T cell immunity. Well-defined phase II and phase III trials involving larger patient populations are cur-rently in progress, with the goal of defining the efficacy of thymic hormones as adjuvant therapy in cancer as well as in a number of other diseases associated with immune abnor-malities.

THYMOSINS AND IMMUNE SENESCENCE

The thymus plays a central role in the ontogenetic development of the immune system. Its major role during embryonic development is to produce mature T cells, a process mediated by thymic hormones. TF5 has been shown to substitute for certain immuno-logic functions of the thymus, and TF5 treatment partially reestablishes immune function in athymic or neonatally thymectomized animals (Low et al., 1978). During the aging process, the thymus undergoes various and profound changes. Thymus weight is greatest in relation to body weight at or near the time of birth and declines progressively with age. At puberty in humans, the gland weighs 30 to 40 grams; by age 60 its weight has fallen to between 10 and 15 grams (Kendall et al., 1980). There is also a significant linear increase in the lipid content of the thymus from birth to age 50 years. These observations are consistent with the hypothesis that age-related changes in the thymus play an important role in the decline of immune function and overall immune surveillance. Thymic involution explains to some degree the generalized T lymphocyte defects that accompany aging: helper T cells fail to provide adequate help for B cell–mediated antibody production and for suppressor T cell activity.

Without adequate regulation of T-suppressor cell activity, autoantibody production may increase, thus leading to an increase in autoimmune disease. Using an indirect fluorescent antibody technique for the localization of $T\alpha_1$ in the human thymus, thymic medullary cells were found to contain $T\alpha_1$. However, these cells progressively declined in number beginning at 13 years of age (Hirokawa et al., 1982). The atrophy of the thymus and the decrease in its $T\alpha_1$ content have far-reaching implications for the maintenance of immune function in the elderly. These changes may directly result in loss of T cell functional activity, reactivity to thymic hormones, and changes in interleukin 2 (IL-2) production and response. Two interleukins have been reported to show age-related changes in activity: IL-1, the lymphocyte activating factor (Inamiza et al., 1985); and IL-2, the T-cell growth factor (Doria and Frasca, 1985). $T\alpha_1$ has been shown to increase IL-2 production by mitogen-stimulated human PBL in vitro (Zatz et al., 1984). These data support a growing body of evidence that relates the age-associated decline in immune function with the potential of thymic hormones to restore lymphokine production in the elderly.

Another theory on senescence postulates an age-related decline in the ability of the brain to produce and control certain key hormones that regulate the function of the endocrine glands (in particular, the adrenals, ovaries, testes, and thyroid). Research in this area indicates that thymic hormonal activity is essential for the regulatory functions of the brain and for the brain-controlled activities of the endocrine glands. Involution of the thymus, which precedes most other manifestations of aging, may be the key to the senescence of the immune and endocrine systems that occurs with aging.

One of the most practical and immediately promising potential applications of thymosin may be its ability to stimulate immunity in the elderly. Influenza, for example, remains a major cause of death in the elderly and in children less than 2 years of age. The inability of current influenza vaccines to induce effective neutralizing antibodies in these two high-risk groups reflects immaturity of the immune system in childhood and depression of immune function in the elderly. In vitro studies in humans (Ershler, 1984) have shown (Fig. 17–8) a dose-related enhancement of specific antibody synthesis by $T\alpha_1$ (100 μg/mL, 200 μg/mL, and 50 μg/mL) following immunization with influenza vaccine in vivo. These findings have provided the rationale for clinical trials in which $T\alpha_1$ is used to increase immune function before immunization. Preliminary clinical trials involving administration of $T\alpha_1$ in conjunction with influenza or hepatitis vaccines are currently under way at the University of Wisconsin Medical Center in Madison, the Cornell Medical Center in New York, and the University of Maryland Medical Center in Baltimore. Restoration of normal immune function in the elderly may decrease the number and severity of illnesses associated with aging and may prolong life itself. Before attempting

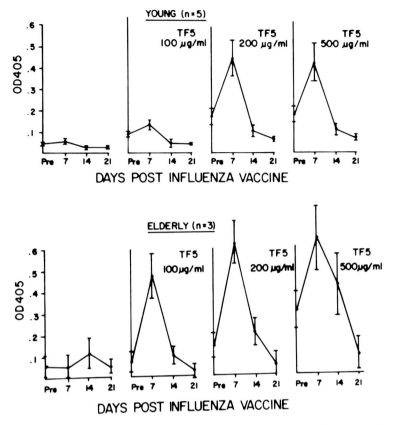

Figure 17–8. Thymosin fraction 5 (TF5) stimulates production of anti-influenza antibodies by lymphocytes from elderly patients after immunization with influenza vaccine.

long-term therapeutic intervention, however, a more sophisticated understanding of the interrelationship between thymus-dependent immunity, endocrine function, and the central nervous system is necessary.

FUTURE PERSPECTIVES

The ultimate clinical application of the thymosins may be to provide a means of safely augmenting specific T lymphocyte functions. Thymic hormones may be used as adjunctive therapy to increase T cell responses to tumor cells and to bacterial pathogens, thus reducing the high incidence of infection that accompanies cancer treatment. The promising results of preliminary clinical trials with TF5 and $T\alpha_1$ provide a strong rationale for expanding basic research programs to purify and characterize the thymosins, with the ultimate goal of broader clinical application.

We anticipate that well-planned clinical trials over the next several years will help determine the optimal conditions for use of thymic hormones in the treatment of a variety of diseases associated with congenital and acquired immune deficiency and with aging.

Basic research into the chemistry and biology of individual thymic factors indicates the existence of a family of peptides that influence different subpopulations of T cells. Purification and characterization of these thymic peptides should result in significant improvements in our ability to modulate the immune system. Clinical trials to date have clearly suggested that the thymosins play a major role in maintaining and restoring immune function and in augmenting specific lymphocyte activities in primary and secondary immunodeficiency diseases, infections, and autoimmune diseases, and cancer.

Acknowledgments

These studies were supported in part by a grant from the National Cancer Institute CA-24974 and gifts from Alpha One Biomedical, Washington D.C.

References

Barrett, D. J., Wara, D. W., Ammann, A. J., and Cowan, M. J. (1980). Thymosin therapy in the DiGeorge syndrome. *J. Pediatr. 97*:66.

Caldarella, J., Goodall, G. Y., Fellix, A. M., Heimer, E. O., Salvin, S. B., and Horecker, B. L. (1983). Thymosin α_{11}: A peptide related to thymosin α_1 isolated from calf thymosin fraction 5. *Proc. Natl. Acad. Sci. USA 80*:7424–7427.

Doria, G., and Frasca, D. (1985). Effects of thymosin α_1 on immunoregulatory T-lymphocytes. In *Thymic Hormones and Lymphokines* (Goldstein, Allan L. ed.). New York: Plenum Press, pp. 445–454.

Erickson-Viitanen, S., and Horecker, B. L. (1984). Thymosin β_{11}: A peptide from trout liver homologous to thymosin β_4. *Arch. Biochem. Biophys. 233*:815–820.

Erickson-Viitanen, S., Ruggieri, S., Natalini, P., and Horecker, B. L. (1983a). Distribution of thymosin β_4 in vertebrate classes. *Arch. Biochem. Biophys. 221*:570–576.

Erickson-Viitanen, S., Ruggieri, S., Natalini, P., and Horecker, B. L. (1983b). Thymosin β_{12} in mammalian tissue. *Arch. Biochem. Biophys. 225*:407–413.

Ershler, W. B. (1984). Augmentation of antibody synthesis in vitro by thymosin fraction 5. In *Thymic Hormones and Lymphokines* (Goldstein, Allan L. ed.). New York: Plenum Press, pp. 297–305.

Folkers, K., and Shiek, H.-M. (1984). The finding and significance of spermidine and spermine in thymic tissue and extracts. In *Thymic Hormones and Lymphokines* (Goldstein, Allan L. ed.). New York: Plenum Press, pp. 89–95.

Garaci, C. R., Favalli, C., Delgobbo, V., Garaci, E., and Jaffee, B. M. (1983). Thymosin action mediated by prostaglandin release. *Science 200*:1163–1165.

Goldstein, A. L., Low, T.L.K., McAdoo, M., McClure, J. M., Thurman, J. B., Rossio, J., Lai, C. Y., Chang, D., Wang, S. S., Harvey, C., Ramel, A. H. and Meienhofer, Jr. (1977). Thymosin α_1: Isolation and sequence analysis of an immunologically active thymic polypeptide. *Proc. Natl. Acad. Sci. USA 74*:725–729.

Goldstein, A. L., Low, T.K.L., Thurman, G. B., Zatz, M., Hall, N. R., McClure, J. E., Hu, S. K., and Schulof, R. S. (1981). Thymosins and other hormonal like factors of the thymus gland. In *Immunological Aspects of Cancer Therapeutics* (Mihich, E., ed.). New York: Wiley, pp. 137–190.

Goldstein, A. L., Slater, F. D., and White, A. (1966). Preparation, assay and partial purification of a thymic lymphocytopoieitic factor (thymosin). *Proc. Natl. Acad. Sci. USA 56*:1010–1017.

Gregoire, C., (1935). Recherches sur la symbiose lymphoepitheliale au niveau thymus de mammifere. *Arch. Biol. Liège 46*:77–722.

Hall, N. R., McGillis, J. P., and Spangelo, B. (1982). In *Current Concepts in Human Immunology*

In contrast to B cells, NK cells do not express surface immunoglobulins, and unlike T cells, they do not bear classical T cell markers (see below). NK cells do not display phagocytic activity and are predominantly nonadherent (Herberman, 1980, 1982a; Lotzová, 1983a; Lotzová and McCredie, 1978). However, some NK cells do adhere to nylon wool, glass, or plastic, especially after activation (Chang et al., 1983; Herberman, 1982a; Lotzová et al., 1984b, 1986c; Savary and Lotzová, 1986). We and others have shown that the majority of NK cells adhere to plastic 24 hours after activation with interleukin-2 (IL-2) (Lotzová et al., 1990; Melder et al., 1988).

Cytochemical analysis has shown that NK cells are acid phosphatase and β-glucuronidase–positive and nonspecific esterase-negative (Ferrarini et al., 1980). Morphologically, NK cells are quite distinct from other lymphocytes; most NK cell activity in humans and other species is closely associated with large granular lymphocytes (LGLs) (Lotzová, 1986c, 1987b; Savary and Lotzová, 1986; Timonen et al., 1981). LGLs are characterized by a relatively large size (15–20 μm in diameter), indented to reniform nucleus, and the presence of cytoplasmic azurophilic granules. However, despite the close correlation of NK cell function and LGL morphology, it is important to note that only 75% of LGLs exhibit NK cell activity. Furthermore, some MHC-nonrestricted T cells also display LGL morphology (Biron and Welsh, 1986; Lanier and Phillips, 1986; Lotzová and Herberman, 1986b). Finally, not all NK cells are LGL—some lymphocytes with agranular morphology also express NK cell activity (Herberman, 1982a). Whether other morphologically distinct lymphocyte types are responsible for NK cell activity remains to be determined.

NK Cell Surface Phenotype

The cell surface phenotypes of human NK cells have been studied extensively by various laboratories and are summarized in Table 18–2. The majority of human NK cells react with Leu-11, B73.1, and 3G8 antibodies (Herberman, 1986), which have been shown to recognize epitopes on the Fc gamma receptor (CD16 antigen). Although treatment with anti-CD16 antibodies abrogates most typical NK cell activity, the CD16 molecule is not NK cell–specific—it is also present on granulocytes and some macrophages (Knapp et al., 1989). Most NK cells are stained by the OKM1 and Leu-15 antibodies which detect the CD11b antigen that is expressed on monocytes/macrophages, on polymorphonuclear leukocytes, and on a subset of T cells (Herberman, 1980, 1982a; Herberman and Ortaldo, 1981; Ortaldo et al., 1981). However, this antigen is expressed at low density on NK cells and does not appear to be expressed on cultured NK cells (Perussia et al., 1987). The CD11a molecule (identified by the LFA-1 antibody)—which is rather ubiquitously expressed on cytotoxic cells, including T cells, monocytes, and polymorphonuclear lymphocytes—is also present on NK cells. The CD56 antigen (NKH1, Leu-19) was recently described on essentially all resting and activated NK cells in human peripheral blood (Lanier et al., 1986; Lanier and Phillips, 1986) and on interleukin 2 (IL-2)–activated NK cells in bone marrow and spleen (Lotzová, 1989; Lotzová and Savary, 1987; Lotzová et al., 1989). This antigen is also expressed on a small proportion of T lymphocytes that mediate MHC-nonrestricted cytotoxicity in vitro (Hercend et al., 1985; Lanier et al., 1986; Lanier and Phillips, 1986; Schmidt and Ritz, 1986).

404   DEVELOPMENT OF THE IMMUNE SYSTEM IN HEALTH AND DISEASE

Table 18–2.   Cell Surface Phenotype of Human NK Cells as Compared
to T Lymphocytes and Other Cells

Antigen[a]	Antibody	Expression on NK cells[b]	Expression on other cells[c]
CD2	T11, Leu-5, 9.6	subset	T cells
CD3	T3, Leu-4	−	T cells
CD4	T4, Leu-3	−	T subset; M
CD5	T1, Leu-1	−	T cells; B subset
CD6	T12	−	T cells; B subset
CD7	3A1, Leu-9	+	T cells; P
CD8	T8, Leu-2	subset	T subset
CD11a	LFA-1	+	T cells; B cells; M; PMN
CD11b	OKM1, Mo-1	+	T cells; M; PMN
CD11c	Leu-M5	+	M; PMN; B subset
CD16 (FcR)	Leu-11, B73.1, 3G8	+	PMN; M; subset
CD18	MHM23; M232; 11H6	+	T cells; M; PMN
CD25 (IL-2R)	Tac	act.	act. T cells; B cells; act. M
CD38	T10, Leu-17	+	T cells; PC
CD56	Leu-19, NKH1	+	T subset; NEDC
CD57	Leu-7, HNK1	subset	T subset; B subset; brain
CD71 (TrR)	T9	act.	act. T cells; B cells; M
HLA-DR (Ia)	anti-HLA-DR	act.	B cells; M; act. T cells
P75	TU27	+	T subset; M

[a]FcR, Fc receptor; IL-2R, IL-2 receptor; TrR, transferrin receptor.
[b]+ = present; − = absent.
[c]M, monocytes; PMN, polymorphonuclear lymphocytes; P, platelets; PC, plasma cells; act., activated; NEDC, neuroectodermal cells.

NK cells also share a variety of other cell surface antigens with subpopulations of T cells. For instance, a high proportion of NK cells express the CD38 antigen, which is recognized by the monoclonal antibodies OKT10 and Leu-17 and is present on thymocytes and activated T lymphocytes (Ortaldo et al., 1981). A certain percentage of NK cells possess low-affinity receptors for sheep red blood cells (CD2 antigen), which can be detected either functionally by rosette formation or phenotypically by staining with one of the monoclonal antibodies OKT11, Leu-5, and 9.6 (Herberman, 1980, 1982a; Herberman and Ortaldo, 1981; Ortaldo et al., 1981). A proportion of NK cells also reacts with the OKT8 and Leu-2 monoclonal antibodies, which recognize the CD8 antigen. This antigen is also present on the subset of T cells that mediate class I–restricted effector functions (Herberman and Ortaldo, 1981; Perussia et al., 1983). Although the CD57 (HNK1) antigen (identified by the Leu-7 antibody) was initially thought to be specific for NK cells (Abo and Balch, 1981), current data indicate that this antibody reacts with only 30%–80% of human peripheral blood NK cells (Herberman, 1986; Trinchieri and Perussia, 1984). In addition, this antigen was also detected on suppressor T cells, on a B cell subset, and on brain cells (Knapp et al., 1989).

Most studies indicate that resting human NK cells do not express substantial levels of class II MHC antigens (Ng et al., 1980; Perussia et al., 1983), which are present primarily on B cells, on macrophages, and on some activated T cells. However, some investigators reported that a proportion of resting NK cells express HLA-DR antigens (Brooks and Moore, 1986; Herberman, 1986; Ortaldo et al., 1981), while activated NK cells are known to express HLA-DR antigens, as well as the transferrin receptor and CD25 antigen (low-affinity IL-2 receptor). Most studies indicate that NK cells do not express the CD3, CD4, or CD5 antigens that are characteristic markers of T-lineage cells (Herberman, 1986; Ortaldo et al., 1981). The majority of human NK cell activity appears to be mediated by an LGL population expressing the $CD3^-$, $CD16^+$, $CD56^+$ phenotype. This population can also mediate antibody-dependent (ADCC) activity. NK activity can also be expressed by an LGL subset that lacks either the CD16 or the CD56 antigen—i.e., $CD3^-$, $CD16^-$, $CD56^+$ or $CD3^-$, $CD16^+$, $CD56^-$. These latter subsets appear to represent a minority of NK cells in human peripheral blood.

Similar to the situation in humans, rat NK cells exhibit LGL morphology, express the asialo GM1 cell surface glycolipid and react with antibodies directed against "suppressor" T cells (OX8) (Herberman, 1982a; Reynolds et al., 1981). The morphology of murine NK cells is less clear. Although some NK cell activity is associated with LGL, it appears that agranular lymphocytes, as well as medium-sized lymphocytes, may contribute to NK cell activity. However, mouse NK cells also express the asialo GM1 glycolipid (Herberman, 1980, 1982a; Lotzová, et al., 1986c), and some NK cells share the Thy-1 and Lyt-1 antigens with T cells (Herberman, 1980, 1982a; Herberman and Ortaldo, 1981; Lotzová et al., 1986c, d). Mouse NK cells also express the NK 1 antigen, whose expression was reportedly restricted to NK cells (Herberman, 1980, 1982a; Lotzová et al., 1982a, 1983b); however, recently this antigen was also detected on a subset of T cells (Sykes, 1990).

Divergence of NK Cells from MHC-Nonrestricted T Cells

It has recently been shown that a minor subset of MHC-nonrestricted T lymphocytes also exhibits the LGL morphology and expresses the CD56 cell surface molecule characteristic of NK cells. After activation with IL-2, this subset can mediate NK-like activity, including lysis of NK-sensitive targets such as K-562 (Lotzová and Ades, 1989). However, much experimental evidence indicates that these two lymphocyte subsets are clearly divergent. In contrast to NK cells, these MHC-nonrestricted T lymphocytes express the CD3, CD4, and CD5 antigens and either α/β or γ/δ T cell receptors (TCR). Functionally, there appears to be a difference in the cytotoxic efficacy of these two lymphocyte subpopulations: NK cells appear in general to be more potent killers than MHC-nonrestricted T lymphocytes. Because of these differences, MHC-nonrestricted T cells should not be classified as NK cells, but should be designated as MHC-nonrestricted cytolytic T lymphocytes.

PHYLOGENY OF NK CELLS

Phylogenetically, NK cell function precedes the development of specific T and B cell–mediated responses; NK cell activity can be detected in primitive invertebrates (Table 18–3) (Savary and Lotzová, 1986). Even though the cytotoxic cells in the latter species have not been characterized precisely, they resemble NK cells functionally and morphologically. To take several examples, these effector cells in annelids are nylon wool-nonadherent (Valembois et al., 1980), while in echinoderms and mollusks they are nonspecific esterase-negative and their cytotoxic function is inhibited by monosaccharides (Decker et al., 1981). Finally, coelomocytes of some earthworms display LGL morphology (Hostetter and Cooper, 1972; Linthicum et al., 1977).

Table 18–3.   NK-Like Cytotoxicity in Various Invertebrate Species

Phylum	Species	Type of Target Cells Killed
Sipunculids	*Sipunculus nudus* (unsegmented worms)	Allogeneic and xenogeneic erythrocytes
Annelids	*Eisenia fetida* (earthworm)	Allogeneic leukocytes
	Lumbricus terrestris (earthworm)	Xenogeneic skin grafts
	Glycera (bloodworm)	Xenogeneic erythrocytes and mastocytoma
Echinoderms	*Pisaster gigantus* (starfish)	Xenogeneic erythrocytes, mononuclear cells, mastocytoma, and B cell line
	Asterias rubens (starfish)	Xenogeneic cell lines
Mollusks	*Megathura crenulata* (keyhole limpet)	Xenogeneic erythyrocytes and mononuclear cells
Arthropods	*Parachaeraps bicarinatus* (crayfish)	Xenogeneic ascites
	Homarus americanus (lobster)	Xenogeneic mastocytoma

Primitive vertebrates, such as fish and amphibians, also display NK-like cytotoxic activity. For instance, cells from fresh water fish, including *Cyprinus carpio* (carp), *Carassius cuvieri* (crucian carp), *Ctenopharyngodon idella* (grass carp), *Misgrunus anguillicandatus* (oriental weather fish), and *Channa argus* (northern snakehead), display cytotoxic activity to established mammalian cell lines of normal or malignant origin (Hinuma et al., 1980). The highest cytotoxic activity in these species was found in the kidney (a hemopoietic organ in fish), with lower and variable reactivity in peripheral blood and spleen. The cytotoxicity was similar to that of mammalian NK cells both in its rapidity (lysis within 6 hours) and lack of species restriction. Natural cytotoxic cells were also identified in the nurse shark, *Ginglymostoma cirratum* (Pettey and McKinney, 1980) and in the amphibian, *Xenopus laevis* (Roder et al., 1981). In amphibia, the splenocytes are the main source of NK-like cells, while bone marrow and thymus exhibited only low levels of cytotoxicity.

In avian species, such as chicken and quail, natural cytotoxic cells, which lyse normal and malignant targets, are found in the spleen and peripheral blood but not in the thymus or the bursa of Fabricius (Fleischer, 1980; Leibold et al., 1980; Sharma and Coulson, 1979; Yamada et al., 1980). NK cells are quite broadly distributed among other species, including rats, hamsters, guinea pigs, cats, dogs, pigs, and nonhuman primates. It is of interest to note the similarity and preservation of NK cell activity throughout phylogeny: a phylogenetically conserved NK receptor-like molecule expressed on fish, mouse, and rat NK cells has recently been described (Harris et al., 1989). This phylogenetic conservation of NK cells supports the important function(s) of these cells for the organism.

ORIGIN AND TISSUE DISTRIBUTION OF NK CELLS

Experimental evidence suggests that NK cells in humans and rodents originate in the bone marrow; studies in mice also indicate that this tissue is necessary for proper differentiation of NK cells (Hackett et al., 1986; Herberman, 1980, 1982a; Kumar et al., 1979; Lotzová and Herberman, 1986a; Lotzová and McCredie, 1978; Lotzová et al., 1986c; Reynolds and Ward, 1986). However, NK cell activity has consistently been found to be low in bone marrow, suggesting that functional NK cells are not present in this tissue (Herberman, 1980, 1982a; Lotzová et al., 1979a,b; Lotzová and Savary, 1987).

The highest levels of NK cell activity in humans, rats, and mice have been found in the spleen and peripheral blood (Herberman, 1980, 1982a; Lotzová, 1989; Lotzová and McCredie, 1978; Reynolds and Ward, 1986), while lower levels are present in the lungs, liver, lymph nodes, and peritoneal cavity (Herberman, 1980; Lotzová et al., 1984a,b; 1986a,c; 1987b; Reynolds and Ward, 1986; Savary and Lotzová, 1986). In some strains of rats, however, peritoneal exudate cells were found to express high levels of NK activity (Lotzová et al., 1984b; Reynolds and Ward, 1986). NK cell cytotoxic activity could also be detected in human tonsils and thoracic duct and in murine intestinal epithelium and lamina propria (Eremin et al., 1978; Herberman, 1982a; Lotzová and McCredie, 1978; Oldham et al., 1978; Reynolds and Ward, 1986). NK activity is reproducibly absent from nonactivated thymus of humans and rodents; furthermore, the thymus is not necessary for generation or expression of NK activity. This latter is evident due to the fact that some strains of athymic mice possess similar or higher cytotoxic potential than the corresponding euthymic animals (Herberman, 1980, 1982a).

REGULATION OF NK CELL ACTIVITY

NK cell activity can be up- and downregulated by various cell populations as well as by a variety of biological or chemical agents. In this section we will briefly review cells and substances that regulate NK cell activity.

Downregulation of NK Cell Activity

NK cell cytotoxic function has been shown to be downregulated by a variety of other cells, which may include macrophages, T lymphocytes, and granulocytes, depending on the system studied. For instance, suppression of peripheral blood NK cell activity was observed in patients with breast carcinoma after surgery (Uchida et al., 1982). The cells active in this suppression were shown to be monocytes, based on their adherence to plastic and Sephadex G-10 columns and their staining for nonspecific esterase. Similarly, macrophage/monocyte-like suppressor cells were identified in carcinomatous pleural effusions of patients with lung cancer (Uchida and Micksche, 1981) and in bronchoalveolar washings from patients undergoing diagnostic bronchoscopy (Bordignon et al., 1982). In another system, human monocytes stimulated by poly I:C were reported to inhibit NK cell activity in vitro (Koren et al., 1981).

Suppressor cells were also detected in ascitic fluid from some ovarian cancer patients (Allavena et al., 1981). However, these cells were plastic-nonadherent, and it was therefore concluded that macrophages or monocytes did not contribute to suppression of NK cell activity in this system. The exact nature of these suppressor cells was not determined. Suppression mediated by T lymphocytes was also reported (DeBoer et al., 1976; Tarkkanen et al., 1983a,b); these cells were identified in the peripheral blood of normal donors and of cancer patients and in cord blood.

Granulocytes represent another cell population that may regulate NK cell function. Kay and Smith (1983) demonstrated that granulocytes from normal donors or from patients with chronic granulomatous disease suppressed the activity of autologous or allogeneic NK cells. NK inhibition in these studies could be mediated by membrane fragments or extracts from sonicated granulocytes, although intact granulocytes had the highest suppressive activity. Regulation of NK activity by granulocytes was also reported independently by Seaman et al. (1982).

Involvement of suppressor cells in regulation of murine NK cell activity was first reported by our laboratory (Lotzová, 1980; Lotzová and Savary, 1986; Savary and Lotzová, 1978). We demonstrated that NK suppressor cells could be identified in infant mice, in *Corynebacterium parvum (C. parvum)*–treated mice, and in F_1 hybrid recipients tolerant to parental bone marrow transplants. After these initial observations, suppressor cells directed against NK activity were reported by various other laboratories (see for review Lotzová and Savary, 1986). Suppressor cells were detected in various mouse strains that displayed low levels of NK activity (Blair et al., 1983; Riccardi et al., 1981), in peritoneal exudates from mice with normal levels of NK activity (Brunda et al., 1982), in aged rats (Bash and Vogen, 1984), and in the thymic tissue of mice and rats (Nair et al., 1981; Zöller and Wigzell, 1982).

In addition to naturally occurring suppressor cells, various agents and procedures have been shown to induce NK cell–directed suppressor cells. These agents include *Bacillus Calmette-Guérin* (BCG), *C. parvum,* pyran copolymer, adriamycin, β-estradiol, carrageenan, hydrocortisone, and irradiation (see for review Lotzová and Savary, 1986; Oehler and Herberman, 1978). Analysis of suppressor cells in various systems indicated considerable phenotypic and functional heterogeneity, depending on the experimental conditions. Macrophages and T cells were found to downregulate NK activity in experimental animals, similar to the situation in humans (Lotzová and Savary, 1986). In addition, NK cell suppressor activity appeared to be present in cell populations whose phenotypes could not be clearly defined using available techniques and/or antibodies. We recently found that erythroblasts can also interfere with murine NK cell cytotoxic function (Pollock and Lotzová, 1987).

The functions of human and animal NK cells can also be inhibited by a variety of agents that do not appear to act through induction of suppressor cell activity. These include corticosteroids, cyclophosphamide, cyclosporin A, silica, prostaglandins, cAMP, EDTA, phorbol esters, protease inhibitors, and some sugars (Brunda and Holden, 1980; Herberman, 1980, 1982a; Introna et al., 1981; Oehler and Herberman, 1978; Ortaldo et al., 1984). Downregulation of NK cells may be expressed at various levels, including decreased NK cell numbers or production; changes in NK cell turnover, migration, or chemotaxis; and direct effects on NK cell lytic activity.

Potentiation of NK Cell Activity

A variety of biological response-modifying agents and chemicals were shown to potentiate NK cell activity in vivo and in vitro (Herberman, 1980, 1982a; Lotzová, 1983a, 1987c; Lotzová and Herberman, 1986b; Lotzová and Savary, 1987; Lotzová et al., 1984a,b; 1986a,c). NK cell-mediated cytotoxicity has been shown to be augmented in the presence of certain bacteria, such as BCG and *C. parvum,* (the latter bacteria augments NK cytotoxicity only soon after its administration), and by viruses and tumor cells. In addition, various species of interferons (IFN), IFN-inducers, and other lymphokines increase NK activity (Henney et al., 1981; Kuribayashi et al., 1981; Lotzová, 1986; Lotzová and Herberman 1986a,b; Lotzová and Savary, 1984; Lotzová et al., 1982a, 1986b,

1987b; Trinchieri and Santoli, 1979). IFN and interleukin 2 (IL-2) were shown to be highly potent stimulators of NK cell activity (Henney et al., 1981; Kuribayashi et al., 1981; Lotzová, 1986; Lotzová and Savary, 1984; Trinchieri and Santoli, 1979). The mechanisms of NK cell potentiation by IFN have been studied quite extensively. Interferon was shown to augment NK cell cytotoxicity through several mechanisms, including increase in NK cell frequency and lytic activity. IFN potentiates binding of tumor cells by NK cells and augments recycling activity; as well, it accelerates the kinetics of target cell lysis (Gustafsson and Lundgren, 1981; Henney et al., 1981; Kuribayashi et al., 1981; Lotzová, 1986; Lotzová et al., 1982a; Lotzová and Savary, 1984; Saksela et al., 1979; Silva et al., 1980; Targan and Dorey, 1980; Trinchieri and Santoli, 1979). It is important to note, however, that high doses or multiple exposures to IFN have a suppressive effect on NK activity (Lotzová et al., 1982a).

Even though the effects of IL-2 on NK cell activity have been extensively described and clinical trials using this substance for treatment of cancer patients have been reported, the mechanisms by which it potentiates NK activity have not been thoroughly defined. We have shown in our laboratory that in vitro culture of human peripheral blood mononuclear cells (obtained from normal donors or leukemic patients) with IL-2 resulted in a generalized potentiation of all components of the NK cell lytic mechanism. These include higher efficiency of NK cell binding to tumor cells, increased recycling of effectors, more rapid lysis, and the production of cytotoxic factors. IL-2 also increased the frequency of cytotoxic NK cells (Lotzová and Savary, 1990c; Lotzová et al., 1987b). Thus, the mechanisms of NK cell potentiation by IL-2 appear multiple, similar to those mediated by IFN.

NK CELLS AS THE MAJOR EFFECTOR POPULATION IN THE LYMPHOKINE-ACTIVATED KILLING (LAK) PHENOMENON

Because of the possible importance of IL-2–generated killer cells in the treatment of cancer, I will summarize recent data from various laboratories on the morphological and phenotypic characteristics of IL-2–activated killers that mediate LAK activity, as well as their precursors.

It was initially reported that IL-2–activated killer cells represent a unique cell population and are distinct from NK cells (Grimm et al., 1982). These cells were subsequently designated as lymphokine-activated killer (LAK) cells and were reported to belong to the T cell lineage. However, analysis of the LAK phenomen by various laboratories has subsequently shown that LAK activity is not mediated by a new and unique cell type. Most investigators agree that NK cells are the principal effector population in blood, bone marrow, and spleen that is involved in LAK activity (Herberman et al., 1987; Lotzová, 1987b; Lotzová and Herberman, 1987; Lotzová and Savary, 1987; Lotzová et al., 1987b, 1989). Interestingly, NK cells were also reported to contribute to the LAK activity from thymus (Ramsdell et al., 1988). Identification of NK cells as the principal mediators of LAK activity is based on the observations detailed below.

The original distinction between LAK and NK cells was based on the apparent ability of the former but not the latter to destroy fresh tumor cells (Grimm et al., 1982). These initial studies were incomplete, however, and did not distinguish between the cytolytic activity of endogenous (unstimulated) NK cells and NK cells stimulated with IL-2. These studies did not address the question of whether IL-2–activated NK cells could lyse fresh tumor cells. However, this point has now been clarified with the demonstration that IL-2–activated NK cells have the ability to kill a wide variety of fresh tumor cells (Herberman et al., 1987; Lotzová and Herberman, 1987; Lotzová et al., 1987b). Furthermore, in contrast to initial studies suggesting that IL-2–activated killer cells were of T cell lineage, more recent work has clearly established that the principal effector cell population involved in LAK activity in humans has LGL morphology and expresses the NK cell surface phenotype (CD16[+], CD56[+]/Leu-19[+], CD3[−], CD4[−], and CD5[−]) (Table 18–4) (Herberman et al., 1987; Lotzová, 1987a,b, 1989a; Lotzová and Herberman, 1987; Lotzová et al., 1987b, 1989). Analysis of precursors for IL-2–induced human killer cells indicated that NK cell–enriched peripheral blood or bone marrow populations were the most efficient in generation of LAK activity; conversely, little or no killer cell activity could be generated from populations depleted of NK cells (Herberman et al., 1987; Itoh et al., 1985; Lotzová, 1987b; Lotzová and Herberman, 1987; Lotzová et al., 1987b).

Murine and rat splenic and peripheral blood LAK activity has also been identified in effector cell populations that display NK cell characteristics (Table 18–4), and high levels of LAK activity could be generated from highly purified LGL, but not from purified T cells.

Most evidence to date thus suggests that LAK activity in peripheral blood, spleen, and bone marrow is mediated primarily by IL-2–activated NK cells. Although T cells can contribute to LAK activity, this appears to be a minor component. It cannot be excluded, however, that future studies looking at a larger panel of tumor cell targets may show that some malignant cells have a higher sensitivity to T cell–derived LAK activity. Because IL-2–activated NK cells appear to play the principal role in LAK activity against a number of tumors, including leukemias and renal, ovarian and head and neck cancers, it is reasonable to suggest that adoptive immunotherapy using effector populations enriched for NK cells may be more effective in generating anticancer responses in these patients than current therapy using unseparated mononuclear cell populations.

Recently, a subset of IL-2–activated human peripheral blood lymphocytes was shown to acquire adherence to plastic early after activation (Lotzová, 1989; Lotzová et al., 1991; Melder et al., 1988). These lymphocytes, designated adherent lymphokine–activated killer cells (A-LAK), display substantially higher oncolytic function and growth inhibitory activity than the conventional LAK. We recently showed that A-LAK can also be derived from patients with acute myelogenous leukemia in remission or relapse (Lotzová, 1990; Lotzová et al., 1991) and that NK cells play an important role in the devel-

Table 18–4.   Major Characteristics of Effector
Cells with LAK Activity

Human	Mouse	Rat
LGL	LGL	LGL
CD3[−]	asialo GM1[+]	asialo GM1[+]
CD4[−]	L3T4[−]	OX8[+]
CD5[−]	Lyt-1[−]	OX6 (Ia)[+]
CD16[+]	Lyt-2[−]	OX6 (Ia)[+]
CD56[+]	Thy-1[+]	pan-T cell[−]
		helper T cell[−]
		laminin[+]

Table 18-5. Cytotoxicity of IL-2–Generated
Tumor-Infiltrating Lymphocytes

| Target | IL-2 | Percentage of Cytotoxicity[a] | |
		TIL	ASC
K-562	−	2.1 ± 0.7	7.1 ± 2.6
	+	68.7 ± 3.2	69.0 ± 1.9
OV-2774	−	0.7 ± 0.4	0.2 ± 0.2
	+	52.3 ± 5.9	66.9 ± 8.8
Autologous	−	0.3 ± 0.2	n.t.
tumor	+	24.4 ± 2.8	n.t.

[a]TIL, tumor-infiltrating lymphocytes; ASC, ascitic fluid-associated lymphocytes. Effector cells were tested before and after seven days in culture with recombinant IL-2 (10^3 U/ml) at a 25:1 effector : target ratio.

opment of A-LAK activity (Lotzová et al., 1991). Consequently, A-LAK may represent a new approach to adoptive therapy of leukemia and perhaps other cancers.

NK CELL INVOLVEMENT IN TIL ACTIVITIES

Tumor-infiltrating lymphocytes (TIL) have shown potential for treatment of cancer patients (Topalian et al., 1988). After activation with IL-2, these cells express lytic activity against a range of fresh tumor cells. TILs were originally studied with the aim of selecting highly specific therapeutically relevant populations of cytotoxic T lymphocytes, which would "home" to the tumor. However, this proved overly simplistic, and TIL have been shown to represent a rather heterogeneous population of cells. We have studied the cytotoxic efficacy and tumor specificity of TIL derived from ascitic and solid human ovarian cancers and analyzed the contributions of NK cells and cytolytic T cells to antitumor effects. The data in Table 18–5 show that high levels of cytotoxicity were expressed by IL-2–activated TIL from both ascitic and solid ovarian cancers but that no specificity in tumor cell killing was observed. TIL showed lytic activity against fresh autologous tumors, as well as the ovarian cell line, OV-2774, and the erythroleukemia cell line, K-562. Phenotypic characterization of TIL showed that the principal cytotoxic population expressed the CD5⁻, CD56⁺ phenotype of NK cells (Table 18–6). Thus NK cells contribute significantly not only to LAK activity, but also to TIL cytotoxic function.

NK CELL ROLE IN DEFENSE AGAINST TUMORS IN VIVO

Increasing experimental evidence suggests a major role for NK cells in in vivo resistance to a variety of malignant cells and in regulating their metastatic properties. The original studies in mice, which implicated NK cells in antitumor defense, were purely correlative and demonstrated impairment of tumor growth in animals with high NK cell activity, as well as progressive tumor growth in mice that exhibited low NK cell cytotoxic potential. These studies were complemented by the observation that mice (or tissues) with high NK

Table 18–6.   Characterization of Cytotoxic
Tumor-Infiltrating Lymphocytes[a]

Lymphocyte Treatment[b]	Percentage of Control Cytotoxicity	
	TIL	ASC
C'	96	98
CD5⁻	91	109
CD56⁻	1	3

[a]Effector cells were cultured with 10^3 U/ml of IL-2 for 11–14 days, then tested for cytotoxicty against OV-2774 in a 3-hour ^{51}Cr-release assay.
[b]The monoclonal antibody treatment was performed as described previously (Lotzová et al., 1987b); C', complement treatment.

cell cytotoxic activity showed rapid in vivo clearance of radioactively labeled transplanted tumors, in comparison to mice or tissues with low NK cell activity (Herberman, 1980, 1982a; Lotzová, 1983a, 1984). Furthermore, a direct correlation was observed between potentiation of NK cell activity and increased resistance to tumor growth (Herberman, 1980, 1982a; Lotzová, 1985; Lotzová and Herberman, 1986a). More direct evidence supporting involvement of NK cells in tumor resistance came from studies with beige mice, which are genetically deficient in NK activity. These investigations showed that leukemias and solid tumors grew more rapidly and were more lethal in homozygous NK-deficient beige mice than in heterozygous littermates with normal NK cell activity (Lotzová, 1984; Pollack, 1983; Talmadge et al., 1980). Similarly, radioactively labeled leukemia cells were cleared more rapidly in vivo by heterozygous than by homozygous beige mice. However, these studies must be interpreted carefully. Beige mice also display defects in the functions of granulocytes, T cells, and B cells (Clark et al., 1982; Saxena et al., 1982).

Additional evidence supporting the important role of NK cells in tumor resistance came from studies in mice experimentally depleted of NK cell activity. Mice treated with NK 1.1 or asialo GM1 antibodies, which abrogate virtually all NK activity, displayed considerably lower resistance to growth of primary tumors in vivo and to the development of metastases (Barlozzari et al., 1985; Gorelik et al., 1982; Habu and Okumura, 1982; Kasai et al., 1981; Kawase et al., 1982; Lotzová, 1984; Lotzová et al., 1986d; Pollack, 1983). We have also shown increased growth of murine ascitic tumors after depletion of NK cell activity with NK 1.1 antibody and, conversely, decreased growth of the same tumors after NK cell activation (Lotzová et al., 1986d). Additional evidence supporting the role of NK cells in tumor resistance has come from the observation that mice inoculated with tumors mixed with NK cells displayed a lower tumor incidence and/or an increase in latent time of tumor appearance, as compared to mice transplanted with tumor cells only or with tumor cells mixed with NK-depleted populations (Kasai et al., 1979).

More direct evidence implicating NK cells in tumor defense came from in vivo reconstitution experiments. Transfer of NK cells or IL-2–dependent NK cell clones to mice or rats with low or no NK activity resulted in the expression of strong tumor resistance (Warner and Dennert, 1982). Given the role of NK cells in tumor defense, it may be significant that high levels of NK activity were detected in liver and lungs, frequently sites of metastasis, after treatment with various biological response modifying agents (Lotzová et al., 1984b, 1986c,d; Wiltrout et al., 1984). This suggests that NK cells in these tissues may play a role in local tumor defense. All of these data strongly support a role of NK cells in defense against tumors and suggest that augmentation of NK cell activity may be therapeutically beneficial. It should be noted that a number of carcinogens, such as urethane and dimethylbenzanthracene, and tumor promoters, including phorbol esters and teleocidin, suppress NK cell cytotoxic function (Herberman, 1980). Suppression of NK cell activity by carcinogens may lead to failure of NK cell antitumor surveillance mechanisms and thus contribute to tumor induction by these agents.

The role of NK cells in containment of human malignant diseases is more difficult to assess. However, several lines of evidence suggest NK cell involvement in defense against human cancers. It has been reported that individuals with defective NK cell cytotoxic activity show high susceptibility to neoplasia. One example of such an association is seen in patients with Chediak-Higashi syndrome, in whom an NK cell deficiency is paralleled by increased susceptibility to development of lymphoma (Roder et al., 1980). A relationship between susceptibility to hemopoietic malignancies and NK cell deficiencies has also been noted in patients with the X-linked lymphoproliferative syndrome and in patients with systemic lupus erythematosus (Hoffman, 1980; Sullivan et al., 1980).

We have shown in our laboratory that patients with leukemia and preleukemic disorders may display low or no NK cell activity, due to heterogeneous defects in NK cell cytotoxic mechanisms (Lotzová et al., 1979a,b, 1983a, 1985, 1986b, 1987b). These include defective NK cell ability to bind or kill tumor cells, failure of NK cell recycling, or to produce cytotoxic factors (Lotzová et al., 1987b). It is possible that the defective NK cell activity observed in patients with preleukemic disorders may predispose to the development of leukemia.

NK cell involvement in resistance to leukemia was initially suggested by the high sensitivity of leukemia-lymphoma targets to NK-mediated killing (Lotzová, 1983a, 1987c; Lotzová and Herberman, 1986b; Lotzová et al., 1979a, 1986b, 1987a). Furthermore, we showed that unstimulated NK cells are able to inhibit the growth of fresh clonogenic leukemic cells in vitro (Lotzová, 1987c; Lotzová and Savary, 1990b) and can kill fresh leukemic cells after activation with IL-2 (Lotzová et al., 1987b). Our investigations also showed that the defective NK cell cytotoxic activity in leukemic patients could be restored by culture of peripheral blood effector cells with IL-2 for one to several weeks (Lotzová, 1989; Lotzová et al., 1987b). The cultured effector cells were found to express normal cytotoxic mechanisms and were able to kill autologous leukemic cells (Lotzová, 1989; Lotzová and Savary, 1990c; Lotzová et al., 1989). Moreover, the effector cells involved in antileukemic activity could be unequivocally characterized as NK cells. Interestingly, cytotoxic NK cells could also be generated from the functionally inert bone marrow and from the spleen of CML patients after culture with IL-2 (Lotzová and Savary, 1987; Lotzová, 1989). These observations suggest that infusions of IL-2–activated NK cells, or NK cell activation by IL-2 treatment in vivo may be of therapeutic value in leukemic patients. It may also be possible to generate NK cells from bone marrow or spleen of CML patients undergoing splenectomy for purposes of adoptive therapy.

A possible role for NK cells in resistance to human solid tumors was suggested by the observed correlation between increased levels of peritoneal NK cell activity induced by viral oncolysates and regression of malignant ovarian ascites (Lotzová, 1986; Lotzová et al., 1984a, 1986a, 1987a). Additionally, our studies showed that IL-2–activated NK cells or their factors were tumoricidal for fresh or cultured ovarian tumor cells. These results indicate that NK cells are also active against solid human tumors and that patients with ovarian carcinomas may benefit from NK cell augmentation or adoptive transfer.

The observation that NK cells are the primary effector cell population that mediates LAK activity suggests that these cells may be responsible for at least some of the antitumor responses observed after adoptive LAK therapy. As discussed above, it is therefore possible that the anticancer effects of LAK therapy will be improved if NK cell-enriched populations are used in adoptive therapy trials. Such an approach would be valid, however, only in tumors that are sensitive to NK cell-mediated lysis. It is also of concern to note that IL-2–activated T cells, which have only marginal anticancer effects, can substantially kill or inhibit proliferation of bone marrow cells and hemopoietic progenitors (Lotzová and Savary, 1990a; Savary and Lotzová, 1990).

OTHER FUNCTIONS OF NK CELLS

Although NK cells were initially described based on their role in defense against tumor cells, accumulating evidence indicates that these cells have a substantially broader range of functional activities. NK cells appear to play an important role in resistance to infections caused by a number of viruses, parasites, and fungi (Welsh, 1986). Among the viruses sensitive to NK cell attack are herpes simplex, influenza, measles, and hepatitis viruses, cytomegalovirus, lymphocytic choriomeningitis virus, Newcastle disease virus, and Sendai virus (Herberman, 1980, 1982a; Lotzová, 1983a, 1985; Lotzová and Ades, 1989; Lotzová and Herberman, 1986a; Welsh, 1986). Other microorganisms that have been reported to be regulated by NK cells include the intraerythrocytic protozoan parasite, *Babesia microtii,* another parasite, *Trypanozoma cruzi,* and a yeast, *Cryptococcus neoformans* (Lotzová, 1985; Welsh, 1986).

Endogenous NK cells also play a role in regulation of the growth and function of hemopoietic and lymphoid cells. Unstimulated NK cells have been shown to inhibit the growth of granulocytic colonies in vitro and to suppress or potentiate the in vitro colony forming capacity of erythroid progenitors (Degliantoni et al., 1985; Hansson et al., 1982; Mangan et al., 1984; Pistoia et al., 1983). Colony forming activity of T cells was also reported to be upregulated by NK cells (Pistoia et al., 1983). Consistent with the regulatory role of NK cells in human hemopoiesis, we have shown that NK cells are one of the effector populations involved in rejection of murine bone marrow allografts (Lotzová and Savary, 1983; Lotzová et al., 1982a, 1983b). NK cells were also reported to be involved in regulation of differentiation and antibody production by B cells (Abruzzo and Rowley, 1983; Arai et al., 1983). NK cells also secrete a variety of biologically important cytokines, including IFN, IL-1, IL-2, IL-4, IL-5, NK cell cytotoxic factor (NKCF), lymphotoxin, and tumor necrosis factor (TNF) (Kasai et al., 1979; Lotzová and Herberman, 1986a, b; Scala et al., 1986; Wright et al., 1985). While some of these factors (e.g., NKCF) are directly involved in NK cell lytic mechanisms, others, including IFN, IL-2, and IL-4, may be involved in regulation of the growth and function of hemopoietic and lymphoid cells and their progenitors.

Less clearly understood, but potentially of great clinical relevance, is the reported involvement of NK cells in the development of graft vs. host disease after allogeneic bone marrow transplantation (Lopez et al., 1980) and in various nonmalignant disorders, chiefly autoimmune in origin (Merrill, 1986). In sum, the multitude of defense and regulatory functions in which NK cells are involved suggests that these cells play a major role in preserving the integrity of the organism.

NK CELL HETEROGENEITY AND MECHANISM OF KILLING

A great amount of information has been obtained during the last decade on various aspects of NK cell antitumor activity and on other NK cell functional activities. It is not presently clear whether all of these NK cell functions are mediated by the same or different NK cell subpopulations, or whether these distinct functional activities are related to the differentiation or activation state of the NK cell. The observation that NK cells acquire killing capacity after only brief stimulation with IL-2 or IFN suggests that some changes in functional activity—in this instance acquisition of tumor-killing properties—may be associated with different stages of NK cell activation. Considerable heterogeneity appears to exist within the NK cell system with regard to cell surface phenotypes, spectrum of target cell recognition and lysis, and physical and morphological characteristics of NK cells (Herberman and Callewaert, 1985; Lotzová, 1983a; Lotzová and Herberman, 1986a). Thus, it remains to be resolved whether the diversity of NK cells at both the functional and phenotypic levels reflects only their stage of differentiation and their degree of activation, or true clonal and functional heterogeneity.

An important point that has lately caused considerable confusion and which must be addressed in any consideration of NK cell heterogeneity is the distinction between NK cells per se and NK-like activity. As indicated earlier in this chapter, other cell populations, including T cells and monocyte/macrophages, can under certain conditions kill NK-susceptible targets such as K-562 and YAC-1 (Herberman, 1982b; Leung et al., 1983). These cells expressing NK-like cytotoxic activity have sometimes been loosely and incorrectly designated as NK cells. Such categorization is quite misleading as it is conceivable that more than one type of effector cell can kill the same target cells. Thus, an effector cell population that kills NK-sensitive targets should not be designated as an NK cell unless it expresses the typical NK cell surface phenotype and other NK-related morphological and functional characteristics.

Despite a very considerable increase in our knowledge of NK lytic mechanisms during the last few years, we still do not clearly know the molecular identity of the target cell structure(s) that trigger NK cell cytotoxicity or of the NK cell receptor involved in target cell recognition. Most data indicate that a diverse range of target cell structures may be recognized by NK cells. Glycoproteins corresponding to a wide range of molecular weights, ganglioside GM2, transferrin receptor, matrix laminin, and hemopoietic histocompatibility (Hh) antigens (Lotzová, 1983b) have all been implicated as NK cell target molecules (Durdik et al., 1980; Hiserodt et al., 1990; Ortaldo et al., 1983; Vodivelich et al., 1983). Furthermore, the LFA-1 molecule, which is expressed on NK cells, and its corresponding ligand I-CAM-1 were suggested to be important in the initial adherence phase of NK-target conjugation and CD2 and its natural ligand LFA-3 in the subsequent activation of NK cell lytic activity (Bierer and Burakoff, 1988). A laminin-like molecule expressed on highly purified LGL also was suggested to play a role in NK cell cytotoxicity (Hiserodt et al., 1990). However, further research is needed to delineate more precisely the role of these structures in NK cytotoxic activity or other functions.

Two groups of investigators recently provided some structural and functional evidence for a putative receptor molecule expressed on NK cells. One of these groups developed a monoclonal antibody (mAb36) to K-562 glycoproteins which also reacted with molecules on other NK-sensitive targets. An anti-idiotypic antibody to mAb36 was also prepared and was found to block binding and lysis of target cells by CD3⁻ LGL. In addition, the anti-idiotypic antibody induced direct lysis of target cells when cross-linked with a second antibody (Anderson, 1991; Anderson, et al., 1989; Ortaldo et al., 1989). The anti-idiotype antibody was reported to immunoprecipitate two proteins of 110 and 150 kD (Anderson, 1992). Another group of investigators prepared monoclonal antibodies against fish cytotoxic cells that were subsequently shown to recognize a receptor-like molecule on mouse, rat, and human NK cells. This receptor-like molecule was found to be active in signal transduction as indicated by stimulation of NK cell proliferation (Harris et al., 1989).

In contrast to these limited studies on the recognition phase of cytotoxicity, greater progress has been made in analysis of post-recognition events of the NK cell cytotoxic mechanism. It has been convincingly demonstrated that an important step in the NK cell lytic mechanism involves rearrangement of NK cell cytoplasmic granules and other organelles toward the target cell, followed by exocytosis of the granules and release of cytotoxic products, which then bind to the target cell membrane and lead to target cell lysis. The nature and characteristics of the cytotoxic molecules have been recently reviewed in detail (Young, 1989).

BIOLOGICAL SIGNIFICANCE OF NK CELLS

The primary function of NK cells appears to be related to homeostatic maintenance of the integrity of the organism. This is expressed as protection against a variety of micro-organisms, which potentially cause infections, and through regulation of growth and function of other cell populations, primarily those of hemopoietic and lymphoid origins. The role of NK cells in mediating these phenomena appears to precede that of defense against malignant cells. This hypothesis is consistent with the observation that NK cells appear early in phylogeny in primitive invertebrates. Malignant transformation is rare in these species.

NK cell involvement in defense against infections, and their role in regulation of differentiation, growth, and function of hemopoietic and lymphoid cells may also under-lie a basic mechanism of homeostatic control in man. Consequently, changes in the num-bers or functions of NK cells may lead to disturbances of the equilibrium of hemopoiesis and lymphopoiesis, to suboptimal or supraoptimal production of various cytokines, or to changes in NK cell-mediated antimicrobial and antitumor defenses. These changes in NK cell numbers or activity may ultimately result in the development of benign or malig-nant diseases of the hemopoietic and lymphoid systems, including leukemia, aplastic ane-mia, and various types of immunodeficiency and autoimmune diseases.

The cytocidal activity of NK cells against leukemias and ovarian cancers, and the more recent evidence that these cells are also able to lyse human renal carcinomas and squamous cell cancers (Lotzová and Savary, 1989; 1990c), suggests that NK cells may have potential therapeutic applications. For instance, in vitro activation and expansion of NK cells with IL-2 and subsequent in vivo adoptive transfer of these cells may provide a useful therapy for patients with NK cell–sensitive leukemias, ovarian cancers, and other tumors. Alternatively, it may be possible to directly activate NK cells in vivo. A similar approach may be used to potentiate NK cell–mediated antimicrobial defenses in individ-uals with certain infectious diseases or in cancer patients treated with chemotherapy. NK cells may also have therapeutic effects in leukemic patients receiving autologous bone marrow transplants. It may be possible to use NK cells to eradicate residual leukemic cells from the marrow before autologous bone marrow transplantation. Furthermore, transfer of NK cells before bone marrow transplantation may be useful in eradication of residual leukemic cells in the prospective marrow recipient.

In summary, there is now substantial evidence that NK cells play an important role in defense against cancer and infection, and that they may be involved in the regulation of various other biological processes. We hope that a greater understanding of the basic immunobiology of NK cells will allow us to more efficiently use these cells for therapeutic purposes.

Acknowledgment

I wish to acknowledge Cetus Corporation, Emeryville, California, for their generous gift of IL-2 and Pamela Baxter, for expert assistance in the preparation of this manuscript. Work from this laboratory is supported by grant CA 39632 from the National Cancer Institute.

References

Abo, T., Balch, B. M. (1981). A differentiation antigen of human NK cells and K cells identified by a monoclonal antibody (HNK-1). *J. Immunol. 127*:1024–1029.

Abruzzo, L. B., and Rowley, D. A. (1983). Homeostasis of the antibody response: Immunoregulation by NK cells. *Science 222*:581–585.

Allavena, P., Introna, M., Mangioni, C., and Mantovani, A. (1981). Inhibition of natural killer activity by tumor-associated lymphoid cells from ascites ovarian carcinomas. *J. Natl. Cancer Inst. 67*:319–325.

Anderson, S. (1992). Molecular cloning of human and murine natural killer cell tumor recognition proteins. In *NK Cell-Mediated Cytotoxicity: Receptors, Signalling and Mechanisms* (Lotzová, E., and Herberman, R. B., eds.). Boca Raton, Fla.: CRC Press.

Anderson, S., Frey, J., Roder, J., Young, H., and Ortaldo, J. (1989). Analysis of putative NK receptor gene product (Abstract). *Nat. Immun. Cell Growth Regul. 8*:118–119.

Arai, S., Yamamoto, H., Itoh, K., and Kumagi, I. (1983). Suppressive effect of human natural killer cells on pokeweed mitogen-induced B cell differentiation. *J. Immunol. 131*:651–657.

Barlozzari, T., Leonhardt, J., Wiltrout, R., Herberman, R. B., and Reynolds, J. W. (1985). Direct evidence for the role of LGL in the inhibition of experimental tumor metastasis. *J. Immunol. 134*:2783–2789.

Bash, J. A., and Vogen, D. (1984). Cellular immunosenescence in F344 rats: Decreased natural killer (NK) cell activity involves changes in regulatory interactions between NK cells, interferon, prostaglandin and macrophages. *Mech. Aging Dev. 24*:49–65.

Bierer, B. E., and Burakoff, S. J. (1988). T cell adhesion molecules. *FASEB J. 2*:2584–2590.

Biron, C. A., and Welsh, R. M. (1986). Antigenic distinctions and morphological similarities between proliferating natural killer and cytotoxic T cells. In *Natural Immunity, Cancer and Biological Response Modification* (Lotzová, E., and Herberman, R. B., eds.). Basel: S. Karger, pp. 289–297.

Blair, P. B., Staskawicz, M. O., and Sam, J. S. (1983). Inhibitor cells in spleens of mice with low natural killer activity. *J. Natl. Cancer Inst. 71*:571–577.

Bordignon, C., Villa, F., Allavena, P., Introna, M., Biondi, A., Avallone, R., and Mantovani, A. (1982). Inhibition of natural killer activity by human bronchoalveolar macrophages. *J. Immunol. 129*:587–591.

Brooks, C. F., and Moore, M. (1986). Presentation of a soluble bacterial antigen and cell surface alloantigens by large granular lymphocytes (LGL) in comparison with monocytes. *Immunology 58*:343–350.

Brunda, M. J., and Holden, H. T. (1980). Prostaglandin-mediated inhibition of natural killer cell activity. In *Natural Cell-Mediated Immunity against Tumors* (Herberman, R. B., ed.). New York: Academic Press, pp. 721–734.

Brunda, M. J., Taramelli, D., Holden, J. T., and Varesio, L. (1982). Suppression of murine natural killer cell activity by normal peritoneal macrophages. In *NK Cells and Other Natural Effector Cells* (Herberman, R. B., ed.) New York: Academic Press, pp. 535–548.

Chang, Z. L., Hoffman, T., Bonvini, E., Stevenson, H. C., and Herberman, R. B. (1983). Spontaneous cytotoxicity of human and mouse tumor cell lines by peripheral blood mononuclear cells: Contributions of adherent and nonadherent NK-like cells. *Scand. J. Immunol. 18*:439–449.

Clark, E. A., Roths, J. B., Murphy, E. D., Ledbetter, J. A., and Clagett, J. A. (1982). The beige *(bg)* gene influences the development of autoimmune disease in SB/Le male mice. In *NK Cells and Other Natural Effector Cells.* (Herberman, R. B., ed.). New York: Academic Press, pp. 301–306.

DeBoer, K. P., Kleinman, R., and Teodorescu, J. (1976). Identification and separation by bacterial adherence of human lymphocytes that suppress natural cytotoxicity. *J. Immunol. 126*:276–281.

Decker, J. M., Elmholt, A., and Muchmore, A. V. (1981). Spontaneous cytotoxicity mediated by invertebrate mononuclear cells toward normal and malignant vertebrate targets: Inhibition by defined mono- and disaccharides. *Cell. Immunol. 59*:161–170.

Degliantoni, G., Perussia, B., Mangoni, L., and Trinchieri, G. (1985). Inhibition of bone marrow colony formation by natural killer cells and by natural killer cell–derived colony-inhibiting activity. *J. Exp. Med. 161*:1152–1168.

Durack, D. T. (1984). Opportunistic infections and Kaposi's sarcoma in homosexual men. *New Engl. J. Med. 305*:1465–1467.

Durdik, J. M., Beck, B. N., Clark, E. A., and Henney, C. A. (1980). Characterization of a lymphoma cell variant selectively resistant to natural killer cells. *J. Immunol. 125*:683–688.

Eremin, O., Coombs, R.R.A., Plumb, D., and Ashby, J. (1978). Characterization of the human natural killer (NK) cell in blood and lymphoid organs. *Int. J. Cancer. 21*:41–50.

Ferrarini, M., Candoni, A., Frazi, T., Ghigliotti, C., Leprini, A., Zicca, A., and Grossi, C. E. (1980). Ultrastructural and cytochemical markers of human lymphocytes. In *Thymus, Thymic Hormones and T Lymphocytes* (Aiuti, F., and Wigzell, H., eds.). New York: Academic Press, pp. 39–48.

Fleischer, B. (1980). Effector cells in avian spontaneous and antibody-dependent cell-mediated cytotoxicity. *J. Immunol. 125*:1161–1166.

Gorelik, E., Wiltrout, R., Okumura, K., Habu, S., and Herberman, R. B. (1982). Acceleration of metastatic growth in anti-asialo GM1-treated mice. In *NK Cells and Other Natural Effector Cells* (Herberman, R. B., ed.). New York: Academic Press, pp. 1131–1337.

Grimm, E. A., Mazumder, A., Zhang, H. Z., and Rosenberg, S. A. (1982). Lymphokine-activated killer cell phenomenon: Lysis of natural killer–resistant fresh solid tumor cells by interleukin-2–activated autologous human peripheral blood lymphocytes. *J. Exp. Med. 155*:1823–1841.

Gustafsson, A., and Lundgren, E. (1981). Augmentation of natural killer cells involves both enhancement of lytic machinery and expression of new receptors. *Cell. Immunol. 62*:367–376.

Habu, S., and Okumura, K. (1982). Evidence for in vivo reactivity against transplantable and primary tumors. In *NK Cells and Other Natural Effector Cells* (Herberman, R. B., ed.). New York: Academic Press, pp. 1323–1330.

Hackett, J., Jr., Bennett, M., and Kumar, V. (1986). Natural killer cell precursors in the bone marrow are distinct from lymphoid and myeloid progenitors. In *Natural Immunity, Cancer and Biological Response Modification* (Lotzová, E., and Herberman, R. B., eds.). Basel: S. Karger, pp. 40–49.

Hansson, M., Beran, M., Andersson, A., and Kiessling, R. (1982). Inhibition of in vitro granulopoiesis by autologous or allogeneic human NK cells. *J. Immunol. 129*:126–132.

Harris, D. T., Friedman, L. J., Devlin, R. B., McKinnon, K. P., Koren, H. S., and Evans, D. L. (1989). Putative antigen receptor on human natural killer cells is a signal-transducing molecule (Abstract). *Nat. Immun. Cell Growth Regul.* 8:120–121.

Henney, C. S., Kuribayashi, K., Kern, D. E., and Gillis, S. (1981). Interleukin-2 augments natural killer cell activity. *Nature* 291:335–338.

Herberman, R. B. (ed.) 1980. *Natural Cell-Mediated Immunity against Tumors.* New York: Academic Press.

Herberman, R. B. (ed.) 1982a. *NK Cells and Other Natural Effector Cells.* New York: Academic Press.

Herberman, R. B. 1982b. Overview and perspective: Natural resistance mechanism. In *Macrophages and Natural Killer Cells: Regulation and Function* (Norman, S. J., and Sorkin, E., eds.). New York: Plenum Press, pp. 799–808.

Herberman, R. B. (1986). Natural killer cells. *Annu. Rev. Med.* 37:347–352.

Herberman, R. B., Balch, C., Bolhuis, R., Golub, S., Hiserodt, J., Lanier, L. L., Lotzová, E., Phillips, J. H., Riccardi, C., Ritz, J., Santoni, A., Schmidt, R. E., Uchida, A., and Vujanovic, N. (1987). Most lymphokine activated killer (LAK) activity mediated by blood and splenic lymphocytes is attributable to stimulation of natural killer (NK) cells by interleukin-2. *Immunol. Today* 8:178–181.

Herberman, R. B., and Callewaert, D. M. (eds.) (1985). *Mechanisms of Cytotoxicity by NK Cells.* Orlando, Fla: Academic Press.

Herberman, R. B., and Ortaldo, J. R. (1981). Natural killer cells: Their role in defense against disease. *Science* 214:24–30.

Hercend, T., Griffin, J. D., Bensussan, A., Schmidt, R. E., Edson, M. A., Brennan, A., Murray, C., Daley, J. F., Schlossman, S. F., and Ritz, J. (1985). Generation of monoclonal antibodies to a human natural killer clone. Characterization of two natural killer-associated antigens, NKH1 and NKH2, expressed on subsets of large granular lymphocytes. *J. Clin. Invest.* 75:932–943.

Hinuma, S., Abo, T., Kumagai, K., and Mata, M. (1980). The potent activity of fresh water fish kidney cells in cell-killing: I. Characterization and species-distribution of cytotoxicity. *Dev. Comp. Immunol.* 4:653–666.

Hiserodt, J. C., van de Brink, M.R.M., and Schwarz, R. E. (1990). Surface structures involved in tumor cell recognition by fresh and IL-2–activated natural killer cells. In *Interleukin-2 and Killer Cells in Cancer* (Lotzová, E. and Herberman, R. B. eds.). Boca Raton, Fla.: CRC Press, pp. 351–361.

Hoffman, T. (1980). Natural killer function in systemic lupus erythematosis. *Arthritis Rheum.* 23:30–35.

Hostetter, R. K., and Cooper, E. L. (1972). Coelomocytes as effector cells in the earthworm immunity. *Immunol. Commun.* 1:155–183.

Introna, M., Allavena, P., Spreafico, F., and Mantovani, A. (1981). Inhibition of human natural killer activity by cyclosporin A. *Transplantation 31:113–116.*

Itoh, K., Tilden, A. B., Kumagai, K., and Balch, C. M. (1985). Leu-11⁺ lymphocytes with natural killer (NK) activity are precursors of recombinant interleukin-2 (IL-2)-induced activated (AK) cells. *J. Immunol.* 134:802–806.

Kasahaya, T., Djeu, J. Y., Dougherty, S. F., and Oppenheim, J. J. 1983. Capacity of human large granular lymphocytes (LGL) to produce multiple lymphokines: interleukin-2, interferon and colony-stimulating factor. *J. Immunol. 131:*2379–2385.

Kasai, M., Leclerc, J. C., McVay-Boudreau, L., Shen, F. W., and Cantor, H. (1979). Direct evidence that natural killer cells in nonimmune spleen cell populations prevent tumor growth in vivo. *J. Exp. Med.* 149:1260–1264.

Kasai, M., Yoenda, T., Habu, S., Maruyama, Y., Okumura, K., and Tokunaga, T. (1981). In vivo effect of anti-asialo GM-1 antibody on natural killer activity. *Nature* 291:334–335.

Kawase, I., Urdahl, D. L., Brooks, C. G., and Henney, C. S. (1982). Selective depletion of NK cell activity in vivo and its effect on the growth of NK-sensitive and NK-resistant tumor cell variants. *Int. J. Cancer* 29:567–574.

Kay, H. D., and Horwitz, D. A. (1981). Evidence by reactivity with hybridoma antibodies for a possible myeloid origin of peripheral blood cells active in natural cytotoxicity and antibody-dependent cell-mediated cytotoxicity. *J. Clin. Invest.* 66:847–851.

Kay, H. D., and Smith, D. L. (1983). Regulation of human lymphocyte-mediated natural killer (NK) cell activity: I. Inhibition in vitro by peripheral blood granulocytes. *J. Immunol.* 130:475–483.

Knapp, W., Rieber, P., Dorken, B., Schmidt, R. E., Stein, H., Boren, A.E.G.Kr.v.d. (1989). Towards a better definition of human leucocyte surface molecules. *Immunol. Today* 10:253–258.

Koren, H. S., Anderson, S. J., Fisher, D. G., Copeland, C. S., and Jensen, P. J. (1981). Regulation of human natural killing: I. The role of monocytes, interferon and prostaglandins. *J. Immunol.* 127:2007–2013.

Kumar, V., Ben-Ezra, J., Bennett, M., and Sonnenfeld, G. (1979). Natural killer cells in mice treated with ^{89}strontium: Normal target binding cell numbers but inability to kill even after interferon administration. *J. Immunol.* 123:1832–1838.

Kuribayashi, K., Gillis, S., Kern, D. E., and Henney, C. S. (1981). Murine NK cell cultures: Effects of interleukin-2 and interferon on cell growth and cytotoxic reactivity. *J. Immunol.* 126:2321–2327.

Lanier, L. L., Le, A. M., Civin, C. I., Loken, M. R., and Phillips, J. H. (1986). The relationship of CD16 (Leu-11) and Leu-19 (NKH1) antigen expression on human peripheral blood NK cells and cytotoxic T lymphocytes. *J. Immunol.* 136:4480–4486.

Lanier, L. L., and Phillips, J. H. (1986). A schema for the classification of cytotoxic lymphocytes based on T cell antigen receptor gene rearrangement and Fc receptor (CD 16) or NKH-1/Leu 19 antigen expression. In *Natural Immunity, Cancer and Biological Response Modification* (Lotzová, E., and Herberman, R. B., eds.). Basel: S. Karger, pp. 10–13.

Leibold, W., Janotte, G., and Peter H. H. (1980). Spontaneous cell-mediated cytotoxicity (SCMC) in various mammalian species and chickens: Selective reaction pattern and different mechanisms. *Scand. J. Immunol.* 11:203–222.

Leung, K. H., Fischer, D. G., and Koren, H. S. (1983). Erythromyeloid tumor cells (K-562) induce PGE synthesis in human peripheral blood monocytes. *J. Immunol.* 131:445–449.

Linthicum, D. S., Stein, E. A., Marks, D. H., and Cooper, E. L. (1977). Electron-microscopic observations of normal coelomocytes from the earthworm, *Lumbricus terrestris. Cell. Tiss. Res.* 185:315–330.

Lopez, C., Kirkpatrick, D., Livnat, S., and Storb, R. (1980). Natural killer cells in bone marrow transplantation. *Lancet* 2:1025.

Lotzová, E. (1980). *C. parvum* mediated suppression of the phenomenon of natural killing and its analysis. In *Natural Cell-Mediated Immunity against Tumors* (Herberman, R. B., ed.). New York: Academic Press, pp. 735–752.

Lotzová, E. (1983a). Function of natural killer cells in various biological phenomena. *Surv. Synth. Path. Res.* 2:41–46.

Lotzová, E. (1983b). Hematopoietic histocompatibility: Genetic and immunological aspects. *Comp. Immunol.* 3:468–493.

Lotzová, E. (1984). The role of natural killer cells in immune surveillance against malignancies. *Cancer Bull.* 36:215–226.

Lotzová, E. (1985). Effector immune mechanisms in cancer. *Nat. Immun. Cell Growth Regul.* 4:293–304.

Lotzová, E. (1986). Therapeutic possibilities of virus-modified tumor cell extract and interleukin-2 in human ovarian cancer. *Nat. Immun. Cell Growth Regul.* 5:277–282.

Lotzová, E. (1987a). Human natural killer cells: Their role and possible therapeutic application in leukemia. *Clin. Immunol. News.* 8:56–60.

Lotzová, E. (1987b). Interleukin-2-generated killer cells: Their characterization and role in cancer therapy. *Cancer Bull. 39*:30–38.

Lotzová, E. (1987c). Possible application of natural killer cells and interleukin-2 in therapy of human leukemia. In *Cellular Immunotherapy of Cancer* (Truitt, R. L., Gale, R. P., and Bortin, M. M., eds.). New York: Alan R. Liss, pp. 259–266.

Lotzová, E. (1989). Cytotoxicity and clinical application of activated NK cells. *Med. Oncol. Tumor Pharmacother. 6*:93–98.

Lotzová E. (1990). Role of human circulating and tumor-infiltrating lymphocytes in cancer defense and treatment. *Nat. Immun. Cell Growth Regul. 9*:253–264.

Lotzová, E., and Ades, E. W. (1989). Natural killer cells: Definition, heterogeneity, lytic mechanism, functions, and clinical application. *Nat. Immun. Cell Growth Regul. 8*:1–9.

Lotzová, E., and Herberman, R. B. (eds.) (1986a). *Immunobiology of Natural Killer Cells*, vols. 1, 2. Boca Raton, Fla: CRC Press.

Lotzová, E., and Herberman, R. B. (eds.) (1986b). *Natural Immunity, Cancer and Biological Response Modification.* Basel: S. Karger.

Lotzová, E., and Herberman, R. B. (1987). Reassessment of LAK phenomenology—a review. *Nat. Immun. Cell Growth Regul. 6*:109–115.

Lotzová, E., and McCredie, K. B. (1978). Natural killer cells in mice and man and their possible biological significance. *Cancer Immunol. Immunother. 4*:215–221.

Lotzová, E., McCredie, K. B., Maroun, J. A., Dicke, K. A., and Freireich, E. J. (1979a). Some studies on natural killer cells in man. *Transplant. Proc. 11*:1390–1392.

Lotzová, E., McCredie, K. B., Muesse, L., Dicke, K. A., and Freireich, E. J. (1979b). Natural killer cells in man: Their possible involvement in leukemia and bone marrow transplantation. In *Experimental Hematology Today* (Baum, S. J., and Ledney, G. S., eds.). New York: Springer-Verlag, pp. 207–213.

Lotzová, E., Pollack, S. B., and Savary, C. A. (1982a). Direct evidence for the involvement of natural killer cells in bone marrow transplantation. In *NK Cells and Other Natural Effector Cells* (Herberman, R. B., ed.). New York: Academic Press, pp. 1535–1540.

Lotzová, E., and Savary, C. A. (1983). Natural resistance to foreign hemopoietic transplants: A possible model of leukemia surveillance. In *13th International Cancer Congress, Biology of Cancer* (Mirand, E. A., Hutchinson, W. B., and Mihich, E., eds.). New York: Alan R. Liss, pp. 125–135.

Lotzová, E., and Savary, C. A. (1984). Stimulation of NK cell cytotoxic potential of normal donors by two species of recombinant alpha interferon. *J. Interferon Res. 4*:201–213.

Lotzová, E., and Savary, C. A. (1986). Regulation of NK cell activity by suppressor cells. In *Immunobiology of Natural Killer Cells*, vol. II. (Lotzová, E., and Herberman, R. B., eds.). Boca Raton, Fla.: CRC Press, pp. 163–177.

Lotzová, E., and Savary, C. A. (1987). Generation of NK cell activity from human bone marrow. *J. Immunol. 139*:279–284.

Lotzová, E., and Savary, C. A. (1989). Function of interleukin-2 activated NK cells in leukemia resistance and treatment. In *Interleukin-2 and Killer Cells in Cancer* (Lotzová, E., and Herberman, R. B., eds.). Boca Raton, Fla.: CRC Press, pp. 41–56.

Lotzová, E., and Savary, C. A. (1990a). Dichotomy in reactivity of IL-2 activated human lymphocyte subsets against normal and malignant hematopoietic cells. In *Proceedings of the 6th International NK Cell Workshop* (Schmidt, R. E., ed.). Basel: S. Karger, pp. 280–284.

Lotzová, E., and Savary, C. A. (1990b). Function of interleukin-2 activated NK cells in leukemia resistance and treatment. In *Interleukin-2 and Killer Cells in Cancer* (Lotzová, E., and Herberman, R. B., eds.). Boca Raton, Fla.: CRC Press, pp. 41–56.

Lotzová, E., and Savary, C. A. (1990c). Growth kinetics, function and characterization of lymphocytes infiltrating ovarian tumors. In *Interleukin-2 and Killer Cells in Cancer* (Lotzová, E., and Herberman, R. B., eds.). Boca Raton, Fla.: CRC Press, pp. 153–161.

Lotzová, E., Savary, C. A., Dicke, K. A., and Jagannath, S. (1989a). Role of NK cells in tumor cell growth and eradication. In *Experimental Hematology Today* (Baum, S. J., Dicke, K. A., Lotzová, E., and Pluznik, D. H. (eds.). New York: Springer-Verlag, pp. 17–21.

Lotzová, E., Savary, C. A., Freedman, R. S., and Bowen, J. M. (1984a). Natural killer cell cytotoxic potential of patients with ovarian carcinoma and its modulation with virus-modified tumor cell extract. *Cancer Immunol. Immunother.* *17*:124–129.

Lotzová, E., Savary, C. A., Freedman, R. S., and Bowen, J. M. (1986a). Natural immunity against ovarian tumors. *Comp. Immunol. Microbiol. Infect. Dis.* *9*:269–275.

Lotzová, E., Savary, C. A., Freedman, R. S., and Bowen, J. M. (1987a). Natural killer cell antitumor activity of patients with ovarian carcinoma. Induction of cytotoxicity by viral oncolysates and interleukin-2. In *Diagnosis and Treatment Strategy for Gynecological Cancer* (Rutledge, F. M., Freedman, R. S., and Gershenson, C. M., eds.). Austin: University of Texas Press, pp. 123–126.

Lotzová, E., Savary, C. A., Freedman, R. S., Edwards, C. L., and Wharton, J. T. (1988). Recombinant IL-2-activated NK cells mediate LAK activity against ovarian cancer. *Int. J. Cancer* *42*:225–231.

Lotzová, E., Savary, C. A., Gutterman, J. U., and Hersh, E. M. (1982b). Natural killer cell-mediated cytotoxicity: Modulation by partially purified and cloned interferon-α. *Cancer Res.* *42*:2480–2488.

Lotzová, E., Savary, C. A., and Herberman, R. B. (1986b). Antileukemia reactivity of endogenous and IL-2 activated NK cells. In *Natural Immunity, Cancer and Biological Response Modification* (Lotzová, E., and Herberman, R. B., eds.). Basel: S. Karger, pp. 177–195.

Lotzová, E., Savary, C. A., and Herberman, R. B. (1987b). Induction of NK cell activity against fresh human leukemia in culture with interleukin-2. *J. Immunol.* *138*:2718–2727.

Lotzová, E., Savary, C. A., Herberman, R. B., and McCredie, K. B. (1987c). The role of natural killer cells in resistance to leukemia: Generation of antileukemia activity in NK-deficient leukemic patients. In *Novel Approaches in Cancer Therapy* (Lapis, K., and Eckhardt, S., eds.). Budapest: Akademiai Kiado, pp. 305–312.

Lotzová, E., Savary C. A., and Keating, M. J. (1983a). Leukemia diseased patients exhibit multiple defects in natural killer cell lytic machinery. *Exp. Hematol.* *10*:83–95.

Lotzová, E., Savary, C. A., Keating, M. J., and Hester, J. P. (1985). Defective NK cell mechanism in patients with leukemia. In *Mechanism for Cytotoxicity by NK Cells* (Herberman, R. B., and Callewaert, D., eds.). New York: Academic Press, pp. 507–519.

Lotzová, E., Savary, C. A., Khan, A., and Stringfellow, D. A. (1984b). Stimulation of natural killer cells in two random-bred strains of athymic rats by interferon-inducing pyrimidinone. *J. Immunol.* *132*:2566–2570.

Lotzová, E., Savary, C. A., Lowlachi, M., and Murasko, D. M. (1986c). Cytotoxic and morphological profile of endogenous and pyrimidinone-activated murine NK cells. *J. Immunol.* *136*:732–740.

Lotzová, E., Savary, C. A., and Pollack, S. B. (1983b). Prevention of rejection of allogeneic bone marrow transplants by NK 1.1 antiserum. *Transplantation* *34*:490–494.

Lotzová, E., Savary, C. A., Pollack, S. B., and Hanna, N. (1986d). Induction of tumor immunity and natural killer cell cytotoxicity in mice by 5-halo-6-phenyl pyrimidinone. *Cancer Res.* *46*:5004–5008.

Lotzová, E., Savary, C. A., Pollock, R. D., and Fuchshuber, P. (1990). Immunologic and clinical aspects of natural killer cells in human leukemia. *Nat. Immun. Cell Growth Regul.* *9*:173–181.

Lotzová, E., Savary, C. A., Totpal, K., Schachner, J., Lichtigen, B., McCredie, K. B., and Freireich, E. J. (1991). Highly oncolytic adherent lymphocytes: Therapeutic relevance for leukemia. *Leukemia Res.* *15*:245–254.

Mangan, K. F., Harnett, M. R., Matis, S. A., Winkelstein, A., and Abo, T. (1984). Natural killer cells suppress human erythroid stem cell proliferation in vitro. *Blood* *63*:260–269.

Melder, R. J., Whiteside, T. L., Vujanovic, N. L., Hiserodt, J. C., and Herberman, R. B. (1988). A new approach to generating antitumor effector cells for adoptive immunotherapy using human adherent lymphokine-activated killer cells. *Cancer Res.* 48:3461–3469.

Merrill, J. E. (1986). The implications of aberrant natural killer (NK) cell activity in nonmalignant chronic diseases. In *Immunobiology of Natural Killer Cells,* vol. 2. (Lotzová, E., and Herberman, R. B., eds.). Boca Raton, Fla.: CRC Press, pp. 75–87.

Nair, M.P.N., Schwartz, S. A., Fernandes, F., Pahwa, R., Ikehara, S., and Good, R. A. (1981). Suppression of natural killer (NK) cells activity of spleen cells by thymocytes. *Cell. Immunol.* 58:9–18.

Ng, A-K., Induiveri, F., Pellegrino, M. A., Molinaro, G. A., Quaranta, V., and Ferrone, S. (1980). Natural cytotoxicty and antibody-dependent cellular cytotoxicity of human lymphocytes depleted of HLA-DR bearing cells with monoclonal HLA-DR antibodies. *J. Immunol.* 124:2336–2340.

Oehler, J. R., and Herberman, R. B. (1978). Natural cell-mediated cytotoxicity in rats: III. Effects of immunopharmacologic treatment on natural reactivity and on reactivity augmented by polyinosinic-polycytidylic acid. *Int. J. Cancer* 21:111–119.

Oldham, R. K., Forbes, J. T., and Niblach, G. D. (1978). Natural killer activity in human thoracic duct lymphocytes. *Proc. Am. Assoc. Cancer Res.* 21:210.

Ortaldo, J. R., Kantor, R., Sega, D., Bolhuis, R. H., and Bino, T. (1989). Identification of a proposed NK receptor. In *Natural Killer Cells and Host Defense* (Ades, E. W., and Lopez, C., eds.). 5th Int. Natural Killer Cell Workshop, Hilton Head, S.C., 1988. Basel: S. Karger, pp. 221–226.

Ortaldo, J. R., Lewis, J. T., Braatz, J., Mason, A., and Henkart, P. (1983). Isolation of target antigens from NK susceptible targets. In *Intercellular Communication in Leucocyte Function* (Parker, J. W., and O'Brien, R. L., eds.). New York: Wiley, pp. 551–558.

Ortaldo, J. R., Sharrow, S. O., Timonen, T., and Herberman, R. B. (1981). Analysis of surface antigens on highly purified human NK cells by flow cytometry with monoclonal antibodies. *J. Immunol.* 126:2401–2409.

Ortaldo, J. R., Timonen, T. T., and Herberman, R. B. (1984). Inhibition of activity of human NK and K cells by simple sugars: Discrimination between binding and post-binding events. *Clin. Immunol. Immunopathol.* 31:439–443.

Perussia, B., Fanning, V., and Trinchieri, G. (1983). A human NK and K cell subset shares with cytotoxic T cell expression of the antigen recognized by antibody OKT8. *J. Immunol.* 131:223–231.

Perussia, B., Ramoni, C., Anegow, I., Cuturi, M. C., Faust, J., and Trinchieri, G. (1987). Preferential proliferation of natural killer cells cocultured with B lymphoblastoid cell lines. *Nat. Immun. Cell Growth Regul.* 6:171–188.

Perussia, B., Starr, S., Abraham, S., Fanning, V., and Trinchieri, G. (1983). Human natural killer cells analyzed by B73.1 monoclonal antibody blocking Fc receptor functions: I. Characterization of lymphocyte subset reactive with B73.1. *J. Immunol.* 130:2133–2141.

Pettey, C. L., and McKinney, E. C. (1980). Effect of decreased environmental temperature on spontaneous cytotoxicity in the nurse shark. *Proc. Fed. Am. Soc. Exp. Biol.* 39:934.

Pistoia, V., Nocera, A., Peralta, A., Leprini, A., Ghio, R., and Ferrarini, M. (1983). Large granular lymphocytes have a regulatory role on the growth of human peripheral blood cells and erythroid colonies. *Surv. Synth. Path. Res.* 2:47–56.

Pollack, S. (1983). In vivo functions of natural killer cells. *Surv. Synth. Path. Res.* 2:93–106.

Pollock, R. E., and Lotzová, E. (1987). Surgical stress related suppression of NK activity: A possible role in tumor metastasis. *Nat. Immun. Cell Growth Regul.* 6:269–275.

Ramsdell, F. J., Gray, J. D., and Golub, S. H. (1988). Similarities between LAK cells derived from human thymocytes and peripheral blood lymphocytes: Expression of NKH1 and CD3 antigens. *Cell Immunol.* 114:209–221.

Reynolds, C. W., Sharrow, S. O., Ortaldo, J. R., and Herberman, R. B. (1981). Natural killer activity in the rat: II. Analysis of surface antigens on LGL by flow cytometry. *J. Immunol.* 127:2204–2208.

Reynolds, C. W., and Ward, J. M. (1986). Tissue and organ distribution of NK cells. In *Immunobiology of Natural Killer Cells*, vol. 1. (Lotzová, E., and Herberman, R. B., eds.). Boca Raton, Fla.: CRC Press, pp. 63–71.

Riccardi, C., Santoni, A., Barlozzari, T., Cesarini, C., and Herberman, R. B. (1981). Suppression of natural killer (NK) activity by splenic adherent cells of low NK-reactive mice. *Int. J. Cancer* 28:811–818.

Roder, J. C., Haliotis, T., Kline, M., Korec, S., Jett, J. R., Ortaldo, J., Herberman, R. B., Katz, P., and Fauci, A. S. (1980). A new immunodeficiency disorder in humans involving NK cells. *Nature* 284:533–536.

Roder, J. C., Karre, K., and Kiessling, R. (1981). Natural killer cells. *Prog. Allergy.* 28:66–159.

Saksela, E., Timonen, T., and Coubel, K. (1979). Human natural killer cell activity is augmented by interferon via recruitment of "pre NK" cells. *Scand. J. Immunol.* 10:257–266.

Savary, C. A., and Lotzová, E. (1978). Suppression of natural killer cell cytotoxicity by splenocytes from *Corynebacterium parvum*–injected, bone marrow–tolerant, and infant mice. *J. Immunol.* 120:239–243.

Savary, C. A., and Lotzová, E. (1986). Phylogeny and ontogeny of NK cells. In *Immunobiology of Natural Killer Cells*, vol. 1 (Lotzová, E., and Herberman, R. B., eds.). Boca Raton, Fla.: CRC Press, pp. 45–61.

Savary, C. A., and Lotzová, E. (1990). Down-regulation of human bone marrow cells and their progenitors by IL-2 activated lymphocytes. In *Interleukin-2 and Killer Cells in Cancer* (Lotzová, E., and Herberman, R. B., eds.). Boca Raton, Fla.: CRC Press, pp. 89–96.

Saxena, R. K., Saxena, Q. B., and Adler, W. H. (1982). Defective T-cell response in beige mutant mice. *Nature* 295:240–241.

Scala, G., Djeu, J. Y., Allavena, P., Kashara, T., Ortaldo, J. R., Herberman, R. B., and Oppenheim, J. J. (1986). In *Immunobiology of Natural Killer Cells,* vol. 2. (Lotzová, E., and Herberman, R. B., eds.). Boca Raton, Fla.: CRC Press, pp. 133–144.

Schmidt, R. E., and Ritz, J. (1986). Clonal analysis of natural killer cell lineage and function. In *Natural Immunity, Cancer and Biological Response Modification* (Lotzová, E., and Herberman, R. B., eds.). Basel: S. Karger, pp. 19–33.

Seaman, W. E., Gindhart, T. D., Blackman, M. A., Dalal, B., Talal, N., and Werb, Z. (1982). Suppression of natural killer activity in vitro by monocytes and polymorphonuclear leukocytes. *J. Clin. Invest.* 69:876–888.

Sharma, J. M., and Coulson, B. D. (1979). Presence of natural killer cells in specific-pathogen-free chickens. *J. Natl. Cancer Inst.* 63:527–531.

Silva, A., Bonavida, B., and Targan, S. (1980). Model of action of interferon-mediated modulation of natural cytotoxic activity: Recruitment of pre-NK cells and enhanced kinetics of lysis. *J. Immunol.* 125:479–484.

Stutman, O. (1975). Immunodepression and malignancy. *Adv. Cancer Res.* 22:261–422.

Stutman, O., Figarella, E., Paige, C. J., and Lattime, E. C. (1980). Natural cytotoxic (NC) cells against solid tumors in mice: General characteristics and comparison to natural killer (NK) cells. In *Natural Cell-Mediated Immunity against Tumors* (Herberman, R. B., ed.). New York: Academic Press, pp. 187–229.

Sullivan, J. L., Byron, K. S., Brewster, F. F., and Purtilo, D. T. (1980). Deficient natural killer cell activity in x-linked lymphoproliferative syndrome. *Science* 210:543–545.

Sykes, M. (1990). Unusual T cell populations in adult murine bone marrow: Prevalence of $CD3^+CD4^-CD8^-$ and $\alpha\beta$ TCR^+ $NK1.1^+$ cells. *J. Immunol.* 145:3209–3215.

Targan, S., and Dorey, F. (1980). Dual mechanism of interferon augmentation of natural killer cytotoxicity (NKCC). *Ann. N.Y. Acad. Sci.* 350:121–129.

Talmadge, J. E., Meyers, K. M., Prieur, D. J., and Starkey, J. R. (1980). Role of natural killer cells in tumor growth and metastasis: C57BL/6 normal and beige mice. *J. Natl. Cancer Inst.* 65:929–935.

Tarkkanen, J., Saksela, F., and Paavolainen, M. (1983a). Suppressor cells of natural killer activity in normal and tumor-bearing individuals. *Clin. Immunol. Immunopathol.* 28:29–38.

Tarkkanen, J., Saksela, E., von Willebrand, E., and Lehtonen, E. (1983b). Suppressor cells of the human NK activity: Characterization of the cells and mechanism of action. *Cell. Immunol.* 79:265–278.

Timonen, T., Ortaldo, J. R., and Herberman, R. B. (1981). Characteristics of human large granular lymphocytes and their relationship to natural killer and K cells. *J. Exp. Med.* 153:569–582.

Topalian, S. L., Solomon, D., Avis, F. P., Chang, A. E., Freerksen, D. L., Linehan, W. M., Lotze, M. T., Robertson, C. N., Seipp, C. A., Simon, P., Simpson, C. G., and Rosenberg, S. A. (1988). Immunotherapy of patients with advanced cancer using tumor-infiltrating lymphocytes and recombinant interleukin-2: A pilot study. *J. Clin. Oncol.* 6:839–853.

Trinchieri, G., and Perussia, B. (1984). Biology of disease. Human natural killer cells: Biologic and pathologic aspects. *Lab. Invest.* 50:489–513.

Trinchieri, G., and Santoli, D. (1979). Antiviral activity induced by culturing lymphocytes with tumor-derived or virus-transformed cells: Enhancement of human natural killer cell activity by interferon and antagonistic inhibition of susceptibility of target cells to lysis. *J. Exp. Med.* 147:1314–1333.

Uchida, A., Kolb, R., and Micksche, M. (1982). Generation of suppressor cells for natural killer activity in cancer patients after surgery. *J. Natl. Cancer. Inst.* 68:735–741.

Uchida, A., and Micksche, M. (1981). Suppressor cells for natural killer activity in carcinomatous pleural effusions of cancer patients. *Cancer Immunol. Immunother.* 11:255–263.

Valembois, P., Roch, P., and Boiledieu, D. (1980). Natural and induced cytotoxicities in sipunculids and annelids. In *Phylogeny of Immunological Memory* (Manning, M. J., ed.). New York: Elsevier/North-Holland, pp. 47–55.

Vodivelich, L., Sutherland, R., Schreider, C., Newman, R., and Greaves, M. (1983). Receptor for transferrin may be a "target" structure for natural killer cells. *Proc. Natl. Acad. Sci. USA* 80:835–839.

Warner, J., and Dennert, F. (1982). In vivo function of a cloned cell line with NK activity: Effects on bone marrow transplants, tumor development and metastases. *Nature 300*:31–34.

Welsh, R. M. (1986). Regulation of virus infections by natural killer cells. *Nat. Immun. Cell Growth Regul.* 4:169–199.

Wiltrout, R. M., Mathieson, B. J., Talmadge, J. E., Reynolds, C. W., Zhang, S. R., Herberman, R. B., and Ortaldo, J. R. (1984). Augmentation of organ-associated natural killer activity by biological response modifiers: Isolation and characterization of large granular lymphocytes from the liver. *J. Exp. Med.* 160:1431–1449.

Wright, S. C., Wilbur, S. M., and Bonavida, B. (1985). Studies on the mechanism of natural killer cell-mediated cytotoxicity: VI. Characterization of human, rat and murine natural killer cytotoxic factor. *Nat. Immun. Cell Growth Regul.* 4:202–220.

Yamada, A., Hayami, M., Yamanouchi, K., and Fujiwara, K. (1980). Detection of natural killer cells in Japanese quail. *Int. J. Cancer 26*:381–385.

Young, J. D. (1989). Killing of target cells by lymphocytes: A mechanistic view. *Physiol. Rev.* 69:250–314.

Zöller, M., and Wigzell, H. (1982). Normally occurring inhibitory cells for natural killer cell activity: II. Characterization of the inhibitory cell. *Cell. Immunol.* 74:27–39.

19

The Immunobiology of Aging

JULIA GREEN-JOHNSON,

ANDREW W. WADE,

MYRON R. SZEWCZUK

We will review developmental changes in the immune system of mice and humans in the context of the aging process. Aging is a relatively poorly defined process involving complex multisystem changes from the fetus to the later years of life. In general, advanced age can lead to diminished functions in both humoral and cell-mediated immunity. The first evidence of age-related change directly related to the immune system is physiological in nature. Thymic involution begins early in life and has far-reaching effects. The most obvious and direct effect is diminished function of T lymphocytes in the aging animal (Fig. 19-1). This, in turn, can influence other branches of the immune system. Because this subject was recently reviewed in depth (Wade et al., 1988), we have taken a new approach by describing the developmental changes that occur with age in the immune system. We will focus on the cellular interactions and physiology of the immune system that are affected by the aging process. The cellular aspects of altered immune responses that we will consider include changes in lymphocyte numbers, subsets, and surface markers, in idiotype expression, and in various aspects of cellular immunity, including proliferative responses, lymphokine production, cytotoxicity, suppression, B cell responses, and antibody production. The role of the thymus and thymic hormones, as well as the influence of microenvironment on lymphocyte function, will also be considered.

AGING AND THE IMMUNE SYSTEM: AN OVERVIEW

Many studies have examined age-related immunological function as a property of entire populations of cells, leading to the conclusion that diminished function is a general phenomenon. However, when lymphocytes are examined on an individual basis, even though a diminished proportion of old cells is able to respond, those that are capable of responding also retain the ability to exhibit full effector function. This applies to several aspects of immune function, including help for antibody production, lymphokine synthesis, and cytolytic activity. It now appears that age-associated reductions in immune reactivity generally reflect diminished numbers of functional cells rather than qualitative changes in cell function.

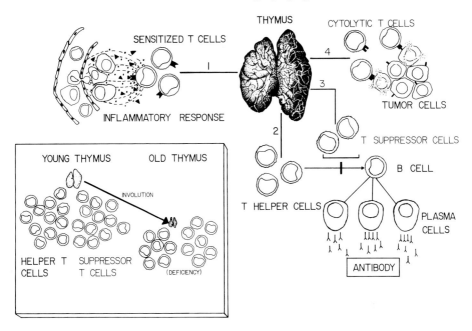

Figure 19–1. A summary of the aging process of the thymus gland and its effects on the immune system. **Step 1:** T cells that mediate delayed hypersensitivity. **Step 2:** Helper T cell activity required for thymus-dependent antibody production. **Step 3:** Suppressor T cell function involved in downregulation of the immune response. **Step 4:** Cytolytic T cells necessary for cell-mediated immunity. (Reprinted by permission from Szewczuk, 1987.)

The decline in immune function may not be equally severe throughout the whole organism. Evidence is accumulating to suggest that the cellular microenvironment plays a major role in age-related influences on cell function. It is also becoming clear that changes in cell function can ultimately be traced back to the basic age-related induction of intracellular defects, especially in the cell's ability to transduce and respond to signals. It appears that the loss of overall function occurs as a result of an increasing number of cells displaying these defects. Some individual cells, however, do remain functional and are capable of responding and reacting in the same way as "young" cells.

Accumulating evidence now suggests that the mucosal immune system is less susceptible to the effects of age by virtue of its distinct cell populations, and that bone marrow stem cells in old individuals retain the potential to produce fully functional cells. Overall, it appears that the microenvironment in which cells mature influences their susceptibility to age-related alterations in immune function.

GENERAL CHANGES IN THE IMMUNE SYSTEM WITH AGING

Age-related changes in serum components, lymphocyte numbers, serum antibody levels, and autoantibody production have been studied extensively. Some reports have found age-associated changes in cell numbers, whereas others have found none. Recent studies such as the Baltimore Longitudinal Study on Aging (Shock, 1984) concluded that age does not lead to significant changes in major cell types or cell subsets. The controversy probably reflects the difficulties in standardizing human populations for aging studies. For example, some studies have used different criteria for defining "old" and "healthy," and these differences contribute to the variations between reports. Changes in minor cell subsets cannot be ruled out, as little information is available in this area to date. In one study, the numbers of resting human T cells, as identified by the monoclonal antibody 3G5, were found to increase with age (Rabinowe et al., 1987). To explain the observed inconsistencies in age-related changes in lymphocyte numbers in mice, it was suggested that circannual and circadian rhythms might be among the contributing factors (Brock, 1987).

Although attempts have been made to correlate the appearance of various autoantibodies in old humans and in mice with autoimmune disease, such correlations do not seem to be definitive. Instead, autoantibodies may indicate a reduced efficiency of the regulatory system that normally controls their expression. Kato and Hirokawa (1988) recently showed that young and old mice have comparable numbers of B cell clones that are capable of producing autoreactive antibodies, supporting the idea that autoantibody production arises from dysregulation of such clones rather than an age-related increase in their frequency.

Levels of antibody produced by elderly humans after immunization with flagellin, influenza, or pneumococcal vaccine or with tetanus toxoid are lower than those observed in young adults. Aged mice also have lower levels of serum antibody after immunization than young mice have, and they show lower numbers of antibody-producing cells in the spleen (see reviews by Wade et al., 1988; Wade and Szewczuk, 1984). Although responses to T-independent antigens remain unaltered until late in life, responses to T-dependent antigens are impaired earlier (Wade and Szewczuk, 1988), indicating that the aging process affects T cells earlier in life than it affects B cells.

There is no clear correlation among changes in serum components, lymphocyte numbers, and changes in immune function with age. Changes in a cell's capacity to function are not necessarily accompanied by obvious changes in surface markers. Thus measurements of cell types, as identified by common surface markers, are not always an accurate indicator of cell functions.

Changes in Cellular Immunity

Most cellular immune functions decline with age (see Fig. 19–2). Individual mice (and humans) show age-related variations in some aspects of cellular immunity, as well as in the extent of their responses. Mice may show alterations in levels of suppression or cytotoxicity, while they retain normal levels of T cell help. As a consequence, large variations are not uncommon when measuring parameters of cellular immunity. To further complicate matters, individual mice also show heterogeneity in their responses to different antigens. Changes in the frequency of T cells responding to a specific antigen may occur with age, while the frequency of cells responding to another antigen may remain unaltered. Thus, conclusions drawn from studying one antigen or type of response may not be universally applicable to other antigens or other cellular immune responses.

In the following sections, we will present a more detailed description of age-associated changes in the major components of cellular immunity.

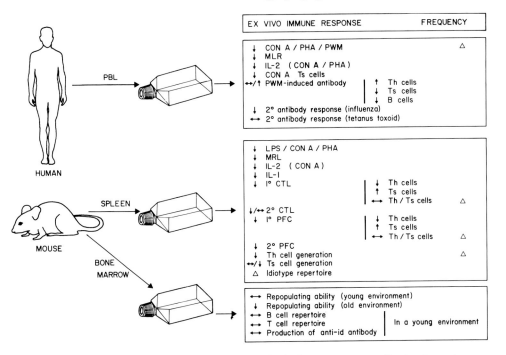

| EX VIVO IMMUNE RESPONSE | | | FREQUENCY |

Figure 19-2. A summary of the changes that occur in the immune system of both mice and humans when responses are examined in culture. Included for selected immune responses is the notation that indicates some of the possible reasons for the change with age. This supplemental information lies either to the right of the vertical lines on the figure or under the column marked FREQUENCY. A listing of three different supplemental items for a response indicates three possible reasons for the response reduction. A delta symbol in the FREQUENCY column indicates that part of the response reduction can be ascribed to an altered frequency of responding cells in the old animal. Upward arrows denote an increase with age in the response or component number; downward arrows denote a decrease; horizontal arrows indicate unchanged reactivity or numbers with age. Symbol descriptions: CON A, PHA, PWM, LPS: proliferation to concanavalin A, phytohemagglutinin, pokeweed mitogen, and lipopolysaccharide, respectively. (PBL, peripheral blood lymphocyte; MLR, mixed lymphocyte reaction; IL-2, interleukin 2; IL-1, interleukin 1; CTL, cytotoxic T lymphocyte; PFC, plaque-forming cells; 1°, primary immune response; 2°, secondary immune response; Id, idiotypic; Th, T helper lymphocytes; Ts, T suppressor lymphocytes.) (Reprinted by permission from Wade et al., 1988.)

Cytotoxic T Lymphocytes (CTL)

Murine cytotoxic T lymphocytes (CTL) show decreased responses to a variety of antigenic stimuli with age. Primary and secondary CTL responses to modified self and allogeneic targets, virus-infected cells, and tumor cells all decline with age.

In spite of the overall decline in CTL activity, there is no age-related change in the lytic ability of those CTL that do remain functional. This has been confirmed in limiting dilution analyses. Furthermore, it has been reported that the CTL affinity for targets is reduced with age. It thus appears that the major defect in murine CTL activity may occur at the lethal hit stage rather than at the level of target binding. Bloom et al. (1988) showed that young and old cells bind similar numbers of target cells, but they found significantly fewer old cells bound to lysed targets, indicating their reduced lytic capacity. Taken together, these results suggest that the frequency of CTL lysing targets decreases with age whereas the frequency of cells that bind targets does not. Thus, a defect in the lytic machinery of aged CTL would cause a reduction in the effective CTL response. Among the several possibilities for the defect in lytic capacity of old CTL cells are alterations in cytoskeletal mobility or reduced production of lytic molecules involved in the lethal hit stage.

As mentioned above, variations in the response to different antigens influence whether or not a decline is detected. In general, old mice have fewer spleen cells triggered by antigen, as determined by limiting dilution analysis. However, the frequencies of precursor CTL cells in older mice vary depending on the alloantigen used for stimulation (Chang and Gorczynski, 1984). The thymic microenvironment may influence generation of the CTL repertoire. Studies have shown that the splenic CTL repertoire changes with age, whereas that of the thymus does not (Chang and Gorczynski, 1984).

Although their relative contributions are difficult to assess, other factors may contribute to the decline in CTL activity. These include the reduced availability of IL-2 in older individuals and inadequate or inefficient antigen presentation. Reduced responsiveness to IL-2 may also be a factor. Addition of exogenous IL-2 to cultures of old CTL cells can often improve levels of cytotoxicity, although not always to the level of activity in young mice.

Natural Killer and Lymphokine-Activated Killer Cells

In humans, natural killer (NK) cell activity seems to be an exception to the general trend to functional decline in cell-mediated immunity with age. Levels of NK activity of human peripheral blood lymphocytes (PBL) actually increase with age (Krishnaraj and Blandford, 1988). Changes are also seen in the numbers of NK cells of different subsets. Surprisingly, this age-associated increase in NK activity does not correlate with an increase in the most active NK subset (Leu-7a$^-$11a$^+$), but instead with increased numbers of cells in the subsets that show weak or variable activity (Leu-7a$^+$11a$^-$ and Leu-7a$^+$11a$^+$). Increased numbers of another NK subset, identified as Leu-7a$^+$11c$^+$, are also seen with age, and this subset is thought to mediate the increase in NK activity. Several possible explanations for the increased NK levels seen with age can be offered: (1) increased NK levels may be predictors of longevity, because they provide protection against cancer and viruses; (2) increased NK activity could reflect accumulated bioinsults with age; (3) the results could be interpreted as showing inefficiency in the NK system because increased numbers are seen chiefly in the NK "weak" subset; (4) increased NK cell numbers might compensate for decreased availability of, or sensitivity to, lymphokines; (5) the increase is apparent, rather than real, and is due to an age-related shift in NK subsets among immune compartments (only PBL were analyzed in this study). Other possibilities are that these findings could reflect an increase in the frequency of active NK cells or a change in activity per NK cell.

Mice differ from humans in that their splenic NK activity decreases with age. While NK activity in bone marrow from young and old mice was found to be similar, splenic NK activity was decreased in old mice. In contrast, mice had generally constant levels of lymphokine-activated killer cell (LAK) activity, although some changes did occur with age. Bone marrow cells of old mice developed levels of LAK activity equivalent to those of young mice. Most LAK precursors (pLAK) are asialo GM$^+$ (ASGM$^+$), and most LAK effector cells express Thy-1, regardless of age. Interferon (IFN) augments the IL-2–mediated generation of LAK activity in cells from young bone marrow, but it inhibits LAK generation from old bone marrow. Although there is no decline in LAK activity with age, there is an age-associated difference in either pLAK numbers or their regulation. Thus, young marrow pLAK cells may be more sensitive to upregulation by IFN than are old pLAK cells. This may be due to the generation of more suppressor activity by IFN in old than in young bone marrow. Alternately, pLAK cells in old marrow may be more sensitive to IFN inhibitory effects than are young pLAK cells.

T_H Lymphocyte Function: Lymphokine Production

Impaired function of T_H (helper T) cells is readily apparent in the decreased levels of lymphokines found in old mice. Defects both in interleukin (IL) synthesis, or the ability of cells to respond to stimuli by producing interleukins, and in the response of T_H lymphocytes to interleukins have been reported.

Thoman and Weigle (1987) reviewed the effects of age on the synthesis and function of IL-1 and IL-2. IL-1 production in rats is not affected by age, and young and old rat T cells activated with Con A (concanavalin A) respond equally well to IL-1. In mice, however, the number of IL-1–producing cells decreases with age. When combinations of old and young murine T cells and macrophages are examined in culture, the decrease in responsiveness seems to reflect defects in the functions of old T cells and macrophages. This decrease in responsiveness may result from defective cell–cell interactions or even from feedback mechanisms. Changes in IL-1 production in humans have not been examined in sufficient detail to justify further discussion.

Aging may alter the expression of IL-1 receptors on lymphocytes. In mice, the number of cells that bind and respond to IL-1 decreases with age, suggesting decreased receptor expression. In addition, decreased signal transducing ability of old cells may be involved. However, when IL-1 is added to cultures of old cells, it has minimal restorative effects, implying that deficient IL-1 production may not be a major factor in the decline of the immune response.

Diminished IL-2 production and IL-2 binding could contribute to the decline of most immune functions. Early studies demonstrated that Con A–stimulated IL-2 production decreased with age (Chang et al., 1982). Experiments involving combinations of old and young adherent cells with T cells have shown that the decrease in IL-2 production is due partly to decreased adherent cell function as well as to defective T cell function. IL-2 synthesis in old mice is reduced in response to mitogens as well as antigens. In contrast, IL-2 synthesis in old rats is reduced in response to antigens, but not after stimulation by mitogens (Gilman et al., 1982). In rats, the defect in IL-2 synthesis may be stimulus-dependent, as synthesis is enhanced by stimulation by PHA and Con A but not with alloantigens. Thus, reduced IL-2 synthesis can be overcome by the concurrent engagement of several cell suface receptors. Mitogen-induced synthesis of IL-2 is also decreased in old humans (Kennes et al., 1983).

At a molecular level, decreased IL-2 production can result from events that occur before IL-2 synthesis or secretion. Decreases of 72% in IL-2 production were found in rats between ages 4 to 30 months by Cheung et al. (1983) and were paralleled by decreased proliferative responses. When ^{3}H-valine uptake was used as a measure of protein synthesis, a 74% decrease over the period from 4 to 30 months was observed, suggesting that an age-related decline in protein synthesis may underlie the decline in IL-2 production. However, other studies have since localized the defect in IL-2 synthesis to a stage preceding the production of IL-2 mRNA (Wu et al., 1986). Levels of IL-2 mRNA expression in aged rats were reduced over those in young rats, decreasing by 85% at 29 months as compared to 5 months. This decrease paralleled decreases in both IL-2 production and T cell proliferation. No age-related changes were seen in the size of the mRNA transcripts, or in degradation or post-transcriptional processing of IL-2 mRNA. Because bulk cultures were used, it is not possible to determine whether decreased mRNA levels resulted from decreased amounts of IL-2 mRNA produced per cell, or from a decreased number of cells that produced IL-2 mRNA transcripts. A more recent study using in situ hybridization showed that average levels of IL-2 mRNA per cell did not differ between activated young and old murine lymphocytes (Fong and Makinodan, 1989), although 40% fewer old lymphocytes than young lymphocytes expressed elevated IL-2 mRNA levels after activation. These results correlated well with data from limiting dilution experiments showing that diminished IL-2 production was not due to a decrease in the capacity of old lymphocytes to produce IL-2, but instead reflected the failure of some old T lymphocytes to produce IL-2. Individual old T lymphocytes that synthesize IL-2 produce amounts comparable to those produced by individual young T lymphocytes, although the precursor frequency of IL-2–producing lymphocytes is diminished by three- to five-fold in aged mice. Decreased levels of mRNA for IL-2, IFN-γ, IL-6, GM-CSF, and IL-1 β have also been found in old humans following lymphocyte activation, suggesting that this defect is widespread and is not unique to a particular cytokine or to a species.

IL-2 receptor expression is also reduced in old murine and human lymphocytes. This likely accounts for some of the results obtained in studies on the restoration of immune responses with exogenous IL-2. Gillis et al. (1981) showed that the responses of old human lymphocytes to PHA and Con A are reduced even in the presence of IL-2, while, under the same conditions, young cells show a 30%–50% increase in proliferation. Kennes et al. (1983) found that addition of IL-2 to aged human lymphocytes increased proliferative responses to basal young levels. In mice there is some variation in the reconstituting effects of IL-2. Proliferation is improved but not completely restored to young levels, whereas other T-lymphocyte responses, including cytotoxicity, suppression, and helper activity for antibody production, may be restored to young levels. The restorative effects of IL-2 cannot be correlated directly with function. For example, Rosenberg et al. (1983) found that addition of IL-2 did not improve proliferative responses, nor did it improve T cell help to B cells for antibody production. The observed differences in the levels of helper activity for antibody production may be explained by variations in the frequency of both T and B cells specific for the antigen used.

In vivo treatment of old mice with IL-2 can reconstitute CTL generation as well as correct the age-related defect in expression of IgD receptors on B lymphocytes. IgD receptor expression is normally induced by exposure to IgD or to IFN-γ, but it is deficient in "old" lymphocytes (Coico et al., 1987). When old mice are treated with IL-2, IgD receptor expression is partially restored. T lymphocytes from old rats bind less IL-2 than lymphocytes from young rats, and, unlike young lymphocytes, they do not respond to exogenous IL-2 with increased proliferation (Gilman et al., 1982). The disparity in the results of IL-2 reconstitution experiments could reflect differences in the levels of IL-2 receptors on T lymphocyte subsets with age, or it could imply that proliferation of lymphocytes is more susceptible to the effects of aging than are other types of responses. Thoman and Weigle (1987) proposed that the age-induced defect in T cell secondary responses is at the expansion stage rather than in the sensitization stage; otherwise, responses would not improve with IL-2 addition.

The influence of age on other interleukins and lymphokines has not been examined in detail. In one study, Chang et al. (1988) reported a two-fold decrease with age in IL-3 production by splenic T cells. By mixing old and young T cells and adherent cells, they found that the reduction reflected diminished T cell rather than diminished adherent cell function. They also noted a strong correlation in the production of IL-2 and IL-3 between young and old mice. The investigators suggest that a single class of T_H cells are responsible for producing IL-2 and IL-3. Presumably, this T_H class is very sensitive to the aging process. With respect to other lymphokines, there is no decline in the production of interferon-γ of the B cell differentiation factor (BCDF) by old human lymphocytes (Hara et al., 1987). In fact, BCDF production by old T cells is three-fold greater than in young cells following stimulation with *Staphylococcus aureus* (Hara et al., 1987). The same group noted an inverse relationship between IL-2 production and BCDF production for both old and young cells.

T_S Lymphocytes

The precise role of suppressor T cells (T_S) in the decline of immune function with age is difficult to assess. Part of the problem is due to variation between individuals as well as variability in responses to different antigens. Analysis of T_S function with age must also take into account the contributions of different types of suppressor activity. These aspects have doubtless contributed to the contradictions in the literature on the role of suppressor activity in age-related changes in immune function, as well as the effects of aging on suppressor activity.

Decreased levels of antibody production by spleen cells of old mice have been attributed to increased levels of suppressor T lymphocyte activity. Globerson et al. (1981) reported an increase in naturally occurring suppressor cell activity in aging mice. In contrast, Kishimoto et al. (1979) reported decreased natural suppressor cell activity in aging humans. Thoman and Weigle (1983) found that old murine spleen cells showed decreased T_S activity, partly due to the decreased availability of IL-2. Overall it appears that the generation of specific T_S is impaired with age, although not to the same extent as are CTL generation and T lymphocyte proliferation.

Idiotypic interactions play an integral role in suppressing some immune responses. Age-related changes in the repertoire of B cell idiotypes may parallel changes in the T_H repertoire recognized by T_S cells. Suppressor cells generated in mice by immunization with TNP-conjugated spleen cells showed age-related restriction of their activity. Suppressor cells induced in young animals inhibited immune responses of young animals only, whereas suppressor cells produced in old animals only inhibited the antibody response of old animals (Hausman et al., 1985). Doria et al. (1987) also showed age-restricted functions for T suppressor inducer cells and T suppressor transducer cells. Although the exact mechanisms responsible for this restriction are unclear, the results clearly show differences in regulatory cell recognition between young and old mice.

B Lymphocyte Responses

Most studies that analyze in vitro B cell responses in humans have used pokeweed mitogen (PWM) as the stimulation. In general, there appears to be no reduction in total IgG synthesis in cultures of cells from old individuals. Some age-related reduction in B cell sensitivity to T cell factors is observed, possibly reflecting reduced expression of receptors for lymphokines. Based on the few studies that have examined in vitro antibody production in response to stimuli other than PWM, it appears that the responses of elderly individuals are not greatly altered. For example, old humans produce only slightly reduced amounts of anti-influenza antibody and unchanged levels of anti-tetanus toxoid antibody, as compared with young subjects (Ershler et al., 1984b).

Splenic B cells from old mice give impaired proliferative responses to bacterial lipopolysaccharide (LPS). Limiting dilution analyses of spleen cell populations from old mice indicate that a reduced frequency of LPS-responsive cells could account, in part, for their lower proliferative responses. This may not hold true, however, for all stimuli or antigen specificities. Zharhary and Klinman (1986) showed an increase in the frequency of phosphorylcholine-specific B cells with age. Still, T and B cells both tend to show a decreased frequency of cells responding to certain stimuli with age. Age-induced changes within spleen cell populations of old mice appear to be permanent and are refractory to environmental influences due to the fact that the impaired function of old B cells could not be restored by transferring them to a young irradiated animal (Averill and Wolf, 1985).

The antibody response to T-dependent or T-independent antigens is depressed, as detected in the serum or by measuring splenic antibody-secreting cells (PFC). Qualitative changes also occur in the immune response, with preferential loss of IgG and high-affinity antibody. This, in turn, may reflect diminished T cell function because T cell help is required for affinity maturation and class switching. Age-related changes in the levels of specific antibody produced depend on the isotype measured, suggesting that aging may also affect events associated with isotype switching (Ponnappan et al., 1987; Wade and Szewczuk, 1988). Limiting dilution analyses using unstimulated and LPS-stimulated B cells indicate that the B cell repertoire for an antigen can either expand, contract, or remain unchanged, depending on the antigen tested. Such changes in cell frequencies were not seen in old nude mice, suggesting that T cells influence the B cell repertoire. This was confirmed by reconstitution experiments using young or old T cells and young B cells, in which the animals receiving old T cells showed a decrease in B cell frequency.

The diminished B lymphocyte response of old mice is not due to defects in the ability of the responding cells to be activated by their respective stimuli. No differences are found in the optimum antigen doses or the kinetics of the responses of old B cells. Individual antibody-producing clones from aged animals are also normal with respect to the amount and relative affinity of the antibody they produce. Snow (1987) found no differences in the responses (PFC) of old and young purified antigen-specific B cells following in vitro stimulation, suggesting that the aged environment plays a major role in diminished B lymphocyte responsiveness.

B cell activity in old mice is also influenced by alterations in T suppressor cells or T helper cells or by defects in antigen presentation, making it difficult to determine the extent to which B cells are influenced by age. For example, the influence of diminished T_H cell function on the responses of aged B cells can be improved by adding IL-2 to cultures (Thoman and Weigle, 1985) or by treating old mice with thymopoietin pentapeptide (Weksler et al., 1978). Alterations in the idiotypic repertoire with age may also play a role by suppression of B cell responses by specific autoanti-idiotypic antibodies (Goidl et al., 1983; Szewczuk and Campbell, 1980).

AGING AND CHANGES IN CELL SURFACE MOLECULES AND INTRACELLULAR MECHANISMS

Development of Cell Surface Molecules

Correlation of cell surface markers with cellular functions is not always reliable. Most earlier studies examined the expression of markers, rather than the density of such markers, on the cell surface. It may be that minor changes in the level of expression of a cell surface marker or receptor may influence the cell's ability to respond to signals or to inter-act with other cells. Recent studies have examined age-related changes in the density of murine lymphocyte surface antigens (see Fig. 19–3).

Earlier studies provided indirect evidence for qualitative changes in cell surface anti-gens. For example, Callard et al. (1979) indirectly showed age-related changes in surface determinants on old mouse lymphocytes. Spleen cells from aged C57Bl/6 mice were shown to proliferate in a mixed lymphocyte reaction (MLR) against spleen cells from young C57Bl/6 mice and vice versa, suggesting that age-related changes in surface anti-gens had occurred. This may reflect changes in density or minor structural alterations of antigens already present on the lymphocyte surface, as no novel age-associated antigen has been shown to account for the response in the MLR. Actually, cell-surface markers specific for aged lymphocytes are relatively rare. It may be noteworthy that a senescence-

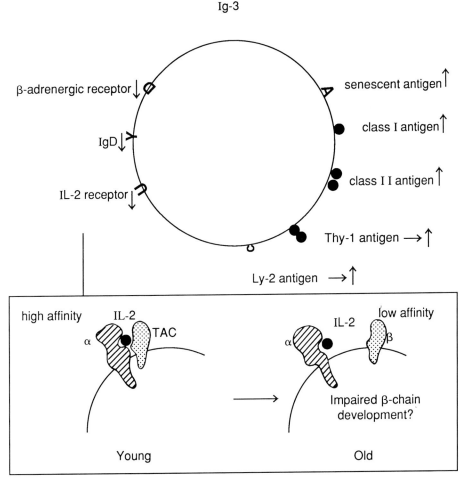

Figure 19–3. A summary of the developmental changes with age in cell surface molecules and interleukin 2 (IL-2) receptors. Receptors for IL-2 consist of two polypeptides [α (75 kD) and β (55 kD) chains]. β chains are recognized by anti-TAC monoclonal antibodies. Upward arrows denote an increase with age in the component number; downward arrows denote a decrease; horizontal arrows indicate unchanged numbers with age.

associated antigen was identified by Kay (1981) and was shown to be expressed on the surface of aged red blood cells, platelets, lymphocytes, and neutrophils. The antigen appears to function in the clearance of senescent cells.

Komatsubara et al. (1986) reported that the stimulatory capacity of aged dendritic cells in syngeneic MLR was higher than that of young dendritic cells. These findings may reflect age-related alterations in dendritic cell membrane antigens. However, these changes in stimulatory capacity do not reflect changes in Ia density, as the same group also measured levels of class II MHC antigens on murine dendritic cells and found no changes with age. Both B and T lymphocytes were found to exhibit age-related increases in the levels of class I and II MHC antigens (Sidman et al., 1987).

Age-associated changes in expression of other cell surface antigens remain controversial. The density of Thy-1 and Ly-2 surface markers on murine lymphocytes has been reported either to decrease or to remain unchanged with age (Sidman et al., 1987; Utsuyama and Hirokawa, 1987). As both studies used flow cytometry to measure antigen density in the same strain of mice, the reasons for these differences are not obvious.

Age-related changes in the densities of receptors for cytokines and other growth factors appear to be somewhat more pronounced than changes in other types of cell surface markers. In early studies, an age-related decrease in the number of IL-2 receptors per cell was found, and this correlated with the reduced ability of old lymphocytes to absorb IL-2 (Chang et al., 1982). This has since been confirmed in humans, as well as in mice. Proust et al. (1988) found a four-fold decrease with age in the frequency of IL-2 responsive cells following Con A activation. Moreover, this correlated with a decrease in the number of high-affinity IL-2 receptors induced on old T cells by Con A. Receptors induced on old T cells were functional and were capable of transducing signals for proliferation. The defect in IL-2 receptor induction could be bypassed by using phorbol dibutyrate, which directly activated protein kinase C, although the receptors induced under these conditions were impaired functionally. These findings suggest an age-induced defect in the ability of membrane receptors to transmit intracellular signals such as those required to induce IL-2 receptor expression. The defect in high-affinity IL-2 receptor expression was seen early in the response to Con A (8–16 hours) and was less marked by 40 hours. The high-affinity IL-2 receptor consists of two chains, each of which can bind IL-2 weakly (Fig. 19–3). Resting T cells do not express the β chain, but they may express up to 50,000 molecules per cell after activation, some of which associate with the α chain to form the high-affinity IL-2 receptor. It would therefore appear that there is an age-related impairment in the induction of β chains due to defects in intracellular signals.

Other receptors also show altered expression with age. Levels of β-adrenergic receptors on murine splenic lymphocytes decrease with age (Kohno et al., 1986). Lymphocytes from aged mice also fail to express IgD receptors following exposure to IgD or IFN-γ, both of which normally induce IgD receptors in young lymphocytes (Coico et al., 1987). Treatment of aged mice with IL-2 leads to an increase in IgD receptor expression, however, suggesting that this receptor deficiency is the result of insufficient IL-2 availability rather than an intrinsic cell defect. The variety of receptors affected by aging implies that such alterations in receptor density may be a generalized phenomenon.

Activation of Senescent Lymphocytes

The first evidence that activation was impaired in old lymphocytes was obtained from studies that measured responses to mitogens or to alloantigens in mixed lymphocyte reac-

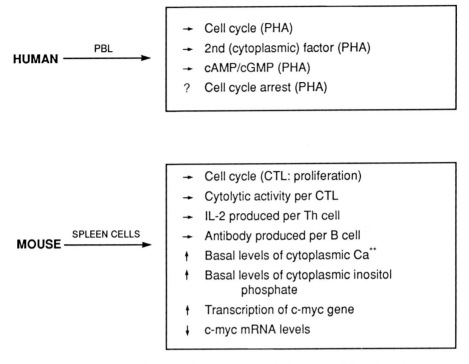

Figure 19–4.   A summary of the age-related changes to the immune cell function when responses are considered on an individual cell basis. Most activity is conserved during aging. See Figure 19–2 for details of symbols. (Reprinted by permission from Wade et al., 1988.)

tions. With very few exceptions, it was found that T lymphocyte proliferation in response to PHA, Con A, PWM, or alloantigen was decreased in aged mice and humans. This defect applies to a wide variety of activating signals, as proliferation of old T or B lymphocytes to other stimuli—including protein A and tuberculin active peptide—and to mitogenic anti-CD3 and anti-Leu-4 antibodies is similarly reduced. Some exceptions include the proliferative responses to tobacco glycoprotein and to anti-Ig by human lymphocytes, which do not decrease with age. On the other hand, there is an age-related increase in the proliferative capacity of murine T cells to the Mls[a] determinant and to allo-MHC class II antigens.

The site of the major defect in proliferation remains controversial. Limiting dilution analysis shows that the number of mitogen-responsive cells is reduced, but that the proliferative ability of individual cells is not diminished by age (see Fig. 19–4). Negoro et al. (1986) concluded that the diminished proliferative response of aged T cells to tuberculin active peptide was due to the inability of stimulated old cells to undergo repeated cycles of replication. These conclusions were supported by the findings of Hara et al. (1987), who showed that the number of B cells proliferating in response to stimulation with protein A from *Staphylococcus aureus* (Cowan strain) was not reduced with age, but that their ability to sustain replication was impaired. Thus, the proliferative kinetics of old lymphocytes can remain unchanged, with no alteration in cell cycle time. However, these results appear to contradict those from limiting dilution analyses, which show a reduced frequency of cells initially capable of responding to the stimuli with no reduction in their proliferative ability (Miller, 1984). It may be that the cells detected in the limiting dilution analyses were those able to successfully undergo repeated replication.

More recent studies have examined defects in the proliferation process at the subcellular level (Fig. 19–4). Old murine T cells show impaired cytosolic-free calcium responses or intracellular levels of protein kinase C (Miller, 1986). Direct activation of protein kinase C and elevation of intracellular calcium concentrations using phorbol esters and ionophores bypasses the signaling defects and produces much higher proliferative responses than does stimulation via cell surface receptors. Old T lymphocytes also require more ionomycin to produce a rise in intracellular calcium levels than do young lymphocytes, indicating that old lymphocytes are refractory to the signals that normally trigger calcium elevation.

Further evidence for defective signal transduction in aged murine lymphocytes was obtained by Proust et al. (1987). Translocation of protein kinase C following Con A stimulation was found to be only 50% of that of young murine T lymphocytes. Furthermore, although cytoplasmic Ca^{2+} and inositol phosphate were increased to the same levels following Con A activation in old and young lymphocytes, the basal levels of both were higher in old lymphocytes, resulting in a smaller net increase. Lerner et al. (1988) also reported that elevation of intracellular calcium is impaired in old lymphocytes, although they found no alterations in the production of inositol phosphates with age. While the conflicting results of these studies do not allow us to define the intracellular signaling mechanisms in T cells that are impaired with age, it appears that the reduced responses may relate to abnormal signal transduction.

Another example of a basic alteration in intracellular signaling processes is seen in the reduction of c-*myc* mRNA levels in stimulated murine lymphocytes (Buckler et al., 1988). The reasons for this defect have not been identified, as the kinetics of c-*myc* mRNA accumulation remain unchanged with age. In addition, mitogenic stimulation produces equivalent increases in transcription of the c-*myc* gene in old and young T cells. The rate of c-*myc* mRNA decay is also unaltered by age. It was postulated that the defect may lie in post-transcriptional processing of c-*myc* mRNA, perhaps at the level of polyadenylation or splicing, or in the export of transcripts to the cytoplasm.

Thus, it is becoming increasingly apparent that age-associated defects in proliferation reflect fundamental changes in the biochemical or subcellular processes that accompany intracellular signaling.

DEVELOPMENTAL ASPECTS OF THYMIC INVOLUTION DURING AGING

Thymic involution begins within a year after birth in humans. Major changes include depletion of cortical lymphoid cells, an increase in the numbers of epithelial cells and macrophages in the medulla, and a decrease in thymic weight and volume (Steinmann, 1986). Steinmann and Müller-Hermelink (1984) also found no major changes in the numbers of immature TdT^+ (terminal deoxynucleotidyl transferase) lymphocytes among cortical thymocytes in thymuses from humans aged 21 to 70 years, even though physiologic thymic involution was obvious. Thus, the correlation of involution with decreased function may not be as direct as was originally thought.

Serum levels of some thymic hormones decrease with age. Thymic hormones normally promote maturation and differentiation of immature T cells, so that the decreased responsiveness of old individuals may, in part, be due to their inability to produce fully differentiated and functional T helper cells from circulating precursors.

Thymectomy in mice can also lead to decreased production of IgG antibodies, in response to T-dependent antigens, and to a loss of high-affinity antibody, both of which require T cell help. Hiramoto et al. (1987) reported on the effects of long-term treatment of mice with thymic hormones or synthetic analogues, including thymopoietin, thymosin, Facteur thymique serique (thymic humoral factor), and an enzyme-resistant variant, thymosin fraction 4 (TM4). Mice were treated at weekly intervals beginning at 2 months of age and extending over their entire life span. Only TM4 had an effect, increasing the median survival time by 25% in one strain of mouse, although it did not increase the maximum life span. TM4-treated mice had fewer morphological abnormalities at autopsy.

The involuted thymus can retain at least some functional capacity, although there is ample evidence for impaired production of thymic hormones in old mice. Furthermore, several other studies have shown that different aspects of murine cellular immunity, including helper T cell function, CTL function, antibody response to T-dependent antigens, and frequencies of T cells responding to mitogens can be partially restored to young levels by in vivo supplementation with thymic extracts or thymic hormones. These have included thymopoietin pentapeptide and thymosin α_1. Transplantation of GH_3 pituitary adenoma cells, which produce growth hormones and prolactin, restores thymic structure and function in rats, leading to improved T lymphocyte function (Kelley et al., 1986). Such studies support the concept that thymic involution with age leads to a loss of mature helper T cells with concomitant decreases in IL-2 production.

Similar findings have been reported regarding the reconstituting effects of thymic hormones in humans. The addition of thymosin fraction 5 (TM5) to cultures of old human lymphocytes following immunization against influenza or tetanus preferentially enhanced the production of antibodies to both antigens (Ershler et al., 1984 a, b). Studies of old humans treated with thymic hormones confirm the physiological relevance of these hormones. For example, treatment of elderly individuals with either thymopoietin pentapeptide, thymostimulin, or synthetic thymopentin led to some improvement in their T cell responses to mitogens in culture (Meroni et al., 1987) and to a reduced rate of infections (Pandolfi et al., 1983) when compared to untreated individuals. Aged individuals with reduced delayed hypersensitivity (DTH) responses to recall antigens showed improvement after treatment with synthetic thymopentin (TP5). The improved DTH responses were accompanied by enhanced in vitro lymphocyte proliferation and IL-2 production after mitogenic stimulation. Interestingly, the levels of immunoglobulin produced after stimulation by PWM did not increase. Because the T cells that produce IL-4 and IL-5 (which are involved in B cell differentiation) do not belong to the same subset as those that produce IL-2, it is possible that thymopentin does not influence this T_H cell subset. These findings suggest that decreased levels of one thymic hormone may not be responsible for all age-related immune defects, and that more complex interactions between thymic hormones and T cells may exist.

In spite of the evidence for gross physiologic and morphologic thymic involution with age, the extent to which these changes influence the ability of the thymus to participate in lymphocyte maturation is still unclear. The decrease in thymic hormone levels is ultimately felt throughout the immune system, as they are necessary for helper T cell function. As discussed previously, impaired T_H function can lead to decreased levels of immunostimulatory and regulatory lymphokines, thus impairing many different branches of the immune system.

MICROENVIRONMENTAL INFLUENCES ON THE DEVELOPMENT OF THE AGED IMMUNE SYSTEM

Bone Marrow Stem Cells

The bone marrow is an important tissue in which to evaluate age-associated changes because it continues to produce precursor cells for the immune system throughout life. Studies of bone marrow stem cells in vivo have suggested that their function is relatively unaffected by aging. Although there is some controversy over the ability of transplanted bone marrow stem cells to successfully repopulate the immune system of irradiated hosts (see review by Wade and Szewczuk, 1984), it appears that stem cells of old mice are as capable as those of young mice in this regard. Recently, Harrison et al. (1989) proposed that excess concentrations of precursors less primitive than primitive stem cells (PSC) are present in old marrow, and that their contribution to the differentiated cell population gradually declines. It was suggested that such precursors, as well as true PSC, proliferate in old donors to compensate for age-dependent deficiencies.

Considerable attention has recently been focused on the capacity of old marrow cells to produce an immune response following transfer into a "young" environment. If impaired functions were observed after cell transfer, this would suggest that the immune response of old animals is at least partly due to intrinsically programmed defects in the animal's stem cell population. On the other hand, if normal "young" function occurred, this would suggest that the reduced immune response in old mice results from downregulatory influences within the aged animals. There is evidence that the number of bone marrow cells doubles in aged mice. The ability of pluripotent stem cells from old marrow to differentiate into discrete foci in the spleen (colony forming units) remains unchanged with age. The repopulating ability of marrow from old mice is apparently normal, even when it is forced to reconstitute the animals repeatedly by cyclic treatment with hydroxyurea (Micklem and Ansell, 1984). In another study where the bone marrow was shielded while the animal was irradiated at a dose sufficient to deplete unshielded immune cells, an immune response occurred without the downregulation by anti-idiotypic antibodies normally seen in old mice (Kim et al., 1985). Thus, production of the anti-idiotypic antibodies responsible for suppression of the immune response in this system is not programmed in stem cells from either young or old mice. Instead, radiation-sensitive cells within the systemic environment of old animals induce the production of this regulatory antibody.

When bone marrow cells from old mice mature in an irradiated young animal they produce antibody-secreting cell (PFC) responses typical of young animals and also show a B cell repertoire indistinguishable from that of young marrow, although there are some differences in the splenic cell repertoire. The influence of the environment can be further shown when marrow stem cells are injected into old recipients. In this case, the B cell repertoires and responses are characteristic of those found in the spleens of old mice (Gorczynski et al., 1984). Thus, the environment in which lymphocytes mature appears to determine whether they show "young" or "old" characteristics.

T cells from irradiated recipients of young or old marrow also show the same patterns of immune characteristics as those in B cells. Transfer of young or old marrow cells to a young recipient results in a "young" proliferative response to PHA (Harrison et al., 1977). The CTL repertoire of young recipients of bone marrow from young or old mice corresponds to that normally found in young mice (Gorczynski and Chang, 1984). Conversely, in an "old" environment, young or old bone marrow cells develop the responsiveness patterns characteristic of old mice, and marrow cells from aged animals show even more impaired responses when they are transferred to an aged environment. Thus, marrow cells of old mice are not completely identical to those of young mice. This implies that young marrow cells are less sensitive than old cells to suppressive effects in the aged environment.

In general, little change occurs in stem cell populations with age, based on comparisons of the functional capacity of marrow lymphocytes from young and old mice. This is in marked contrast to the changes seen in lymphocyte populations from other tissues such as the spleen. Bone marrow cells from old mice, even though not completely identical to marrow cells from young mice, do retain the potential to respond normally to a number of stimuli once they are removed from the negative influences of the aged environment. The contribution of intrinsic age-related defects in this case is difficult to assess. What these studies do indicate is that the microenvironment influences the extent to which the development of lymphocyte function is altered with age.

Mucosal Immune System

The mucosal immune system, which encompasses over a third of the body's lymphoid tissue and forms the first line of defense against infection and foreign antigen exposure, has largely been ignored in aging studies. This system performs a critical regulatory role by preferentially allowing access of selected antigens to the systemic immune system while it prevents access to other antigens. The unique mucosal environment is capable of sustaining the growth and maturation of cells distinct from those in other compartments, supporting the idea that this immune system, while in communication with the rest of the body, also remains largely distinct. It is made up of unique populations of cells with distinct gut-homing properties and characteristics. For example, it contains T cells that express γ/δ receptors, as well as typical α/β T cells (Eldridge et al., 1987). Peyer's patches of X-linked immunodeficient *(xid)* CBA/N mice contain mature B lymphocytes, in contrast to the immature splenic B cell populations of these immunodeficient mice (Eldridge et al., 1983). The Peyer's patches of *xid* mice also contain T lymphocytes capable of providing help to normal B lymphocytes, while systemic T lymphocytes are unable to provide help. Furthermore, the gut-associated tissue has also been implicated as a primary lymphoid area in some animals (Reynolds et al., 1981). Whether the presence of such unique lymphocyte populations in the mucosa may account for the age-related developmental differences seen in this microenvironment is an intriguing question.

Little information about age-associated changes in the mucosal immune system exists (Wade et al., 1988). Szewczuk et al. (1981) measured the production of IgM, IgG, and IgA in the spleen and mucosal tissues following intraperitoneal immunization with antigen and found that, whereas responses in the spleen and peripheral lymph nodes

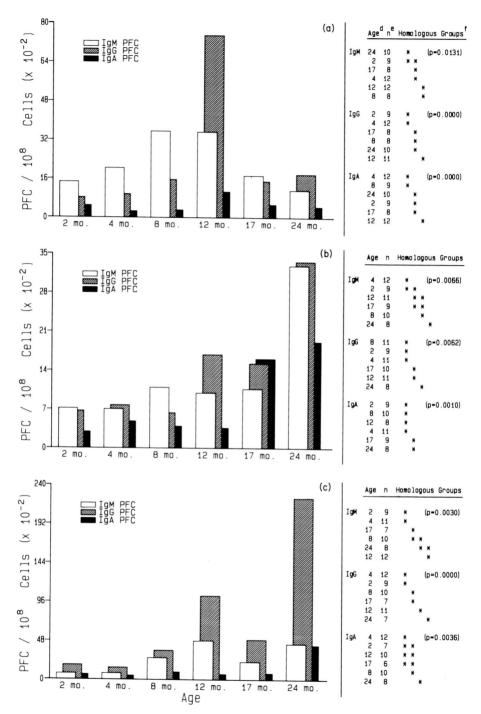

Figure 19–5. Change with age in the primary plaque-forming cell (PFC) response of the spleen, mesenteric lymph node (MLN), and bronchial lymph node (BLN) following intraperitoneal immunization with TNP-KLH. Primary IgM, IgG, and IgA PFC responses were determined in the (**a**) spleen, (**b**) MLN, and (**c**) BLN 14 days after primary immunization with 500 μg TNP-

decreased with age, mucosal responses (mediastinal and mesenteric lymph nodes) were unaltered for all three isotypes. Rivier et al. (1983) also found that in vitro IgA production induced by dextran B1355 in mesenteric lymph node (MLN) cells increased dramatically with age. Sullivan et al. (1987) detected an increase in the amount of IgA found in tears with age, reflecting increased IgA production by lacrimal plasma cells.

Even less information is available on changes in mucosal T lymphocyte function with age. Ernst et al. (1987) found that lymphocytes from Peyer's patches of aged mice retained their capacity to proliferate in response to Con A, PHA, and alloantigen, along with their ability to generate alloimmune cytotoxic cells, whereas aged splenic lymphocytes showed reduced responses. The lack of age-associated immune dysfunction in Peyer's patches was not due to altered response kinetics or to proportions of T cells in mucosal tissue. Our preliminary results, obtained from studies on MLN lymphocytes, support those of Ernst and co-workers with regard to responses to PHA, but not to Con A. Nonetheless, these studies suggest that the mucosal immune system is less susceptible to age-related decline in function.

There is now considerable evidence that supports the hypothesis that the functional activities of systemic and mucosal lymphocytes do not undergo parallel age-associated changes. Unimpaired B cell function in the mucosa of aged mice can be shown by measuring the heterogeneity of avidity of antibody produced by B cells. The heterogeneity of the primary anti-TNP antibody-secreting cell (PFC) response in the spleen and peripheral lymph nodes (PLN) becomes restricted during aging, with a preferential loss of high-affinity clones. However, "old" mucosal lymphocytes produced antibody with a full range of heterogeneity of avidity, without loss of high-avidity clones (Szewczuk and Campbell, 1981a, b). The restricted heterogeneity of avidity of the response in the spleens of old mice could reflect defective T cell help. Our findings suggest that T_H cell function remains unimpaired in the mucosal tissues of old mice.

The primary and secondary immune responses of aged C57BL/6J mice show the same dichotomy between systemic and mucosal responses (Wade and Szewczuk, 1988). Splenic B cell production of anti-TNP antibody declines after 8 months of age, while PFC responses in the MLN and bronchial (or mediastinal) lymph nodes (BLN) did not show significant changes (see Fig. 19–5). Thus, the primary and secondary mucosal-associated lymph node responses were found to be essentially unaffected by age, in contrast to responses in the spleens of aged mice.

Intragastric immunization results in IgM, IgG, and IgA PFC responses in the spleen and MLN that remain elevated or peak at the oldest time point examined (Wade and Szewczuk, 1988). The enhanced PFC response in the MLN of old mice primed intragastrically may therefore result from activation, trafficking, and subsequent seeding of gut tissue by mucosal cells, which are only slightly affected by age. Systemic seeding by these mucosally stimulated cells would result in elevated PFC responses in the spleen as well.

KLH in complete Freund's adjuvant. The height of each bar represents the geometric mean (PFC/10^8 cells) of the particular response. The statistical analysis is presented on the right side of the figure. Listed for each tissue are (**d**) the ages of the animals ranked by response magnitude (lowest at the top); (**e**) the number of animals in each age group; and (**f**) the statistical grouping of animals into homologous groups by Fisher's LSD procedure. The P values from one-way ANOVA are listed in brackets on the far right. (Reprinted by permission from Wade and Szewczuk, 1988.)

Szewczuk and Campbell (1981a) showed that antigen-specific B cells within the systemic immune system of old mice recognize, bind, and are reversibly inhibited by anti-idiotypic antibodies on their surface. No such antibody could be demonstrated on mucosal lymphocytes from these same aged animals. These findings indicate that the idiotype repertoires expressed by systemic and mucosal lymphocytes of aged mice are different and non-overlapping. The inductive signals or repertoire changes responsible for the development of augmented systemic anti-idiotyic regulation during senescence are apparently not triggered, or are not effective, in the mucosal immune system.

To further examine age-related changes in the idiotype repertoire expressed in the spleen and MLN of C57BL/6J mice, sera from young and old C57BL/6J mice previously immunized with TNP-BGG were tested for their ability to reversibly inhibit the PFC responses of spleen and MLN cells from young and old mice (Wade and Szewczuk, 1986). Reversible suppression in this system is mediated by circulating anti-idiotypic antibody (Goidl et al., 1984). It was found that serum from old immune mice inhibited PFC responses from both old and young animals to a greater extent than did serum from young mice. The greater inhibition of splenic PFC responses in old mice than in young animals by serum from old immune animals is compatible with the work of Goidl et al. (1980) and suggests that the idiotype repertoire stimulated by antigen undergoes a change with age. In contrast, immune serum from old mice has little effect on the PFC response of cells from the MLN of young or old animals (see Fig. 19–6). Thus, the idiotype repertoire produced in response to TNP-BGG is not the same as that in B cells in the spleen and MLN of old mice.

Intragastric (ig) immunization induces high levels of anti-idiotypic suppression of PFC in young but not in old mice (Wade and Szewczuk, 1988). These findings suggest that direct mucosal stimulation not only allows for elevated systemic and mucosal PFC responses in aged animals, but also allows their escape from the suppressive anti-idiotypic antibody regulation normally induced by other routes of stimulation.

Other evidence for cell populations that express unique idiotypic repertoires is found in the work of Gearhart and Cebra (1979). They found that the frequency of phosphorylcholine-specific cells did not differ between the spleen, MLN, and Peyer's patches, but that expression of the T15 idiotypic marker was different between systemic and mucosal tissues. Most spleen cells expressed the T15 idiotype, whereas the majority of Peyer's patch cells did not. This lends further support to the idea that Peyer's patches contain a unique population of B cells.

In general, the mucosal immune system contains entirely separate cell populations from the systemic immune system. These include distinct subpopulations of T and B cells, as well as lymphocyte subpopulations that express unique receptor and idiotype repertoires. The reasons for the existence of such distinct cell populations remain to be determined. The continuous exposure of lymphocytes in the gut environment to antigens derived from diet and microbial flora may influence the maintenance of these unique cell populations and require that their activity be unimpaired even at a time when the systemic immune system shows a decline in function.

CONCLUSIONS

The developmental changes occurring in the immune system with age are subtle and are not readily assessed by measurements of cell numbers or concentrations of serum com-

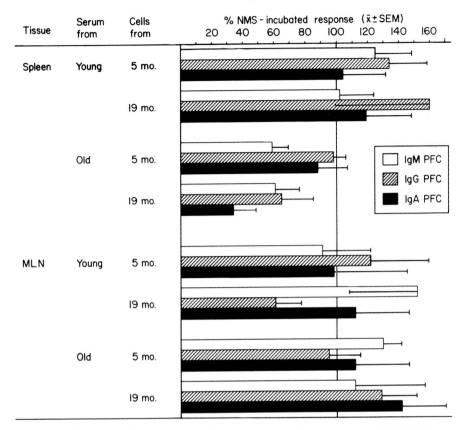

Figure 19–6.   The effect of immune serum from young and old mice on the anti-TNP PFC response in the spleen and MLN of young and old mice immunized with TNP-BGG. Serum was removed on day 14 from 5- and 20-month-old TNP-BGG–immune mice and later depleted of anti-TNP antibody by (2×) batch absorption on TNP-BGG sepharose (as confirmed by hemagglutination with TNP-conjugated sheep red blood cells). The inhibitory effect of this serum was tested on immune cells from the spleen and MLN of young (4-mo) and old (20-mo) mice. After pretreatment of each target cell population with 10^{-8} M TNP-EACA to remove endogenously bound anti-idiotypic antibody, a cell aliquot was incubated (10 minutes on ice) with 0.02 ml of the immune serum, NMS or HBSS medium, washed, and tested in a normal PFC assay. The arithmetic means from six independent experiments are presented. (Reprinted by permission from Wade et al., 1988.)

ponents. More likely, an amplification effect is in operation, with slight changes in levels of thymic hormones having significant effects on T cell function, particularly in T_H function. In turn, impaired lymphokine production by aged T_H cells has a major effect on many other aspects of the immune system. Underlying this is a basic cellular defect in the ability to respond to activating and stimulating signals, so that even when lymphokines are produced, the cells that normally respond to them are unable to or can do so only at reduced levels. Because some responses can be restored to a large extent by addition of IL-2, the major defect seems to be in the loss of cells that respond to stimulation by IL-2 production.

Mice and humans show similar changes in their immune systems with age. Old animals retain relatively normal levels of stem cells, serum proteins, lymphocytes, and other cell types, but, because of regulatory changes, environmental influences, or intracellular defects, these cells respond submaximally to extrinsic stimuli. Cells in the aged environment can also promote the production of suppressive anti-idiotypic antibodies. These antibodies can bind and inhibit the normal functioning of cells within the systemic immune system.

Restoration or improvement of immune responses in old individuals by modulators of T cell function (thymic hormones and IL-2) suggests that changes in T helper cells account for a major portion of these immune deficits. One characteristic of murine responses seen in aged individuals is the heterogeneity of the response. Examination of individual responses frequently yields old individuals who fall within the "young" range, so that identification of age-associated changes requires large test populations and careful statistical analysis.

The reduction in the proliferative potential of T and B cells with age can be attributed to a reduction in the number of cells able to respond to signals and to a shortage of T helper signals, especially IL-2. In CTL and B cell responses that require some degree of cellular interaction and cooperation, the age-associated alterations are more complex. The combined effects of a repertoire change, diminished IL-2 production, altered receptor expression, and augmented suppressor activity all serve in differing degrees to downregulate the responses of old animals. Once cells have matured in an aged environment, supplementation with IL-2 or thymic hormone can restore responses to only a limited extent. Thus, changes in populations and in repertoires cannot be corrected.

It appears that for a limited number of old cells there is no change in the intrinsic ability to respond to stimuli. The decline in immune function may therefore reflect a reduced frequency of cells able to respond—whether it is to interleukins, growth factors, or antigens. This is reflected throughout the total cell population as decreases are seen in proliferation, cytolytic T cell activity, and lymphokine production.

Proliferative responses of lymphocytes from humans and mice to mitogens decrease with age. This is probably due to similar age-associated biochemical and structural changes in the lymphocytes. Murine lymphoid stem cells remain fairly normal into old age, but they are subject to elevated suppressive and regulatory influences when they enter the aged systemic environment. Cells within the systemic immune system of old mice show depression of immune reactivity as a consequence of regulatory changes that are remarkably similar for all stimuli. If cells from old animals are stimulated with Con A, PHA, T-dependent antigens, or allogeneic cells, age-dependent changes in responding cell frequency and IL-2 production are universally found.

Thymic involution probably contributes to depressed immune responses by causing insufficient production of various thymic hormones. The augmentation of most responses seen with exogenous IL-2 or thymic hormones implies that thymic involution leads to failure of aged mice to fully differentiate T helper cells. The demonstration of a normal CTL repertoire within the thymus of old mice implies that its ability to shape the T cell repertoire may continue into old age, suggesting that thymic involution with age does not correlate with the loss of all aspects of thymic function.

The multitude of effects exerted by age on cell function makes it difficult to assess its effects on any distinct cell type. The immune system is an interrelated network of cells that require complex interactions to operate at full functional capacity. Thus any age-induced change can have far-reaching repercussions, making the original defect difficult to pinpoint. When they are examined on an individual basis, the cells that remain able to respond, albeit diminished in number, retain the ability to exhibit full effector function. It follows that at least some of the age-associated reduction in immune reactivity is mediated at the level of cellular interactions, and that the decline in overall immune responsiveness cannot be solely attributed to loss of cell metabolism. The microenvironment in which lymphocytes develop also influences the extent to which the aging process affects their functional capacity. The challenge in the next few years will be to determine the exact mechanisms operating to maintain unique cell populations refractory to the aging process.

Acknowledgments

This work was supported in part by grants-in-aid of research from the Medical Research Council of Canada, MA-7347, and the E. K. Dawson Foundation. M. Szewczuk is a Career Scientist of the Ontario Ministry of Health.

References

Averill, L. E., and Wolf, N. S. (1985). The decline in murine splenic PHA and LPS responsiveness with age is primarily due to an intrinsic mechanism. *J. Immunol. 134*:3859–3863.

Bloom, E. T., Kubota, L. F., and Kawakami, K. (1988). Age-related decline in the lethal hit but not the binding stage of cytotoxic T cell activity in mice. *Cell. Immunol. 114*:440–446.

Brock, M. A. (1987). Age-related changes in circannual rhythms of lymphocyte blastogenic responses in mice. *Am. J. Physiol. 252*:R299–305.

Buckler, A. J., Vie, H., Sonenshein, G. E., and Miller, R. A. (1988). Defective T lymphocytes in old mice. Diminished production of mature c-*myc* RNA after mitogen exposure not attributable to alterations in transcription or RNA stability. *J. Immunol. 140*:2442–2446.

Callard, R. E., Basten, A., and Blanden, R. V. (1979). Loss of immune competence with age may be due to a qualitative abnormality in lymphocyte membranes. *Nature 281*:218–220.

Chang, M.-P., and Gorczynski, R. M. (1984). Peripheral (somatic) expansion of the murine cytotoxic T lymphocyte repertoire: I. Analysis of diversity in recognition repertoire of alloreactive T cells derived from the thymus and spleen of adult or aged DBA/2J mice. *J. Immunol. 133*:2375–2380.

Chang, M.-P., Makinodan, T., Peterson, W. J., Strehler, B. L. (1982). Role of T cells and adherent cells in age-related decline in murine interleukin 2 production. *J. Immunol. 129*:2426–2430.

Chang, M.-P., Utsuyama, M., Hirokawa, K., and Makinodan, T. (1988). Decline in the production of interleukin-3 with age in mice. *Cell. Immunol. 115*:1–12.

Cheung, H. T., Twu, J.-S., and Richardson, A. (1983). Mechanism of the age-related decline in lymphocyte proliferation: Role of IL-2 production and protein synthesis. *Exp. Gerontol. 18*:451–460.

Coico, R. F., Gottesman, S. R., Siskind, G. W., and Thorbecke, G. J. (1987). Physiology of IgD: VIII. Age-related decline in the capacity to generate T cells with receptors for IgD and partial reversal of the defect with IL 2. *J. Immunol. 138*:2776–2781.

Doria, G., Mancini, C., Frasca, D., and Adorini, L. (1987). Age restriction in antigen-specific immunosuppression. *J. Immunol. 139*:1419–1425.

Eldridge, J. H., Beagly, K. W., and McGhee, J. R. (1987). Immunoregulation in the Peyer's patch microenvironment: Cellular basis for the enhanced responses by the B cells of X-linked immunodeficient CBA/N mice. *J. Immunol. 139*:2255–2262.

Eldridge, J. H., Kiyono, J. H., Mickalek, S. M., and McGhee, J. R. (1983). Evidence for a mature B cell population in Peyer's patches of young adult *xid* mice. *J. Exp. Med. 157*:789–794.

Ernst, D. N., Weigle, W. O., and Thoman, M. L. (1987). Retention of T cell reactivity to mitogens and alloantigens by Peyer's patch cells of aged mice. *J. Immunol. 138*:26–31.

Ershler, W. B., Moore, A. L., Hacker, M. P., Ninomiya, J., Naylor, P., and Goldstein, A. L. (1984a). Specific antibody synthesis in vitro: II. Age-associated thymosin enhancement of antitetanus antibody synthesis. *Immunopharmacology 8*:69–77.

Ershler, W. B., Moore, A. L., and Socinski, M. A. (1984b). Influenza and aging: Age-related changes and the effects of thymosin on the antibody response to influenza vaccine. *J. Clin. Immunol. 4*:445–454.

Fong, T. C., and Makinodan, T. (1989). In situ hybridization analysis of the age-associated decline in IL-2 mRNA expressing murine T cells. *Cell. Immunol. 118*:199–207.

Gearhart, P. J., and Cebra, J. J. (1979). Differentiated B lymphocytes: Potential to express particular antibody variable and constant regions depends on site of lymphoid tissue and antigen load. *J. Exp. Med. 149*:216–227.

Gillis, S., Kozak, R., Durante, M., and Weksler, M. E. (1981). Immunological studies of aging: Decreased production of and response to T cell growth factor by lymphocytes from aged humans. *J. Clin. Invest. 67*:937–942.

Gilman, S. C., Rosenberg, J. S., and Feldman, J. D. (1982). T lymphocytes of young and aged rats: II. Functional defects and the role of interleukin-2. *J. Immunol. 128*:644–650.

Globerson, A., Abel, L., and Umiel, T. (1981). Immune reactivity during ageing: III. Removal of peanut-agglutinin binding cells from ageing mouse spleen cells leads to increased reactivity to mitogens. *Mech. Aging Dev. 16*:275–284.

Goidl, E. A., Choy, J. W., Gibbons, J. J., Weksler, M. E., Thorbecke, G. J., and Siskind, G. W. (1983). Production of auto-antiidiotypic antibody during the normal immune response: VII. Analysis of the cellular basis for the increased auto-antiidiotypic antibody production by aged mice. *J. Exp. Med. 157*:1635–1645.

Goidl, E. A., Samarut, C., Schneider-Gadicke, A., Hockwald, N. L., Thorbecke, G. J., and Siskind, G. W. (1984). Production of auto-anti-idiotypic antibody during the normal immune response: IX. Characteristics of the auto-anti-idiotype antibody and its production. *Cell. Immunol. 85*:25–33.

Goidl, E. A., Thorbecke, G. J., Weksler, M. E., and Siskind, G. W. (1980). Production of auto-anti-idiotypic antibody during the normal immune response: Changes in the auto-anti-idiotypic antibody response and the idiotype repertoire associated with aging. *Proc. Natl. Acad. Sci. USA 77*:6788–6792.

Gorczynski, R. M., and Chang, M.-P. (1984). Peripheral (somatic) expansion of the murine cyto-toxic T lymphocyte repertoire: II. Comparison of diversity in recognition repertoire of allo-reactive T cells in spleen and thymus of young or aged DBA/2J mice transplanted with bone marrow from young or aged donors. *J. Immunol. 133*:2381–2389.

Gorczynski, R. M., Chang, M.-P., Kennedy, M., MacRae, S., Benzing, K., and Price, G. B. (1984). Alterations in lymphocyte recognition repertoire during aging: I. Analysis of changes in immune response potential of B lymphocytes from non-immunized aged mice, and the role of accessory cells in the expression of that potential. *Immunopharmacology 7*:179–194.

Hara, H., Negoro, S., Miyata, S., Saiki, O., Yoshizaki, K., Tanaka, T., Igarashi, T., and Kishimoto, S. (1987). Age-associated changes in proliferative and differentiative response of human B cells and production of T cell–derived factors regulating B cell functions. *Mech. Aging Dev. 38*:245–258.

Harrison, D. E., Astle, C. M., and Doubleday, J. W. (1977). Stem cell lines from old immunodefi-cient donors give normal responses in young recipients. *J. Immunol. 118*:1223–1227.

Harrison, D. E., Astle, C. M., and Stone, M. (1989). Numbers and functions of transplantable prim-itive immunohematopoietic stem cells: Effects of age. *J. Immunol. 142*:3833–3840.

Hausman, P. B., Goidl, E. A., Siskind, G. W., and Weksler, M. E. (1985). Immunological studies of aging: XI. Age-related changes in idiotype repertoire of suppressor T cells stimulated dur-ing tolerance induction. *J. Immunol. 134*:3802–3807.

Hiramoto, R. N., Ghanta, V. K., and Soong, S. (1987). Effect of thymic hormones on immunity and life span. In *Aging and the Immune Response: Cellular and Humoral Aspects* (Goidl, E. A., ed.). New York and Basel: Marcel Dekker, pp. 177–198.

Kato, K., and Hirokawa, K. (1988). Qualitative and quantitative analysis of autoantibody production in aging mice. *Aging: Immunol. Infect. Dis. 1*:177–190.

Kay, M.M.B. (1981). Isolation of the phagocytosis-inducing IgG-binding antigen on senescent somatic cells. *Nature 289*:491–494.

Kelley, K. W., Brief, S., Westly, H. J., Novakofski, J., Bechtel, P. J., Simon, J., and Walker, E. B. (1986). GH3 pituitary adenoma cells can reverse thymic aging in rats. *Proc. Natl. Acad. Sci. USA 83*:5663–5667.

Kennes, B., Brohee, D., and Neve, P. (1983). Lymphocyte activation in human aging: V. Acquisition of response to T cell growth factor and production of growth factors by mitogen-stimulated lymphocytes. *Mech. Aging Dev. 23*:103–113.

Kim, Y. T., Goidl, E. A., Samarut, C., and Weksler, M. E. (1985). Bone marrow function: I. Peripheral T cells are responsible for the increased auto-anti-idiotypic response of older mice. *J. Exp. Med. 161*:1237–1242.

Kishimoto, S., Tomino, S., Mitsuya, H., and Fujiwara, H. (1979). Age-related changes in suppressor functions of human T cells. *J. Immunol. 123*:1586–1593.

Kohno, A., Seeman, P., and Cinader, B. (1986). Age-related changes of beta-adrenoceptors in aging inbred mice [published erratum appears in *J. Gerontol.* 1986 *41*:698]. *J. Gerontol. 41*:439–444.

Komatsubara, S., Cinader, B., and Muramatsu, S. (1986). Functional competence of dendritic cells of ageing C57BL/6 mice. *Scand. J. Immunol. 24*:517–525.

Krishnaraj, R., and Blandford, G. (1988). Age-associated alterations in human natural killer cells: 2. Increased frequency of selective NK subsets. *Cell Immunol. 114*:137–148.

Lerner, A., Philosophe, B., and Miller, R. A. (1988). Defective calcium influx and preserved inositol phosphate generation in T cells from old mice. *Aging: Immunol. Infect. Dis. 1*:149–157.

Meroni, P. L., Barcellini, W., Frasca, D., Sguotti, C., Borghi, M. O., De Bartolo, G., Doria, G., and Zanussi, C., (1987). In vivo immunopotentiating activity of thymopentin in aging humans: Increase of IL-2 production. *Clin. Immunol. Immunopathol. 42*:151–159.

Micklem, H., and Ansell, J. (1984). The self-renewal capacity of murine haematopoietic stem cells: A non-aging population. In *Lymphoid Cell Functions in Ageing* (DeWeck, A. L., ed.). City: EURAGE, pp. 53–58.

Miller, R. A. (1984). Age-associated decline in precursor frequency for different T cell–mediated reactions, with preservation of helper or cytotoxic effect per precursor cell. *J. Immunol. 132*:63–68.

Miller, R. A. (1986). Immunodeficiency of aging: Restorative effects of phorbol ester combined with calcium ionophore. *J. Immunol. 137*:805–808.

Negoro, S., Hara, H., Miyata, S., Saiki, O., Tanaka, T., Yoshizaki, K., Igarashi, T., and Kishimoto, S. (1986). Mechanisms of age-related decline in antigen-specific T cell proliferative response: IL-2 receptor expression and recombinant IL-2 induced proliferative response of purified Tac-positive T cells. *Mech. Aging Dev. 36*:223–241.

Pandolfi, F., Quinti, I., Montella, F., Voci, M. C., Schipani, A., Urasia, G., and Aiuti, F. (1983). T-dependent immunity in aged humans: II. Clinical and immunological evaluation after three months of administering a thymic extract. *Thymus 5*:235–240.

Ponnappan, U., Cinader, B., Gerber, V., and Blaser, K. (1987). Age-dependent changes in isotype expression and down-regulation of C57BL/6 mice. *Scand. J. Immunol. 25*:45–54.

Proust, J. J., Filburn, C. R., Harrison, S. A., Buchholz, M. A., and Nordin, A. A. (1987). Age-related defect in signal transduction during lectin activation of murine T lymphocytes. *J. Immunol. 139*:1472–1478.

Proust, J. J., Kittur, D. S., Buchholz, M. A., and Nordin, A. A. (1988). Restricted expression of Mg-induced high affinity IL-2 receptors in aging mice. *J. Immunol. 141*:4209–4216.

Rabinowe, S. L., Nayak, R. C., Krisch, K., George, K. L., and Eisenbarth, G. S. (1987). Aging in man: Linear increase of a novel T cell subset defined by antiganglioside monoclonal antibody 3G5. *J. Exp. Med.* *165*:1436–1441.

Reynolds, J. D., Cahill, R.N.P. and Trnka, Z. (1981) Peyer's patches as a bursal equivalent: A new look at some old arguments. In: *Aspects of Development and Comparative Immunology* (Solomon, J. B., ed.). Oxford and New York: Pergamon Press, pp. 265–272.

Rivier, D. A., Trefts, P. E., and Kagnoff, M. F. (1983). Age-dependence of IgA anti-(1–3) dextran B1355 response in vitro. *Scand. J. Immunol.* *17*:115–121.

Rosenberg, J. S., Gilman, S. C., and Feldman, J. D. (1983). Effects of aging on cell cooperation and lymphokine responsiveness to cytokines. *J. Immunol.* *130*:1754–1758.

Shock, N. W. (1984). Normal human aging: The Baltimore longitudinal study of aging. Washington, D.C.: Department of Health and Human Services.

Sidman, C., Luther, E., Marshall, J., Ngujen, K.-A., Roopenian, D., and Worthen, S. (1987). Increased expression of major histocompatibility complex antigens on lymphocytes from aged mice. *Proc. Natl. Acad. Sci. USA* *84*:7624–7628.

Snow, E. C. (1987). An evaluation of antigen-driven expansion and differentiation of hapten-specific B lymphocytes purified from aged mice. *J. Immunol.* *139*:1758–1762.

Steinmann, G. G., (1986). Changes in the human thymus during aging. *Curr. Top. Pathol.* *75*:43–88.

Steinmann, G. G., and Müller-Hermelink, H.-K. (1984). Immunohistological demonstration of terminal transferase (TdT) in the age-involuted human thymus. *Immunobiology* *166*:45–52.

Sullivan, D. A., Hann, L. E., and Allansmith, M. R. (1987). The influence of age on the ocular secretory immune system of the rat. *Adv. Exp. Med. Biol.* *216*:1395–1407.

Szewczuk, M. R. (1987). Aging and the future of preventive medicine. *Persp. Geriatrics* *3*:29–38.

Szewczuk, M. R., and Campbell, R. J. (1980). Loss of immune competence with age may be due to auto-anti-idiotypic antibody regulation. *Nature* *286*:164–166.

Szewczuk, M. R., and Campbell, R. J. (1981a). Differential effect of aging on the heterogeneity of the immune response to a T-dependent antigen in systemic and mucosal-associated lymphoid tissues. *J. Immunol.* *126*:472–477.

Szewczuk, M. R., and Campbell, R. J. (1981b). Lack of age-associated auto-anti-idiotypic antibody regulation in mucosal-associated lymph nodes. *Eur. J. Immunol.* *11*:650–656.

Szewczuk, M. R., Campbell, R. J., and Jung, L. K. (1981). Lack of age-associated immune dysfunction in mucosal-associated lymph nodes. *J. Immunol.* *126*:2200–2204.

Thoman, M. L., and Weigle, W. O. (1983). Deficiency in suppressor T cell activity in aged animals: Reconstitution of this activity by interleukin 2. *J. Exp. Med.* *157*:2184–2189.

Thoman, M. L., and Weigle, W. O. (1985). Reconstitution of in vivo cell-mediated lympholysis responses in aged mice with interleukin 2. *J. Immunol.* *134*:949–952.

Thoman, M. L., and Weigle, W. O. (1987). Age-associated changes in the synthesis and function of cytokines. In *Aging and the Immune Response* (Goidl, E. A., ed., New York: Marcel Dekker, pp. 199–224.

Utsuyama, M., and Hirokawa, K. (1987). Age-related changes of splenic T cells in mice—a flow cytometric analysis. *Mech. Aging Dev.* *40*:89–102.

Wade, A. W., Green-Johnson, J., and Szewczuk, M. R. (1988). Functional changes in systemic and mucosal lymphocyte repertoires with age: An update review. *Aging: Immunol. Infect. Dis.* *1*:65–97.

Wade, A. W., and Szewczuk, M. R. (1984). Aging, idiotype repertoire shifts, and compartmentalization of the mucosal-associated lymphoid system. *Adv. Immunol.* *36*:143–188.

Wade, A. W., and Szewczuk, M. R. (1986). Changes in the mucosal-associated B cell response with age. In *Aging and the Immune Response* (Goidl, E. A., ed.). New York: Marcel Dekker, pp. 95–121.

Wade, A. W., and Szewczuk, M. R. (1988). Evidence for the divergence of systemic and mucosal B cells with age in the primary and secondary immune responses to T-dependent antigens:

Route of priming influences isotype profile of antigen-specific memory and antibody-secreting B cells. *Aging: Immunol. Infect. Dis. 1*:99–119.

Weksler, M. E., Innes, J. B., and Goldstein, G. (1978). Immunological studies of aging: IV. The contribution of thymic involution to the immune deficiencies of aging mice and reversal with thymopoietin. *J. Exp. Med. 148*:996–1006.

Wu, W., Pahlavani, M., Cheung, H. T., and Richardson, A. (1986). The effect of aging on the expression of interleukin 2 messenger ribonucleic acid. *Cell. Immunol. 100*:224–231.

Zharhary, D., and Klinman, N. R. (1986). A selective increase in the generation of phosphorylcholine-specific B cells associated with aging. *J. Immunol. 136*:368–370.

20

Development of Normal and Abnormal Immune Functions in Humans

HANS-MICHAEL DOSCH,

MAY F. HUI, PAUL J. DOHERTY

Human primary immunodeficiency diseases can provide unique insights into the processes that direct the acquisition and expression of immunocompetence (Gelfand and Dosch, 1983; Komiyama et al., 1988; Matricardi et al., 1984; Royer and Reinherz, 1987). Few of these diseases have natural counterparts in experimental animals, and inborn immune defects in animals are largely without analogues in humans (e.g., Bosma et al., 1988; Dosch et al., 1985, 1991a; Kincade, 1987). This overview will refer to such species differences and discuss human primary immunodeficiency syndromes within the framework of normal lymphocyte ontogeny.

Despite great progress in our understanding of the molecular basis of normal immune function, the majority of primary immune deficiency disorders remain poorly understood. The more severe defects involve primary developmental blocks in either B or T lymphocyte maturation, and only rarely are both cell lineages affected such as the deficiency of adenosine deaminase (ADA, see below). Severe combined immunodeficiency (SCID) describes the absence of both humoral and cellular lymphocyte function. In most cases the combined defect is due to a primary block in T cell development (Gelfand and Dosch, 1982). The presence in such patients of inherently normal B-lineage cells underlines the essentially complete T cell dependency of the human humoral immune system (Gelfand and Dosch, 1983).

A number of congenital multiorgan diseases affect the immune system. Examples include DiGeorge syndrome, Wiscott-Aldrich syndrome, ataxia telangiectasia, and α-1-antitrypsin deficiency (Cooper, 1987; Gelfand et al., 1979; Gelfand and Dosch, 1983). Three X-chromosomal abnormalities are associated with severe, yet heterogeneous immune disorders: X-linked lymphoproliferative disease (XLP), X-linked agammaglobulinemia (XLA), and an X-linked form of SCID (Rousset et al., 1986; Seemayer, 1987; Tatsumi and Purtilo, 1986). The XLA and X-SCID defects have been mapped (Conley and Puck, 1988; Hendriks et al., 1989; Mensink et al., 1987). The vast majority of human immune defects are not global but affect functions of one or more B or T cell subsets selectively. The clinical pictures are highly heterogeneous. Although only a minority are directly life threatening, even clinically minor defects (e.g., IgA deficiency) carry significant long-term risks, in particular involving autoimmune and malignant diseases (Cunningham, 1989; Kersey et al., 1988).

Structural definitions of immune defects are still scant; progress has been slowed by the rarity of their occurrence. A recent report identified IL-2 deficiency as the underlying cause of a new form of SCID, and the affected patient responded well to chronic IL-2 therapy (Pahwa et al., 1989). Abnormal expression of class II MHC molecules may occur as the result of a transacting regulatory gene defect outside of the MHC locus and may result in SCID (dePreval et al., 1985). While defective expression of CD8 similarly leads to SCID (Roifman et al., 1989), the absence of $CD4^-8^+$ cells in β_2-m–deficient mice does not (Koller et al., 1990; Zijlstra et al., 1990). Patients with defects in purine salvage enzyme or lymphocyte 5′-nucleotidase function also develop life-threatening immunodeficiency syndromes (Cohen et al., 1980a,b, 1981).

HUMAN T CELL ONTOGENY

Development of the T Cell Compartment

As the primary site for rearranging T cell receptor (TCR) α, β and TCR γ, δ genes, the thymus is critical for the production of mature circulating T cells. The human thymic microenvironment develops early: a primitive thymic rudiment forms at 4 weeks of gestation. MHC antigens critical to formation of the expressed TCR repertoires similarly appear early in ontogeny, and their abnormal expression leads to SCID (Berg et al., 1989, 1990; Reith et al., 1988). Epithelial cells in the thymic rudiment express class II MHC antigens by 7 weeks, and both epithelial cells and macrophage-like cells express class I and class II MHC antigens at 8½ weeks (Lobach and Haynes, 1987).

Pluripotent stem cells derived successively from the yolk sac (60 days), from the fetal liver (50–150 days), and from bone marrow (80 days) colonize the thymus (Strominger, 1989). At 8–9 weeks of gestation 50% of the thymocytes are $CD7^+2^+$, while the remainder are $CD7^+2^-$. No T cell differentiation antigens other than CD7 and CD2 are seen this early in development (Haynes et al., 1988).

At 10 weeks of fetal development the encapsulating fibrous connective tissue penetrates the thymus, giving rise to discrete lobules. The T cell markers CD3, CD4, CD5, and CD8 are all expressed by this time (Lobach and Haynes, 1987). The presence of CD3 suggests that T cell receptors may also be expressed (Deibel et al., 1983; Haynes et al., 1988; Osmond, 1985; Royo et al., 1987; Swerdlow et al., 1988).

Both CD1 and CD3 surface antigens are present on T cells throughout the thymus after 12 weeks of gestation. At 14 weeks each lobule consists of an inner medulla, clearly demarcated from a subcapsular and an inner cortex. Characteristic of the medulla at 15 to 16 weeks are Hassal's bodies (keratinized epithelial cell swirls). Hassal's bodies are absent in most patients with the intrathymic form of SCID, who make up the largest single group of patients with this disease (Dosch et al., 1978; Gelfand and Dosch, 1983; Pyke et al., 1975). After 15 weeks, when partitioning of the thymus into the cortex and the medulla occurs, CD1 is expressed only on cortical cells (Lobach and Haynes, 1987). By 16 weeks the non-lymphoid component of the thymus already has a postnatal appearance (Lobach and Haynes, 1987).

In mice the emergence of the TCR repertoire is a late event in fetal ontogeny. TCR β becomes detectable at low levels on the surface of rodent thymocytes by day 17 of fetal life, and high-density staining is not observed until after birth (Crispe et al., 1986; Richie et al., 1988). In striking contrast, the emergence of a diversified TCR repertoire is an early event in human fetal ontogeny. We have studied the diversity of expressed TCR genes as early as 17 weeks of gestation (Doherty et al., 1991). In these experiments we amplified RNA from various fetal tissues using variable-region (V) gene family–specific oligonucleotide primers for cDNA amplification by the polymerase chain reaction (Doherty et al., 1989).

Table 20–1 summarizes the expression of the 20 V_β TCR gene families in fetal spleen tissue. It is clear that the entire spectrum of these 20 families is expressed at this point in development and resembles adult repertoires with respect to the relative proportions of V_β region expression (% usage) (Doherty et al., 1990). The expressed TCR V_β repertoire at 17 weeks gestation appears as diverse as that in thymus from donors at several years of age. The presence of the T cell markers CD3, CD4, and CD8 prior to 17 weeks of gestation suggests that the repertoire may be formed even earlier. Thus, fetal T cell development may occur concurrently (or somewhat earlier) than fetal B cell diversification (see below) (Dosch et al., 1989; Hui et al., 1989).

Enzymes Critical to Lymphocyte Ontogeny

The maturation of T cells in the thymus is characterized by changing patterns of enzyme activities that are critical to thymic differentiation. These include the purine metabolic enzymes adenosine deaminase (ADA) and purine nucleoside phosphorylase (PNP), terminal deoxynucleotidyl transferase (TdT), and the nucleoside kinases (Cohen et al., 1983a, 1984; Ma et al., 1982).

In cortical thymocytes high levels of deoxynucleotidetriphosphates (dNTP) result from high levels of deoxynucleoside kinases and low levels of 5'-nucleotidase activity. Together with maximal levels of TdT, high levels of dNTP in immature T (and B) cells are likely essential for N-region addition in rearranging TCR (and Ig) genes (Cohen et al., 1984). In addition, an inverse relationship exists between the levels of ADA and PNP, with ADA levels higher in more immature cells, decreasing in amount up the T cell maturation pathway. The relative proportions of ADA and PNP activities shift the relative sizes of the four dNTP pools. The direct consequence of this shift in maturing T cells is not clear, but the absence of ADA results in SCID, and a deficiency in PNP results in less complete but still severe T cell defects. High cytosolic levels of deoxynucleotides due to absent ADA or PNP activity are selectively toxic to T-lineage cells in vivo and in vitro, and this selective toxicity has been used to therapeutically manipulate T cell function in

Table 20–1.   TCR V_β Usage in Human Fetal Spleen at 17 Weeks Gestation

TCR β Family	$V_\beta 1$	$V_\beta 2$	$V_\beta 3$	$V_\beta 4$	$V_\beta 5$	$V_\beta 6$	$V_\beta 7$	$V_\beta 8$	$V_\beta 9$	$V_\beta 10$
% Usage	5.62	3.71	4.93	2.99	6.45	6.67	10.19	4.76	2.69	2.99
SD	1.38	2.08	1.21	1.14	1.47	2.23	2.08	1.26	0.53	0.60

TCR β Family	$V_\beta 11$	$V_\beta 12$	$V_\beta 13$	$V_\beta 14$	$V_\beta 15$	$V_\beta 16$	$V_\beta 17$	$V_\beta 18$	$V_\beta 19$	$V_\beta 20$
% Usage	3.63	7.41	9.10	3.72	2.16	1.57	4.85	2.81	8.34	4.81
SD	0.69	1.77	2.11	1.27	0.78	0.47	1.82	0.38	4.16	1.53

rodents and humans (Cohen et al., 1980a, 1983a,b, 1984; Dosch et al., 1980; Ruers et al., 1987; Urba et al., 1989).

Thymic Selection

The process of thymic selection has been studied in depth, primarily in the mouse. A combination of positive and negative selection destroys about 5×10^7 thymocytes/day in the weanling mouse; most of these cells are cortical, and full replacement of the 1.5×10^8 cortical cells requires 3–4 days (Rothenberg, 1990). It remains a paradox that both positive and negative selection depend on the recognition, at least in part, of class I and class II MHC antigens in the thymus by the T cell antigen receptor. Because random rearrangement of TCR genes is more likely to result in a failure of positive selection (failure to recognize self MHC antigens) rather than negative selection (e.g., following high-affinity recognition of self MHC antigens), most of the destruction of thymocytes is likely due to a failure of positive selection, which prevents exit from a programmed process leading to cell death (apoptosis) (von Boehmer et al., 1989).

The most compelling evidence that T cells are positively selected for recognition of class I or class II MHC in the thymus comes from studies of transgenic mice and from antibody blocking experiments. Selection requires coexpression of the T cell receptor and CD8 or CD4 on the surface of the selected thymocyte (Sha et al., 1988; Teh et al., 1988; von Boehmer et al., 1989). CD8 coexpression is invariably observed on T cells restricted in their reactivity to peptides presented by class I MHC antigens (Zijlstra et al., 1990); CD4 coexpression is associated with restriction by class II MHC antigens.

Teh et al. (1988) described mice that expressed transgenes coding for a TCR recognizing the male-specific H-Y antigen in the context of class I H-2D^b molecules. It is clear from experiments with female mice that both the TCR and thymic MHC antigens are involved in the selection of mature CD4$^-$8$^+$ cells from immature CD4$^+$8$^+$ T cells. This positive selection occurs in the absence of the nominal antigen (H-Y is absent in females), and the data are consistent with an MHC–TCR interaction, with MHC antigens determining the CD4/CD8 phenotype of peripheral T cells. In this example class I MHC-recognition rescues CD4$^-$8$^+$ T cells from apoptosis (von Boehmer et al., 1989). Although class I- and II-dependent selection processes in humans are less well understood, a "rescue from apoptosis model" by positive recognition would be consistent with the observation in humans that aberrant MHC expression results in SCID (dePreval et al., 1985; Reith et al., 1988).

Severe Combined Immunodeficiency (SCID)

Patients with SCID represent a heterogeneous group of primary T cell defects that lead to a secondary failure of B cell immunity (see below). Both B and T cell lineages are affected in only a subset of patients—i.e., those patients with abnormalities in the purine salvage enzymes adenosine deaminase, purine nucleoside phosphorylase, or 5′-nucleotidase (Cohen et al., 1983a, 1984; Gelfand and Dosch, 1983; Gelfand et al., 1984). There are no animal models for these disorders, but ADA deficiency has now become the first acknowledged target of attempts at gene therapy (Geiger and Nagy, 1986; Willis et al., 1987).

The major T cell sublineages are distinguished by CD4 and CD8 expression. Congenital abnormalities in the expression of CD4 are associated with considerable immune defects (Akahonai et al., 1986; Levinson et al., 1985). The selective, intrathymic elimination of $CD8^+$ cells also leads to SCID, implying a critical role for $CD8^+$ T cell sublineages in the development of T cell competence (Roifman et al., 1989).

Isolated interleukin deficiencies are at best rare (Matheson and Green, 1987; Pahwa et al., 1989), and replacement therapies analogous to intravenous gamma globulin are therefore not indicated or available in B cell defects. Therapeutic strategies must aim at replacement of the defective function, balancing the severity and risk of the defect with those of therapy. SCID is a difficult clinical problem, and treatment today relies on bone marrow transplantation (Gelfand and Dosch, 1983; Gelfand et al., 1984).

In the largest subset of SCID patients, a block in intrathymic T cell development reflects the failure of thymic organ development. Transplantation of (genetically unrelated) fetal thymic tissue or of in vitro–grown thymic epithelium from normal (unrelated) donors can successfully promote the differentiation of functional host-derived T-lineage cells and lead to lasting, albeit often incomplete, immune reconstitution (Gelfand and Dosch, 1983; Goldmann et al., 1979). With improvements in techniques of marrow transplantation, these therapeutic strategies are now rarely used; they are logistically difficult, and in vitro culture of thymic epithelium may select for some but not other functional sublineages of this tissue. Even after removal of immunocompetent, mature T cells from a bone marrow cell graft, replacement of host lymphoid cells by related or unrelated, MHC-matched donor cells is sufficient for long-term immune reconstitution despite the absence of a functional thymus in these patients.

As illustrated in Table 20–1, TCR V_β genes expressed in fetal thymus and in the spleen by the second trimester show V-region gene family usage similar to that observed in adult tissue (Doherty et al., 1991). In patients with the intrathymic form of SCID, the absence of functional thymic epithelium prevents this expansion, selection, and diversification process (Gelfand and Dosch, 1982). For example, the expression of ADA in both pro-thymocytes and the thymic epithelium is a prerequisite for successful interaction of marrow-derived T progenitor cells with thymic epithelium, and this initial cell–cell interaction is the first identified event that is critical for further maturation of these cells into variously defined precursor T cells (Shore et al., 1981).

The exact function of ADA in this process is uncertain, and it may affect more than one function. We have previously used limiting dilution analysis to estimate the diversity of B cell repertoires in SCID patients, monitoring the frequency of cells specific for a given antigen and which are able to differentiate into an (IgM) antibody-secreting (plaque-forming) cell during in vitro culture in the presence of allogeneic T cell help (Dosch and Gelfand, 1977, 1979; Dosch et al., 1978; Shore et al., 1978, 1981). As shown in Table 20–2, patients with the ADA^+ form of SCID had rather normal B cell repertoires, whereas the two cases of ADA deficiency that were investigated showed extremely narrow repertoires among the few B cells that managed to mature and appear in the circulation.

Many ADA^+ SCID patients generate normal-sized B lymphocyte pools that show normal functionality when examined in vitro (see Table 20–2) (Dosch et al., 1978; Gelfand and Dosch, 1982). The observation that the absence of functional T lymphocytes results in *combined* (B and T cell) immunodeficiency emphasizes that human B lymphocyte function is T cell–dependent to a much higher (almost absolute) degree than that in mice (Dosch and Gelfands 1979). The finding in such patients of intact B cell ontogeny up to the stage of $sIgM^+/D^+$ cells implies that there is little T cell requirement during earlier stages of B cell development (Dosch et al., 1989).

Table 20-2.   Frequency[a] of Antigen/Hapten-Specific PFC Precursor B Cells

B Cell Donor	Ovalbumin	TNP[b]	Sheep RBC	FITC
Normal[c]	1/3460	1/212	1/1100	1/4620
ADA⁻SCID	<1/10⁶	**1/9**	1/14870	<1/10⁶
ADA⁻SCID	<1/10⁶	1/5780	**1/22**	<1/10⁶
ADA⁺SCID[c]	1/4210	1/180	1/1320	1/4040

[a]Number of B cells containing on average one B cell specific for the antigen or hapten indicated.
[b]Trinitrophenyl (hapten).
[c]Average of results in four donors.

HUMAN B CELL ONTOGENY

Human lymphocyte ontogeny begins early during embryogenesis and by 8–10 weeks gestational age, definitive B and T cell populations are present in several tissues, notably yolk sac, fetal liver, and bone marrow (Antin et al., 1986; Dosch et al., 1989; Edwards et al., 1986; Haynes et al., 1988, 1989; Hofman, et al., 1984; Lobach and Haynes, 1987; Rainaut et al., 1987; Royo et al., 1987; Spencer et al., 1986; Westerga and Timens, 1989). These cells have undergone at least some Ig or TCR rearrangement events, respectively (Haynes et al., 1989; Hui et al., 1989; Royo et al., 1987). and it is likely that lymphocyte ontogeny is indicated and perhaps initiated by the tissue-specific induction of recombinase function (Schatz et al., 1989; Sen and Baltimore, 1986; Weaver and Baltimore, 1987). Abnormalities in this process underlie the development of SCID in the *scid/scid* mutant mouse (EMBO Workshop, 1989). A human recombinase defect analogous to the mouse mutation has so far not been found in patients with SCID and would thus at best be rare.

Until about 16 weeks gestational age the populations of B-lineage cells are small, and a majority express the pre-B cell phenotype (reviewed in Dosch et al., 1989). The sole expressed heavy chain is IgM. Among the small population of fully developed (surface Ig⁺) B lymphocytes, κ and λ light chains are expressed in the same relative proportions (~60:40) as are seen in adult human sera and B lymphocytes. This implies that light chain choice is not a developmentally regulated function.

The human fetal bone marrow subserves a function somewhat akin to that provided for the T cell lineage by the thymus (Dosch et al., 1989). Beginning at approximately 16 weeks gestational age, the marrow B cell compartment undergoes a dramatic proliferative expansion until over 25% of the nucleated marrow cells are of B cell lineage (Fig. 20–1). Because still over half of these cells are pre-B cells, this massive expansion likely derives from the proliferation of immature precursors. This numeric expansion is directly coupled to a similarly massive differentiative event that generates full (adult-type) isotype diversity over as short a time as 1 week. The B lymphocytes that are generated express surface Ig with expected (adult) $\kappa:\lambda$ ratios. Perhaps more interestingly, all isotypes (including IgE) are expressed in the relative proportions typical of adult lymphoid tissues (Dosch et al., 1989). The rapidity with which these events occur suggest that they may be triggered by some external signal. The reasons for the failure of these newly generated B lymphocytes to develop into secretory cells in vivo, and for the staggered pattern of Ig heavy chain expression over the first postnatal years, are unknown (see below).

As the fetus matures beyond 20 weeks gestation, the B cell compartment in marrow comes to resemble that of the adult quantitatively and functionally, except that the pools of cells that secrete Ig at very high rates without further in vitro stimulation (Hibi and

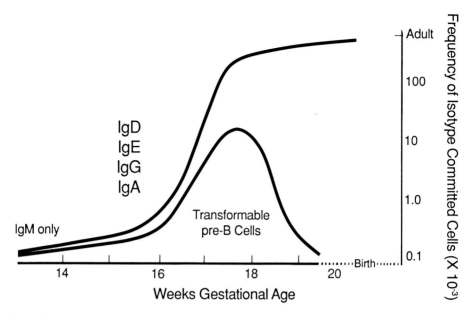

Figure 20–1. Ontogeny of Ig isotype diversity in humans: development of human B lymphocytes in fetal bone marrow. Marrow mononuclear cells of various gestational ages were transformed by Epstein-Barr virus, and limiting dilution cultures were analyzed for the frequencies of cells able to express a given Ig isotype. The dramatic expansion of B cell pools by 16 weeks gestational age is accompanied by the rapid, full isotype diversification that includes all four heavy chain classes. The large population of transformable pre-B cells disappears by ~20 weeks when the frequencies of secretory transformants reach adult levels. The relative proportions of cells committed and able to express a given heavy chain is similar after 16 weeks gestation and resembles adult proportions.

Dosch, 1986a, b) accumulate only slowly during the first years of life. Fetal marrow B and pre-B cells are highly transformable (>60%) by Epstein-Barr virus (EBV), and this constitutes one of the earliest B-lineage markers. These B cell transformants secrete Ig at rates typical of adult EBV-transformed lines (~10^6 Ig molecules/cell/hr) (Dosch and Gelfand, 1987; Dosch et al., 1985). As shown in Figure 20–1, the large pools of EBV-transformable pre-B cells gradually disappear entirely and become replaced by a smaller pool of pre-B cells, which, like pre-B cells in adult marrow cannot be transformed by EBV (Hibi et al., 1986). We have used this observation to argue that the stringent cellular prerequisites that dictate transformability by EBV include developmentally regulated genes (Dosch et al., 1990).

   B cell lines generated during the expansion phase of the B cell compartment have been useful in studies of the phenotypic and genomic changes that characterize B cell ontogeny (Fig. 20–2) (Barankiewicz et al., 1989b; Gregory et al., 1987). We recently identified four stages of B cell differentiation and maturation that have been characterized on the basis of cell surface phenotype, Ig gene rearrangements, and protein expression patterns (Hui et al., 1989).

   B cell progenitors in the first group constitute the earliest stage of cells in the B lineage defined so far. These B-progenitor cells are difficult to classify because of their lack of expression of most established markers. Heavy and light chain genes in B progenitors are retained in germline configuration, but they express CD21 (Dosch et al., 1989), low levels of CD9 (BA1), and low but significant ectonucleotidase levels (Barankiewicz et al., 1988). As well, they display B cell type volume regulatory function (Hui et al., 1989). Volume regulation in T and B cells differs dramatically: hypoosmotic challenge induces an immediate cell swelling phase in all cells, but even very immature T-lineage cells respond within minutes with restoration of the original cell volume, whereas B cells regulate volume very slowly and never return to their original cell volume (Cheung et al., 1982). Thus, cell physiological differences between T- and B-lineage cells are already present early in development.

B cell ectonucleotidases (ecto-ATPase, ecto-ADPase, and ecto-AMPase) generate extracellular adenosine from free systemic ATP, which is present at significant levels. This function may be important in B→T cell communication, as only T cells have adenosine receptors that are able to regulate numerous cell functions (Barankiewicz et al., 1989a, b). At least within lymphoid cell lineages, these ectoenzyme activities represent early B cell–specific markers whose activities increase during the course of B cell maturation (Barankiewicz et al., 1989b).

The second group of immature B-lineage cells encompasses early pre-B cells. These cells remain negative for surface, cytoplasmic, and secretory Ig (sIg$^-$, cIg$^-$, and Ig$^-$), but otherwise they express typical B lymphocyte surface markers. Along with the surface markers described above, early pre-B cells express class II MHC (Ia) antigens, CD19 (B4), and often (but not invariably) CD20 molecules (Fig. 20–2). Both heavy and light chain genes remain in germline configuration. CD19 is an early B cell differentiation marker that disappears as the cell matures (Hui et al., 1989).

Cells in the third group are classical pre-B lymphocytes (Cooper, 1981), character-

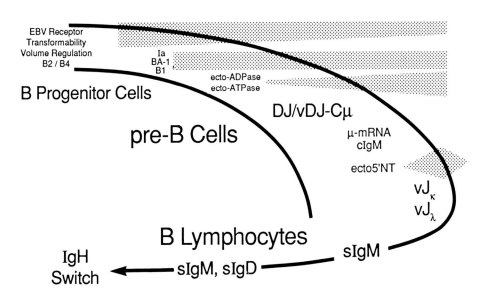

Figure 20–2. Phenotypic, genomic, and functional markers of human B cell development. The earliest markers of commitment to the B cell lineage include B cell–type volume regulatory response to hypoosmotic challenge and EBV transformability. The surface markers indicated and the acquisition of ectonucleotidases (shaded areas indicate variations during maturation) accompany stepwise genomic rearrangements that lead to Ig expression.

ized by the presence of cytoplasmic IgM heavy chain and the $CD20^+/CD21^+/MHC$-class II^+/sIg^- cell surface phenotype. Productive heavy chain gene rearrangement of VDJ-C_μ has occurred, but light chain genes remain unrearranged. C_μ heavy chain RNA is transcribed beyond the C_μ locus to include the next downstream Ig H locus, IgD in only a proportion of such cells, perhaps suggesting that transcription as well as expression (e.g., Fig. 20–1) of IgD is developmentally regulated (Hui et al., 1989). Cytosolic δ heavy chain protein is not found in these cells. Immunoglobulin heavy and light chain gene rearrangements occur in an ordered and hierarchical fashion, in which heavy chain gene rearrangement precedes that of light chain genes, while κ gene rearrangements generally precede those of λ genes. This latter event is usually but not always (Hui et al., 1989) accompanied by κ deletion. In contrast to the mouse immune system, immunoglobulin variable-region gene rearrangements show a strong 3′–5′ bias life long, a fact that reflects the interspersed location of V-region genes with the bulk of all V genes mapped to the 3′ third of the V gene locus (Walter et al., 1991). Parallel to gene rearrangement events, the enzyme terminal deoxynucleotidyl transferase has its peak expression in these cells and may be detected by cytoplasmic staining (Loken et al., 1988).

The last group of developing B cells that have been delineated includes immature and mature virgin B lymphocytes, which are characterized by the presence of surface IgM and surface IgM/IgD, respectively. In both stages, productive light chain gene rearrangements have occurred, thus enabling the assembly and export of complete Ig molecules. The dual expression of surface IgM and IgD occurs by alternative RNA splicing of a long C_μ-C_δ primary transcript because classical switch sequences are absent in the C_μ-C_δ intron (Milstein et al., 1984). The majority of these cells will either die, secrete IgM, or switch to downstream heavy chain loci, while a minority ($<10\%$) produces secretory IgD in addition to IgM. Approximately 1% of the latter cells can go on to produce IgE (Chan and Dosch, 1989; MacKenzie and Dosch, 1989).

Development of Mature B Cell Compartments

In the human T cell lineage the crucial role of thymic function is essentially completed during fetal life, and even early thymectomy produces few, if any, immune abnormalities (see above). Considering the defined period of time in which B-lineage expansion and diversification occurs, and the amount of expansion achieved, fetal marrow may play a somewhat thymus-equivalent role for the B cell lineage. The adult-like B cell repertoires that are generated at around 16 weeks gestation (Fig. 20–1) are exported to peripheral tissues (Antin et al., 1986; Edwards et al., 1986; Hofman et al., 1984; Hui et al., 1989; Spencer et al., 1986; Westerga and Timens, 1989). During postnatal life, peripheral tissues such as the lymph nodes contain the bulk of virgin, competent B-lineage cells, which recirculate to the marrow upon antigen-driven recruitment (Hibi and Dosch, 1986a, b).

Table 20–3 summarizes the functions of B cell pools in the neonate and the adult. About 70% of the systemic Ig production in the adult derives from nonproliferating, surface Ig^+ cells in bone marrow, which secrete their antibody at rates of approximately 10^8 molecules/cell/hour (Hibi et al., 1986; Hibi and Dosch, 1986a, b). The bulk of antibody deficiency syndromes involve the generation and function of these cells; indeed, pre-B cell development appears intact even in patients with XLA (congenital, X-linked agammaglobulinemia) (Buckley, 1986; Cooper, 1981, 1987). Because obligate female XLA carriers show a complete bias in B cell X chromosome inactivation pattern such that they

Table 20-3.   Isotype Commitment of Ig Secretors

Tissue Source	Spontaneous Ig Secretors	EBV Activated	PWM Activated
Fetal Marrow	none	IgG/A/M/D/E	some IgM
Cord Blood	none	(IgG/A/M/D/E)[a]	IgM only
Adult Marrow	IgG/A/M/D	IgM only	very few IgM
Adult Blood	Very few (IgD/E in atopics, IgM/D/(E)/G/A in SLE)	IgG/A/M/D/E	IgG/A/M, ± IgD, no IgE

[a]*Limiting dilution cultures only.*

use exclusively the unaffected chromosome, it appears that the defect in these patients involves a B cell gene or the regulation of a gene required during or after the pre-B to B cell transition (Conley, 1985; Conley and Puck, 1988; Hendriks et al., 1989; Mensink et al., 1986, 1987).

The presence of high-rate Ig-secreting cells in the blood of patients with systemic lupus erythematosus (SLE) is consistent with abnormal trafficking of B-lineage cells in this condition (Becker et al., 1981; Fauci and Moutsopoulos, 1981; Hoch et al., 1983; Okubo et al., 1982). As only bone marrow contains populations of highly efficient, specialized regulatory cells that are able to modulate the secretory functions of these high-rate Ig producers, such abnormal trafficking may contribute to the abnormal antibody production in SLE (Dosch et al., 1985b; Mills et al., 1985). The coexistence of SLE and antibody deficiency (i.e., hypogammaglobulinemia) has rarely been reported (Peters et al., 1984).

Normal human immune ontogeny and its failures, which lead to the development of primary immunodeficiency syndromes, differ significantly from immune development in rodents where lymphocyte diversification is a perinatal process and is thus much more likely to be influenced by extrinsic antigens (see above). Given the early presence of large pools of cells able to express all Ig isotypes (e.g., Table 20-3), it is unclear why the full isotype diversity of serum immunoglobulins is acquired only over a prolonged period of up to several years. This staggered in vivo expression of Ig genes in newborn children does not appear to represent a maturation process defined solely by limiting precursor cell pool sizes, as is seen following marrow transplantation (Mitus et al., 1989). Instead, immunoregulatory events may interfere with expression of the generated B cell repertoires during the first years of life.

In this regard, we have previously studied patients with IgA deficiency, the most common inherited immune abnormality in humans (Carneiro et al., 1989; Cunningham, 1989). Our finding of near normal numbers of cells that are able to secrete IgA in vitro but not in vivo implies that in at least some patients the disorder may not reflect a primary inability of B cells to express such isotypes but, rather, a failure to discontinue the regulatory restriction in isotype expression that persists from the perinatal stages of development.

This is illustrated in Figure 20-3, which shows Ig isotype diversity among blood B cells activated by Epstein-Barr virus (EBV). The figure summarizes the results of limiting dilution experiments (Dosch et al., 1989; Hibi et al., 1986; Hibi and Dosch, 1986a, b; Stein et al., 1986). Our culture system uses multiple very low density cultures (1–1000 cells) and irradiated feeder cells to provide for optimal B cell growth and to minimize regulatory interactions that can severely reduce the numbers of responding B cells (Hibi

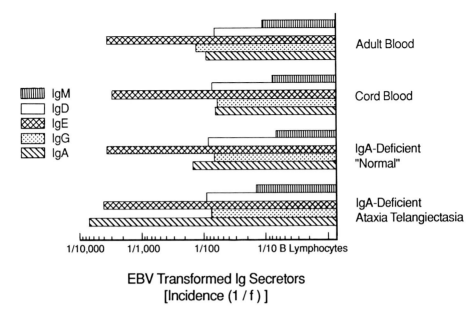

Figure 20–3.   Isotype diversity in B cells from normal patients and from patients with IgA deficiency, or ataxia telangiectasia. Results of limiting dilution experiments with EBV-transformed peripheral blood B cells are expressed as the incidence—that is, the number—of B cells that on average contain one responsive cell committed to the secretion of the Ig isotype indicated. Essentially normal numbers of IgA-secreting cells are present in most donors with in vivo selective IgA deficiency, whereas such cells are rare in IgA-deficient patients with ataxia telangiectasia.

and Dosch, 1986; MacKenzie and Dosch, 1989; Stein et al., 1986). Approximately 10% of all B cells are activated by EBV and secrete IgM, and 5%–10% of these cells also secrete IgD. These figures are similar in adult and cord blood (Dosch and Hibi, 1987). Under limiting dilution conditions, the frequencies of IgG- and IgA-committed cells in cord blood and adult peripheral blood are comparable, although these isotypes are not normally coexpressed in vivo beyond the first year of age. Consistent with a possible regulatory role for T lineage cells, T cell–dependent B cell activators such as pokeweed mitogen (PWM) are unable to activate such cells in cord blood, and, more importantly, EBV fails to activate these cells in bulk cultures of 10^5–10^6 or more cells (Ahmad et al., 1982; Andersson, 1985; Conley and Cooper, 1981; Cooper, 1981; Eibl et al., 1980; Kuritani and Cooper, 1982) (Table 20–3).

Antibody Deficiency Disorders

IgA deficiency is the most common humoral immune defect, and although it is usually of little or no direct clinical consequence, autoimmune and atopic disorders are found at higher frequencies among IgA-deficient patients, particularly females (Andersson, 1985; Carneiro et al., 1989; Conley and Cooper, 1981; Cunningham, 1989; Ferreira et al., 1988). As shown in Figure 20–3, the number of IgM-, IgM/IgD-, or IgG-secreting cells is normal in IgA deficiency following in vitro infection with EBV. In addition, we consistently observe near-normal numbers of IgA-secreting, EBV-activated B cells in limiting dilution cultures of clinically "normal" patients with IgA deficiency. In this respect, and as discussed elsewhere (Dosch and Hibi, 1987), lymphocytes from such patients behave like cord blood cells in that special culture conditions can unmask the presence of cells able to secrete an isotype that is not expressed in vivo. Patients with ataxia telangiectasia (AT) frequently have IgA deficiency in addition to other immune abnormalities, and this humoral immunodeficiency may require therapeutic intervention during their frequent episodes of bacterial infection (Cunningham, 1989; Paganelli et al., 1988; Ryser et al., 1988). Consistent with an intrinsic B cell abnormality, the frequency of cells able to secrete IgA is severely reduced (if they can be detected at all) in AT patients with IgA deficiency (Fig. 20–3).

The spectrum of human B cell immunodeficiency diseases covers a wide range of selective and global defects that involve one or more immunoglobulin isotypes (Cunningham, 1989; Pollack et al., 1985; Tedder et al., 1985). The common clinical feature is reduced host resistance particularly to bacterial infections. The spectrum of clinically overt antibody deficiency can overlap with autoimmunity in a subset of patients with SLE (Cronin et al., 1989; Goldstein et al., 1985; Peters et al., 1984). There are few phenotypic or functional similarities between the many humoral immunodeficiency disorders, which are often accompanied by subtle or overt abnormalities in the functions of other immune cells. Although it might be expected a priori, defects in specific cytokines have not been shown to underlie antibody deficiency states to date (Matheson and Green, 1987). At a clinical level, it is possible to distinguish between patients who require (and do well with) IV IgG replacement therapy, and those who can mount sufficient host resistance to need clinical monitoring only. Given the tremendous heterogeneity in clinical presentation and the absence of acceptable insights into pathogenetic mechanisms of B cell immunodeficiency, we will focus below on what is generally viewed as an absolute block in B cell development: infantile/congenital, X-linked (or Bruton-type) agammaglobulinemia (XLA) (Cooper, 1987).

The disease is surprisingly heterogeneous in its biological and clinical manifestations, and its inheritance (Leickley and Buckley, 1986) has been mapped (Conley and Puck, 1988; Hendriks et al., 1989; Mensink et al., 1987). In patients with XLA, at least the initial diversification process is probably intact, and such patients generate normal numbers of marrow pre-B cells (Conley, 1985; Cooper, 1981, 1987). However, the pre-B to B cell transition and the emigration of marrow B-lineage cells to the periphery appear blocked in these patients, and only a very few B-lineage cells (<5% of normal) escape into the periphery (Conley, 1985). Although these cells can be transformed into Ig-secreting B cell lines in most patients, there may be subsets of patients and/or subsets of B-lineage cells that show abnormal rearrangement processes (Conley, 1985; Schwaber and Chen, 1988; Schwaber et al., 1988; Schwaber and Rosen, 1984). However, despite the fact that small but still considerable numbers of B cells may escape the XLA defect and show normal diversity and rearrangements (Anker et al., 1989), these cells seem unable to express secretory functions in vivo. In this sense, these populations resemble fetal and perinatal diversified B cell populations that seem unable to achieve differentiation into secretory cells in vivo but can do so under special circumstances in vitro.

References

Ahmad, R., Zielinski, C. C., Mannhalter, J. W., Zlabinger, G., Rockenschaub, A., and Eibl, M. M. (1982). Plaque-forming cells in human cord blood: A soluble factor suppressing differentiation but not proliferation of B cells. *Int. Arch. Allergy Appl. Immunol.* 69:132–136.

Akahonai, Y., Aoki, S., Takei, T., Mori, M., Kobayashi, T., and Yachi, A. (1986). A patient with IgD (lambda)-multiple myeloma accompanied by OKT4 epitope deficiency. *Rinsho Ketsueki.* 27:2304–2308.

Andersson, U. (1985). Development of B lymphocyte function in childhood. *Acta Paediatr. Scand.* 74:568–573.

Anker, R., Conley, M. E., and Pollok, B. A. (1989). Clonal diversity in the B cell repertoire of patients with X-linked agammaglobulinemia. *J. Exp. Med.* 169:2109–2119.

Antin, J. H., Emerson, S. G., Martin, P., Gadol, N., and Ault, K. A. (1986). Leu-1$^+$ (CD5$^+$) B cells. A major lymphoid subpopulation in human fetal spleen: Phenotypic and functional studies. *J. Immunol.* 136:505–510.

Barankiewicz, J., Dosch, H. M., Cheung, R., and Cohen, A. (1989a). Relationship between extracellular and intracellular nucleotide metabolism in human lymphocytes. *Adv. Exp. Med. Biol.* 253B:475–479.

Barankiewicz, J., Dosch, H. M., and Cohen, A. (1988). Extracellular nucleotide catabolism in human B and T lymphocytes: The source of adenosine production. *J. Biol. Chem.* 263:7094–7098.

Barankiewicz, J., Hui, M., Cohen, A., and Dosch, H. M. (1989b). Differential expression of ecto-nucleotide metabolic enzymes during immunoglobulin gene rearrangements in human pre-B-cells. *Adv. Exp. Med. Biol.* 253B:455–461.

Becker, T. M., Lizzio, E. F., Merchant, B., Reeves, J. P., and Steinberg, A. D. (1981). Increased multiclonal antibody-forming cell activity in the peripheral blood of patients with SLE. *Int. Arch. Allergy Appl. Immunol.* 66:293–303.

Berg, L. J., Frank, G. D., and Davis, M. M. (1990). The effect of MHC gene dosage and allelic variation on T cell receptor selection. *Cell.* 60:1043–1053.

Berg, L. J., Pullen, A. M., Fazekas de St Groth, B., Mathis, D., Benoist, C., and Davis, M. M. (1989). Antigen/MHC-specific T cells are preferentially exported from the thymus in the presence of their MHC ligand. *Cell* 58:1035–1046.

Bosma, M., Schuler, W., and Bosma, G. (1988). The *scid* mouse mutant. *Curr. Top. Microbiol. Immunol.* 137:197–202.

Buckley, R. H. (1986). Humoral immunodeficiency. *Clin. Immunol. Immunopathol.* 40:13–24.

Carneiro, S.M.M., Carbonare, S. B., Rozentraub, R. B., de Araujo, M. N., Riberiro, M. A., and Porto, M. H. (1989). Frequency of selective IgA deficiency among Brazilian blood donors and healthy pregnant women. *Allergol. Immunopathol. (Madrid)* 17:213–215.

Chan, M. A., and Dosch, H.-M. (1989). Human IgE response: Virus-activated IgE secretors are interleukin-2-dependent cells. *Int. Arch Allergy Appl. Immunol.* 89:90–97.

Cheung, R. K., Grinstein, S., Dosch, H. M., and Gelfand, E. W. (1982). Volume regulation by human lymphocytes: characterization of the ionic basis for regulatory volume decrease. *J. Cell. Physiol.* 112:189–196.

Cohen, A., Barankiewicz, J., Lederman, H. M., and Gelfand, E. W. (1983a). Purine and pyrimidine metabolism in human T lymphocytes: Regulation of deoxyribonucleotide metabolism. *J. Biol. Chem.* 258:12334–12340.

Cohen, A., Barankiewicz, J., Lederman, H. M., and Gelfand, E. W. (1984). Purine metabolism in human T lymphocytes: Role of the purine nucleoside cycle. *Can. J. Biochem. Cell. Biol.* 62:577–583.

Cohen, A. Dosch, H. M., and Gelfand, E. W. (1981). Induction of ecto 5′-nucleotidase activity in human thymocytes. *Clin. Immunol. Immunopath.* 18:287–290.

Cohen, A., Lee, J. W., Dosch, H. M., and Gelfand, E. W. (1980a). The expression of nucleoside toxicity during T-cell differentiation. *J. Immunol.* 124:1578–1582.

Cohen, A., Lee, J. W., and Gelfand, E. W. (1983b). Selective toxicity of deoxyguanosine and ara-binosyl guanine for T-leukemic cells. *Blood.* 61:660–666.

Cohen, A., Mansour, A., Dosch, H. M., and Gelfand, E. W. (1980b). Absence of lymphocyte ecto-5′-nucleotidase in familial reticuloendotheliosis and combined immunodeficiency. *Clin. Immunol. Immunopath.* 15:245–250.

Conley, M. E. (1985). B cells in patients with X-linked agammaglobulinemia. *J. Immunol.* *134*:3070–3074.

Conley, M. E., and Cooper, M. D. (1981). Immature IgA B cells in IgA-deficiency patients. *N. Engl. J. Med.* *305*:495–497.

Conley, M. E., and Puck, J. M. (1988). Carrier detection in typical and atypical X-linked agammaglobulinemia. *J. Pediatr.* *112*:688–694.

Cooper, M. D. (1981). Pre-B cells: Normal and abnormal development. *J. Clin. Immunol.* *1*:81–89.

Cooper, M. D. (1987). B Lymphocytes: Normal development and function. *N. Engl. J. Med.* *317*:1452–1456.

Crispe, E. N., Husmann, L. A., and Bevan, M. J. (1986). T cell receptor expression and receptor-mediated induction of clonal growth in the developing mouse thymus. High surface β-chain density is a requirement for functional maturity. *Eur. J. Immunol.* *16*:1283–1288.

Cronin, M. E., Balow, J. E., and Tsokos, G. C. (1989). Immunoglobulin deficiency in patients with systemic lupus erythematosus. *Clin. Exp. Rheumatol.* *7*:359–364.

Cunningham, R. C. (1989). Clinical and immunologic analyses of 103 patients with common variable immunodeficiency. *J. Clin. Immunol.* *9*:22–33.

Deibel, M.R.J., Riley, L. K., Coleman, M. S., Cibull, M. L., Fuller, S. A., and Todd, E. (1983). Expression of terminal deoxynucleotidyl transferase in human thymus during ontogeny and development. *J. Immunol.* *131*:195–200.

dePreval, C., Lisowska, G. B., Loche, M., Griscelli, C., and Mach, B. (1985). A trans-acting class II regulatory gene unlinked to the MHC controls expression of HLA class II genes. *Nature* *318*:291–293.

Doherty, P. J., Huesca-Contreras, M., Dosch, H. M., and Pan, S. (1989). Rapid amplification of complementary DNA from small amounts of unfractionated RNA. *Anal. Biochem.* *177*:7–10.

Doherty, P. J., Roifman, C. M., Pan, S., Cymerman, U., Ho, W. K., Thompson, E., and Cohen, A. (1991). Expression of the human T-cell receptor V_β repertoire. *Mol. Immunol.* *28*:607–612.

Dosch, H.-M., Cochrane, D.M.G., Cook, V. A., Leeder, S. L., and Cheung, R. K. (1991). Exogenous but not endogenous EBV induces lymphomas in beige/nude/xid mice carrying human lymphoid xenografts. *Int. Immunol.* *3*:731–735.

Dosch, H. M., and Gelfand, E. W. (1977). Generation of human plaque-forming cells in culture: Tissue distribution, antigenic and cellular requirements. *J. Immunol.* *118*:302–308.

Dosch, H. M., and Gelfand, E. W. (1979). Specific in vitro IgM response of human B-cells: A complex regulatory network modulated by antigen. *Immunol. Rev.* *45*:242–274.

Dosch, H. M., and Gelfand, E. W. (1987). Poly-L-lysine (PLL) plaque assay for the measurement of antigen-activated human B-lymphocytes. *Meth. Enzymol.* *150*:223–244.

Dosch, H.-M., and Hibi, T. (1987). Ontogeny of bone marrow B cell function in man. In *Clinical Immunology* (Pruzanski, W., and Seligmann, M., eds.). Amsterdam: Elsevier, pp. 311–318.

Dosch, H.-M., Lam, P., and Guerin, D. (1985a). Differential regulation of activation, clonal expansion, and antibody secretion in human B cells. *J. Immunol.* *135*:3808–3816.

Dosch, H. M., Lam, P., Hui, M. F., and Hibi, T. (1989). Concerted generation of Ig isotype diversity in human fetal bone marrow. *J. Immunol.* *143*:2464–2469.

Dosch, H.-M., Lam, P., Hui, M. F., Hibi, T., and Cheung, R. K. (1990). EBV utilizes a unique activation pathway for the transformation of human B cells. *Int. Immunol.* *2*:833–848.

Dosch, H.-M., Ledgley, C. J., White, D., Lam, P., and Mills, G. B. (1985b). Lymphocyte function in human bone marrow: II. Characterization of an interleukin 2–sensitive T precursor–cell population. *J. Clin. Immunol.* *5*:345–356.

Dosch, H.-M., Lee, J. W., Gelfand, E. W., and Falk, J. A. (1978). Severe combined immunodeficiency disease: A model of T-cell dysfunction. *Clin. Exp. Immunol.* *34*:260–261.

Dosch, H. M., Mansour, A., Cohen, A., Shore, A., and Gelfand, E. W. (1980). Inhibition of suppressor T-cell development following the administration of deoxyguanosine. *Nature.* *285*:494–496.

Dosch, H. M., White, D., and Grant, C. (1985). Reconstitution of nude mouse T cell function in vivo: IL 2–independent effect of human T cells. *J. Immunol. 134*:336–342.

Edwards, J. A., Durant, B. M., Jones, D. B., Evans, P. R., and Smith, J. L. (1986). Differential expression of HLA class II antigens in fetal human spleen: Relationship of HLA-DP, DQ, and DR to immunoglobulin expression. *J. Immunol. 137*:490–497.

Eibl, M., Zielinski, C. C., Ahmad, R., Steurer, F., and Rockenschaub, A. (1980). Plaque-forming cells in human cord blood: Studies on T and B cell function. *Clin. Exp. Immunol. 41*:176–181.

EMBO Workshop (1989). The *scid* mouse: Characterization and potential uses (European Molecular Biology Organization, Basel, February 20–22, 1989, Proceedings). *Curr. Top. Microbiol. Immunol. 152*:1–263.

Fauci, A. S., and Moutsopoulos, H. M. (1981). Polyclonally triggered B cells in the peripheral blood and bone marrow of normal individuals and in patients with systemic lupus erythematosus and primary Sjögren's syndrome. *Arthritis Rheum. 24*:577–583.

Ferreira, A., Garcia, R.M.C., Lopez, T. M., Pascual, S. D., and Fontan, G. (1988). Anti-IgA antibodies in selective IgA deficiency and in primary immunodeficient patients treated with gamma-globulin. *Clin. Immunol. Immunopathol. 47*:199–207.

Geiger, J. D., and Nagy, J. I. (1986). Lack of adenosine deaminase deficiency in the mutant mouse wasted. *FEBS Lett. 208*:431–434.

Gelfand, E. W., Cox, D. W., Lin, M. T., and Dosch, H.-M. (1979). Association of severe combined immune deficiency disease and α1-antitrypsin deficiency. *Lancet. 2*:202–204.

Gelfand, E. W., and Dosch, H. M. (1982). Differentiation of precursor T lymphocytes in man and delineation of the selective abnormalities in severe combined immune deficiency disease. *Clin. Immunol. Immunopath. 25*:303–315.

Gelfand, E. W., and Dosch, H. M. (1983). Diagnosis and classification of severe combined immunodeficiency disease. *Birth Defects 19*:65–72.

Gelfand, E. W., McCurdy, D., Rao, C. P., and Cohen, A. (1984). Absence of lymphocyte ecto-5′-nucleotidase in infants with reticuloendotheliosis and eosinophilia (Omenn's syndrome). *Blood 63*:1475–1480.

Goldmann, S. F., Niethammer, D., Flad, H. M., Belohradsky, B. H., Colombani, J., Dieterle, U., Dosch, H. M., Gelfand, E. W., Tollner, U., Fliedner, T.F.M., and Kleihauer, E. (1979). Hemopoietic and lymphopoietic split chimerism in severe combined immunodeficiency disease (SCID). *Transplant. Proc. 11*:225–229.

Goldstein, R., Izaguirre, C., Smith, C. D., Mierins, E., and Karsh, J. (1985). Systemic lupus erythematosus and common variable panhypogammaglobulinemia: A patient with absence of circulating B cells. *Arthritis Rheum. 28*:100–103.

Gregory, C. D., Kirchgens, C., Edwards, C. F., Young, L. S., Rowe, M., Forster, A., Rabbitts, T. H., and Rickinson, A. B. (1987). Epstein-Barr virus–transformed human precursor B cell lines: Altered growth phenotype of lines with germ-line or rearranged but nonexpressed heavy chain genes. *Eur. J. Immunol. 17*:1199–1207.

Haynes, B. F., Denning, S. M., Singer, K. H., and Kurtzberg, J. (1989). Ontogeny of T-cell precursors: A model for the initial stages of human T-cell development. *Immunol. Today 10*:87–91.

Haynes, B. F., Martin, M. E., Kay, H. H., and Kurtzberg, J. (1988). Early events in human T cell ontogeny: Phenotypic characterization and immunohistologic localization of T cell precursors in early human fetal tissues. *J. Exp. Med. 168*:1061–1080.

Hendriks, R. W., Mensink, E. J., Kraakman, M. E., Thompson, A., and Schuurman, R. K. (1989). Evidence for male X chromosomal mosaicism in X-linked agammaglobulinemia. *Hum. Genet. 83*:267–270.

Hibi, T., Chan, M. A., Petsche, D., and Dosch, H. M. (1986). Phenotype, frequency, and EBV responsiveness of human marrow B and pre-B cells. *J. Immunol. 136*:3211–3218.

Hibi, T., and Dosch, H.-M. (1986a). Limiting dilution analysis of the B cell compartment in human bone marrow. *Eur. J. Immunol. 16*:139–145.

Hibi, T., and Dosch, H.-M. (1986b). Lymphocyte function in human bone marrow: III. Isotype commitment, metabolic and secretory characteristics of immunoglobulin producing cells. *Cell. Immunol. 98*:34–45.

Hoch, S., Schur, P. H., and Schwaber, J. (1983). Frequency of anti-DNA antibody producing cells from normals and patients with systemic lupus erythematosus. *Clin. Immunol. Immunopathol. 27*:28–37.

Hofman, F. M., Danilovs, J., Husmann, L., and Taylor, C. R. (1984). Ontogeny of B cell markers in the human fetal liver. *J. Immunol. 133*:1197–1201.

Hui, M. F., Lam, P., and Dosch, H. M. (1989). Properties and heterogeneity of human fetal pre-B cells transformed by EBV. *J. Immunol. 143*:2470–2479.

Jordan, S. C. (1989). Intravenous gamma-gobulin therapy in systemic lupus erythematosus and immune complex disease. *Clin. Immunol. Immunopathol.* S164–169.

Kersey, J. H., Shapiro, R. S., and Filipovich, A. H. (1988). Relationship of immunodeficiency to lymphoid malignancy. *Pediatr. Infect. Dis. J. 5*:S10–S12.

Kincade, P. W. (1987). Experimental models for understanding B lymphocyte formation. *Adv. Immunol. 41*:181–267.

Koller, B. H., Marrack, P., Kappler, J. W., and Smithies, O. (1990). Normal development of mice deficient in β_2-m, MHC class I proteins and $CD8^+$ T cells. *Science 248*:1227–1230.

Komiyama, A., Ichikawa, M., Kanda, H., Aoyama, K., Yasui, K., Yamazaki, M., Kawai, H. Miyagawa, Y., and Akabane, T. (1988). Defective interleukin-1 production in a familial monocyte disorder with a combined abnormality of mobility and phagocytosis-killing. *Clin. Exp. Immunol. 73*:500–504.

Kuritani, T., and Cooper, M. D. (1982). Human B cell differentiation: II. Pokeweed mitogen-responsive B cells belong to a surface immunoglobulin D–negative subpopulation. *J. Exp. Med. 155*:1561–1566.

Leickley, F. E., and Buckley, R. (1986). Variability in B cell maturation and differentiation in X-linked agammaglobulinemia. *Clin. Exp. Immunol. 65*:90–99.

Levinson, A. I., Hoxie, J. A., Kornstein, M. J., Zembryki, D., Matthews, D. M., and Schreiber, A. D. (1985). Absence of the OKT4 epitope on blood T cells and thymus cells in a patient with thymoma, hypogammaglobulinemia, and red blood cells aplasia. *J. Allergy Clin. Immunol. 76*:433–439.

Lobach, D. F. and Haynes, B. F. (1987). Ontogeny of the human thymus during fetal development. *J. Clin. Immunol. 7*:81–97.

Loken, M. R., Shah, V. O., Hollander, Z. and Civin, C. I. (1988). Flow cytometric analysis of normal B lymphoid development. *Pathol. Immunopathol. Res. 7*:357–370.

Ma, D.D.F., Sylwestrowicz, T. A., Granger, S., Massaia, M., Franks, R., Janossy, G., and Hoffbrand, A. V. (1982). Distribution of terminal deoxynucleotidyl transferase and purine degradative and synthetic enzymes in subpopulations of human thymocytes. *J. Immunol. 129*:1430–1435.

MacKenzie, T., and Dosch, H. M. (1989). Clonal and molecular characteristics of the human IgE-committed B cell subset. *J. Exp. Med. 169*:407–430.

Matheson, D. S., and Green, B. J. (1987). Defect in production of B cell differentiation factor-like activity by mononuclear cells from a boy with hypogammaglobulinemia. *J. Immunol. 138*:2469–2472.

Matricardi, P. M., Capobianchi, M. R., Paganelli, R., Facchini, J., Sirianni, M. C., Seminara, R., Dianzani, F., and Aiuti, F. (1984). Interferon production in primary immunodeficiencies. *J. Clin. Immunol. 4*:388–394.

Mensink, E. J., Schuurman, R. K., Schot, J. D., Thompson, A., and Alt, F. W. (1986). Immunoglobulin heavy chain gene rearrangements in X-linked agammaglobulinemia. *Eur. J. Immunol. 16*:963–967.

Mensink, E. J., Thompson, A., Schot, J. D., Kraakman, M. E., Sandkuyl, L. A., and Schuurman, R. K. (1987). Genetic heterogeneity in X-linked agammaglobulinemia complicates carrier detection and prenatal diagnosis. *Clin. Genet. 31*:91–96.

Mills, G. B., Ledgley, C. J., Hibi, T., White, D., Lam, P., and Dosch, H. M. (1985). Lymphocyte function in human bone marrow: I. Characterization of two T cell populations regulating immunoglobulin secretion. *J. Immunol.* *134*:3036–3041.

Milstein, C. P., Deverson, E. V., and Rabbitts, T. H. (1984). The sequence of the human immunoglobulin μ-δ intron reveals possible vestigial switch segments. *Nucleic Acids Res.* *12*:6523–6535.

Mitus, A. J., Stein, R., Rappeport, J. M., Antin, J. H., Weinstein, H. J., Alper, C. A., and Smith, B. R. (1989). Monoclonal and oligoclonal gammopathy after bone marrow transplantation. *Blood 74*:2764–2768.

Okubo, M., Kamata, K., and Uchiyama, T. (1982). Pseudoplaque formation by cytophilic immunoglobulins in the sera of lupus patients as measured using reverse haemolytic plaque assay. *Clin. Exp. Immunol. 48*:499–504.

Osmond, D. G. (1985). The ontogeny and organization of the lymphoid system. *J. Invest. Dermatol. 1*:2S–9S.

Paganelli, R., Capobianchi, M. R., Ensoli, B., Doffizi, G. P., Facchini, J., Dianzani, F., and Aiuti, F. (1988). Evidence that defective gamma interferon production in patients with primary immunodeficiencies is due to intrinsic incompetence of lymphocytes. *Clin. Exp. Immunol. 72*:124–129.

Pahwa, R., Chatila, T., Pahwa, S., Paradise, C., Day, N. K., Geha, R., Schwartz, S. A., Slade, H., Oyaizu, N., and Good, R. A. (1989). Recombinant interleukin 2 therapy in severe combined immunodeficiency disease. *Proc. Natl. Acad. Sci. USA. 86*:5069–5073.

Peters, J., Astrup, L., and Andersen, V. (1984). Hypogammaglobulinaemia in systemic lupus erythematosus: Report of a case with evidence for spontaneously activated T suppressor cells. *Clin. Exp. Rheumatol. 2*:145–149.

Pollack, S., Reisner, Y., Koziner, B., Good, R. A., and Hoffman, M. K. (1985). B-cell function in common variable immunodeficiency: Suppression of in vitro anti-sheep erythrocyte antibody production by T cells and monocytes. *Immunology 54*:89–96.

Pyke, K. W., Dosch, H. M., Ipp, M. M., and Gelfand, E. W. (1975). Demonstration of an intrathymic defect in a case of severe combined immunodeficiency disease. *N. Engl. J. Med. 293*:424–428.

Rainaut, M., Pagniez, M., Hercend, T., Daffos, F., and Forestier, F. (1987). Characterization of mononuclear cell subpopulations in normal fetal peripheral blood. *Hum. Immunol. 18*:331–337.

Reith, W., Satola, S., Sanchez, C. H., Amaldi, I., Lisowska, G. B., Griscelli, C., Hadam, M. R., and Mach, B. (1988). Congenital immunodeficiency with a regulatory defect in MHC class II gene expression lacks a specific HLA-DR promoter binding protein, RF-X. *Cell 53*:897–906.

Richie, E. R., McEntire, B., Crispe, N., Kimura, N., Lanier, L. L., and Allison, J. P. (1988). α/β T-cell antigen receptor gene and protein expression occurs at early stages of thymocyte differentiation. *Proc. Natl. Acad. Sci. USA 85*:1174–1178.

Roifman, C. M., Hummel, D., Martinez, V. H., Thorner, P., Doherty, P. J., Pan, S., Cohen, F., and Cohen, A. (1989). Depletion of CD8[+] cells in human thymic medulla results in selective immune deficiency. *J. Exp. Med. 170*:2177–2182.

Rothenberg, E. V. (1990). Death and transfiguration of cortical thymocytes: A reconsideration. *Immunol. Today. 11*:116–119.

Rousset, F., Souillet, G., Roncarolo, M. G., and Lamelin, J. P. (1986). Studies of EBV–lymphoid cell interactions in two patients with the X-linked lymphoproliferative syndrome: Normal EBV-specific HLA-restricted cytotoxicity. *Clin. Exp. Immunol. 63*:280–289.

Royer, H. D., and Reinherz, E. L. (1987). T lymphocytes: Ontogeny, function, and relevance to clinical disorders. *N. Engl. J. Med. 317*:1136–1142.

Royo, C., Touraine, J. L., and de Bouteiller, O. (1987). Ontogeny of T lymphocyte differentiation in the human fetus: Acquisition of phenotype and functions. *Thymus 10*:57–73.

Ruers, T. J., Buurman, W. A., and van der Linden, C. J. (1987). 2′-Deoxycoformycin and deoxy-adenosine affect IL 2 production and IL 2 receptor expression of human T cells. *J. Immunol. 138*:116–122.

Ryser, O., Morell, A., and Hitzig, W. H. (1988). Primary immunodeficiencies in Switzerland: First report of the national registry in adults and children. *J. Clin. Immunol. 8*:479–485.

Schatz, D. G., Oettinger, M. A., and Baltimore, D. (1989). The V(D)J recombination activating gene, RAG-1. *Cell 59*:1035–1048.

Schwaber, J., and Chen, R. H. (1988). Premature termination of variable gene rearrangement in B lymphocytes from X-linked agammaglobulinemia. *J. Clin. Invest. 81*:2004–2009.

Schwaber, J., Payne, J., and Chen, R. (1988). B lymphocytes from X-linked agammaglobulinemia: Delayed expression of light chain and demonstration of lyonization in carriers. *J. Clin. Invest. 81*:514–522.

Schwaber, J., and Rosen, F. S. (1984). Lymphoid cell lines from patients with "non-secretory" agammaglobulinemia produce glycosylated heavy chains which are reduced in molecular weight. *J. Mol. Cell. Immunol. 1*:279–291.

Seemayer, T. A. (1987). Molecular basis of selected primary immunodeficiency disorders. *Arch. Pathol. Lab. Med. 111*:1114–1117.

Sen, R., and Baltimore, D. (1986). Multiple nuclear factors interact with the immunoglobulin enhancer sequences. *Cell 46*:705–716.

Sha, W. C., Nelson, C. A., Newberry, R. D., Kranz, D. M., Russel, J. H., and Loh, D. Y. (1988). Selective expression of an antigen receptor on CD8-bearing T lymphocytes in transgenic mice. *Nature. 335*:271–274.

Shore, A., Dosch, H. M., and Gelfand, E. W. (1978). Induction and separation of antigen-dependent T-helper and T-suppressor cells in man. *Nature 274*:586–587.

Shore, A., Dosch, H. M., and Gelfand, E. W. (1981). Role of adenosine deaminase in the early stages of precursor T-cell maturation. *Clin. Exp. Immunol. 44*:152–155.

Spencer, J., MacDonald, T. T., Finn, T., and Isaacson, P. G. (1986). The development of gut associated lymphoid tissue in the terminal ileum of fetal human intestine. *Clin. Exp. Immunol. 64*:536–543.

Stein, L. D., Chan, M. A., Hibi, T., and Dosch, H. M. (1986). Epstein-Barr virus–induced IgE production in limiting dilution cultures of normal human B cells. *Eur. J. Immunol. 16*:1167–1170.

Strominger, J. L. (1989). Developmental biology of T cell receptors. *Science 244*:943–950.

Swerdlow, S. H., Angermeier, P. A., and Hartman, A. L. (1988). Intrathymic ontogeny of the T cell receptor associated CD3 (T3) antigen. *Lab Invest. 58*:421–427.

Tatsumi, E., and Purtilo, D. T. (1986). Epstein-Barr virus (EBV) and X-linked lymphoproliferative syndrome (XLP). *Aids Res. 1*:S109–113.

Tedder, T. F., Crain, M. J., Kubagawa, H., Clement, L. T., and Cooper, M. D. (1985). Evaluation of lymphocyte differentiation in primary and secondary immunodeficiency diseases. *J. Immunol. 135*:1786–1791.

Teh, H. S., Kisielow, P., Scott, B., Kishi, H., Uematsu, Y., Bluthman, H., and von Boehmer, H. (1988). Thymic major histocompatibility complex antigens and the α/β T-cell receptor determine the CD4/CD8 phenotype of T cells. *Nature. 335*:229–233.

Urba, W. J., Baseler, M. W., Kopp, W. C., Steis, R. G., Clark, J. W., Smith J. W., Coggin, D. L., and Longo, D. L. (1989). Deoxycoformycin-induced immunosuppression in patients with hairy cell leukemia. *Blood 73*:38–46.

von Boehmer, H., Teh, H. S., and Kisielow, P. (1989). The thymus selects the useful, neglects the useless and destroys the harmful. *Immunol. Today. 9*:57–61.

Walter, M. H., Dosch, H.-M., and Cox, D. W. (1991). A deletion map of the human immunoglobulin heavy chain variable region. *J. Exp. Med. 174*:335–349.

Weaver, D., and Baltimore, D. (1987). B lymphocyte–specific protein binding near an immunoglobulin κ-chain gene J segment. *Proc. Natl. Acad. Sci. USA. 84*:1516–1520.

Westerga, J., and Timens, W. (1989). Immunohistological analysis of human fetal lymph nodes. *Scand. J. Immunol. 29*:103–112.

Willis, E. H., Carson, D. A., and Shultz, L. D. (1987). Adenosine deaminase activity in recipients of bone marrow from immunodeficient mice homozygous for the wasted mtuation. *Biochem. Biophys. Res. Commun. 145*:581–585.

Zijlstra, M., Bix, M., Simister, N. E., Loring, J. M., Raulet, D. H., and Jaenisch, R. (1990). β_2-microglobulin deficient mice lack CD4$^-$8$^+$ cytolytic T cells. *Nature 344*:742–746.

INDEX